2015 International Conference on IC Design & Technology

(ICICDT 2015)

Leuven, Belgium
1-3 June 2015

IEEE Catalog Number: CFP15412-POD
ISBN: 978-1-4799-7670-6

**Copyright ©2015 by the Institute of Electrical and Electronic Engineers, Inc
All Rights Reserved**

Copyright and Reprint Permissions: Abstracting is permitted with credit to the source. Libraries are permitted to photocopy beyond the limit of U.S. copyright law for private use of patrons those articles in this volume that carry a code at the bottom of the first page, provided the per-copy fee indicated in the code is paid through Copyright Clearance Center, 222 Rosewood Drive, Danvers, MA 01923.

For other copying, reprint or republication permission, write to IEEE Copyrights Manager, IEEE Service Center, 445 Hoes Lane, Piscataway, NJ 08854. All rights reserved.

***This publication is a representation of what appears in the IEEE Digital Libraries. Some format issues inherent in the e-media version may also appear in this print version.

IEEE Catalog Number: CFP15412-POD
ISBN 13:   978-1-4799-7670-6

Additional Copies of This Publication Are Available From:

Curran Associates, Inc
57 Morehouse Lane
Red Hook, NY 12571 USA
Phone: (845) 758-0400
Fax: (845) 758-2633
E-mail: curran@proceedings.com
Web: www.proceedings.com

2015 International Conference on IC Design & Technology (ICICDT 2015)

Leuven, Belgium
1-3 June 2015

IEEE Catalog Number: CFP15412-POD
ISBN: 978-1-47997-670-6

TABLE OF CONTENTS

SESSION A: LOW POWER MEMORY TECHNOLOGY AND CIRCUITS
Session Chair: Hideto Hidaka

Invited - Selectors for High Density Crosspoint Memory Arrays: Design Considerations, Device Implementations and Some Challenges Ahead 1
Bogdan Govoreanu, Leqi Zhang, Malgorzata Jurczak

Current Pulse Generator for Multilevel Cell Programming of Innovative PCM 5
Athanasios Kiouseloglou, Gabriele Navarro, Alessandro Cabrini, Guido Torelli, Luca Perniola

Design Technology Co-optimization for Enabling 5nm gate-all-around Nanowire 6T SRAM 9
Trong Huynh-Bao, Sushil Sakhare, Julien Ryckaert, Dmitry Yakimets, A. Mercha, D. Verkest, Aaron Voon-Yew Thean, Piet Wambacq

Assessment of SiGe Quantum Well Transistors for DRAM Peripheral Applications 13
R. Ritzenthaler, T. Schram, G. Eneman, A. Mocuta, N. Horiguchi, A. V.-Y. Thean, A. Spessot, M. Aoulaiche, P. Fazan

Low Standby Power Capacitively Coupled Sense Amplifier for Wide Voltage Range Operation of Dual Rail SRAMs 17
Anuj Grover, Promod Kumar, Mohammad Daud, G. S. Visweswaran, Chittoor Parthasarathy, Jean-Philippe Noel, David Turgis, Bastien Giraud, Guillaume Moritz

A 6T-SRAM in 28nm FDSOI Technology with Vmin of 0.52V Using Assisted Read and Write Operation 21
Ashish Kumar, Vinay Kumar, Dhori Kedar Janardan, G.S.Visweswaran, Kaushik Saha

SESSION B – EMERGING TECHNOLOGIES
Session Chairs: Thomas Ernst And Denis Flandre

Invited - NEMS Switches: Opportunities and Challenges in Emerging IC Technologies 25
Philip X.-L. Feng

Wide Band Study of Silicon-on-Insulator Photodiodes on Suspended Micro-Hotplates Platforms 31
N. André, G. Li, P. Gérard, O. Poncelet, Y. Zeng, S.Z. Ali, F. Udrea, L.A. Francis, D. Flandre

Evaluation of 32-Bit Carry-Look-Ahead Adder Circuit with Hybrid Tunneling FET and FinFET Devices 35
Tse-Ching Wu, Chien-Ju Chen, Yin-Nien Chen, Vita Pi-Ho Hu, Pin Su, Ching-Te Chuang

Area and Routing Efficiency of SWD Circuits Compared to Advanced CMOS 39
Odysseas Zografos, P. Raghavan, Yasser Sherazi, Adrien Vaysset, Florin Ciubatoru, Bart Soree, Rudy Lauwereins, Iuliana Radu, Aaron Thean

SESSION C – IIIGH FREQUENCY BUILDING BLOCKS
Session Chairs: Stefano D'amico And Andrea Scarpa

Low-Phase Noise Variation VCO Implementing Resistorless Digitally Controlled Varactor 43
Mohammed Aqeeli, Abdullah Alburaikan, Xianjun Huang, Zhirun Hu

SESSION D: ADVANCED TRANSISTORS AND MATERIALS
Session Chairs: Bich-Yen Nguyen And Dina H. Triyoso

Invited - Dimensioning for Power and Performance Under 10nm: The Limits of FinFETs Scaling 47
M. Garcia Bardon, P. Schuddinck, P. Raghavan, D. Jang, D.Yakimets, A. Mercha, D. Verkest, A. Thean

Impact of Fin Shape Variability on Device Performance Towards 10nm Node 51
Kazuyuki Tomida, Keizo Hiraga, Morin Dehan, Geert Hellings, D. Jang, Kenichi Miyaguhi, Thomas Chiarella, Minsoo Kim, A. Mocuta, Naoto Horiguchi, A. Mercha, D. Verkest, Aaron Thean

Modeling FinFET Metal Gate Stack Resistance for 14nm Node and Beyond 55
Kenichi Miyaguchi, Bertrand Parvais, Lars-Åke Ragnarsson, Piet Wambacq, P. Raghavan, A. Mercha, A. Mocuta, D. Verkest, Aaron Thean

Static and Dynamic Power Management in 14nm FDSOI Technology 59
O. Weber, E. Josse, J. Mazurier, M. Haond

Lateral NWFET Optimization for Beyond 7nm Nodes 63

D. Yakimets, D. Jang, P. Raghavan, G. Eneman, H. Mertens, P. Schuddinck, A. Mallik, M. Garcia Bardon, N. Collaert, A. Mercha, D. Verkest, A. Thean, K. De Meyer

Nonparabolicity and Confinement Effects of IIIV Materials in Novel Transistors 67

M. Ali Pourghaderi, A. Mocuta, Aaron Thean

SESSION E – ADVANCED CMOS DEVICE RELIABILITY
Session Chairs: Koji Eriguchi And Yuichiro Mitani

Invited - PBTI for n-type Tunnel FinFETs 70

W. Mizubayashi, T. Mori, K. Fukuda, Y. X. Liu, T. Matsukawa, Y. Ishikawa, K. Endo, S. O'uchi, J. Tsukada, H. Yamauchi, Y. Morita, S. Migita, H. Ota, M. Masahara

Reliability Impact of Advanced Doping Techniques for DRAM Peripheral MOSFETs 74

A. Spessot, R. Ritzenthaler, T. Schram, M. Aoulaiche, M. Cho, Maria Toledano Luque, Naoto Horiguchi, P. Fazan

Impact of Random Telegraph Noise on Ring Oscillators Evaluated by Circuit-level Simulations ... 78

Azusa Oshima, P. Weckx, B. Kaczer, Kazutoshi Kobayashi, Takashi Matsumoto

Simple Technique for Prediction of Breakdown Voltage of Ultrathin Gate Insulator Under ESD Testing 82

Yuichiro Mitani, Kazuya Matsuzawa

Off-State Stress Degradation Mechanism on Advanced *p*-MOSFETs 86

M. Cho, A. Spessot, B. Kaczer, M. Aoulaiche, R. Ritzenthaler, T. Schram, P. Fazan, Naoto Horiguchi, Dimitri Linten

SESSION F - CAD AND RELIABILITY CHALLENGES IN ADVANCED TECHNOLOGIES
Session Chair: Wei Guo And Thuy Dao

Invited - Deadspace-aware Power/Ground TSV Planning in 3D Floorplanning 90

Shengcheng Wang, Farshed Firouzi, Fabian Oboril, Mehdi B. Tahoori

Impact of Device and Interconnect Process Variability on Clock Distribution 94

Nathalie Fievet, P. Raghavan, Rogier Baert, Frederic Robert, A. Mercha, D. Verkest, Aaron Thean

Impact of Time-dependent Variability on the Yield and Performance of 6T SRAM Cells in an Advanced HK/MG Technology 98

P. Weckx, B. Kaczer, J. Roussel, F. Catthoor, G. Groeseneken

Countering Early Propagation and Routing Imbalance of DPL Designs in a Tree-based FPGA 102

Emna Amouri, Shivam Bhasin, Yves Mathieu, Tarik Graba, Jean-Luc Danger

FinFET Stressor Efficiency on Alternative Wafer and Channel Orientations for the 14 nm Node and Below 106

G. Eneman, A. De Keersgieter, A. Mocuta, N. Collaert, A. Thean

SESSION G – POWER DEVICE RELIABILITY AND PLASMA-INDUCED DAMAGE
Session Chairs: Yuichiro Mitani And Koji Eriguchi

Invited - Trapping Induced Parasitic Effects in GaN-HEMT for Power Switching Applications 110

Gaudenzio Meneghesso, Matteo Meneghini, Enrico Zanoni, Piet Vanmeerbeek, Peter Moens

Plasma Induced Damage Investigation in the Fully Depleted SOI Technology 114

M. Akbal, G. Ribes, M. Guillermet, L. Vallier

Plasma-induced Photon Irradiation Damage on Low-k Dielectrics Enhanced by Cu-line Layout 118

Taro Ikeda, Akira Tanihara, Nobuhiko Yamamoto, Shigeru Kasai, Koji Eriguchi, Kouichi Ono

Surface Orientation Dependence of Ion Bombardment Damage During Plasma Processing 122

Yukimasa Okada, Koji Eriguchi, Kouichi Ono

SESSION H - ANALOG AND MIXED-SIGNAL TECHNIQUES
Session Chair: Stefano D'amico And Andrea Scarpa

Invited - High-Speed Analog-to-Digital Converters in Downscaled CMOS 126

Annachiara Spagnolo, Bob Verbruggen, Stefano D'Amico, Piet Wambacq

Overview of Methods to Increase Linearity of High-performance ADC 130

Hua Fan, Kehong Liu, Airong Liu, Lishan Lv, Zhiliang Qiao, Qiang Li

Optimal Design to Maximize Efficiency of Single-Inductor Multiple-Output Buck Converters in Discontinuous Conduction Mode for IoT Applications 134

Yoshitaka Yamauchi, Yuki Yanagihara, Hiroshi Fuketa, Takayasu Sakurai, Makoto Takamiya

SESSION I - HIGH-POWER / HIGH-VOLTAGE
Session Chairs: Jan Ackaert And Thuy Dao

Invited - Thermal Experimental and Modeling Analysis of High Power 3D Packages 138
  H. Oprins, V. Cherman, G. Van der Plas, F. Maggioni, J. De Vos, E. Beyne

Metallization Scheme Optimization of Plasticencapsulated Electronic Power Devices 142
  Jan Ackaert, Tony Colpaert, Aditi Malik, Mario Gonzalez

**I/O Thick Oxide Device Integration Using Diffusion and Gate Replacement (D&GR) Gate Stack
Integration** 146
  *R. Ritzenthaler, T. Schram, M. Cho, A. Mocuta, N. Horiguchi, A. V.-Y. Thean, A. Spessot, C. Caillat, M. Aoulaiche,
  P. Fazan*

**Characterization of Onset Tunneling Voltage (Vonset) Walkout in High-Voltage Deep Trench
Isolation on SOI** 150
  Thuy Dao, Mu-Ling Ger, Jiangkai Zuo

SESSION J – I/O CIRCUITS AND ESD PROTECTION
Session Chairs: Philippe Galy And Lorenzo Cerati

**Invited - Integrated Front-End/Back-End Simulation of Electromagnetic Fields, Lorentz Force
Effects and Fast Current Surges in Microelectronic Protection Devices** 154
  Wim Schoenmaker, P. Galy

Impact of Local Interconnects on ESD Design 158
  Mirko Scholz, Shih-Hung Chen, Geert Hellings, Dimitri Linten, Roman Boschke

ESD Protection Diodes in Optical Interposer Technology 162
  Roman Boschke, G. Groeseneken, Mirko Scholz, Shih-Hung Chen, Geert Hellings, Peter Verheyen, Dimitri Linten

**Preliminary 3D TCAD Electro-thermal Simulations of BIMOS Transistor in Thin Silicon Film for
ESD Protection in FDSOI UTBB CMOS Technology** 166
  S. Athanasiou, S. Cristoloveanu, P. Galy

A High-Speed 2xVDD Output Buffer With PVTL Detection Using 40-nm CMOS Technology 170
  Chua-Chin Wangy, Tsung-Yi Tsai, Wei Lin

SESSION K – 3D INTEGRATION AND LOWER POWER PROCESSORS
Session Chairs: Bich-Yen Nguyen And Juergen Pille

Invited - 3D Monolithic Integration: Stacking Technology and Applications 174
  Ionut Radu, Bich-Yen Nguyen, Gweltaz Gaudin, Carlos Mazure

Through Silicon Via to FinFET Noise Coupling in 3-D integrated Circuits 177
  *A. Rouhi Najaf Abadi, W. Guo, X. Sun, K. Ben Ali, J.P Raskin, M. Rack, C. Roda Neve, M.Choi, V. Moroz, G. Van
  der Plas, I. De Wolf, E. Beyne , P. Absil*

Simple Wafer Stacking 3D-FPGA Architecture 181
  Motoki Amagasaki, Qian Zhao, Masahiro Iida, Morihiro Kuga, Toshinori Sueyoshi

Design of a Low-power Fixed-point 16-bit Digital Signal Processor Using 65nm SOTB Process 185
  Duc-Hung Le, Nobuyuki Sugii, Shiro Kamohara, Xuan-Thuan Nguyen, Koichiro Ishibashi, Cong-Kha Pham

**Power Measurements and Cooling of the Dome 28nm 1.8GHz 24-thread ppc64 μServer Compute
Node** 189
  Ronald P. Luijten, Matteo Cossale, Rolf Clauberg, Andreas Doering

Author Index

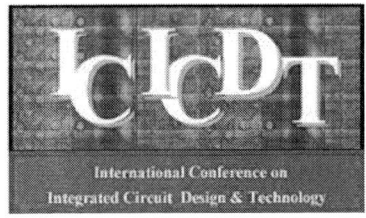

International Conference on IC Design and Technology (ICICDT)

June 1st – 3rd, 2015
Leuven, Belgium
www.icicdt.org

CONFERENCE PROGRAM

Conference Chair:	Dina H. Triyoso
General Chair:	Keith A. Bowman
Executive Chair:	Thuy B. Dao
Keynote Chair:	Aaron Thean
Local Arrangement Chair:	Wei Guo
Tutorial Chair:	Koji Eriguchi
Publicity & Award Chair:	Mariam Sadaka
Publication Chair:	Ben Kaczer
Treasurer:	Yuichiro Mitani
Secretary:	Andrea Scarpa
IEEE EDS Student Chapter Chair:	Denis Flandre

UCLouvain EDS Chapter

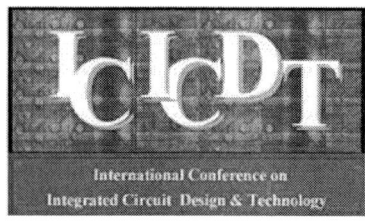

International Conference on IC Design and Technology (ICICDT)

June 1st – 3rd, 2015
Leuven, Belgium
www.icicdt.org

Technical Program Committee

Jan Ackaert, ON Semiconductors

Serge Bardy, NXP

Keith A. Bowman, Qualcomm

David Burnett, GLOBALFOUNDRIES

Lorenzo Cerati, ST Microelectronics

Xi Chen, Qualcomm

Kin P. (Charles) Cheung, NIST

Stefano D'Amico, University of Salento

Thuy B. Dao, Freescale

Marcello De Matteis, Univ. of Milan

Koji Eriguchi, Kyoto University

Thomas Ernst, CEA/LETI

Bao Fang, Qualcomm

Philippe Galy, ST Microelectronics

Wei Guo, imec

Mark Hall, Freescale

Steve Heinrich-Barn, Texas Instruments

Hideto Hidaka, Renesas

Ben Kaczer, imec

Rouwaida Kanj, American U. of Beirut

Bao Liu, University of Texas

Yuichiro Mitani, Toshiba

Bich-Yen Nguyen, Soitcc

Fabio Pellizzer, Micron

Dac Pham, Freescale

Juergen Pille, IBM

Shivam Priyadarshi. Qualcomm

Mariam Sadaka, Soitec

Ashoka Sathanur, Philips

Andrea Scarpa, NXP

Rick Shen, eMemory

Jayakumaran Sivagnaname, Freescale

Akif Sultan, GLOBALFOUNDRIES

Dina Triyoso, GLOBALFOUNDRIES

Michiel van Duuren, NXP

Peiyuan Wang, Qualcomm

Wenke Weinreich, Fraunhofer

UCLouvain EDS Chapter

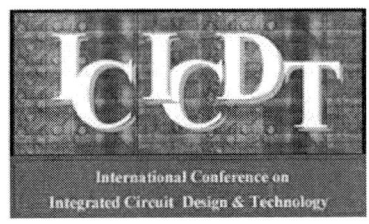

International Conference on IC Design and Technology (ICICDT)

June 1st – 3rd, 2015
Leuven, Belgium
www.icicdt.org

This meeting is sponsored by:

UCLouvain EDS Chapter

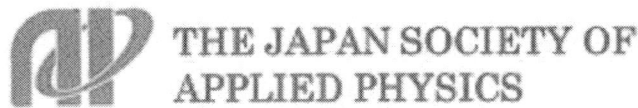

UCLouvain EDS Chapter

Selectors for High Density Crosspoint Memory Arrays: Design Considerations, Device Implementations and Some Challenges Ahead

(invited paper)

Bogdan Govoreanu[*), Leqi Zhang[$), Malgorzata Jurczak

imec, Kapeldreef 75, 3001 Leuven, Belgium
[$)also with KU Leuven, Kasteelpark Arenberg 10, 3001 Leuven, Belgium;
[*)E-mail: bogdan.govoreanu@imec.be, Phone: +32-16-281337

Abstract – Significant progress has been made in recent years in the research and development of Resistive RAM as future nonvolatile memory. To achieve even higher integration densities, the resistive switching element needs a two-terminal selector device, in a one-selector one-resistor (1S1R) serial cell, which enables suppression of the parasitic leakage paths in the memory array. In this paper, review a top-down approach to derive selector requirements, considering a worst-case crosspoint array model and resistive switching element characteristics. The requirements that a selector needs to meet are derived by systematic analysis of various bias schemes, considering a basic set of figures of merit, to describe array functionality and power efficiency. We furthermore briefly discuss several selector implementations to date, outlining their strengths and weaknesses. In the end, we point out selector variability impact on cell operation, which calls for even more stringent selector requirements to be met.

Index Terms — **selector, Resistive RAM (RRAM), crosspoint memory array, bias scheme, current drive, nonlinearity, selector variability.**

I. INTRODUCTION

R ESISTIVE Random Access Memory (RRAM) is the most scalable resistive switching structure experimentally proven to date. Functional devices on a 10nm-scale and below have been reported, which are based on either anion [1] or cation [2] switching species. Implementation of high-density RRAM crosspoint arrays depends critically on the availability of a two-terminal selecting device, able to cut off the sneak leakage paths [3] occurring during memory cell operation. A bipolar resistive switching element (RSE, 1R) and a bidirectional, highly-nonlinear selector device (1S), serially connected in a one-selector one-resistor (1S1R) two-terminal memory cell, enable achieving the smallest 4F^2 cell size (F-minimum feature size). The selector device must act as a rectifier, suppressing the current through the cell at low biases, while turning on at larger biases. Next to enabling higher density, a two-terminal selector device can potentially be fabricated in a back-end-of-line CMOS process. This allows to stack memory on top of logic and may offer important technological and cost benefits for several applications, ranging from stand-alone to embedded memory. Consequently, selector requirements are multiple and derived from circuit, device and process compatibility constraints.

In this paper, we introduce a crosspoint memory array model, to be used for array simulations, and a number of figures of merit, related to array performance and power efficiency. These enable array sizing simulations, as well as derivation of selector requirements that ensure compatibility with available resistive switching elements. We furthermore review various selector device reports, which exploit nonlinear transport mechanisms for implementing the selectivity function. We propose a selector benchmarking metric that enables to assess, to first order, the cell scalability and the memory array size that a particular selector can sustain. Finally, we point out the importance of the selector variability when considering array operation and discuss its impact on device requirements.

II. CROSSPOINT MEMORY ARRAY: MODEL & METHODOLOGY

A. Crosspoint array model

A crosspoint memory array consists of memory cells defined by metallic bitlines (BL) and wordlines (WL) that run on directions normal to each other. In the forthcoming discussion, we adopt a worst-case crosspoint array model [4], considering the longest signal path and the worst data pattern, for assessing its performance and efficiency.

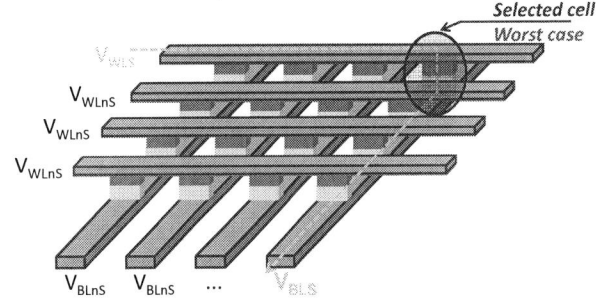

Fig.1: Cross-point array schematic: worst-case cell operation implies addressing the cell situated on the longest signal path, and the worst-case data pattern, which depends on the addressed cell state and on the operation to be performed.

The longest signal path translates into the operation of the farthest cell, in relation to the WL/BL applied/sensed signals. Accessing a particular cell in the array, requires application of specific voltages on the selected and non-selected BL/WL,

978-1-4799-7670-6/15 $31.00 © 2015 IEEE

resulting in selected cell (S), half-selected (HS) BL/WL cells and nonselected (NS) cells. In a worst-case situation, the biasing configuration is captured by a corresponding array model (Fig.2).

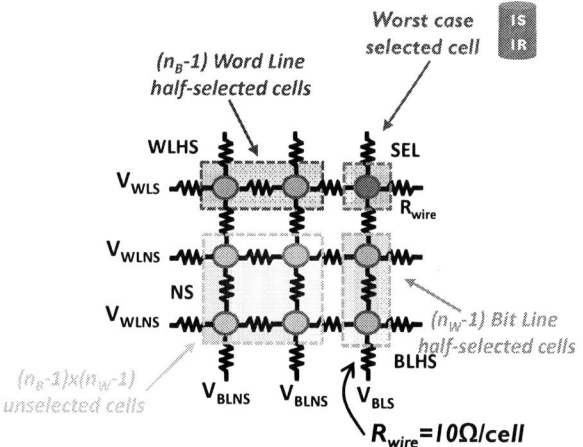

Fig.2: Crosspoint array model, emphasizing the (worst-case) selected cell, SEL, WL and BL half-selected cells, as well as nonselected (NS) cells. The array has a number n_B of BL, and n_W of WL. For square arrays $n_B = n_W$. Assumed wire resistance: 10Ohm/cell.

The worst data pattern depends on the operation being performed as well as on the state of the addressed cell, and is summarized in Table I.

	Set	Reset	Read HRS	Read LRS
SEL	HRS	LRS	HRS	LRS
WLHS	LRS	LRS	LRS	LRS
BLHS	LRS	LRS	LRS	HRS
NS	LRS	LRS	LRS	LRS

Table I: Worst data pattern case, for cell operation (Set, Reset) and cell read-out.

B. Array sizing: analysis methodology and results

The size of a crosspoint array consisting of 1S1R memory cells is estimated, to first order, taking into account a static assessment of its functionality, as well as of its power efficiency. The following performance factors are taken into account to assess the array functionality:

- **Read Margin (RM)**, which is a worst case measure of the separation between the sensed-out on- and off-state levels. Typically, a current-based definition is adopted for the RM:

$$RM = \frac{\Delta I_{BLS}}{I_{ref}} = \left(1 - \frac{\max\{I_{BLS} | SEL = HRS\}}{\min\{I_{BLS} | SEL = LRS\}}\right) \cdot 100[\%] \quad (1)$$

- **Write margin (WM)**, which is a worst case measure of the access voltage transfer to the selected cell. Commonly, a voltage-based definition is employed for the WM:

$$WM = \frac{V_{SEL}}{V_{appl}} = \frac{V_{SEL}}{V_{BLS} - V_{WLS}} \cdot 100[\%] \quad (2)$$

Array power efficiency is assessed based on the power consumed in order to perform read ($\mathbf{P_R}$) and write ($\mathbf{P_W}$) operations, respectively.

For an array to be functional, certain design criteria must be fulfilled, expressed in constraints on the parameters mentioned above. Typical values, further considered here, are summarized in Table II.

	Margin	Power
READ	≥25%	≤10uW/bit
WRITE	≥90%	≤1mW/bit

Table II: Target design region for hte Read and Write margins, as well as for the array efficiency.

A particular biasing scheme needs to be employed in order to read-out or to operate a memory array, for accessing the target cell, while limiting the leakage through unaddressed cells. The most common schemes are either ½-bias or ⅓-bias. These are both particular cases of a general, partial ("x") bias scheme, as summarized in Table III.

Bias Scheme	Line Applied Biases				Cell Biases			
	WLS	BLS	WLNS	BLNS	SEL	WLHS	BLHS	NS
Half-Bias	V	0	V/2	V/2	V	V/2	V/2	0
1/3-Bias	V	0	V/3	2V/3	V	V/3	V/3	-V/3
x-Bias	V	0	x.V	(1-x).V	V	x.V	x.V	(2x-1).V

Table III: WL/BL applied biases and corresponding cell biases, for selected, half-selected and nonselected cells, as illustrated in Fig.2, and considering different biasing schemes.

Dependence of the RM/WM on the array size allows to estimate the maximum array size that still allows having sufficient operating margin, as well as the efficiency of the R/W operations. Assuming a typical resistive switching element (RSE) in a 1S1R cell with an operating (switching) current of 10uA and selector characteristics derived from experimental results [5], SPICE simulation results show that both WM and RM degrade faster for ½-bias schemes, relative to ⅓-bias schemes (Fig.3). However, ⅓-bias schemes are more power hungry, since all unaddressed cells are biased at ±V/3, unlike ½-bias scheme, where only the half-selected cells are biased at V/2.

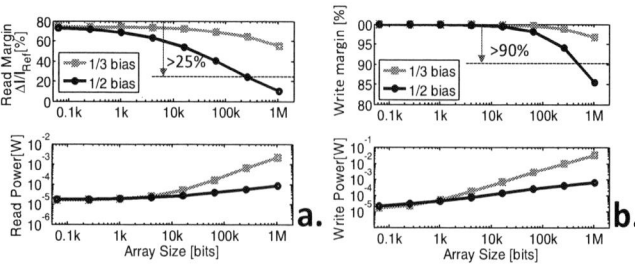

Fig.3: RM/WM and the power consumption, respectively, as a function of the array size, considering a RSE operating at 10uA, with 50kOhm/500kOhm on/off window and a selector with characteristics as experimentally reported in [5].

III. SELECTOR REQUIREMENTS

A. Performance factors

Selector requirements can be derived by considering circuit, device and process constraints, most important of which are summarized in Table IV. In this paper, we limit to the most important performance factors, that allow a first order selector benchmarking: i) *the current drive* (I_0)– i.e. the maximum current that can flow through a selector of a given area; this current must be high enough so as to enable RSE switching. ii) *the operating voltage* (V_{op}), which is the voltage at which the current flowing through selector reaches the current drive, $I_0 = I(V_{op})$. iii) *the selector nonlinearity* (NL) – which is a measure of the selector ability to rectify at program or read-out conditions. Although there are different ways to express this nonlinearity, we will use here the half-bias definition, which, similar to that of the full 1S1R cell, may be expressed as: $\mathbf{NL_{1/2}} = I(V=V_{op})/I(V=V_{op}/2)$, where $I(V=V_{op})=I_0$.

Requirement	Parameter	Remarks
High selectivity at Read-out, Set and Reset (S/R)	Nonlinearity factor, $NL_{1/2}$	Limits the array size
Large current for S/R switching	Current drive $J_{S/R}$	$n \cdot 10^6$ A/cm² for n uA S/R @10nm cell size
Compatible with RSE operation	Operating voltage, V_{op}	Nonlinearity relates to operating voltage
Withstand Forming	BD voltage, V_{BD}	Forming-free RSE preferred
High operating speed	Turn on/off times, t_{on}, t_{off}	Respond faster than the RSE
Do not alter RSE reliability	Cycling endurance, N_{cy}	$N_{cy}(Sel) > N_{cy}(RSE)$
Scalable	Minimum feature size, F_{min}	$F_{min} < F_{min}(RE)$
BEOL process compatibility	BEOL thermal budget & BEOL thermal stability	Must not impact BEOL and be not impacted by BEOL

Table IV: Most important selector requirements and parameter best quantifying them, including circuit (top, red bar), device and process (bottom, blue bar) constraints.

B. Design considerations

We use a top-down approach to derive selector requirements, by formulating the inverse problem, i.e. to find selector characteristics that meet the design constraints (Table II), for a known array size (e.g. 1Mbit) and RSE characteristics. To this goal and in order to ensure a computationally efficient simulation method, while not sacrificing the overall accuracy, the selector's IV characteristics are parametrized, using a simple analytical expression (Fig.4). For practical reasons, this analysis is carried out at a predefined current drive level, set by the RSE characteristics, assumed to be known.

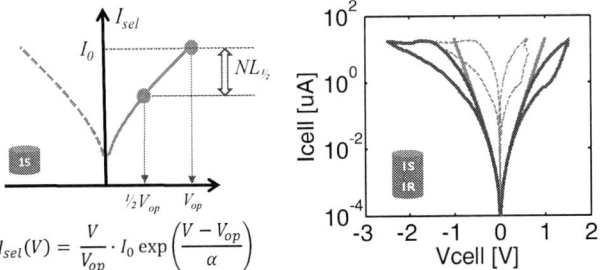

$$I_{sel}(V) = \frac{V}{V_{op}} \cdot I_0 \exp\left(\frac{V - V_{op}}{\alpha}\right)$$

Fig.4: (left): Selector parametric form, illustrating current drive (I_0), and ½-bias NL. (right): 1S1R characteristics (blue solid line) constructed assuming experimentally known RSE characteristics (thin gray dashed line) [6] and a selector parametric form (red solid line), serially connected (inset).

Simulation results for a ½-bias scheme show that the RM increases with selector NL, while the operating voltage has only limited impact. The design region for the RM is compatible with that of the read power (Fig.5), requiring a selector with a $NL_{1/2}$ of about 3000, for a 1Mbit array size, at 10uA operating current. When considering a write operation, the design region is delimited by the WM constraints and both high selector $NL_{1/2}$ and high V_{op} are required (Fig.6).

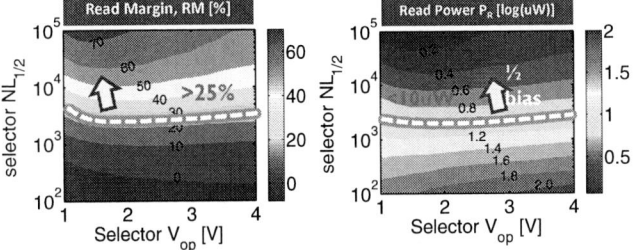

Fig.5: Read Margin and Read Power dependence on selector parameters for a 10uA switching current and an ½–bias scheme. Design region is selected in line with design constraints (Table II).

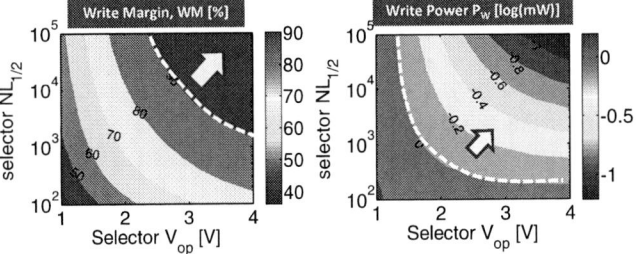

Fig.6: Write Margin and Write Power dependence on selector parameters for a 10uA switching current and an ½–bias scheme. Design region is selected in line with design constraints (Table II).

In ⅓-bias schemes, unlike in an ½-bias scheme, the unselected cells are under a bias of ±V/3 (ignoring the impact of the line resistance). As a consequence, the read power increases considerably and in order to comply with the design constraints, a higher selector $NL_{1/2}$ is required, close to 10^5. Likewise, the write power constraint activates first and, consequently, a selector $NL_{1/2}$ of over 10^4 and V_{op} of about 3V are required (not shown). In fact, an optimal bias scheme can be derived for both read and write operations (Fig.7), which allows deriving the minimum selector $NL_{1/2}$ that meet all constraints. In both cases, a minimal value of about 2000 is required, for an operating voltage of 3V, and for a switching current of 10uA. If the switching current reduces (to e.g. 1uA), the requirements become less stringent, especially since the WM is less affected (e.g. $NL_{1/2}$ of about 800).

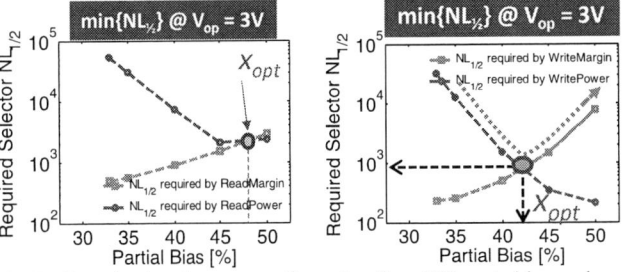

Fig.7: Required selector nonlinearity for different bias schemes, derived from read and write constraints. The optimum partial bias correspond to minimal $NL_{1/2}$ satisfying both constraints.

978-1-4799-7670-6/15 $31.00 © 2015 IEEE

IV. STATUS AND OUTLOOK

A. Implementations

There is a large number of selector implementations proposed to date [7]. The most common category is that of **bidirectional diodes**. These can be realized using Si-based punchthrough-diodes [8], back-to-back Schottky diodes [9], amorphous-Si [10], or oxide-based MIM tunnel diodes [11]. Junction based diodes require thermal budget that exceeds the BEOL. Schottky diodes have limited nonlinearity, while oxide-based diodes have limited drive current. Among these, amorphous-Si based MSM selector has been shown to reach a $NL_{\frac{1}{2}}$ exceeding 5.10^3, with a current drive of about $1MA/cm^2$. Although a drive current-NL trade-off exists, this may eventually be broken out by barrier engineering [12].

Another important category is that of **volatile switches**. These may have different implementations, exploiting phenomena such as *insulator-metal transition* [13], induced by Joule heating or electric field, or *ovonic threshold switching* specific to chalcogenide materials [14]. Such selectors are appealing for their high drive current ability. However, the demonstrated NL remains rather modest. A notable exception is that of *volatile conductive bridges* [15], which may have both high drive and NL. However, the success of this implementation depends on the bridging species diffusivity.

Other **nonlinear mechanisms** used to implement selectors are notably *mixed ionic-electronic conduction* (MIEC) [16], which have high current drive and nonlinearity, however limited voltage operation range, making them suitable particularly for low-voltage RRAM.

B. Benchmarking

Considering the heterogenous nature of the selector constraints (Table IV), it is not easy to benchmark them. In fact, picking up a best selector depends on the application it is meant for. Nevertheless, the key selector performance may be captured in the benchmarking plane (current drive – NL), which, to first order, translates into scalability-array size attainable by a particular selector (Fig.8).

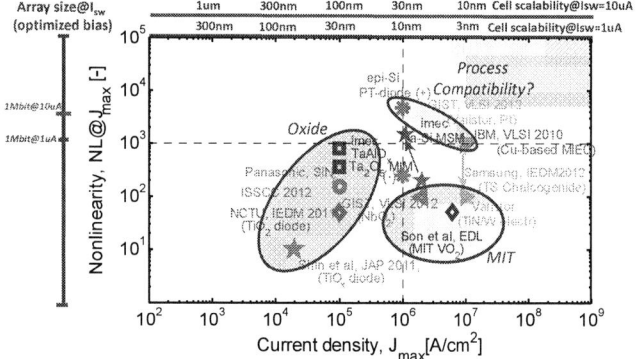

Fig.8: (Current drive – NL) benchmarking metric for selector performance estimation. A NL of 10^3 is required for a 1Mbit array operating at 1uA and a current drive of $1MA/cm^2$ would enable 10nm cell scalability (at 1uA operating current). For a 10uA switching current, 10nm scalability in a 1Mbit array translates into $10MA/cm^2$ current drive and a $NL_{\frac{1}{2}}$ close to 10^4. The target region is indicated in the top right area of the plot.

C. Challenges

It appears that several selectors reported to date cluster in either a high NL region or high current drive region. Fulfilling both remains a challenge, in spite of a few reports situating well in the target region, however typically difficult to manufacture. As the switching current of the RSE can hardly be reduced below 10uA without unbearable reliability penalty, the realistic current drive density is about $10MA/cm^2$, while the NL should remain in the order of $>10^3$ to 10^4, depending on the selector nature and operating voltage. Furthermore, the selector needs to be compatible with the RSE, which means the voltage drop over the whole 1S1R cell needs to be properly distributed between the selector and RSE. For a RSE with typical switching voltages of about ±1.5V, the selector should operate over at least 2V voltage range.

With cell size scaling, variability becomes an increasingly important issue. Unlike RSE, variability of which is strongest in the off-state, selector variability is most critical for the on-state [10]. This further reduces the on-off read window (Fig.9) and, in order to compensate for, selector variability control and/or an even higher NL are necessary.

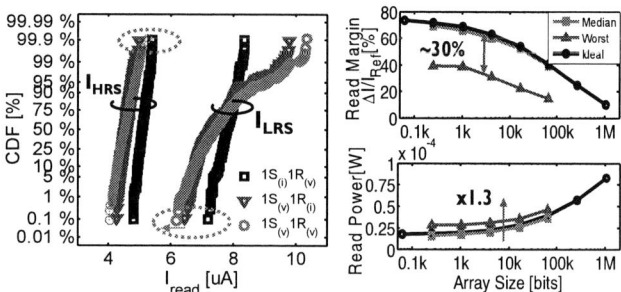

Fig.9: Consideration of all main variability sources (Selector, RSE, data pattern) shows that the on/off-window shrinks, resulting in a RM penalty (e.g.) of about 30%. To compensate for, an even higher selector NL is required.

V. SUMMARY

In summary, we presented a crosspoint array model and analysis methodology to estimate selector device requirements, considering read and write operations, as well as power efficiency constraints. We reviewed several selector implementations, proposing a benchmarking metric that easily translates into attainable cell scalability – array size. To enable dense 1S1R array implementations, current drive of $10MA/cm^2$ or higher might be required, with a NL close to 10^4. Selector variability may come into picture for aggressively scaled 1S1R cells, affecting mostly the on-state. An even better selector performance may be needed to compensate for.

REFERENCES

[1]. B. Govoreanu et al, IEDM Tech Dig. 2011; [2]. J. Park et al, IEDM Tech. Dig, 2011; [3]. M.-J. Lee et al, IEEE IEDM Tech. Dig, 2007; [4]. L.Zhang et al, Proc. IMW, 2014; [5]. L.Zhang et al, El. Dev. Lett, **35**(2), 2014; [6]. B. Govoreanu et al, Proc. IMW, 2013; [7]. ITRS Roadmap, Emerging Devices chapter, 2013; [8]. S. Srinivasan et al, Edl. Dev. Lett. **33**(10), 2012; [9]. A. Kawahara et al, ISSCC 2012; [10]. L.Zhang et al, IEDM Tech. Dig, 2014; [11]. B. Govoreanu et al, El. Dev. Lett, **35**(1), 2014; [12]. B. Govoreanu et al, Proc. IMW, 2015; [13]. M. Son et al, El. Dev. Lett. **32**(11), 2011; [14]. M.-J. Lee et al, IEDM Tech. Dig, 2012; [15]. S.H. Jo et al, IEDM Tech. Dig. 2014; [16]. Gopalakhrisnan et al, VLSI Tech. Symp. 2010.

Current Pulse Generator for Multilevel Cell Programming of Innovative PCM

Athanasios Kiouseloglou[*†], Gabriele Navarro[*], Alessandro Cabrini[†], Guido Torelli[†] and Luca Perniola[*]

[*]CEA, LETI, MINATEC Campus, 17 rue des Martyrs, 38054 Grenoble Cedex 9, France.
[†]Dipartimento di Ingegneria Industriale e dell'Informazione, University of Pavia, via Ferrata 5, 27100 Pavia, Italy.
Email: athanasios.kiouseloglou@cea.fr

Abstract—Multilevel Cell programming, i.e. storing multiple bits per memory cell, is a promising way to increase storage density in Phase Change Memory (PCM). In this paper, it is shown that it is possible to program a PCM device to multiple intermediate resistance states by using a single-pulse programming approach, as opposed to time-consuming iterative write algorithms previously reported in the literature. A circuit that generates current programming pulses with characteristics suitable for the specific target resistance state, is presented and simulated. The programmed resistance variation due to the variations in current is also studied and the programmed resistance states distributions are shown to be adequately spaced from each other, thus providing a viable programming solution for obtaining multiple resistance levels per memory cell.

I. Introduction

Storage Class Memory (SCM) is envisioned as a fast, inexpensive and power efficient non-volatile solid state memory solution that will potentially replace disk drives and large parts of power-hungry volatile working memory, thus providing an effective storage system [1]. Among the emerging non-volatile resistive memory technologies, Phase Change Memory (PCM) is the most promising candidate to support the development of an SCM in energy demanding, high-performance servers. PCM offers a wide variety of features, such as fast read and write access, excellent scalability potential, baseline CMOS technology compatibility and exceptional high-temperature data retention and endurance performances, and can therefore pave the way for applications not only in memory devices, but also in computing systems [2].

Bit storage in PCM relies on the reversible transition between a low-resistive crystalline phase (SET state) and a high-resistive amorphous phase (RESET state) of a chalcogenide alloy, typically $Ge_2Sb_2Te_5$ (GST). Cell programming is driven by current induced Joule heating. More specifically, in order to perform a RESET operation, a current pulse with high amplitude and fast quenching time is forced through the cell, which raises the temperature of the phase change material above its melting point. The active volume of the memory cell is first locally melted and then amorphised by the swift temperature decrease. For the SET operation, a current pulse of intermediate amplitude, which is capable of bringing the active region of the device above the crystallization temperature, is considered adequate [3]. Alternatively, a high-amplitude pulse with a long fall time, which allows enough time for crystallization during quench, can be used.

Since PCM demonstrates a programming window where the resistance ratio between the SET and the RESET state is very high (usually higher than a factor of 100), Multilevel Cell (MLC) programming, i.e. storing multiple bits in a single memory cell, can be used to achieve higher memory density and, hence, lower cost per bit. In order to store N bits per cell, 2^N distinct resistance levels are required. Key issues to address MLC PCM are the programming methodology as well as the stability and the retention of the intermediate resistance levels [4]. Several programming techniques to achieve an intermediate resistance state have been proposed in the literature. These programming schemes are based on time-consuming, iterative write-and-verify algorithms which are utilized to gradually approach the target resistance level and ensure that the desired resistance value is achieved [5], [6].

In this paper, we propose a novel technique for MLC programming of an innovative phase change material, namely N-doped Ge-rich GST, which demonstrates performances promising for MLC applications. Section II presents the recrystallization cartographies of the considered material and highlights the possibility to program a PCM device to a desired resistance state by using a single-pulse procedure. Section III proposes a circuit able to generate the required programming current pulse. Section IV studies the variability of the current pulse characteristics and the ensuing impact over programmed resistance variability. The proposed pulse generator is capable of achieving accurately programmed resistance states, which demonstrate modest variation, thus enabling 2-bit MLC programming for PCM.

II. Recrystallization Cartographies

In order to achieve reliable MLC programming in PCM, we need to ensure that the transition between the RESET and the SET state does not occur abruptly. An ideal candidate for this application should therefore be a material that demonstrates low crystallization speed, thus being capable of gradually passing from the RESET to the SET state and providing the possibility to generate intermediate resistance states. In this respect, GeTe or GST are not good candidates for MLC storage, since the application of a pulse with a constant decrease rate rapidly leads to a SET state of minimum resistance value [7].

Therefore, the material we selected to work with, is an N-doped Ge-enriched GST alloy, which demonstrates low crystallization speed [8], being an ideal candidate for MLC

978-1-4799-7670-6/15 $31.00 © 2015 IEEE

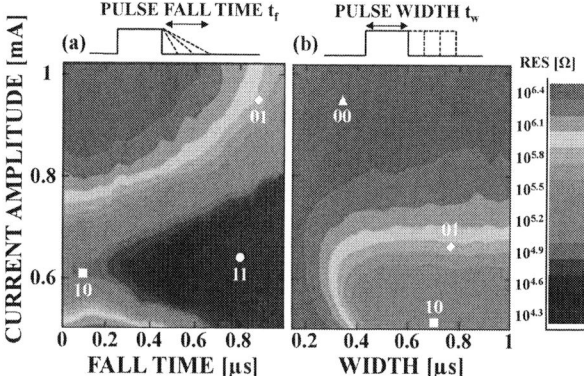

Fig. 1. Recrystallization cartographies of N-doped Ge-rich GST as a function of (a) pulse fall time t_f (t_w = 300 ns) and (b) pulse width t_w (t_f = 5 ns) (mean values from ten tests). Starting from a RESET state, it is possible to reach a desired resistance level by means of single pulses, by adjusting the fall time or the pulse width of the current pulse. White marks correspond to the best choice for programming the four resistance levels for 2-bit-per-cell storage.

applications. The smooth transition between resistance states allows the occurrence of intermediate states even when the pulse amplitude or other pulse characteristics, such as the time width, t_w, or the fall time, t_f, are affected by some variability.

For this specific target, we performed a set of characterization measurements on state-of-the-art 1R devices based on N-doped Ge-rich GST, to determine which combination of current amplitude and pulse fall time is necessary for a desired resistance level with a single programming pulse. A series of electrical tests was performed, during which we first programmed our cells to the RESET state and then applied single programming pulses with increasing amplitude, thus obtaining the recrystallization cartographies shown in Fig. 1. As it has been shown in [8], the application of an initial RESET pulse allows to screen lower current values, enabling a more accurate control of the programmed resistance. The cartographies were then obtained by reading the resistance value after the application of a programming pulse with a fixed t_w (300 ns) and an increasing t_f (Fig. 1(a)), or a pulse with a fixed t_f (5 ns) and an increasing t_w (Fig. 1(b)).

The distinct resistance regions of the recrystallization car-

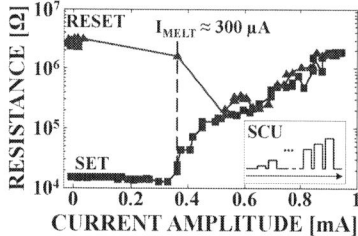

Fig. 2. Programmed resistance as a function of the programming current. A sequence of rectangular pulses (t_w = 50 ns) with increasing current amplitude is applied to the cells, which are initially programmed in the SET or the RESET state. Once a specific current amplitude is reached, the same resistance value is achieved regardless of the previously stored resistance state.

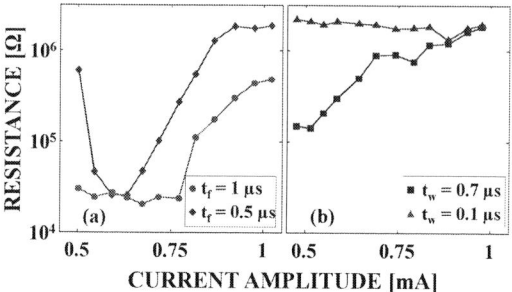

Fig. 3. Programmed resistance as a function of the programming current amplitude for different values of (a) pulse fall time t_f (t_w = 300 ns) and (b) pulse width t_w (t_f = 5 ns). An increased t_f is necessary to reach a stable low-resistance state starting from a RESET state. Increasing t_w enables the programming of intermediate resistance states at lower programming currents. A short t_w at higher current amplitudes ensures a high-resistance state that remains substantially constant in a wide programming current range.

tographies provide evidence that any resistance level can be achieved with a single-pulse programming procedure using a specific current amplitude and a controlled fall time. The effect of the current amplitude on the final resistance can be seen in Fig. 2. Our cells were programmed by means of a Staircase Up (SCU) sequence starting from a fully SET and a fully RESET state. An SCU consists of a sequence of rectangular pulses with a fixed t_w (50 ns) and increasing current amplitude. When starting from a SET state, a current pulse of 300 μA is already capable of melting part of the active region, thus altering the resistance state previously stored in the cell. Starting from a RESET state, once the voltage applied to the cell is higher than the threshold voltage which is required to switch the cell [3], causes the two curves of Fig. 2 to overlap, resulting in the same resistance value regardless of the initially programmed resistance state.

Even when a device is initially programmed in the RESET state, it is possible to end up in a fully SET state if a programming pulse with a sufficiently long t_f (1 μs) is applied, as observed in Fig. 3(a). If a pulse with a smaller t_f is applied, a fully SET state can still be reached, but the current range that is favorable for this transition is limited with respect to the previous case. Fig. 3(b) demonstrates that a pulse with an abrupt t_f (5 ns) and an increased t_w (700 ns in Fig. 3(b)) can still be employed in order to bring the cell to an intermediate resistance level. It is thus possible to reach a specific intermediate resistance when a rectangular pulse of low current is applied to the cell. For our material, the application of a pulse with a long fall time is necessary in order to reach a fully SET state. However, when programming a cell to the RESET state, the width of the applied pulse has a negligible effect on the final resistance value, especially when the applied current pulse has a high amplitude.

III. MLC PROGRAMMING

For reliable MLC storage, it is important that the programmed resistance distributions are kept sufficiently spaced in order not to result in decoding errors during resistance read-

978-1-4799-7670-6/15 $31.00 © 2015 IEEE

out. On that account, we divided our resistance programming window in three equally log-spaced resistance subwindows. The upper bound of the overall programming window corresponds to a fully RESET state (state 00), which can be achieved by a programming pulse with high amplitude and short fall time. The lower bound of the programming window corresponds to the fully SET resistance value (state 11), which is obtained by applying a programming pulse with lower amplitude and long fall time. The programming conditions for states 11 and 00 can be easily derived from the recrystallization cartographies (deep purple region in Fig. 1(a) and deep red region in Fig. 1(b), respectively). Intermediate states 01 and 10 correspond to the upper and the lower bound of the intermediate subwindow, respectively.

We then investigated the cartographies for regions around the target intermediate resistance values, aiming at finding the programming conditions and the actual target resistance values that ensure the minimum impact of the variation of the programming pulse parameters over the obtained resistance.

To this end, we estimated which resistance level close to the target value shows the least sensitivity to current amplitude variation, and for which programming conditions this resistance value can be achieved. Once this resistance level and the corresponding programming conditions were found out, a similar calculation was carried out for the sensitivity of the programmed resistance to fall time variations. The obtained states and the corresponding programming conditions are highlighted by white marks in Fig. 1.

The choice of the best programming conditions to obtain a state 01 with low sensitivity to programming condition variations is straightforward, when a pulse with high amplitude and long fall time is applied to the memory cell (diamond mark in Fig. 1(a)). A similar state can be obtained by the cartography in Fig. 1(b), but in this case, the current range necessary to achieve the desired resistance is narrower, thus requiring highly accurate current amplitude control.

In contrast, a state 10 with the same resistance value and similar sensitivity to current variations can also be found in both cartographies (square mark in Fig. 1), but when an abrupt pulse fall is employed (Fig. 1(b)), the targeted resistance can be achieved at a lower current. Moreover, in this case, the resistance presents a smaller sensitivity in current and is thus preferred. States 00 and 10 can be achieved with an abrupt pulse fall, while an increased fall time may be used for the programming pulses for states 11 and 01.

In order to program the cells to the target resistance levels, the on-chip current generator in Fig. 4 is proposed. The circuit generates a voltage pulse, V_A, which is converted to a current by means of a resistor (R_1 or R_2) and is then provided to the PCM cell selected by MOSFET transistor M_S.

Capacitor C is first precharged to a given voltage V_P through a switch that is enabled when control signal V_W is high and is then discharged at a constant rate (V_W low) by current I_{DCH}, which is obtained by replicating bias current I_{BIAS} by means of current mirror M_1-$M_{2a,2b}$. The voltage across capacitor C, which corresponds to signal V_A, is therefore a pulse with

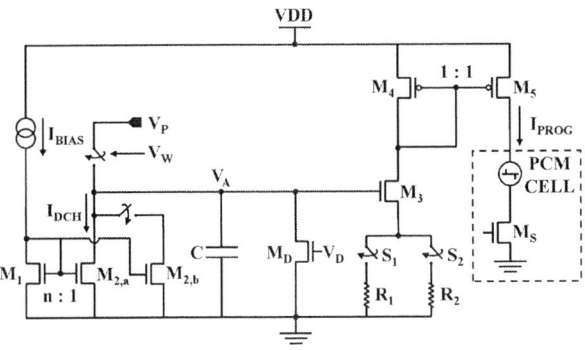

Fig. 4. Circuit schematic of the proposed current pulse generator.

amplitude V_P, time width controlled by the duration of signal V_W, and falling slope controlled by current I_{DCH}.

The two different values of falling slope required by the programming conditions in Fig. 1(a) (states 11 and 01) are achieved by selecting the mirror factor for I_{DCH}, which is obtained by enabling or not transistor M_{2b}. To provide abrupt fall of the pulse when required (programming conditions in Fig. 1(b), states 00 and 10), an additional discharge path for the capacitor was included, namely transistor M_D, which is turned on or off by means of signal V_D.

The generated pulse V_A is then converted to a current pulse by means of the source follower made up by transistor M_3 and a resistor (R_1 or R_2). Switches S_1 and S_2 select the value of the load resistance depending on the required value of the current pulse amplitude. The obtained current, I_{PROG}, is finally fed to the selected PCM cell through current mirror M_4-M_5.

When transistor M_D is enabled to discharge capacitor C, a fall time on the order of a few hundreds of picoseconds is achieved for the current, which enables a good RESET state (state 00) when the amplitude of the generated current pulse is sufficiently high to melt the active portion of the phase change material, or an intermediate (10) state when the current pulse provided has a lower amplitude and a larger pulse width. When capacitor C is discharged by current I_{DCH}, if the current pulse provided to the cell has a low amplitude, a fully SET state (state 11) is achieved whereas, if the amplitude of I_{PROG} is higher, the intermediate resistance state 01 is obtained.

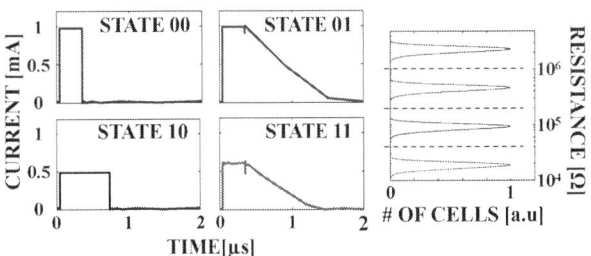

Fig. 5. Simulated current pulse waveforms (left) for each desired resistance state. Resistance distributions (right) considering a maximum resistance variation of 50% (3σ) for each state are also shown.

The circuit was simulated in Cadence environment and the obtained current pulses are shown in Fig. 5 together with the corresponding resistance distributions for each of the four states. According to the recrystallization cartographies of the material, the presented pulses fit the material requirements and are capable of bringing the cell to the desired specific resistance levels.

IV. PROGRAMMED RESISTANCE VARIATION

A key issue of any current pulse generator is the impact of fabrication process spreads over the parameters of the obtained pulse. For this reason, Montecarlo simulations of the pulse generator were performed in order to estimate the effect of programming current variations on the programmed resistance distributions. For resistance states 00 and 10, the fall time of the pulse has substantially no effect, provided that its value is sufficiently short to ensure fast quench. The effects of pulse width variations are also negligible, since control signal V_W is generated by standard digital circuitry (the simulated pulse width variations were on the order of hundreds of ps). The only factor that affects the programmed resistance for these two states is therefore the current pulse amplitude.

In contrast, the programmed resistance of states 11 and 01 is affected by the current amplitude, as well as by the pulse fall time. Fig. 6(a) illustrates the simulated current amplitude variation for all states and the pulse fall time variation for states 11 and 01 (Fig. 6(b)). The observed resistance variation cannot be calculated by simply observing the resistance variation when only the current amplitude or only the pulse fall time varies. In our analysis, we estimated the effect of all variations in the programming pulse characteristics over the programmed resistance.

To this end, we moved across the y-axis of the recrystallization cartographies of Fig. 1(b) (considering a constant programming pulse width) and observed the programmed resistance values for states 00 and 10. For the case of states 11 and 01, where the fall time and the current amplitude variations vary independently, we calculated the resistance value for each simulated combination of pulse amplitude and fall time. After reporting the resistance values, we were able to evaluate the overall resistance variation.

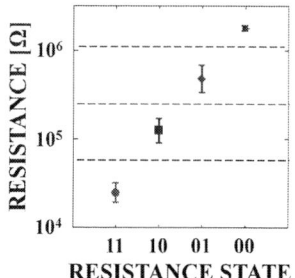

Fig. 7. Estimated variation of the programmed resistance for each resistance state. The estimated resistance variation ensures a safe margin between resistance states, thus enabling 2-bit MLC programming with the proposed pulse generator.

The estimated variation in the programmed resistance for each state can be seen in Fig. 7. The largest resistance variation ($\Delta R = 37.4\%$) is observed for state 10. For states 11 and 01, the programmed resistance varies by $\Delta R = 22.9\%$ and $\Delta R = 30.5\%$, respectively. RESET state 00 shows the least variation ($\Delta R = 7.1\%$). Novel memory cell structures can be utilized in order to address the drift problem and ensure that the programmed resistance distributions do not overlap [9]. From the above, we can conclude that the proposed programming circuit is capable of generating 4 distinct resistance states, which are sufficiently spaced from each other thus allowing accurate resistance state programming for 2-bit-per-cell PCM.

V. CONCLUSION

In this paper, we presented a current pulse generator for programming multiple bits per cell for Phase Change Memory based on N-doped Ge-rich GST. This chalcogenide alloy demonstrates a low crystallization speed and a gradual transition between different resistance levels, thus being an ideal candidate for MLC storage in PCM applications. Studying the recrystallization cartographies of this material, we were able to determine the optimal characteristics of the programming current pulse that are required to bring the memory cell to a target resistance level. A current pulse generator capable of providing current programming pulses with the required characteristics has been presented and simulated. The circuit is capable of providing multiple resistance levels which show modest variation when the applied programming pulses vary due to fabrication process spreads. The accuracy of the programmed resistance levels highlight the effectiveness of the presented technique, thus making the proposed circuit a viable solution for programming multiple resistance levels per cell.

REFERENCES

[1] R.F. Freitas, *et al.*, *IBM J. Res. Dev.*, vol. 52, no. 4/5, pp. 439–447, 2008.
[2] A.L. Lacaita *et al.*, *Microelectron. Eng.*, vol. 109, pp. 351–356, 2013.
[3] A.L. Lacaita, *Solid-State Electron.*, vol. 50, pp. 24–31, 2006.
[4] H. Pozidis *et al.*, *4th IEEE IMW*, 2012, pp. 1–4.
[5] F. Bedeschi *et al.*, *ISSCC Tech. Dig.*, 2008, pp. 428–429.
[6] T. Nirschl *et al.*, *IEDM Tech. Dig.*, 2007, pp. 461–464.
[7] H.Y. Cheng *et al.*, *Proc. EPCOS*, 2013, pp. 1–6.
[8] G. Navarro *et al.*, *IEDM Tech. Dig.*, 2013, pp. 21.5.1–21.5.4.
[9] S. Kim *et al.*, *IEDM Tech. Dig.*, 2013, pp. 30.7.1–30.7.4.

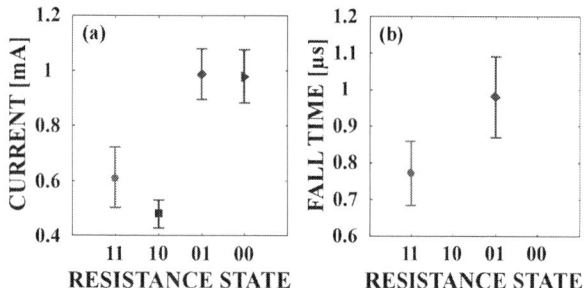

Fig. 6. Simulated (a) current amplitude and (b) corresponding fall time variation of the pulses for the target resistance states. The two pulse characteristics vary independently, therefore, the effect of both of them has to be taken into account to estimate the correct resistance variation.

Design Technology Co-optimization for Enabling 5nm gate-all-around Nanowire 6T SRAM

Trong Huynh-Bao[1,2], Sushil Sakhare[1], Julien Ryckaert[1], Dmitry Yakimets[1,3], Abdelkarim Mercha[1],
Diederik Verkest[1,2], Aaron Voon-Yew Thean[1], Piet Wambacq[1,2]

[1]Imec, Kapeldreef 75, 3001 Leuven, Belgium;
[2]ETRO, Vrije Universiteit Brussel, Brussel, Belgium; [3]ESAT, Katholieke Universiteit Leuven, Heverlee, Belgium
Email: Trong.HuynhBao@imec.be

Abstract—**This paper presents a comprehensive benchmarking and co-optimization of 6T SRAM bitcells designed with 5nm vertical and lateral gate-all-around nanowire FET technology for the first time. A variety of 6T SRAM bitcells configurations combined with different device integration scenarios will be discussed. Our results show that an ultra-dense SRAM bitcell (0.01 um2) can be achieved with vertical FET architecture. The bitcell designed with vertical FET are preferably targeted for low power applications while the lateral FET-based SRAM bitcells could provide 4.5× higher in performance, but resulting in a penalty of 17× increasing in the leakage current compared to the vertical designs. A Vmin of 0.45 V could be obtained for 122 SRAM bitcells implemented with vertical devices.**

Keywords—*5nm, CMOS scaling, DTCO, embedded memory, gate-all-around FETs, nanowire, on-chip variation, parametric yield, 6T SRAM, vertical FET, Vmin.*

I. INTRODUCTION

The exponential growth of the portable and wearable devices such as smart phones and personal healthcare assistants has put an increasing demand for low power and tiny form-factor applications. In several applications, static random access memory (SRAM) occupies a majority proportion of chip area with millions of transistors. Meanwhile, CMOS scaling has taken SRAM into the nanoscale regime where it is difficult to simultaneously achieve the expected density, performance and energy constraints. The 50% area reduction of the SRAM bitcell at every technology node has made it more prone to process variability and reliability issues such as charge trapping, bias temperature instability. A compromise needs to be made for the contradict requirements on the pass-gate for read and write operation in column interleaved memory architectures [1]. The operating voltage of SRAM has not scaled as much as bitcell size as it is strictly difficult to achieve a sufficient parametric yield at low voltage [2]. It becomes challenging to apply a low voltage to both the SRAM and logic in SoC for energy saving. Consequently, the ability of SRAM scaling will rely on the co-optimization of novel transistor architectures, materials, and assist circuit schemes to overcome the intrinsic limitation of SRAM bitcell designs.

Starting with anticipated 5nm technology node, this paper will investigate the scaling prospect and robustness of a disruptive transistor architecture so-called vertical gate-all-around nanowire FET (VNWFET) and lateral nanowire gate-all-around FET (LNWFET) as depicted in Fig. 1 for 6T SRAM bitcell designs. The SRAM bitcells designed with 5nm node

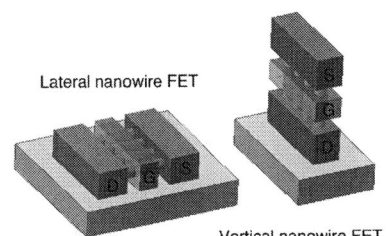

Fig. 1. The lateral nanowire transistor (LNWFET) and vertical naowire transistor (VNWFET). The gate length of a VNWFET device is defined by the thickness of high-κ/metal-gate and not confined by the device footprint area.

design rules are presented. Different configurations of SRAM bitcells implemented with these advanced architecture will be designed and statistically benchmarked using Monte Carlo simulation.

The rest of this paper is organized as follow. Section II introduces the key features and process assumptions for each transistor architecture. The size and layout of SRAM bitcells are also represented in this section. Section III discusses the process variation requirements and details the electrical results of different bitcells configurations for both device architectures.

II. SRAM ARCHITECTURE

A. Process assumption

The specifications for 5nm node technology is briefly listed in TABLE I. For lateral transistor architecture, to achieve the density scaling the gate length (Lg) need to compete with the source/drain (S/D) contact size to maintain an optimized electrostatic control and access resistance. On the contrary, the vertical architecture could allow us to relax the gate length without increasing the device footprint area. Regarding the remaining pitches and dimensions, e.g. contacted gate pitch (CGP), nanowire pitch and metal pitch (MP), both architectures will follow the same guide lines as shown in TABLE I.

A predictive technology model (PTM) based on BSIM-CMG [3] has been built and calibrated with TCAD simulation which takes into account the ballistic transport of the carriers [4]. To achieve the leakage constraints, a high threshold voltage (V_T) is optimized by tuning the gate metal work-function. The leakage current (Ioff) of the NFET and PFET for both architectures are targeted at the same level. The nominal supply voltage is 0.5 V for this technology node.

978-1-4799-7670-6/15 $31.00 © 2015 IEEE

TABLE I.   PROCESS ASSUMPTIONS FOR 5NM NODE

	VNWFET	LNWFET
CGP [nm]	32	32
NW pitch [nm]	18	18
NW diameter [nm]	7	7
Metal pitch [nm]	24	24
Lg [nm]	15	12
Spacer thickness [nm]	5	3

B. SRAM bitcell designs

The configuration of a SRAM bitcell is described by the number of nanowires (NWs) per device and conventionally noted as PU:PG:PD, with PU, PG and PD are the pull-up, pass-gate and pull-down devices. The three typical configurations of SRAM bitcells has been widely adopted by the industry are high density (HD), low voltage (LV) and high performance (HP) [1], [2], [5]. The detail definition for each configuration of SRAM bitcells is represented in TABLE II. These configurations will be employed for benchmarking the SRAM bitcells designed with LNWFET and VNWFET.

TABLE II.   AREA OF SRAM BITCELLS

	VNWFET			LNWFET		
Cell type	HD	LV	HP	HD	LV	HP
Configurations	111	112	122	111	112	122
Area [μm²]	0.010	0.013	0.013	0.014	0.016	0.016

The layout of a 6T-SRAM bitcell implemented with LNWFET technology using 5nm node design rules is illustrated in Fig. 2. In this layout, two dummy nanowires are required for N-PLUS and P-PLUS separation. For an ultimate scaling scenario, the nanowires can be also vertically stacked on top of each other for each device resulting in an increasing number of nanowires per device without any change in bitcell area. This is a major advantage for LNWFET with vertically stacked nanowire.

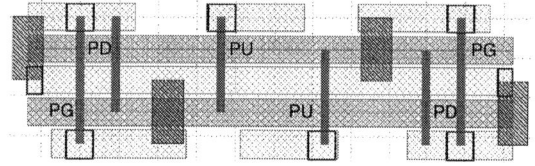

Fig. 2. Layout representation of a low voltage (LV) SRAM bitcell (112) implemented with lateral gate-all-around nanowire FET architecture (LNWFET).

Fig. 3 indicates the layout of the LV SRAM bitcell implemented with VNWFET. The channel is formed vertically and the source/drain is located on top of each other. Thanks to the vertically stacked structure of the gate and source/drain, the P-PLUS and P-PLUS separation could be avoided and resulting in an about 23% of area reduction in the size of SRAM bitcells (TABLE II). The ultra-dense bitcell (0.01 μm²) can be achieved with the VNWFET and provides 29% smaller than the HD bitcell implemented with LWNFET. An additional advantage of VNWFET is the gate length relaxation for improving electrostatic control and not affecting the footprint area.

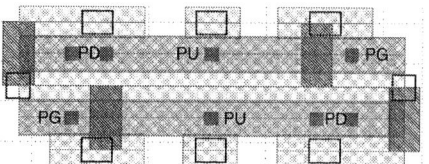

Fig. 3. Layout representation of a low voltage (LV) SRAM bitcell (112) implemented with vertical gate-all-around nanowire FET architecture.

III. ELECTRICAL SIMULATIONS

A. Process variations

A major concern for SRAM design is process variations. The mismatch is the key limiter for the Vmin operation of SRAM bitcells. The analytical equation for σV_T due to random dopant variation is given to be

$$\sigma V_T = q\left(\sqrt{\frac{W_{si}\cdot H_{si}}{W_{eff}}}\right)\cdot \frac{T_{ox}}{\varepsilon ox}\cdot \frac{\sqrt{N_{tot}}}{\sqrt{L_{eff}W_{eff}}} \qquad (1)$$

where q is electron charge, T_{ox} is oxide thickness, ε_{ox} is oxide permittivity, N_{tot} is total channel doping concentration, L_{eff} is physical gate length, and W_{eff} is effective width. For a fully depleted device structure as VNWFET and LNWFET with intrinsic channel, V_T does not have a strong dependence on doping level due to a negligible $T_{ox}\cdot N_{tot}$ quantity. However, the variations from other sources (e.g. work-function, extension resistance...) could have a strong impact on V_T variations. From a circuit perspective, these variations could be lumped together following Pelgrom's rule as

$$\sigma V_T = \frac{A_{vt}}{\sqrt{2L_{eff}W_{eff}}} \qquad (2)$$

where A_{vt} is the slope of the Pelgrom plot [6].

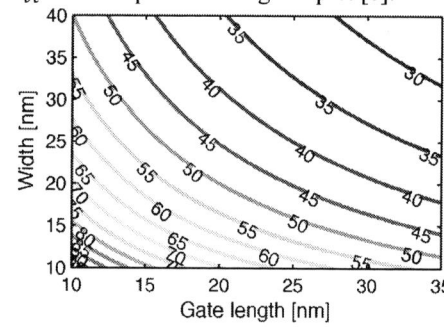

Fig. 4. Transistor random variation as a function of effective width and gate length.

The contour plot of σV_T as a function of device width (W) and gate length (L) with A_{vt} of 1 $mV.\mu m$ is represented in Fig. 4. In order to improve the random V_T variation, we could upsize the bitcell by either increasing the device width or the gate length which lead to a significant penalty in bitcell area. Therefore, for vertical transistor architecture, the gate length is extended without increasing bitcell area. With an ultimate scaling for LNWFET architecture, the V_T variation can be improved by increasing the number of nanowire layers vertically.

978-1-4799-7670-6/15 $31.00 © 2015 IEEE

B. Original SRAM bitcells

The SRAM bitcells are firstly designed with nominal gate length (i.e Lg=15nm for VNWFET, and Lg=12nm for LNWFET) and there is no stacking of nanowire layers. The SRAM read stability (RSNM), read current (Iread), and leakage current (Ileak) is described in TABLE III. Thanks to the gate-all-around device structure, we could obtain a good Iread/Ileak ratio with a short gate length. The bitcells designed with VNWFET could obtain 1.8× higher in Iread/Ileak ratio. The 122 SRAM bitcell achieves the best read and leakage current ratio. However, the read stability margin (μ_{RSNM}/σ_{RSNM}) or access disturb margin (ADM) are too poor for both architecture due to random variation. The ADM of all bitcells configurations are below the 6σ yield target.

TABLE III. READ AND LEAKAGE CURRENT OF SRAM BITCELLS

	VNWFET		LNWFET	
	μ_{RSNM}/σ_{RSNM}	I_{read}/I_{leak}	μ_{RSNM}/σ_{RSNM}	I_{read}/I_{leak}
111	3.3	0.42 x10⁶	2.7	0.19 x10⁶
112	4.4	0.4 x10⁶	3.4	0.23 x10⁶
122	4.2	0.46 x10⁶	3.3	0.28 x10⁶

Fig. 5 indicates the trade-off between the read and leakage current for different bitcell configurations. The longer gate length of VNWFET has made it superior in DIBL and sub-threshold slope and possible for designing the SRAM bitcells with the same read current and 1.5× lower in leakage current with respect to the LNWFET SRAM bitcell. In addition to the short channel effect (SCE), the vertical-based SRAM bitcells are also less susceptible to Vdd variation. Increasing the supply voltage from nominal value to 0.65 V, the leakage of LNWFET SRAM bitcells increase 12% higher than the amount of increment from VNWFET bitcells. The optimized regime as indicated in Fig. 5 are the place where we should target the SRAM bitcells. The results show that the VNFET SRAM bitcells is closer to this regime than the lateral-based design.

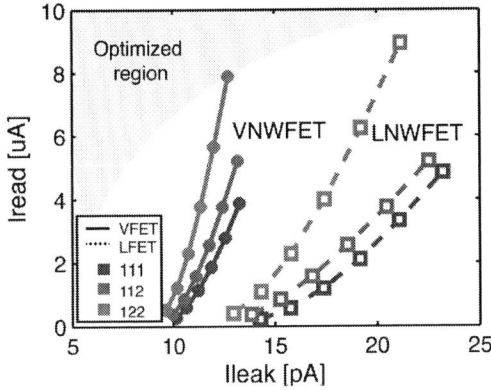

Fig. 5. The scatter plot of read and leakage current for original design of SRAM bitcells

The read stability of SRAM bitcells designed with VNWFET is 14% higher than the bitcells of LNWFET due to its superior in electrostatic control as shown in Fig. 6. Due to vertical orientation of the crystal lattice, the PFET is slightly stronger than NFET device. Consequently, the write trip point

(WTP) for VNWFET is slightly below the LNWFET design. There is a strong contention between the write margin (WRM) and access disturbance margin (ADM). Our results show that, the best write margin could be achieved with 122 SRAM bitcells, and the best ADM could be obtained with the 112 SRAM bitcells.

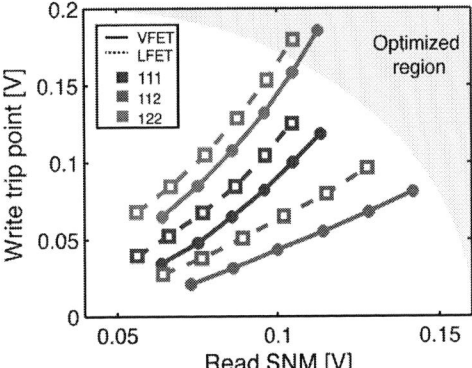

Fig. 6. The scatter plot of read stability and writability for original SRAM bitcell design.

C. Improved SRAM bitcell designs

As the poor the access disturbance margin yield level achieved with the first design, an improved design has been proposed for increasing the yield target. To maintain the bitcell area for density scaling and reach the level of the random ΔV_T variation of sub-40 mV, the gate length for VNWFET is extended to 30 nm. For LNWFET SRAM bitcells, the nanowire will be vertically stacked with three layers which will require an extra cost for process integrations.

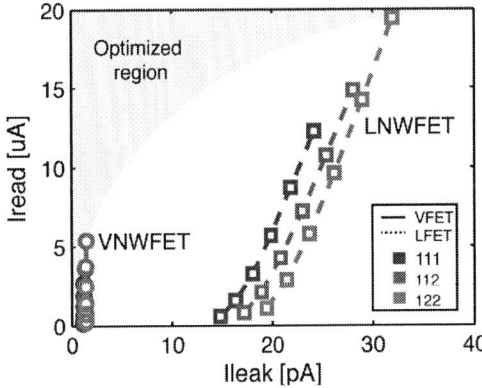

Fig. 7. The scatter plot read and leakage current for SRAM bitcells with stacked nanowire and gate length enhancement.

The read and leakage current for the improved SRAM bitcells is shown in Fig. 7. The consequence of increasing gate length has degraded the Iread of VNWSRAM bitcells by 50%. On the other hand, the Iread/Ileak ratio is improved by more than 5× for VNWFET SRAM bitcells. The leakage current of LNWFET is about 17× higher than the VNWFET design due to the increasing of number of nanowires per devices. It can be observed from the Fig. 7 that the bitcells leakage of VNWFET are insensible to the supply voltage variation. The Ileak of

978-1-4799-7670-6/15 $31.00 © 2015 IEEE

LNWFET design is increased by 2.5× compared to the original designs, and 4.5× higher than vertical bitcells. The inflation of the number nanowire layers could help the LNWFET to boost the Iread current and tolerate to the Vdd and random variation.

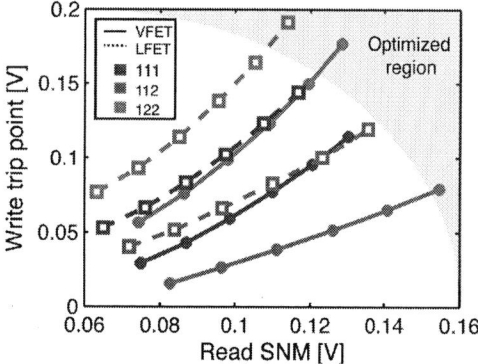

Fig. 8. The scatter plot of read stability and writability for SRAM bitcells with stacked nanowire and gate length enhancement.

The read stability and WTP of all bitcells are slightly improved as in Fig. 8. The RSNM and WTP are improved by 16% for both architecture. But we found that more than 40% and 100% of improvement in the access disturbance margin for VNFET and LNWFET respectively due to the reduction of V_T variability.

D. Vmin of improved SRAM bitcells

Both 112 and 122 SRAM bitcells could achieve a good margin of read stability. For 6σ yield target, the 122 SRAM bitcells of VNWFET could be operated at a minimum supply voltage (Vmin) of 0.45 V as depicted in Fig. 9. The 122 bitcells from LNWFET require 35 mV more in the Vmin to reach the same yield target. The 111 SRAM bitcell need to a minimum supply voltage of 0.6 V in order to reach 6σ yield target for ADM.

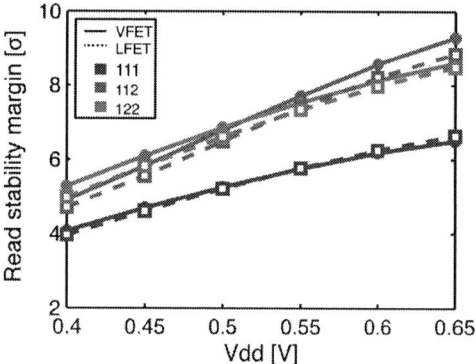

Fig. 9. Cell margin for read stability of SRAM bitcells.

With the Vmin of 0.45 V, the required voltage for bitline (BL) in order to achieve 6σ yield targeting is illustrated in Fig. 10. For the HP bitcells, the required BL voltages are negative 100 mV and 60 mV for designs with VNWFET and LNWFET architecture respectively at the Vmin supply voltage. Therefore, to enable the HP bitcells implemented with nanowire

architecture and operated at Vmin will require write assist techniques (e.g. negative bitline, wordline underdrive). The 122 bitcell designed with LNWFET could operate at 0.65 V without a requirement of negative BL voltage. The rest of SRAM bitcells need to write assist circuit to target 6σ bitcell yield.

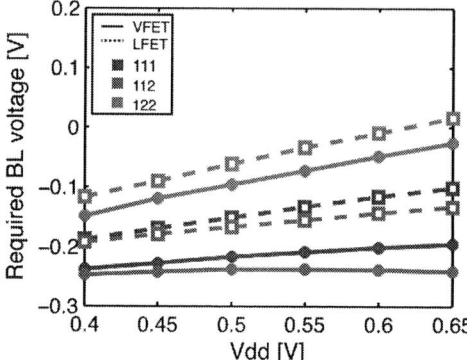

Fig. 10. The required bitline (BL) voltage for 6σ yield.

IV. CONCLUSION

A comprehensive benchmarking of SRAM bitcells designed with vertical and lateral nanowire FET technology has been presented. The VNWFET architecture is shown to be superior in electrostatic control and resulting in a 17× reduction of leakage current compared to the LNWFET architecture and more immune to the Vdd variations. However, due to the limitation of the drive current scalability, the VNWFET is appropriate for low voltage and low power applications. On the other hand, the LNWFET technology should be targeted for high performance applications due to its ability of increasing drive current. A write assist circuit is required for enabling the 6σ yield of SRAM bitcells designed with both architecture. The read stability Vmin of 0.45 V could be obtained for HP bitcells implemented with VNWFET. SRAM bitcells designed with VNWFET can provide 20% reduction in area and operate at about 35 mV lower to achieve the same yield target.

REFERENCES

[1] E. Karl et al., "A 4.6 GHz 162 Mb SRAM Design in 22 nm Tri-Gate CMOS Technology With Integrated Read and Write Assist Circuitry," *IEEE Journal of Solid-State Circuits*, vol. 48, no. 1, pp. 150–158, Jan. 2013.

[2] T. Song et al., "A 14nm FinFET 128Mb 6T SRAM with VMIN-enhancement techniques for low-power applications," in *2014 IEEE International Solid-State Circuits Conference Digest of Technical Papers (ISSCC)*, 2014, pp. 232–233.

[3] Compact Modeling for Multi-Gate Device Structure (BSIM-CMG). [Online]. Available: http://www-device.eecs.berkeley.edu/bsim.

[4] Synopsys Sentaurus Technology Computer-Aided Design (TCAD). [Online]. Available: http://www.synopsys.com/Tools/TCAD.

[5] Y.-H. Chen et al., "A 16nm 128Mb SRAM in high-κ metal-gate FinFET technology with write-assist circuitry for low-VMIN applications," in *2014 IEEE International Solid-State Circuits Conference Digest of Technical Papers (ISSCC)*, 2014, pp. 238–239.

[6] M. J. M. Pelgrom, A. C. J. Duinmaijer, and A. P. G. Welbers, "Matching properties of MOS transistors," *IEEE Journal of Solid-State Circuits*, vol. 24, pp. 1433–1440, 1989.

Assessment of SiGe Quantum Well transistors for DRAM peripheral applications

R. Ritzenthaler, T. Schram, G. Eneman, A. Mocuta,
N. Horiguchi, A. V.-Y. Thean
imec, Kapeldreef 75,3001 Leuven, Belgium
contact: romain.ritzenthaler@imec.be

A. Spessot, M. Aoulaiche, P. Fazan
Micron Technology Belgium, imec Campus,
Kapeldreef 75,3001 Leuven, Belgium

K.B. Noh, Y. Son
Assignee at imec from SK-Hynix

Abstract— **In this work, the potential of $Si_{1-x}Ge_x$ Quantum Wells (SiGe QW) for future DRAM periphery transistors and more generally for Low Power applications is investigated. It is shown that an increase of Ge content in the channel leads to a significant reduction of threshold voltage and to an increase of long channel mobility. However, an increase of external resistance is observed for $Si_{1-x}Ge_x$ Quantum Well devices, which is attributed to junction induced defects creation at the SiGe/Si buffer layer interface. This highlights the need for a dedicated junction solution in SiGe QW devices. The junction leakages are also investigated, and it is found that Band to Band Tunneling is the dominant mechanism setting the minimum Off-state leakage current. Band to Band Tunneling is increasing when the Ge content is increased, and it may effectively cap the allowed Ge channel content for Low Power Applications. The minimum Off state leakage requirement for DRAM peripheral applications is still obtained for a Ge concentration of 45%.**

Keywords— *DRAM, DRAM periphery circuitry, Junction leakage, Low Power applications, SiGe channels*

I. INTRODUCTION (HEADING 1)

Apart from the memory element, Dynamic Random Access Memory (DRAM) technologies need transistors in their peripheral circuitry to perform different operations (address decoding, I/O circuits, etc. [1]). Even if at the moment all the commercially available DRAM devices use standard poly-SiO_2/SiON transistor, in order to keep pace with the growing requirement of the semiconductor industry these DRAM periphery transistors are pushed toward the introduction of High-K/Metal Gate (HKMG) for the next technology nodes [2]. In order to target Low Power mobile applications, their performance must also be close enough to Low STand-by Power (LSTP) logic transistors (low gate and junction leakages are an absolute necessity), which justifies a thicker Equivalent Oxide Thickness (EOT) compared to High Performance Logic applications. DRAM periphery transistors must also stay compliant with DRAM process compatibility requirements and should sustain a high thermal budget (called DRAM Anneal(s) in this work). The DRAM Anneals are linked to the DRAM cell fabrication (storage capacitor, access transistor, BEOL), and should be seen as a <u>constraint</u> for periphery transistors design.

Multi Threshold voltage (V_{TH}) devices are also required for DRAM periphery applications, given the wide range of transistors used in the DRAM periphery. In a MIPS (Metal Inserted Poly Silicon, or "Gate First") flow, a low PMOS V_{TH} is difficult to obtain using work function shifters only, especially without compromising the gate leakage and channel mobility. It has been shown that a Replacement Metal Gate approach could provide a significantly wide range of V_{TH} [3]. Another possibility to reduce the threshold voltage of PMOSFETs is to use the valence band shift with the introduction of $Si_{1-x}Ge_x$ channel [4][5]. In this work, the potential of SiGe Quantum Wells is investigated for DRAM periphery transistors, and more generally for Low Power applications. The Process and drive current performance are presented in part II, while part III is dedicated to the off-state leakage control.

II. PROCESS AND DRIVE PERFORMANCE

The $Si_{1-x}Ge_x$ QW pFET devices were fabricated on 300 mm (100) Si-wafers using a DRAM peripheral device process flow. After STI formation, the compressively strained $Si_{1-x}Ge_x$ layers (Ge content being 25% or 45%) are grown by epitaxy on a 2 nm Si buffer layer. Subsequently, a thin Si cap is grown, followed by the gate stack (SiO_2 interfacial layer, HfO_2, TiN metal gate, PolySilicon). The gate stack features an EOT ranging between 1.4 and 1.6 nm, well within the DRAM periphery transistors requirements [2]. The rest of the flow is standard (gate patterning, extension/halo implants, spacer deposition and etch, HDD implants, activation with RTA annealing) up to the above mentioned DRAM anneal, inserted to mimic the real fabrication of the full memory chip. Final standard BEOL (up to Metal 1) completes the processing. Conventional Si channel devices were also fabricated as references, featuring identical junction scheme. **Fig.1** illustrates the fabricated devices, with inset showing the gate stack (SiGe channel, Si cap, Interfacial layer (IL), high-k gate oxide and Metal gate).

On long channel devices ($L_G=W_G=1\mu m$, **Fig. 2**), the threshold voltage (V_{TH}) shift due to the bandgap narrowing in SiGe QW transistors is clearly visible. An aggressive V_{TH} down to -0.15V is obtained with a channel germanium content of 45%. The long channel mobility boost obtained with SiGe channels is also clearly visible (**Fig. 2**), with an enhanced effect when increasing the Ge content in the channel. The DRAM

978-1-4799-7670-6/15 $31.00 © 2015 IEEE

anneal(s) impact is limited to differences in inversion thicknesses, while subthreshold slopes and drive current identical performance point toward a marginal effect of DRAM anneal(s) on SiGe channel relaxation and Ge out-diffusion in the buffer layer or Si cap [6]. The NBTI performance on SiGe QW devices are also very good, and attributed to a favorable energy decoupling between SiGe channel and gate dielectric defects due to valence band offset [6,7].

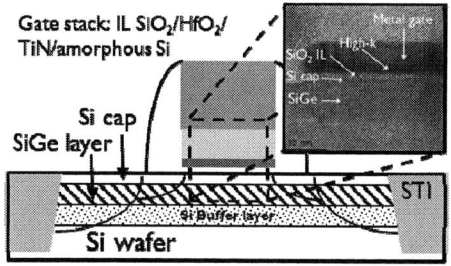

Fig. 1: Sketch of fabricated SiGe QW transistors. Inset shows a TEM cross section of the gate stack.

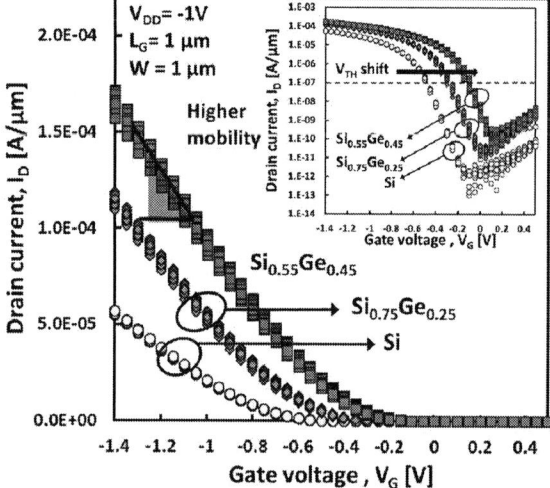

Fig. 2: Long channel saturation drain current I_D vs. gate bias V_G for SiGe channels (Ge concentration of 25% (diamonds) and 45% (squares)) and Si control wafer (circles). Inset shows currents in logarithmic scale. W_G=1 µm, V_{DD}=-1V. EOT (t_{INV} - 4Å)=1.2 nm for Si reference and 1.5 nm for $Si_{1-x}Ge_x$ wafers.

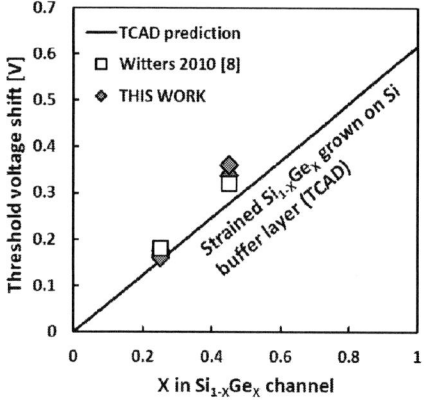

Fig. 3: Comparison of the threshold voltage shift vs. Ge concentration in the channel given by TCAD predictions (line), Logic devices (squares) and DRAM periphery devices (diamonds). V_{DD}= -1V.

The threshold voltage shift, already apparent in **Fig.2**, is plotted against Ge content and compared to TCAD predictions made for SiGe layers grown on Si (and therefore compressively

strained) (**Fig.3**). A good agreement is obtained both with High Performance Logic transistors (featuring a thinner EOT for higher drive current and less stringent gate/junction leakages requirements [8]) and with TCAD predictions.

Regarding the short channel performance, R_{ON} (extracted at $V_{TH,LIN}$-1V and V_D=20mV, **Fig. 4**) demonstrates that high field short channel mobility in both $Si_{0.75}Ge_{0.25}$ and $Si_{0.55}Ge_{0.45}$ is higher than the silicon reference featuring identical junctions scheme. $Si_{0.75}Ge_{0.25}$ and $Si_{0.55}Ge_{0.45}$ exhibit a similar $R_{ON}(L_G)$ slope, showing that they have comparable short channel mobilities. In other words, the long channel performance indicating a clear mobility improvement when the Ge content is increased is still observed for short channels, but with a loss of advantage for $Si_{0.55}Ge_{0.45}$ over $Si_{0.75}Ge_{0.25}$. The fact that mobility is not higher for the 45% germanium content suggests that a channel relaxation (because of LDD and HALO implantation) or a partial Ge out-diffusion might be occurring. Regarding external resistance R_{EXT}, there is a clear increase when the Ge content is increased (**Fig. 4**). This is particularly clear for $Si_{0.55}Ge_{0.45}$, where R_{EXT} is practically twice the value obtained in the Si reference. This strong R_{EXT} increase is correlated to the presence of damage in the SiGe/Si buffer interface, as shown in the TEM image/cross section obtained on $Si_{0.75}Ge_{0.25}$ after RTA Spike annealing (**Fig. 5**; the wafer was processed up to silicidation step to observe implant related damages only).

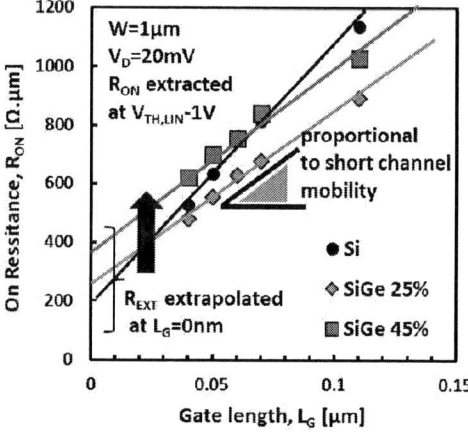

Fig. 4: On resistance R_{ON} vs. gate length L_G obtained for $Si_{0.75}Ge_{0.25}$ (diamonds), $Si_{0.55}Ge_{0.45}$ (squares) and Si control wafer (circles). W_G=1 µm, V_D=-20mV.

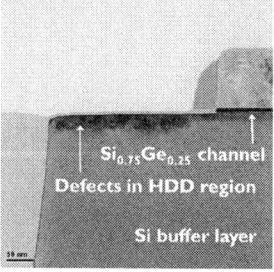

Fig. 5: TEM illustrating implant related damages. Wafer is withdrawn between HDD implants and silicidation, and RTA thermal budget applied.

The increase in short channel mobility dominates the higher external resistance, which is evidenced by a drive current boost at fixed I_{OFF} for $Si_{0.75}Ge_{0.25}$ over Si reference (**Fig. 6**). Due to the increased R_{EXT} over the short channel mobility improvement, devices with a Ge content of 45% are however showing a clear current penalty with regard to the Silicon control wafer. These results demonstrate that for transistors where the SiGe channel is left in the Source/Drain region (i.e. no recess), a specific

978-1-4799-7670-6/15 $31.00 © 2015 IEEE   14

junction solution is required to avoid defect injection and SiGe channel relaxation which will be detrimental to the short channel drive current.

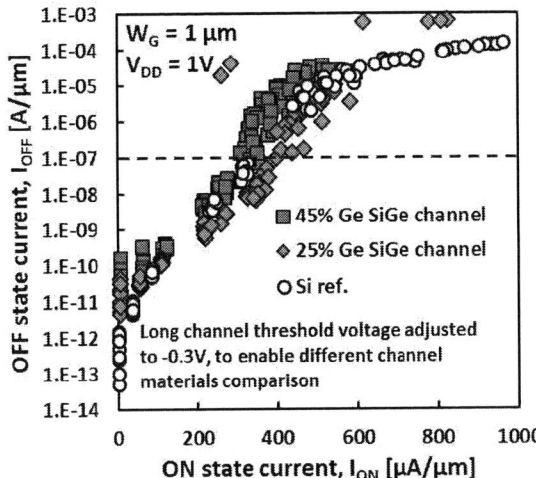

Fig. 6: Intrinsic Transistor Performance (ITP) obtained for $Si_{0.75}Ge_{0.25}$ (diamonds), $Si_{0.55}Ge_{0.45}$ (squares) and Si control wafer (circles). The Long channel threshold voltage V_{TH} is adjusted to -0.3V in order to enable different channel material comparison. W_G=1μm, V_{DD}=-1V.

III. ANALYSIS OF JUNCTION LEAKAGES

Off leakages in a transistor may come from many sources, such as degraded subthreshold slope (diffusion current), gate leakage, extension leakages (possibly enhanced by halo profile [9]), diode leakage in the HDD region, and Gate Induced Drain Leakage (GIDL) [10]. GIDL is created by the band bending in the drain extension due to the electric field issued from the gate; it can be purely Band to Band Tunneling (BTBT), or be enhanced by Trap Assisted Tunneling depending on the energy distribution of interface states. As presented in the introduction, the Low Power transistors are more stringent in terms of I_{OFF} control than High Performance application. Therefore, it is very important to understand the intrinsic limitations of a given technology.

Following the methodology presented in [11,12], a parallel devices array (6000 devices in this study) and area diodes (featuring a surface of 5000 μm²) were used to extract junction leakages. First, the current in area diodes is measured (**Fig. 7**). It is noteworthy that no difference is obtained with regard to the Ge content of the channel, indicating that the HDD/substrate junction is located in the Si substrate well below the SiGe channel, as sketched in **Fig. 1**. Then, the parallel devices array is biased in two configurations in order to decorrelate GIDL from other components (**Fig. 8**). By normalizing considering the layout dimensions of the parallel devices array, HDD leakages are compared to the current in the configuration 1 (i.e. recombination current, extension leakage and a negligible gate contribution). It is found that the HDD leakage component is 2 to 3 orders of magnitude lower than extension + recombination current component (**Fig. 9**). It is therefore a negligible contribution for the setting of the minimal off leakage.

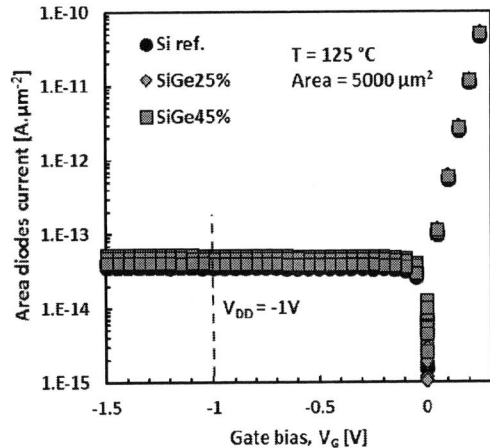

Fig. 7: HDD P/N junction diode current (reverse) obtained for $Si_{0.75}Ge_{0.25}$ (diamonds), $Si_{0.55}Ge_{0.45}$ (squares) and Si control wafer (circles). Surface=5000μm², T=125°C.

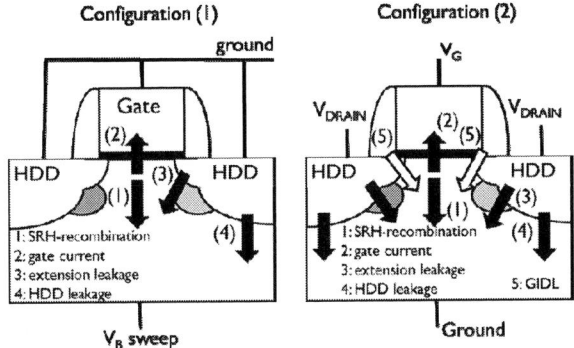

Fig. 8: Configurations used in parallel devices to extract junction leakages (Configuration 1) and GIDL current + junction leakages (Configuration 2). 5916 devices are used in parallel.

Then, by subtracting the current obtained in the configuration (2) with the gate leakage component and the current obtained in the configuration (1), the real GIDL current can be extracted (**Fig. 10**). Comparing the current at matched drain to gate bias V_{DG} obtained in parallel devices and stand-alone transistors, an excellent overlay is found for both 25% and 45% Ge content, as shown in **Fig.10.a** and **10.b**. Furthermore, temperature measurements (from T=25°C to 125°C) were used to extract the activation energy (E_A) of the GIDL current. It is found that E_A is smaller than 0.1 eV for drain to gate bias V_{DG}<-1V (**Fig. 11**). Therefore, the leakage currents in the transistor accumulation regime are not due to a junction implant induced Trap Assisted Tunneling but only to a Band to Band Tunneling mechanism. Hence, the minimum off leakage that can be obtained will be determined by the short channel control (subthreshold slope) and Band to Band Tunneling. In correct agreement with previously published simulations based on BTBT [13], the bandgap narrowing obtained by increasing the Ge content in the channel is yielding an increase of $I_{OFF,MIN}$. These results demonstrate that the specifications generally accepted by the industry for DRAM peripheral applications (I_{OFF}=10^{-10} A/μm, Fig. 10) are reached for a Ge content of 45%.

978-1-4799-7670-6/15 $31.00 © 2015 IEEE

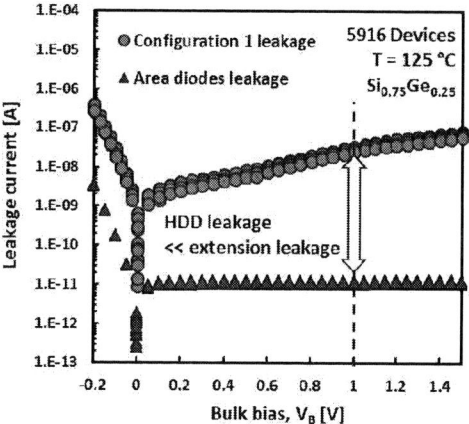

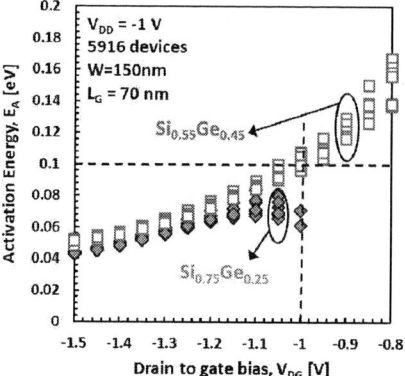

Fig. 11: Activation energy E_A measured for $Si_{0.75}Ge_{0.25}$ (diamonds) and $Si_{0.55}Ge_{0.45}$ (squares) vs. drain to gate bias V_{DG}.

Fig. 9: Comparison of the HDD current measured on area diodes (and matched to the layout dimension of parallel devices) with the current extracted in the configuration 1 of Fig. 8 (HDD leakages + extension leakages + SRH current). HDD leakages are negligible compared to the other components.

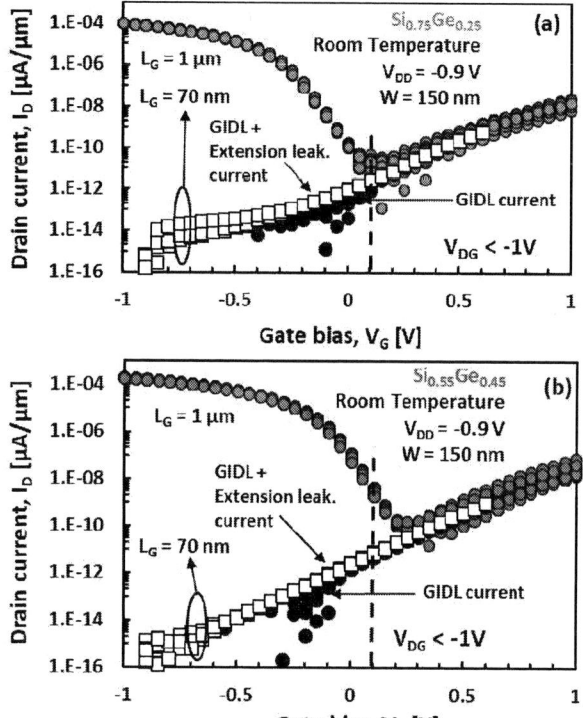

Fig. 10: Comparison of GIDL current and total leakage current (configuration 2 of Fig 8) obtained on parallel devices and transfer characteristics on transistors for $Si_{0.75}Ge_{0.25}$ (a) and $Si_{0.55}Ge_{0.45}$ (b) SiGe channels. V_{DD}=-0.9V. the gate bias corresponding to V_{DG}=-1V (where the BTBT predominates) is indicated by the dashed line.

IV. CONCLUSIONS

$Si_{1-x}Ge_x$ channel technology is a promising candidate for future DRAM applications with demonstrated V_{TH} shift and long channel mobility boost, but with a couple of challenges to be tackled. The first one is the control of the junctions implant and RTA induced channel relaxation and defect injection at the $Si_{1-x}Ge_x$/Si buffer interface, if the SiGe channel is not recessed in the Source/Drain region. The other challenge, true for any Low Power applications, is the control of the minimal I_{OFF} that can be obtained. It is shown in this work that the minimum reachable I_{OFF} is fixed by the short channel control (subthreshold slope) and Band to Band Tunneling. Narrow bandgap materials, used to obtain lower threshold voltage and higher mobility, exhibit higher Band to Band Tunneling. Low Power off-leakage requirements may then cap the Ge content that can be used in the channel, and a careful trade-off considering the supply voltage and the I_{ON} at the Low Power I_{OFF} target has to be made for a given Ge channel concentration. These results demonstrate however that the off-leakage specification for DRAM periphery applications is still reachable with a Ge concentration up to 45%.

REFERENCES

[1] T. Vogelsang, "Understanding the Energy Consumption of Dynamic Random Access Memories," *Microarchitecture (MICRO) 43rd Annual IEEE/ACM International Symposium on*, 2010.

[2] S.Y. Cha, "DRAM Technology - History & Challenges", *IEDM 2011 short course*, 2011.

[3] R. Ritzenthaler et al., "Low-power DRAM-compatible Replacement Gate High-k/Metal Gate Stacks ", proc. of the *42th European Solid-State Device Research conference (ESSDERC)*, 2012.,

[4] S. Krishnan et al., *2011 International Electron Devices Meeting*, 2011, pp. 28.1.1–28.1.4.

[5] H. C.-H. Wang et al.,in *proc. IEDM*, pp.161 -164, 2004 .

[6] Y. Son, K. B. Noh, M. Aoulaiche, R. Ritzenthaler, T. Schram, A. Spessot, P. Fazan, M. Cho, J. Franco, N. Horiguchi and A. V.-Y. Thean., *International Memory Workshop'2014* proceedings, 2014.

[7] J. Franco et al., in *2010 International Electron Devices Meeting*, 2010, no. 7, pp. 4.1.1–4.1.4.

[8] L. Witters et al., in proc. *Symp. VLSI technology*, pp. 181-182, 2010.

[9] G. Eneman et al., *IEEE Transactions on Electron Devices*, vol. 56, no. 12, pp. 3115–3122, Dec. 2009.

[10] K. Roy et al., "Leakage current mechanisms and leakage reduction techniques in deep-submicrometer CMOS circuits," *Proceedings of the IEEE*, vol. 91, no. 2, pp. 305–327, Feb. 2003.

[11] G. Roll et al., in 2010 Proceedings of the *European Solid State Device Research Conference*, pp. 329–332, 2010.

[12] D. Rideau et al., in Proceedings of the *2004 International Conference on Microelectronic Test Structures*, pp. 149–154, 2004.

[13] T. Krishnamohan et al., in *2006 International Electron Devices Meeting*, 2006.

978-1-4799-7670-6/15 $31.00 © 2015 IEEE

Low Standby Power Capacitively Coupled Sense Amplifier for Wide Voltage Range Operation of Dual Rail SRAMs

Anuj Grover,
Promod Kumar,
Mohammad Daud
STMicroelectronics
Plot # 1, Knowledge Park-3,
Greater NOIDA, UP, India
anuj.grover@st.com

G. S. Visweswaran
Department of Electrical
Engineering,
Indian Institute of
Technology Delhi
Hauz Khas, Delhi, India
gswaran@ee.iitd.ac.in

Chittoor Parthasarathy,
Jean-Philippe Noel,
David Turgis
STMicroelectronics,
850 rue Jean Monnet,
38920, Crolles Cedex France
chittoor.parthasarathy@st.com

Bastien Giraud,
Guillaume Moritz
CEA, LETI, MINATEC
Campus,
38054, Grenoble Cedex 9,
France
bastien.giraud@cea.fr

Abstract— **Dual Rail SRAMs are widely used to enable Dynamic Voltage and Frequency Scaling (DVFS) in SRAMs where array voltage cannot be scaled down. DVFS operating points are limited by maximum differential supported between two supplies of the SRAM. To extend gains of DVFS, we propose a Low Standby Power - Capacitively Coupled Sense Amplifier (LSTP-C2SA) that enables further lowering of periphery supply in Dual Rail SRAMs without leading to SRAM cell instability. We present a design method to optimally size the coupling capacitance in LSTP-C2SAs. Designs with LSTP-C2SA are shown to consume 43% lesser read power in DVFS operation at 0.4V in 28nm UTBB FD-SOI when compared to an implementation with standard latch sense amplifier. Silicon measurements confirm LSTP-C2SA functionality at 0.35V.**

Keywords— *Sense Amplifier, Dual Rail SRAMs, DVFS, Wide Voltage Range*

I. INTRODUCTION

Technology scaling has enabled lowering of supply levels for digital logic, but is unable to offer comparable reduction in minimum operational voltage of SRAM cells [1]. Dual Rail memories are widely used to maintain area scaling of the SRAM cell while extracting gains of DVFS by lowering supply levels of only the periphery circuits in the SRAM [2]. When DVFS is applied in dual rail SRAMs, the cell array is maintained at a higher voltage (say VDDA), while the periphery (including control and decoder logic; sense amplifiers; write drivers etc.) is scaled to a lower voltage level (say VDDP).

Memory cell array consumes limited power from the supply, VDDA. The circuits on periphery that run on VDDP supply consume most of the memory power. Therefore, almost entire gains of DVFS accrue even if only VDDP supply is lowered.

Fig. 1 is a functional view of a memory bit-slice where memory cells are coupled with a sense amplifier (SA) and

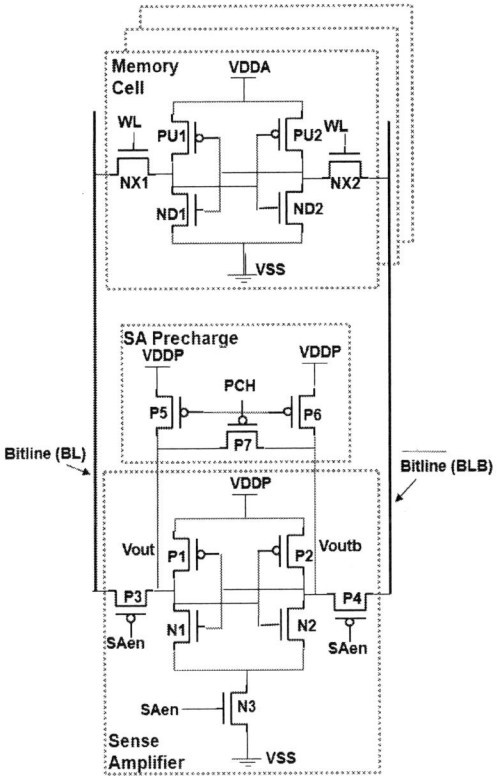

Fig. 1. Conventional latch type Sense Amplifier (LSA) in a memory slice. At reset (WL=0, SAen=0, PCH=0), BL and BLB are connected to SA supply (VDDP).

SA precharge circuits. In memory reset state, P3-P7 are ON and the bitlines are maintained at VDDP level.

In a dual rail SRAM, it is essential to drive wordline (WL) to VDDA to maintain write-ability of the SRAM cell and ensure high read current. Precharging BLs to VDDP (which is lower than VDDA in DVFS mode), improves cell

978-1-4799-7670-6/15 $31.00 © 2015 IEEE

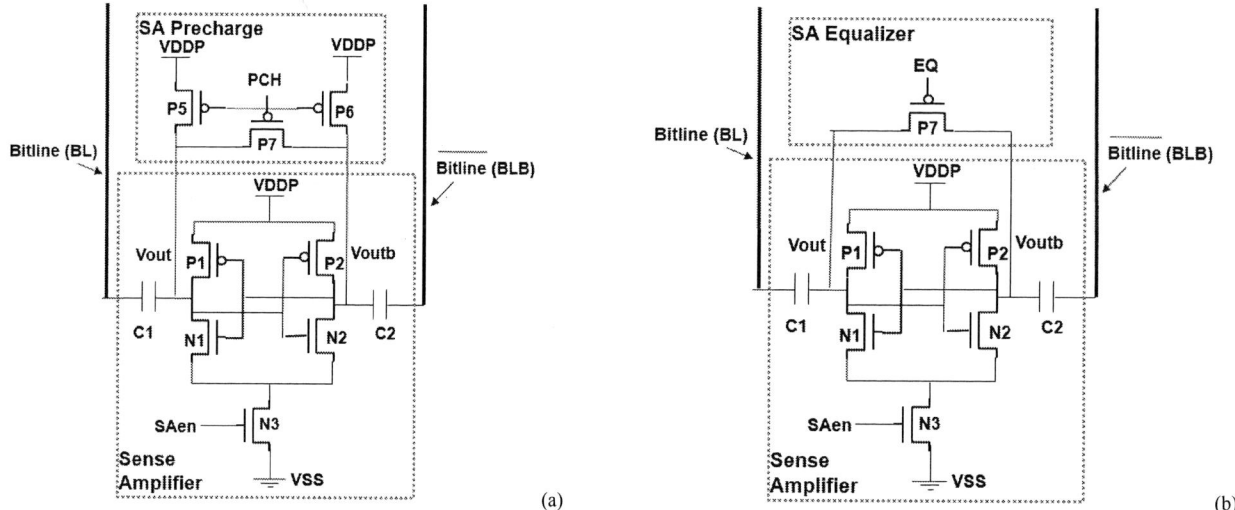

Fig. 2. Proposed Low Standby Power Capacitively Coupled Sense Amplifier (LSTP-C2SAs): (a) VDDP-precharged; (b) Q-point equilibrated.

stability during read mode by reducing the noise injected in the SRAM cell when WL is selected. However, beyond a certain level of lowering on VDDP, uninitiated write process is triggered and stability of memory cell degrades. This sets a bound on VDDP lowering [3]. Further lowering of VDDP beyond this bound can be done only if BLs are precharged to VDDA. From Fig.1, it is evident that BLs connect to SA supply in reset state. Therefore, it becomes necessary to put the SAs also on VDDA. Thus, power consumed by SAs does not scale with VDDP. SA power is a significant part of memory read power and while DVFS to lower VDDP levels is enabled, SRAM power consumption does not reduce.

Capacitively coupled SAs (C2SAs) have been proposed in [4] and [5]. In C2SAs, BLs couple with SA through capacitors instead of transistors P3 and P4 (Fig. 1) and BL supply is not shorted with SA supply in reset mode. This allows SA to be maintained on VDDP supply and reduce power consumption in low frequency read operation, while BLs are charged to VDDA. The proposed C2SAs also enable high performance. However, both suffer from high static current. This can offset dynamic power gains in low frequency modes and defeat the paradigm of DVFS and low power design.

In this work we propose Low Standby Power Capacitively Coupled Sense Amplifiers (LSTP-C2SAs) that utilize low standby power features of conventional latch type sense amplifier (LSA) with benefit of independent BL and SA supplies of C2SAs. This enables extended lowering on VDDP and offers additional DVFS modes to SoC designers.

Rest of the paper details the solution. In Section II, we present the design and operation of LSTP-C2SA. In Section III, we describe the experimental setup and method to characterize the SA and size the coupling capacitors of C2SA. In Section IV, we present the comparison between conventional SA based design and proposed SA based design. Finally, we sum up the results in Conclusions.

II. LOW STANDBY POWER CAPACITIVELY COUPLED SENSE AMPLIFIER (LSTP-C2SA)

We propose two types of Low Standby Power Capacitively Coupled Sense Amplifiers (LSTP-C2SAs) in Fig. 2. In the proposed SAs, P3 and P4 devices of LSA (Fig. 1) are replaced with capacitances, C1 and C2. In the SA of Fig. 2(a), like in LSA, SA nodes are precharged to VDDP level. It is evident that in reset mode, before read operation is triggered, N1 and N2 will be in saturation, while P1 and P2 will be cut-off. Therefore, P1 and P2 have very limited effect on offset of both LSTP-C2SA of Fig. 2(a) and also LSA.

In LSTP-C2SA of Fig. 2(b), SA nodes are equalized at the Q-point of the SA. In Q-point equilibrated LSTP-C2SA, N1, N2, P1, and P2, will all be in saturation. So, P1 and P2 also influence SA offset. They have to be designed slightly larger than in LSA to minimize offset and improve response time.

In Q-point equilibrated LSTP-C2SA, N3 prevents high static current in reset mode and overcomes a key limitation of using the C2SA of [5] for DVFS centric designs. VDDP precharged LSTP-C2SA does not suffer from high static current, but the offset of the SA due to device mismatch is not auto-cancelled as in [5], and therefore N3 is needed to trigger read when sufficient differential voltage has been generated. The LSTP-C2SAs proposed here are well suited for wide range DVFS designs.

It is important to understand the operation of C2SAs and how it is different from LSA to estimate the impact of using LSTP-C2SAs in a SRAM. Differential generated on Vout/Voutb pair is dependent on read current of the memory cell, leakage current of unselected cells on the same BL, BL capacitance, and time-lapse between WL activation and SAen arrival. In LSA, charge transfer efficiency for a given discharge time is dependent on linear resistance of P3 and P4 devices. This resistance increases in DVFS mode when BLs and LSA are at VDDP supply. So, time-lapse between WL activation and SAen arrival has to be designed to be dependent on level of VDDP. However, in LSTP-C2SAs,

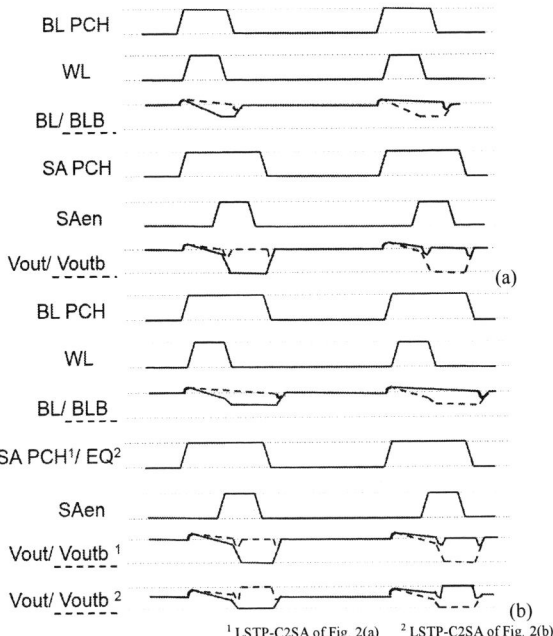

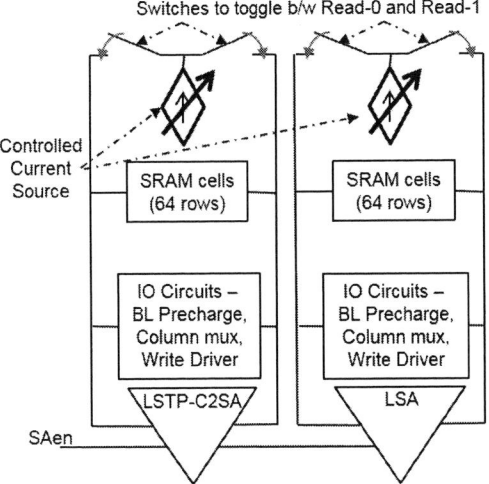

Fig. 3. Read-0 and Read-1 operation in (a) LSA (b) LSTP-C2SAs.

[1] LSTP-C2SA of Fig. 2(a)   [2] LSTP-C2SA of Fig. 2(b)

Fig. 4. Bitlines (BLs) are connected to 64 inactive SRAM cells, BL precharge, write driver, I/O mux, and SAs. Both SAs are enabled simultaneously by SAen signal. Switches connect two controlled current sources to discharge BL/ BLB for Read-0/ Read-1.

charge coupling efficiency depends on the ratio of BL capacitance and C1 and C2 that can be designed to remain voltage independent.

Additionally, in cases where time-lapse between WL activation and SAen arrival is large and leakage currents are not negligible, leakage on non-discharging BL reduces differential voltage created at LSA internal nodes. LSTP-C2SAs are not impacted because C1 and C2 offer high impedance to slow moving leakage signal. Consequently, LSTP-C2SAs achieve faster read at low voltage.

In LSA, the cross-coupled inverters (P1-N1 and P2-N2) need to only drive the output circuit (latch and driver) to complete the read operation. In C2SAs, these inverters also drive C1 and C2 in addition to the output circuits. Consequently, LSTP-C2SAs have slower response time than LSA for same differential at internal nodes and require larger activation pulse on SAen [6]. This offsets some gains discussed above.

Thus, the proposed SAs achieve similar performance as LSA and enable additional DVFS modes with lower VDDP to improve energy efficiency.

III. DESIGN AND CHARATERIZATION

We designed LSTP-C2SAs and LSA in 28nm Ultra-Thin Body and Box Fully Depleted Silicon On Insulation (UTBB FD-SOI) technology. From the discussion in Section II, it is evident that C1 and C2 strongly influence charge coupling efficiency in C2SAs and have to be sized carefully. To be able to read with similar read current, LSA and C2SAs should have similar dispersion of offset requirement.

The setup to measure dispersion of offset current is shown in Fig. 4. A set of 64 memory cells is coupled with LSTP-C2SA and another with LSA. To emulate real case loading on BLs, precharge devices, multiplexer gates and write driver are also connected. Controlled current sources emulate worst case read current in 1000 Monte-Carlo simulations. Simulations are run with different values of C1 and C2. Dispersion of the current at which each SA latches correct output is measured. To ensure that C1 and C2 don't change with voltage of operation, they are implemented through metal capacitance (Fig.5). They are sized such that dispersion of LSTP-C2SAs is same as LSA (Fig. 6).

The SAs are also designed into a dual rail SRAM instance (1024 words, 32 bits, column mux 4, 4 banks) and power consumption of the instance is characterized for implementations with LSA and BLs at VDDA and VDDP and also with LSTP-C2SA at VDDP and BLs at VDDA. The results are presented in Section IV.

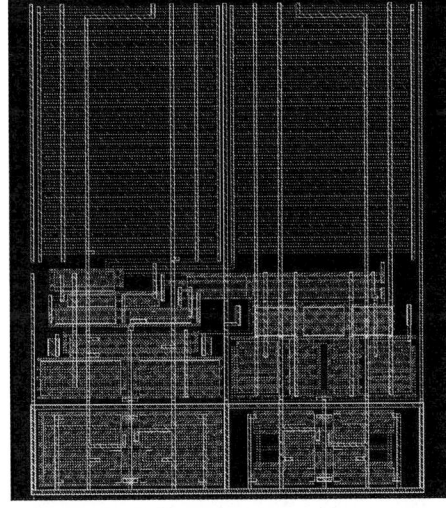

Fig. 5. Layout of LSTP-C2SA and LSA embedded in SRAM. Coupling capacitance has been implemented in M1 (shown) and M3.

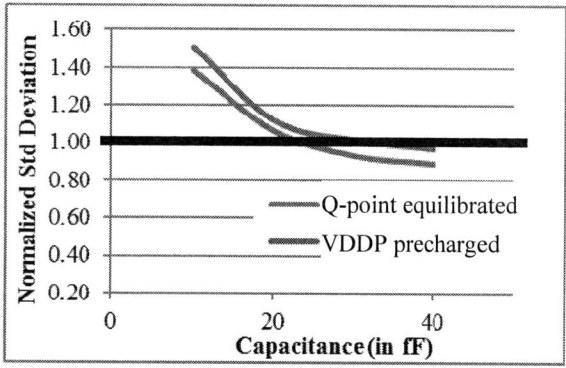

Fig. 6. Results of Monte-Carlo simulations to identify the optimal coupling capacitance for two types of LSTP-C2SA.

IV. RESULTS

Results of the experiment to size C1 and C2 are presented in Fig. 6. While VDDP precharged C2SA requires C1 and C2 to be sized at about 30fF, Q-point equilibrated SA has same offset requirement as LSA with 20fF. In Q-point equilibrated design, mismatch between SA devices is canceled by auto-calibration and dispersion reduces. Therefore C1 and C2 can be sized smaller. When MIM capacitor is available in a technology, LSTP-C2SA can be designed with same area as LSA. In absence of MIM capacitance, C1 and C2 are implemented in M1 (Fig. 5) and M3 layers with 4% impact on instance area.

We compare read power consumed by implementations based on LSA and LSTP-C2SAs in Table I in which power consumption from VDDA and VDDP supplies is presented for designs with LSA and BLs on VDDP (Case-A), LSA and BLs on VDDA (Case-B), and Q-point equilibrated LSTP-C2SA on VDDP and BLs on VDDA (Case-C). Three operating conditions are characterized for system operation at 2GHz (VDDP=VDDA=1.1V), 800MHz (VDDP=0.7V, VDDA=1.0V), and 200MHz (VDDP=0.4V, VDDA=0.8V). The most energy efficient operating point (200MHz) is forbidden in Case-A, because BLs (at VDDP) are 400mV below VDDA and cell stability in read mode is degraded. It is evident that at 200MHz, design with LSTP-C2SA consumes about 43% lower read power than the design with conventional SA. As expected, share of power consumed from VDDA supply increases at low speed operating points when VDDP scales to lower voltages than VDDA.

We benchmark implementations based on Case-B, C2SA of [5], and LSTP-C2SA (Case-C) in Table II for different DVFS scenarios. We use power delay product (PDP) as a Figure of Merit to compare energy consumed by each design. C2SA proposed in [5] is faster than LSA, while LSTP-C2SAs in this work have been optimized for same speed as LSA. Static power is also included in estimating power consumption. It is evident that the implementation with LSTP-C2SA has better power delay product (PDP) than LSA based design and also C2SA of [5] across a wide range of DVFS scenarios. When system is operated primarily at high voltages and high frequency, all the implementations are similar and in-line with observations from Table I – B and C at VDDA=VDDP=1.1V. However, as the probability

TABLE I. READ POWER CONSUMPTION OF A DUAL RAIL SRAM IN DIFFERENT SCENARIOS

Normalized Power consumed	1.1V (VDDP) 1.1V (VDDA)			0.7V(VDDP), 1.0V(VDDA)			0.4V(VDDP), 0.8V(VDDA)		
	(A)	*(B)*	*(C)*	*(A)*	*(B)*	*(C)*	*(A)*	*(B)*	*(C)*
VDDA	1.0	25.0	12.8	0.9	27.2	10.9	-	18.1	7.2
VDDP	25.4	2.2	16.8	13.0	0.8	7.8	-	0.6	3.3
Total	26.4	27.2	29.6	13.9	28.1	18.7	-	18.7	10.5

(A): BL, precharge, SA and write driver on VDDP
(B): BL, precharge, SA and write driver on VDDA
(C): BL, precharge, write driver on VDDA; LSTP-C2SA on VDDP

TABLE II. BENCHMARK OF ENERGY EFFICIENCY IN DIFFERENT DVFS SCENARIOS

Probability of use at Operating Point			LSA (Case-B)	[5]	LSTP-C2SA
2 GHz	**800 MHz**	**200 MHz**			
0.8	0.1	0.1	1.0	0.97	1.01
0.6	0.2	0.2	1.0	0.93	0.92
0.4	0.4	0.2	1.0	0.89	0.87
0.2	0.4	0.4	1.0	0.86	0.76
0.2	0.2	0.6	1.0	0.87	0.71
0.1	0.1	0.8	1.0	0.86	0.64

Power Delay Product (PDP) is used as a measure of energy consumption and is estimated by also including static power. Improved SA delay in [5] has been used in calculating PDP.

of operating at lower frequencies increases, LSTP-C2SA based implementation is more energy efficient.

Both the LSTP-C2SAs have been realized on 28nm UTBB FD-SOI silicon and are tested to be functional in DVFS mode at 0.397V [7]. Dedicated testing of SRAMs confirmed functionality down to 0.35V.

V. CONCLUSIONS

Low Standby Power Capacitively Coupled Sense Amplifiers are proposed to extend lowering of periphery supply without degrading stability of SRAM cells in dual rail SRAMs. At SRAM instance level, LSTP-C2SA based design consumes up to 43% lower power in DVFS operation at 0.4V with improved energy efficiency across a range of DVFS scenarios.

REFERENCES

[1] International Technology Roadmap for Semiconductors (ITRS), http://www.itrs.net/

[2] J. Kulkarni et al., "Dual-VCC 8T-bitcell SRAM Array in 22nm tri-gate CMOS for energy-efficient operation across wide dynamic voltage range," *Symposium on VLSI Circuits 2013*, pp.C126,C127

[3] M. Khellah et al., "Wordline and bitline pulsing schemes for improving SRAM cell stability in low-Vcc 65 nm CMOS designs," *Symposium on VLSI Circuits 2006*, pp.9,10

[4] N. Verma, A.P. Chandrakasan, "A High-Density 45 nm SRAM Using Small-Signal Non-Strobed Regenerative Sensing," *Solid-State Circuits, IEEE Journal of*, vol.44, no.1, pp.163,173, Jan. 2009

[5] B. Giridhar et al., "13.7 A reconfigurable sense amplifier with auto-zero calibration and pre-amplification in 28nm CMOS," *ISSCC 2014*, pp.242,243

[6] G. Moritz et al., "Optimization of a voltage sense amplifier operating in ultra wide voltage range with back bias design techniques in 28nm UTBB FD-SOI technology," *ICICDT 2013*, pp.53,56

[7] R. Wilson et. al., "27.1 A 460MHz at 397mV, 2.6GHz at 1.3V, 32b VLIW DSP, embedding FMAX tracking," *ISSCC 2014*, pp.452,453

A 6T-SRAM in 28nm FDSOI Technology with Vmin of 0.52V Using Assisted Read and Write Operation

Ashish Kumar, Vinay Kumar, Dhori Kedar Janardan
STMicroelectronics Pvt. Ltd.
Greater Noida, India
ashish.kumar@st.com

G.S.Visweswaran, Kaushik Saha
Indian Institute of Technology
New Delhi, India
gswaran@ee.iitd.ac.in

Abstract— **A low Vmin, 6T-SRAM is realized in 28nm FDSOI technology using read and write assist methods. We could reduce the Vmin of SRAM cell to 0.52V for the 0.120um2 high density 6T-SRAM. Reduced read margin of the SRAM cell is recovered using a transient rise in cell supply level through word-line coupling. Write assist is realized using application of PVT selective negative bit-line approach. Bit-line is pulled to a required negative value in order to provide sufficient assistance for the write operation. Required power line overshoot is 50mV to ensure correct read operation. For write assist, the undershoot requirement is within 100mV. We could achieve a performance improvement of 50 percent with no power penalty for a 288Kb capacity SRAM with 2K words of 144bits width. The area overhead of the read and write assist scheme is 1.4 percent and 2.5 percent respectively.**

Keywords— *Low power, SRAM, low Vmin, low voltage, write assist, read assist. SNM, write margin.*

I. INTRODUCTION

As we move into deep nanometer CMOS technology, associated challenges are also increasing. Increased integration density along with large size of SoC (system on chip) is resulting in very high power density. This requires reduction of supply voltage to make the integration feasible. Low voltage operation, however, results in near subthreshold device operation, resulting in large statistical variation in current. Very small geometry adds to the parametric variation of the device. The 6T-SRAM cell using smallest geometry and packed with the highest density, is affected most by the statistical variation. In order to enable the SoC to work at lower voltage, all components must be ensured to function correctly at reduced supply. Ensuring correct functionality of SRAM at low voltage becomes a challenge, putting a limitation on minimum operational voltage (Vmin) for SoC. This is due to reduced static noise margin (SNM) and write margin (WM) of the SRAM cell at low supply voltage, illustrated in Fig.1 and Fig.2. Six-sigma robustness of the design is not achieved at low supply voltage. To recover the loss in stability, specifically SNM and WM, there is need for assist scheme that can help for the correct read and write operations. Several assist schemes has been explored and implemented to provide the solution [1-10].

The assist schemes implemented so far have used methods such as wordline (WL) lowering to improve SNM at low

---

This work is completely sponsored by STMicroelectronics Pvt. Ltd. India

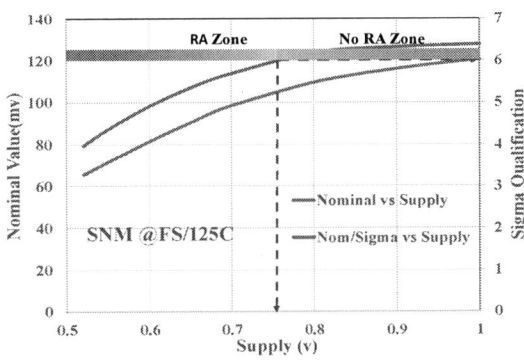

Fig.1. SNM variation with supply voltage (120um² 6T-cell in 28nm FDSOI)

voltage [1][4], negative bitline, column supply lowering and WL boosting as write assist [3-4][6][8-10]. WL lowering results in performance loss and also reduces the efficiency of write assist (WA). Schemes, such as partial suppression of WL [2] are proposed to reduce this loss. Write assist using negative bitline results in excessive undershoot at higher supply levels and its application is limited by reliability concerns. This work presents a WL coupled row supply rise as a read assist (RA) scheme, which also helps to improve the SRAM performance at low voltage. The negative bitline approach is used as WA augmented with another detection scheme to activate the WA only for the required PVT (process, voltage and temperature) condition. This helps to avoid excessive electrical stress at higher voltages, ensuring reliability of the device for its operational lifetime.

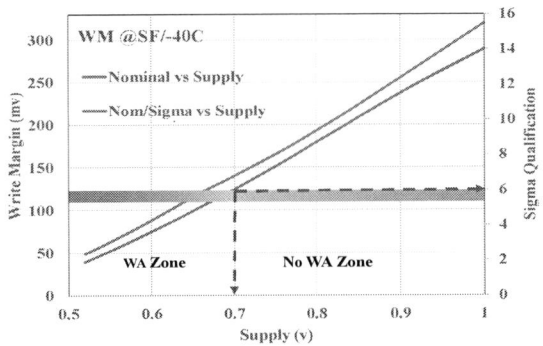

Fig.2. WM Variation with supply voltage. (120um² 6T-cell in 28nm FDSOI)

II. IMPLEMENTATION

A. Read Assist Using RVDD Raise

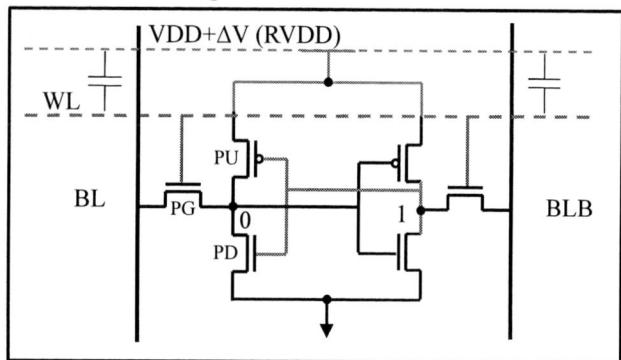

Fig.3. Cell level illustration of RA using RVDD raise

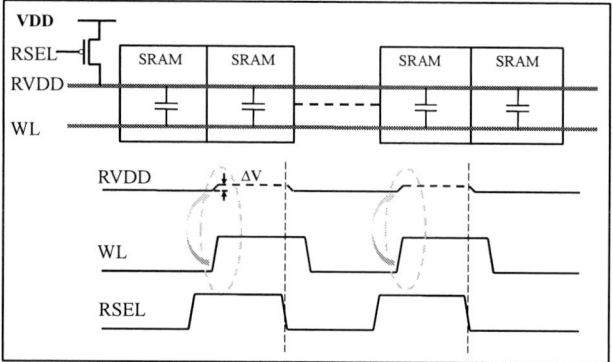

Fig.4. RA scheme using RVDD raise

Assistance to the read operation is provided through supply voltage rise (VDD+ΔV) for the memory cell, while keeping wordline voltage and bitline precharge at VDD level. Fig.3 illustrates the scheme at memory cell level. Trip point for the cross-coupled inverters increases due to rise in RVDD. Also, since the WL is kept at VDD, WL lowering phenomena comes into effect in a natural way. This would make access transistor (PG) weak compared to the pull-down (PD) transistor. Both mechanisms help to increase the SNM. Since the WL level is not going below VDD and PD is driven by RVDD, cell current increases.

Implementation of the scheme requires the cell supply to run in a horizontal direction. Each row of the memory core has its separate supply line. While accessing the row, RVDD is made floating by disconnecting from VDD, as illustrated in Fig.4. The row select signal (RSEL) is used to disconnect the RVDD line from VDD. With RVDD line floating, when WL goes high, the RVDD line gets a boost (ΔV) through capacitive coupling. To augment this coupling, boost capacitors are used to feed additional charge to the RVDD line. Once the RVDD is raised and WL is made high, BL starts discharging through the access transistor.

The read operation is illustrated in Fig.5. Once the read operation is initiated, the floating RVDD starts dropping. This drop is due to memory cell leakage current and coupling from the discharging bitline. By the time RVDD loses a good portion of its rise, BL has discharged enough and reached a

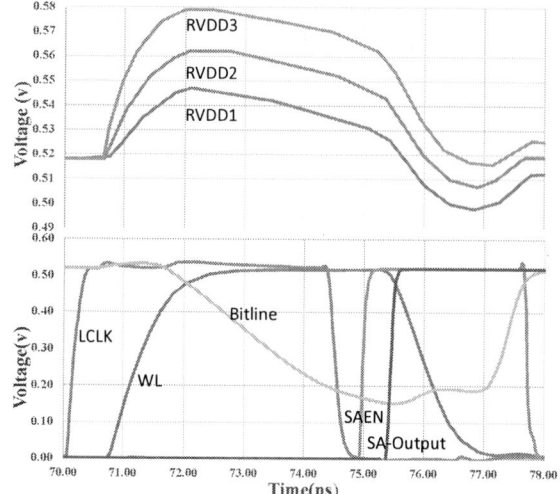

Fig.5. Read operation illustrated using RVDD raise

much lower voltage, injecting lesser current through cell. Hence, the SNM of cell increases. As the bitline voltage reaches point A (Fig.6), the memory cell goes to high SNM state with six-sigma robustness. The RA scheme also needs to be activated during write operation to ensure the stability of half-selected cells, which are connected to the same WL but do not belong to the addressed column for write. Hence, the discharge of RVDD should not be excessive due to simultaneous fast discharge of write bitlines. This has to be ensured by design using additional capacitors to raise the RVDD, if required. However, due to very low overlapping coupling area between bitlines and RVDD, impact of BL discharge during write operation is not prohibitive for the implementation of the scheme. We have used nmos capacitors to feed additional charge on RVDD to have multiple optional voltages namely RVDD1, RVDD2 and RVDD3 (Fig.5).

Read assist, when applied to recover the SNM at FS (fast nmos, slow pmos) process corner, deteriorates the WM of the cell. Although WM is good at FS, reduces significantly due to RVDD raise or WL lowering and becomes critical. As illustrated in Fig.7, 80mV WL lowering is required for a six-sigma SNM robustness compared to the required 50mV of RVDD raise. When these assists are applied, WM deteriorates by 3.8 sigma and 1.3 sigma respectively for WL lowering and RVDD raise.

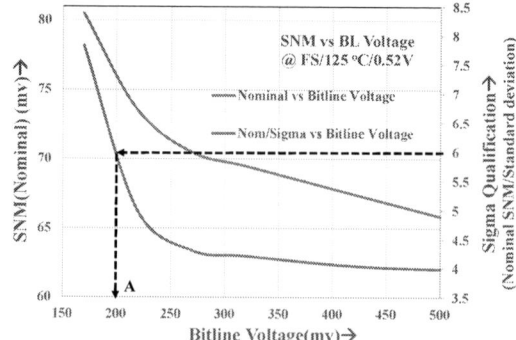

Fig.6. Rise in SNM with BL discharge. (120um² 6T-cell in 28nm FDSOI)

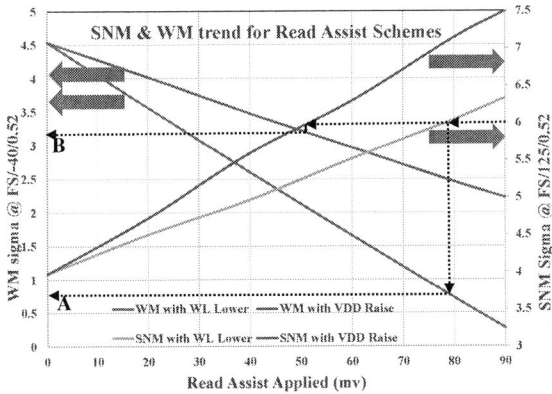

Fig.7. SNM and WM variations for RA schemes. (Using 120um^2 6T-cell in 28nm FDSOI)

B. Write Assist Using Negative Bitline

To assist the write operation, negative bitline or bitline undershoot has been used. This has been evaluated as the most effective way to assist the write operation [2][4][8-10]. Write failures are problem for SRAM at low voltage. An undershoot, designed to recover write failure, ensures correct operation at low voltage. However, this undershoot grows as the supply voltage increase, contrary to the hypothesis that less or no assist is required at higher operating voltages. This undesired undershoot creates higher than allowed electric field stress within the device, posing a threat to the reliability of the device and an accelerated ageing effect (Fig.8). Attempts have been made to limit this undershoot through suppressed coupling signal [8]. However, this does not eliminate the undesired stress completely. In this work, we have used a low write margin PVT detector to disable the write assist at PVTs, where there is no issue with the write operation. Fig.8 illustrates the concept behind the implemented scheme. In conventional WA, the undershoot increases along the direction AA'. In this work, write assist is disabled at lower voltage along AB or AB'. There is a spread of BB' for the detection point due to statistical variation within detection scheme. Write assist is disabled at higher voltage within a spread of BB'.

To realize the scheme, a write margin detector is used, as illustrated in Fig.9. The detector uses a 6T cell modified by using two pass transistors (PG) in series. This cell is kept at targeted RVDD to assess the write margin at similar conditions. This is done to make the detection cell weak for write operation, so as to ensure the statistical coverage of detection over the entire memory array. An array of detector cells is used to reduce statistical variation of detection. Using a pulsed activation of the detector by using signal EN carries out detection. Voltage at PNODE is sampled and evaluated using a half latch sense amplifier. A VDD/2 reference is used for the sense operation.

Fig.10 illustrates the detection of PVTs that require WA. Voltage at PNODE is close to VDD for the cases when access transistor (PG) is relatively much weaker and is close to zero when this is stronger. We observe that at FS/-40C, detector is signaling the need for WA activation. This is due to the fact

that WM is degraded due to RVDD raise. However, magnitude of WA is not needed more than what is required at SF(slow nmos, fast pmos)/-40C, the worst write condition. This detector is activated at regular time intervals to decide the activation of WA, as the PVT conditions do not change abruptly.

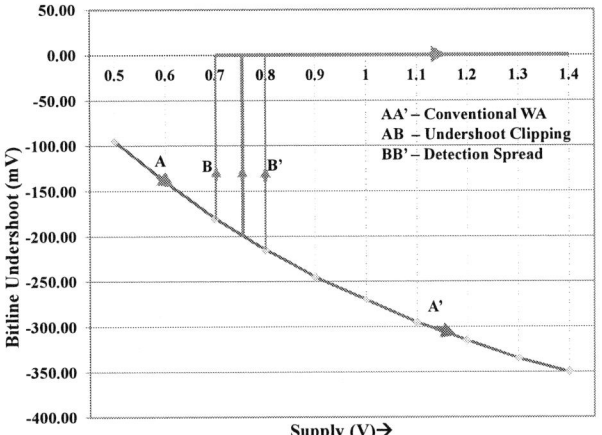

Fig.8. BL undershoot increase with supply. Clipping is applied at higher voltage. (Clipping range demonstrated for SF/-40C case. Clipping will be earlier for other cases)

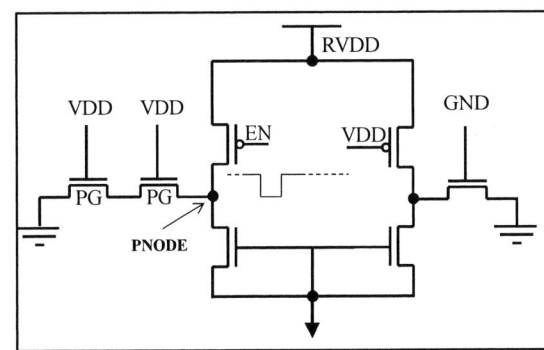

Fig.9. Write margin status detection using array of modified 6T-SRAM Cell.

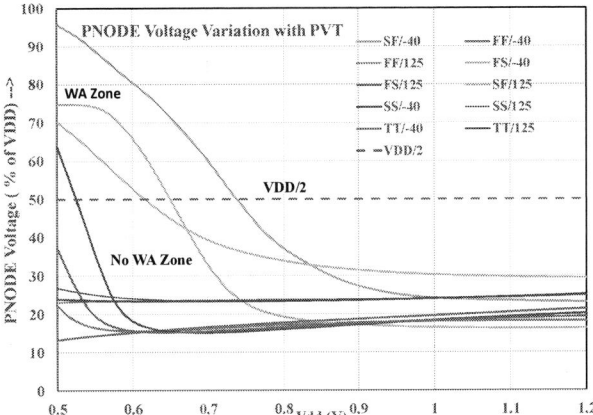

Fig.10. PNODE Voltage is evaluated and WA zone is filtered for WA activation. (For the array of 64 detector cells) (120um^2 6T-cell in 28nm FDSOI)

III. Discussion and Conclusion

TABLE-1

Comparison of RA Schemes			
6-sigma Vmin(mV)	RVDD raise needed(mV)	WL lowering needed(mV)	Boost capacitor need[†]
520	50[*]	80[**]	Yes
550	42	65	Yes
600	30	45	Yes
650	18	28	No
700	12	17	No
740	0	0	No

* Cell current increases by 108% (0.72uA→1.5uA @SS/0.52V/-40C)
** Cell Current decreases by 42% (0.72uA → 0.42uA @SS/0.52V/-40C)
† Boost capacitor needed with RVDD raise scheme

TABLE-2

Comparison with Related Work					
	[1][α]	[4][α]	[9][β]	[10][γ]	This work[δ]
Technology(nm)	20	20	22	32	28
Vmin(mV)	600	750	700	700	520
Area Overhead (%)	NA	9.5	3.4	2.7	3.9

α Wordline lowering as RA and wordline overdrive as WA used.
β Wordline lowering as RA and bitline undershoot as WA used.
γ Reduced Bitline precharge (< Vdd) as RA and bitline undershoot as WA used.
δ RVDD raise as RA and bitline undershoot as WA used.

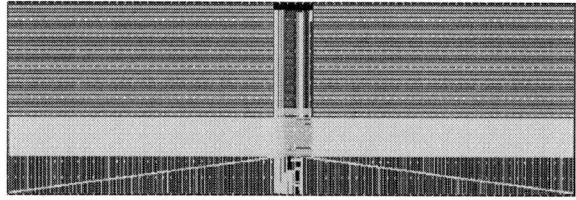

Fig.11. Layout of memory instance (2K words, 144 bits)

We have designed a low voltage SRAM memory instance of 288Kb having 2K words of 144bits width. 6T-SRAM cell with a cell area of $0.120um^2$ in 28nm FDSOI technology is used. Analyses have been done using silicon extracted spice models. Using RVDD rise, we have ensured the read stability at low voltage. We have used 50mV of RVDD raise to achieve Vmin of 0.52V (Table-1). Stability of cell increases from 3.8 sigma to six-sigma, ensuring 99% yield for a capacity of 64Mbits. The RA scheme is implemented with an area overhead of only 1.4 percent. RVDD rise resulted in cell current increase of 108% that provides a worst case performance gain of approximately 50 percent. We could achieve an operational frequency of 6MHz at 0.52V and 50MHz at 0.6V. Dynamic read and write power of the memory are 11uA/MHz and 18uA/MHz respectively at 0.52V. Natural coupling of WL with RVDD is enough to support a Vmin of 650mV, while boost capacitors are needed to enable lower Vmin (Table-1).

Conventional schemes, realized with WL lowering approach [1][2][4][9], could achieve the same Vmin lowering with a performance loss ranging from 10 to 40 percent. The extent of WL lowering required is also much higher compared to RVDD raise. Table-2 provides a comparison of the scheme with related work.

Write assist for this work uses a negative bitline approach. We have used 100mV of BL undershoot [8] to enable the Vmin of 0.52V. Write assist is disabled in a voltage range of 0.70V-0.8V for SF/-40, and is never allowed to be activated above 0.8V. We have implemented write assist scheme including the PVT detection block, within an area overhead of 2.5%. There is no increase in read access power. This is due to the fact that the extra power used for the rise of RVDD through capacitors is compensated with the power saving due to faster bitline discharge with reduced statistical variation. This results in reduced BL discharge required to satisfy six-sigma robustness.

Requirement of write assist indicates the relative robustness of cell with respect to SNM. However, for extremely low Vmin (520mV), when WM is poor even in FS corner, both RA and WA are needed. At moderately high Vmin (600mV), where the need for WA signals sufficient SNM, RA can be disabled during write operation. Deterioration of WM due to RVDD raise (RA) is avoided, reducing the amount of WA needed and saving on dynamic power during write. Fig.11 is a snapshot of the memory instance layout.

Acknowledgement

The authors are indebted to Tushar Sharma, Atul R Menon, Deepak Dahiya and Shawahique Khan for the layout design of memory instance and to Ashima Singh for providing simulation data of write margin status detector.

References

[1] Makoto Yabuuchi et al, "20nm High-Density Single-Port and Dual-Port SRAMs with Wordline-Voltage-Adjustment System for Read/Write Assists", ISSCC 2014, p.p. 234-235.

[2] Jonathan Chang et al, "A 20nm 112Mb SRAM in High-κ Metal-Gate with Assist Circuitry for Low-Leakage and Low-VMIN Applications", ISSCC 2013, p.p. 316-317.

[3] Robert Aitken et al, "On the Efficacy of Write-Assist Techniques in Low Voltage Nanoscale SRAMs",DATE 2010.

[4] Mudit Bhargawa et al, "Low VMIN 20nm Embedded SRAM with Multi-voltage Wordline Control based Read and Write Assist Techniques", Symposium on VLSI Circuits Digest of Technical Papers,2014.

[5] R.Ranica et al, "FDSOI Process/Design full solutions for Ultra Low Leakage, High Speed and Low Voltage SRAMs", Symposium on VLSI Technology Digest of Technical Papers, 2013.

[6] Vivek De et al, "Capacitive Coupling Wordline Boosting with Self-Induced Vcc Collapse for Write Vmin Reduction in 22-nm 8T SRAM" ISSCC 2012, p.p. 234-235

[7] Pramod Kolar et al, "A 32 nm High-k Metal Gate SRAM With Adaptive Dynamic Stability Enhancement for Low-Voltage Operation", IEEE Journal of Solid-State Circuits, Vol. 46, No. 1, Jan 2011, p.p. 76-84.

[8] Yen-Huei Chen et al, "A 16nm 128Mb SRAM in High- κ Metal-Gate FinFET Technology with Write-Assist Circuitry for Low-VMIN Applications", ISSCC 2014, p.p. 238-239.

[9] Eric Karl et al., "A 4.6GHz 162Mb SRAM Design in 22nm Tri-Gate CMOS Technology with Integrated Active Vmin Enhanced Assist Circuitry," ISSCC 2012, pp 230-231

[10] H. Pilo, et al., "A 64Mb SRAM in 32nm High-k metal-gate SOI technology with 0.7V operation enabled by stability, write-ability and read-ability enhancements," ISSCC 2011, pp. 254-256.

NEMS Switches: Opportunities and Challenges in Emerging IC Technologies

(*Invited Paper*)

Philip X.-L. Feng[*]

Department of Electrical Engineering and Computer Science, Case School of Engineering
Case Western Reserve University, Cleveland, OH 44106, USA
[*]Email: philip.feng@case.edu

Abstract—The active search for candidates of an ideal switching device for low-voltage logic and ultralow-power applications has stimulated new explorations of contact-mode switches (relays) based on micro/nanoelectromechanical systems (MEMS/NEMS). This has been driven by the fundamental advantages that mechanical devices offer, such as ideally abrupt switching with zero off-state leakage, suitable for harsh and extreme environments, and very small footprints (*e.g.*, particularly with NEMS). This digest paper is focused upon reviewing the state-of-the-art NEMS switching devices (particularly Si and SiC NEMS), and the potential and possibilities towards enabling logic circuit components and integration with Si and SiC circuits, respectively. We also discuss the opportunities and challenges in other technical aspects, including materials, processes, device design, nanoscale contacts, lifetime and reliability, scaling, and the perspective of monolithic 3D integration with the advancing 3D transistors.

Keywords—logic switch; power consumption; leakage current; subthreshold swing; nanoelectromechanical systems (NEMS); relay; nanoFET; NEMS-CMOS integration; 3D integration

I. INTRODUCTION

Power consumption is a key challenge in today's nanoscale silicon (Si) MOSFETs (metal-oxide-semiconductor field-effect transistors), which are the basic building blocks of virtually all computing and communication devices that have become indispensable in daily lives. While Si MOSFET scaling with the Moore's law has been the underpinning enabler for CMOS ICs' rapid growth and phenomenal achievements for several decades, in recent years the scaling has slowed down or been stagnant in some key aspects (*e.g.*, clock speed, voltage, power, *etc.*). Following the International Technology Roadmap for Semiconductors (ITRS), Si MOSFETs have been scaled from 45nm to 32nm, 22nm, and 14nm technology node from 2008 to 2014; and 10nm and 7nm technology nodes have been planned. At module and chip levels, the standby power consumption has grown comparable to dynamic computing power consumption, and increased faster with down scaling of gate length.

The fundamental reason for the standby power crisis is rooted at the device level – the increasing leakage in Si MOSFET is plaguing the transistors as they shrink in size. Among the major sources of leakage (Fig. 1a), while gate tunneling leakage has been alleviated by adoption of high-k dielectrics, the subthreshold leakage persists as a fundamental limiting factor. In fact, due to its nature of thermal activation [1], Si MOSFET is a far-from-ideal switch (especially at low voltages) for its large subthreshold swing (S~70 to 100mV/dec, or even larger) and ineffective switching off (Fig. 1b). The on-

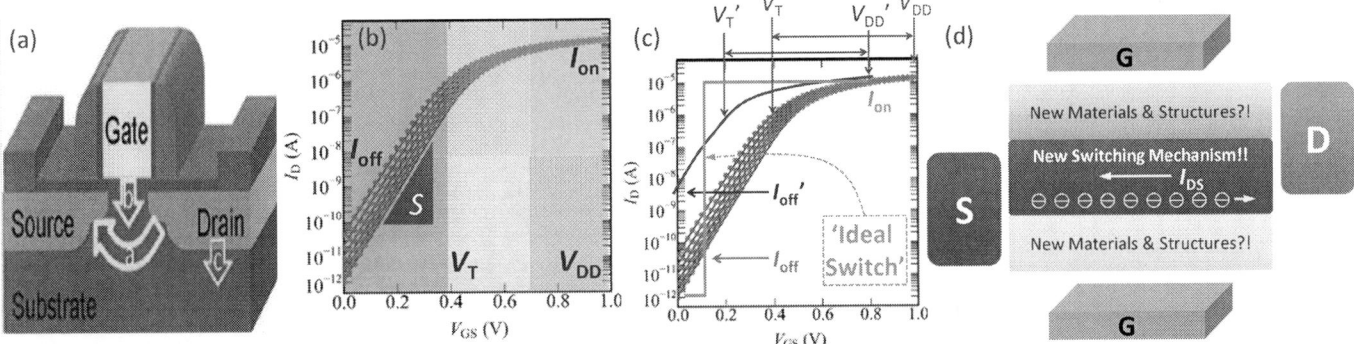

Fig. 1. The quest for a new switching device beyond the Si MOSFET. (a) Illustration of the classical Si MOSFET, with the major transistor leakage sources indicated in the arrows 'a', 'b', and 'c'. (b) Typical transfer characteristics I_D (*i.e.*, I_{DS}) versus V_{GS}, with typical levels of V_T, V_{DD}, I_{on}, and I_{off} noted. The subthreshold swing, S, is also indicated (usually S>60mV/dec), with subthreshold regime coded in a different color than the above-threshold regime. (c) Illustration of the fundamental challenging issue due to the finite subthreshold swing. With voltage scaling (the red curves shift to the blue curve, if V_{DD}-V_T=V_{DD}'-V_T'), I_{off} becomes I_{off}' with exponential increase for linear lowering of voltage. The green line illustrates the desired I_D versus V_{GS} transfer characteristics of an 'ideal switch'. Note that in such case the switching transition is desired to be ideally abrupt (with subthreshold swing S=0 mV/dec), and the subthreshold or off-state leakage should be zero (the green curve indicates the off-state current may be limited by the noise floor of the measurement tool, in such cases). (d) Conceptual and speculated illustration of the intriguing possibilities of the future charge-based electronic switch that could offer abrupt transition and ideal switching characteristics.

978-1-4799-7670-6/15 $31.00 © 2015 IEEE

chip power density limit is ~300 W/cm² due to thermal cooling limit, which severely constrains the IC design. To increase the transistor density without increasing power density, the supply voltage V_{DD} must be decreased to lower the active power dissipation, which results in a reduction in performance. As V_{DD} is lowered, the delay time t_{delay} of the computing circuits (proportional to V_{DD}/I_{on}) increases, and the leakage energy (proportional to $t_{delay} \times I_{off} \times V_{DD}$) also increases. Reducing the transistor threshold voltage V_T along with V_{DD} can alleviate the increase in t_{delay}, but the transistor off-state leakage current I_{off} increases exponentially with reducing V_T (Fig. 1c).

Towards pursuing the 'ideal switch' (with its characteristics depicted by the green curve in Fig. 1c), a number of alternative 'post-CMOS' device technologies have been proposed and are being actively explored [2]. These include approaches that one could view in such categories: **(i)** embedding new concepts or effects (may demand new materials) into today's MOSFETs to modify switching behavior (to achieve $S<60$mV/dec), such as tunneling FET (TFET), impact ionization MOS (I-MOS), negative-capacitance (NC) FET (requiring ferroelectric NC gate), and feedback FET; **(ii)** adopting novel switching mechanisms derived from new structures (different from MOSFET) but are still charge based and using an electrical current to discern on/off states, such as NEMS switches (relays), and piezoelectronic transistors (require special materials and stacks); **(iii)** departing from the paradigm of using electrical charge and current as the logic information token or carrier, such as devices proposed to exploit the states of spins, photons, and other exotic excitons for switching. The choice or preference one could have among all these possible concepts and options may largely depend on how close (and how soon) one would like these 'post-CMOS' technologies to be related to, and/or transformable from, the latest nanoscale MOSFETs; in other words, what is the observer's interests of timeframe and expectation between CMOS and 'post-CMOS'.

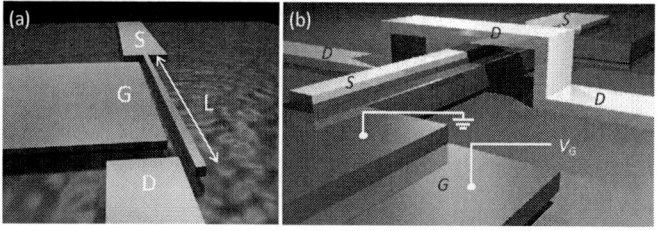

Fig. 2. Basic ideas of the NEMS logic switches and main actuation schemes. (a) Illustration of an in-plane (lateral) *electrostatic* NEMS switch that features a single conducting material for making the structure and contact (on an insulator sacrificial layer). (b) Illustration of an out-of-plane NEMS switch that features *piezoelectric* actuation and composite materials, with piezoelectric active layer sandwiched between metallic layers (ground, gate G) for making the structure, and extra metallic layers (S, D) for making the contact electrodes.

If we focus on an electronic switch that is still charge-based and uses electrical current as on/off state variable, we can see that, towards the 'ideal switch', a new switch good candidate shall offer the following: **(i)** abrupt switching behavior, *i.e.*, small subthreshold swing and steep subthreshold slope; **(ii)** high on/off ratio ($I_{on}/I_{off} > 10^4$ or 10^6); and **(iii)** high enough I_{on} or small on-state resistance (R_{on}). We note that, several today's proposed emerging 'post-CMOS' devices actually only seek to defy the $S>60$mV/dec limit ('Boltzmann tyranny') by attaining $S<60$ mV/dec (but not ideally steep, $S\rightarrow0$mV/dec).

Mechanical switches offer some fundamentally attractive attributes: **(i)** zero off-state leakage; **(ii)** abrupt switching with ideally steep slope (zero subthreshold swing, $S\sim0$mV/dec); **(iii)** high on/off ratio; **(iv)** the switch footprint area (top-view) can be smaller than that of a modern Si MOSFET; **(v)** mechanical switching can operate at high temperatures (plus other harsh environments) where Si MOFETs fail; **(vi)** scaled down to genuinely nanoscale, mechanical switches can operate very fast (as suggested by demonstrations of RF/microwave NEMS resonators with scalable operating frequencies) [3]. These unique features and immense potential have been inspiring and driving researchers to investigate various NEMS switches. Here we briefly summarize some aspects at today's frontiers of NEMS devices. For the length limit, we confine this digest paper with the following considerations: **(i)** We focus on genuinely nanoscale devices with only one non-sub-micron lateral dimension (*i.e.*, only length can be on ~1µm order and all other dimensions are on sub-100nm to ~100nm order). NEMS instead of MEMS are our focus because they are comparably more scaled in dimensions, lower volume (mass), higher speed, and are thus more relevant in terms of scaling and relation to latest MOSFETs. **(ii)** We focus on Si and SiC NEMS enabled either directly by wafer-scale manufacturing technologies, or by processes transferrable to such.

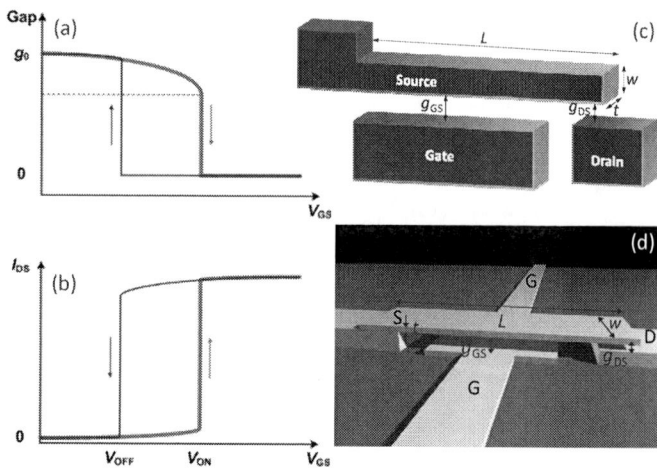

Fig. 3. Basic principles and expected characteristics of *electrostatic* NEMS switches. (a) Modulation of the gap due to actuation and deflection of the NEMS. (b) Expected I_{DS} versus V_{GS} transfer characteristics with electrostatic NEMS switching. (c) 3D illustration of an in-plane device that can be made by single material as the device layer. (d) 3D illustration of an out-of-plane device that would require multilayer surface nanomachining (*e.g.*, structural layers before and after the sacrificial layer).

II. BASIC PRINCIPLES AND SCHEMES

The basic idea is to reinvent the electromechanical relays at nanoscale, via aggressive miniaturization; and also, to engineer electromechanical coupling effects so that distinct on/off states can be realized with high I_{on}/I_{off} values. Off state is defined by an opening air (or vacuum) gap (could be on ~10nm or ~100nm order), thus there is no sub-threshold leakage. The off-state current picked up at an instrument shall be limited by other sources (*e.g.*, leaking substrate which NEMS are sitting on, noise floor of the instrument, *etc.*). On state is defined by a contact which conducts and carries electrical current. In this new 'redux' of the vintage electromechanical switches [4-6]

978-1-4799-7670-6/15 $31.00 © 2015 IEEE

(that had dominated computing before 1950 and only later were replaced by transistors), we can gain and appreciate the following points: **(i)** the fundamental properties of zero off-state leakage, abrupt switching characteristics, and tolerance to high temperatures (or other harsh environmental factors), remain the same; **(ii)** the benefits of aggressively shrinking the relays down into NEMS switches are in the greatly reduced dimensions, unprecedentedly boosted operating speeds and lowered power consumption; **(iii)** the challenges on the mechanical contacts are renewed, now at the nanoscale – compared to conventional macroscopic relays, the contacts in NEMS switches possess new contact materials, areas, surfaces, electrical and mechanical properties, thus demanding new mesoscopic studies. Considering the key metrics for nanoscale contacts in NEMS, there is no obvious fundamental reason suggesting that the deeply scaled small contacts should not outperform their macroscopic counterparts in classical relays.

Figure 2 illustrates representative concepts of in-plane and out-of-plane NEMS switches, combined with engineering of either electrostatic or piezoelectric actuation schemes. Electrostatic NEMS switches can be made on a device layer of single material (*e.g.*, metallic, highly conductive doped Si). Piezoelectric NEMS switches usually require multilayer stacks of materials, including both active piezoelectric crystals to provide efficient electromechanical coupling for actuation, and conductive coatings for actuation and contact electrodes. A number of piezoelectric films (*e.g.*, PZT, AlN, GaN, ZnO, *etc.*) widely explored in various MEMS/NEMS transducers can be considered for adoption. As shown in Figs. 2 & 3, in-plane electrostatic NEMS may be the most convenient to fabricate (possible for a single lithography step). Out-of-plane NEMS switches require additive and subtractive nanomachining of the sacrificial layer, to create the electrostatic coupling gap. Figure 3a&b illustrate the expected switching characteristics of electrostatic NEMS switches. The mechanical hysteresis, its occurrence and magnitude (V_{on}-V_{off}), depend on the properties of the contact (*e.g.*, adhesion force, contact area) and their interplay with the electromechanical designs of NEMS. For 3-terminal switches (Fig. 3c&d), g_{GS}>g_{DS}; and it is desired to have very small actuation gap (g_{GS}) for low-voltage operation. Attainable switching speed is related to the eigen-frequency of mode corresponding to the designed deflection of the structure, which is set by both material properties (*e.g.*, elastic modulus E_Y) and device dimensions (*e.g.*, for cantilever, $L \times w \times t$).

III. SILICON (SI) NEMS SWITCHES

The impetus for Si NEMS switches is natural and huge. In academic research labs, Si NEMS have emerged as attractive enabling components for new transducers, through both top-down nanofabrication and bottom-up synthesis and assembly. Meanwhile, embracing movable Si devices in CMOS platform has been a major theme in the "More-than-Moore" trend. The technology fusion has been stimulating in pushing the envelope of new devices and platforms. Specifically, from NEMS side, ultrathin Si nanowires (NWs) and arrays have been engineered into high-speed vibrating devices such as very-high-frequency resonators and piezoresistive transducers [7,8]; while from advanced Si transistors side, suspended Si NWs and arrays are the mid-way products toward making Si NW gate-all-around (GAA) 3D nanoFETs [9]. The fusion of these technological ingredients has thus quickly led to 8″-wafer-scale VLSI of Si NW NEMS, in state-of-the-art SOI technology [9,10]. This helps prepare for investigation of Si NEMS logic.

A. Low-Voltage Si NEMS Switches in SOI Technology

As shown by the example in Fig. 4, using chips from very thin SOI wafers, low-voltage Si NEMS switches have been demonstrated [11,12]. Such lateral switches can be enabled by electron beam lithography (EBL) at both chip and wafer levels. Similar devices have been attained by using both highly doped SOI or metalized SOI thin films as the device layer [11,12].

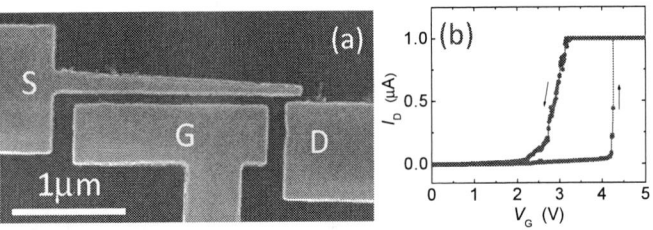

Fig. 4. One representative example of very small Si NEMS switches enabled by very thin SOI technology and electron beam lithography (EBL, not at wafer scale, but on cm-scale samples in university labs). (a) SEM image of a lateral 3-terminal Si NEMS switch made in ~160nm-thick doped SOI film, with one-step EBL. (b) Measured abrupt switching event at low voltage, with expected hysteresis phenomena (see Fig. 3) also observed.

B. Si NEMS Switches Enabled by VLSI of NEMS

VLSI of Si NEMS at 8″-wafer-scale using state-of-the-art

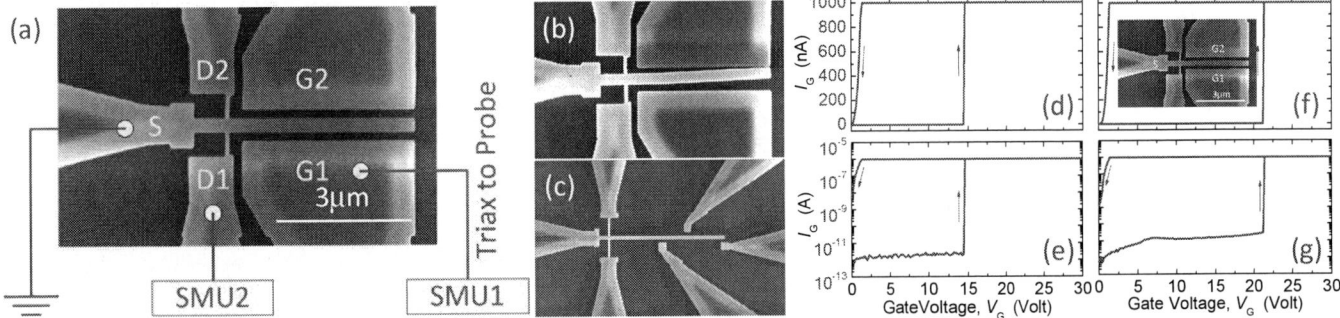

Fig. 5. Representative examples of versatile lateral Si NEMS switching devices, enabled by an 8″-wafer-scale SOI technology for VLSI of NEMS [13,14]. (a) SEM image and measurement circuit for a Si nanowire (NW)-cantilever mechanically-coupled 2-terminal switch, with additional readout by the piezoresistive Si NWs. (b) SEM image showing an actuated device in contact. (c) A device with multiple lateral gates and contact electrodes. (d) & (e) Data (I_{G1-S} vs V_{G1}) of first 'pull-in' switching cycle of a device as shown in (a), in linear and logarithmic scales, respectively. (f) & (g) Second G1-S 'pull-in' switching after overcoming the stiction to G1 by applying a sweeping voltage on G2 (*i.e.*, 'pull-off' from G1 by actuation via G2).

978-1-4799-7670-6/15 $31.00 © 2015 IEEE

SOI technology has enabled unprecedented numbers of high-quality Si NEMS devices with excellent uniformity and overall yield (as compared to EBL devices at chip levels, Fig. 4). Various designs of Si NEMS switching devices have been enabled (see Fig. 5 for examples). We have also explored some of such devices with embedded very thin Si NWs for additional piezoresistive readout of the switching events, by exploiting the piezoresistive effects in thin Si NWs [14].

IV. SILICON CARBIDE (SiC) NEMS SWITCHES

We also briefly introduce and highlight SiC NEMS, not only because SiC is a technologically important material endowed with a wide spectrum of compelling properties (photonic, electronic, thermal, and mechanical), but also because of important latest advances made by SiC NEMS. SiC has recently emerged as the material that has enabled some of the best and robust NEMS switches and logic building blocks with abrupt switching, zero leakage, and long lifetime. Our devices achieve $>10^7$ to 10^{10} switching cycles recorded without device failure at both room and high temperatures (up to 500°C) and *in ambient air* (such lifetime and reliability performance far surpass the lifetimes in all other genuinely nanoscale NEMS switches reported to date) [15-17], while featuring ultrasmall motional volume ($\sim 1\mu m^3$) [15-17], low switching-on voltage ($V_{on,dc} \sim 1V$) for quasi-static switching [6], very high speed (switching time approaching $t_s \sim 1$ to $\sim 10 ns$) [3,6,11], and *milli-Volt switching* in *resonant-mode operations* ($V_{on,res} \sim 10$ to $\sim 100 mV$ level) for $>10^{10}$ to $>10^{12}$ cycles without failure [18], with versatile geometric configurations and signal readout schemes [6,11,18], scalability and manufacturability including hybrid NEMS-transistor integration.

A. Low-Voltage SiC NEMS Switches

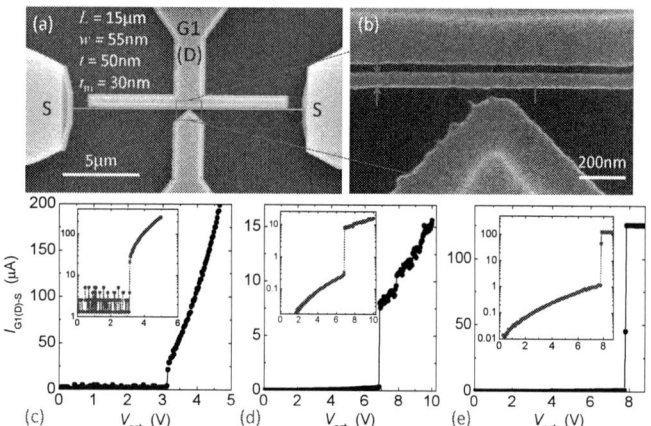

Fig. 6. Examples of low-voltage lateral electrostatic NEMS switches enabled by very thin SiC nanowires (NWs) defined by top-down lithographical processes. (a) A 15μm-long, 55nm-wide SiC NW metallized with 30nm Al on top. (b) Close-in view showing the 55nm width and the 30nm gap enabled by high-resolution EBL, lift-off, and dry etching processes. (c) & (d) NEMS switching events from two 20μm-long devices (same thickness and similar width and gap), with $V_{on} \approx 3.1V$ and 7.0V, respectively. (e) Measured switching event from a 15μm-long device, with $V_{on} \approx 7.8V$. Insets in (c)-(e) are the data shown with measured current in logarithmic scale.

SiC has proven itself as one of the best crystalline materials for making high-performance, robust NEMS [3] (in extensive NEMS fabrication runs through the past decade, given same

designs and same lithographical processes, we almost always enjoy SiC NEMS outperforming their counterparts in Si, SiN, *etc.*). SiC also has major advantages over diamond, as high-quality SiC-on-Si and SiC-on-SiO$_2$ are established. Even without an insulating layer, very high aspect ratio, ~50nm-thin SiC NWs made of SiC-on-Si, with ~30nm coupling gaps, can make low-voltage 2-terminal NEMS switches (Fig. 6) owing to the very low pull-in voltages engineered in such structures. By engineering with the trade-offs, there are pathways toward sub-1V, nanosecond (ns) abrupt SiC NEMS switches [6,11,18].

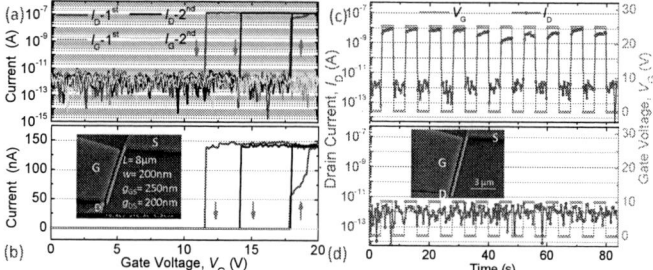

Fig. 7. One representative example of clear and repeatable switching behavior of 3-terminal SiC NEMS switches. (a) & (b) First two switching cycles with hysteresis, measured in another SiC NEMS switch, with currents in logarithmic scale (a) and linear scale (b), respectively. Inset of (b) is an SEM image of the actual device. (c) & (d) Representative data selected from a long-cycle high-precision measurement (recorded over consecutive days). (c) The waveforms show SiC NEMS switch operating multiple cycles following the applied gate voltage waveform ($V_{G,pk} > V_{on}$). (d) The clear control experiment where no mechanical switching happens at a sub-threshold gate voltage ($V_{G,pk} < V_{on}$).

B. Long Cycles of SiC NEMS Switching in Ambient Air

SiC nanocantilevers with in-plane coupled gate (G) & drain (D) electrodes make excellent 3-terminal switches. Sweeping V_G from 0 volt up to slightly above V_{on}, but well below the pull-in-to-gate $V_{PI,G}$ (both values are estimated by modeling with measured device dimensions), and then sweep V_G back to 0V, we record the I_D-V_G and I_G-V_G behavior in quasi-dc, high-precision measurements (current noise floor can be down to ~10fA, limited by settings and cabling). Typical data in Fig. 7a&b clearly show the ideally abrupt switching and expected hysteresis. We note that these switches can be very 'quiet'; and low-noise measurement is key to capturing the details.

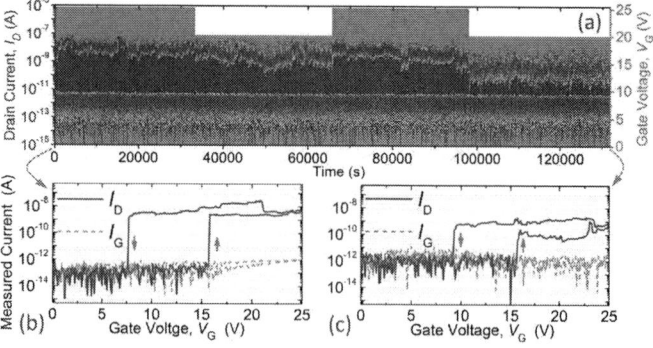

Fig. 8. A representative example of robust, long-lifetime 3-terminal SiC NEMS switches operated in ambient air. (a) Real-time recorded long operating cycles of SiC NEMS switching. The data are taken over 7 days of operation in air, with all details and evolution during the dwelling time of each switching cycle recorded. (b) & (c) Switching characteristics and on-current changes *before* and *after* 7 days of long-cycle testing. Hysteresis *I-V* curves tested at (b) the beginning and (c) the end of the long trace recorded over 7 days (after >14,000 switching cycles; sometimes the device was resting during the 7 days).

978-1-4799-7670-6/15 $31.00 © 2015 IEEE

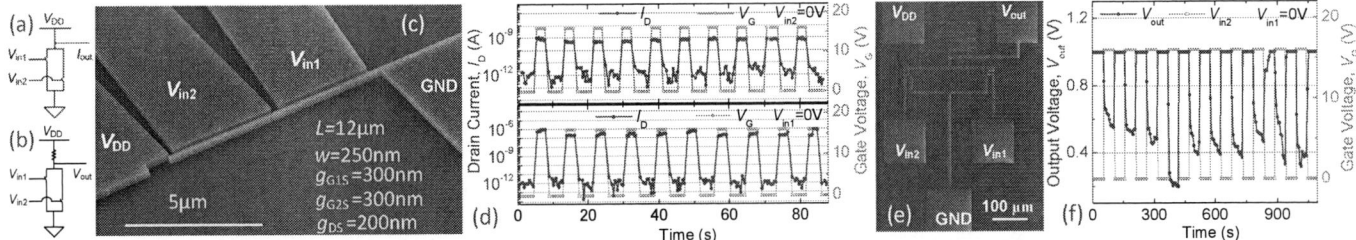

Fig. 9. SiC NEMS logic building blocks and simple circuits. (a) & (b) Illustration of OR and NOR logic functions and schematics of circuits. (c) SEM image of a dual-gate lateral NEMS switch for demonstrating the basic logic function of an OR gate. (d) Time-domain data showing the measured OR function, the output is high when *either V_{in1} or V_{in2}* is above the threshold voltage V_{on}. (e) SEM image of the actual device for demonstrating the NOR gate logic function, with a pull-up resistance $R\sim1M\Omega$. (f) Measured output voltage V_{out} versus input gate voltage $V_G=V_{in2}$, when $V_{in1}=0V$.

As shown in Fig. 7c&d, by driving the gate with a square-wave function with peak voltage $V_{G,pk}>V_{on}$, we actuate the switch on and off while monitoring the currents at drain (I_D) and gate (I_G). Recorded data demonstrate clear abrupt nanomechanical switching with no measurable gate leakage.

The number of switching cycles (*i.e.*, lifetime) is a critical metric and key challenge in engineering NEMS switches. To date, SiC NEMS has demonstrated the best lifetime among all genuinely nanoscale electromechanical switches, with long-cycle experimental measurements performed in ambient air, both at room temperature and at ~500°C. For the first time in NEMS switches, we have carefully measured and *recorded the complete time evolution of all the switching events in real time*, as we switch on/off the devices throughout many cycles over days. Long-cycle data from a representative robust SiC NEMS switches are shown in Fig. 8a. Fig. 8b&c show the switching characteristics *before* and *after* the 7 days testing respectively, confirming the device is still alive with almost the same V_{on} and V_{off}, testifying stable and reliable performance over long testing period in ambient air. We have programmed resting intervals between hot-switching cycles, simulating realistic device operations. Such long cycles also allow for studying *I-V* curve hysteresis over the lifetime, to probe the evolution of switching parameters (*e.g.*, V_{on}, V_{off}, I_{on}, I_{off}, R_{on}).

C. Logic Building Blocks in SiC NEMS

Simple logic functions and building blocks have been realized using SiC NEMS switches. Results in Fig. 9a&c&d demonstrate one configuration for an OR gate employing a novel dual-gate NEMS switch [19], with each gate providing a logic input, V_{in1} and V_{in2}. Fig. 9b&e show an NOR gate based

on SiC NEMS switches in conjugation with a pull-up resistor. Fig. 9f shows the output of the NOR gate $V_{out}= \{V_{in1}$ NOR $V_{in2}\}$. The NOR gate operates at a large output voltage range from 0.2V to 1V at $V_{DD}=1V$.

V. OPPORTUNITIES TOWARDS 3D CO-INTEGRATION

To enable scale-up from individual NEMS switches to NEMS-enriched integrated logic circuits, architectures, and system-level designs, it requires a high level of optimization, robustness, and reproducibility at the individual device level. The fundamental advantages of NEMS logic are ultralow operating power levels and the complete elimination of leakage and standby power consumption. We envision this to generate huge improvements and benefits by using 3D architectures. In modern CMOS, performance and power issues arise from the interconnects that address and connect the active logic devices and route logic signals. Local and global interconnects can account for up to ~80% of the total signal delay and up to ~85% of total power consumption. By increasing the number of active device layers (from 1 to N) and making 3D ICs, ideally the interconnect wire length can drop by a factor on the order $N^{1/2}$, and interconnect-induced parasitic resistances and capacitances would decrease proportionally. This would lead to, in the best scenario, a power drop by $\sim N^{1/2}$, and reduction of delay by $\sim N$. Such enhancements (reduced volume and power, and increased bandwidth) are attractive. However, there could be significant thermal penalty in forming 3D ICs by stacking CMOS device layers with intervening interconnects. This arises from both the increased heat density in active layers and the compromised heat transfer between 3D CMOS stacks. As such, completely extinguishing off-state leakage in 3D ICs (see Fig. 10) would be of critical importance, and highly desired.

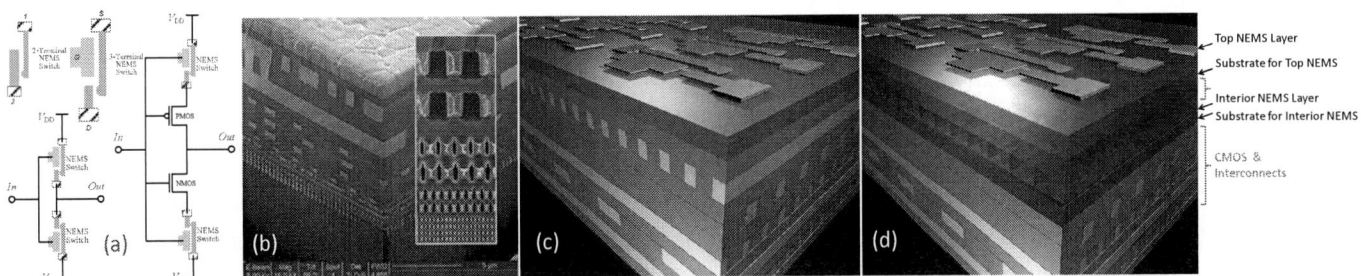

Fig. 10. Illustrations of the basic concepts of 3D co-integration of advanced NEMS switches with emerging transistors. (a) Symbols of generic 2- & 3-terminal NEMS switches, and using *inverter* as an example, illustrations of all-NEMS logic and NEMS-nanoFET co-integrated inverter. (c) An example of the cross-sectional SEM of the 'IBM air/vacuum-gap microprocessor' technology (released in May 2007) [20], featuring air/vacuum gaps in between the interconnects. (c) Illustration of a possible co-integration architecture where one NEMS device layer is made on top of the transistor chip. (d) Another possible co-integration architecture where two or more NEMS device layers could be integrated in the 3D stacks of interconnects. Wherever suited in the air/vacuum-gap layers, the NEMS can be either horizontally or vertically oriented (see Fig. 11 for examples of vertical NW NEMS and vertical NW GAA nanoFETs).

978-1-4799-7670-6/15 $31.00 © 2015 IEEE

NEMS switches and all-NEMS logic building blocks are well suited for realizing 3D co-integration architectures. The temperature-insensitivity and ultralow operational power levels of NEMS minimize issues with thermal budgets compared to those faced by 3D ICs with only CMOS. In principle, all-NEMS logic holds the promise to reliably operate at up to ~700°C. A further advantage inherent to 3D stacking is that the NEMS elements can effectively be encapsulated in vacuum microcavities (see Fig. 10). Vertical vias (*e.g.*, surrounded by low-*k* materials) can wire up both nanoFETs in the bottom layer (*e.g.*, in a CMOS-first process) and NEMS devices in other layers. In a given layer, lateral lines connect devices and functional blocks in the conventional 2D manner. Engineered thermal vias (*e.g.*, with high thermal conductivity) could be incorporated into the 3D stacks to improve thermalization.

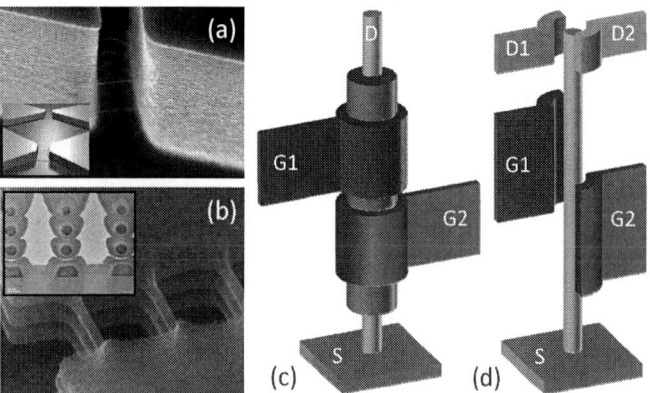

Fig. 11. Technology fusion towards 3D NEMS-nanoFETs co-integration. (a) Example of horizontal, sub-50nm suspended Si nanowires (NWs) for enabling Si NW resonators at RF/microwave frequencies [7,8]. Such Si NWs are made by a combination of top-down lithographic and bottom-up synthesis processes. (b) Example of suspended Si NWs and arrays on the way of making Si NW gate-all-around (GAA) nanoFETs, enabled by large-wafer-scale VLSI processes. (c) Illustration of a generic 3D GAA nanoFET using a vertical NW. (d) Illustration of a generic 3D NEMS switch exploiting a vertical NW.

We envision the first 3D NEMS-nanoFETs co-integration should be realized in a hybrid architecture that incorporates NEMS devices into an existing 3D CMOS architecture with air/vacuum microcavities (Fig. 10b&c). We shall then advance to embed two or more NEMS layers (with enhanced and tested lifetimes) into the 3D stack (Fig. 10d). Further, once adequate reliability and lifetime are attained, it would become possible to explore all-NEMS 3D ICs where no conventional CMOS devices are involved. Such 3D microprocessors would provide ideal temperature/harsh-environment insensitivity and even smaller device footprints; thus this route eventually permits 3D ICs with exceptionally high density. In fact, as shown in Fig. 11, as technology fusion (of horizontal Si NW NEMS resonators and suspended Si NW arrays by GAA nanoFETs process) has led to 8″-wafer-scale VLSI of NEMS, this inspires us that it is natural to exploit vertical 3D GAA nanoFETs processes to make arrays of vertical NEMS switches (Fig. 11c&d), toward high-density 3D ICs with NEMS-nanoFETs.

VI. CHALLENGES AND STRATEGIES

The ideally abrupt switching characteristics and additional outstanding attributes of NEMS endorse their advantages and promises. Meantime, each emerging switch candidate faces its own challenges. For NEMS, to realize their full potential and transform the prototyped devices into large-scale applications, at the device level, one key challenge we shall focus on is the understanding and control of nanoscale contacts. We envision these viable strategies: **(i)** deepen the studies of nanocontacts combined with a 'materials genome' approach (*e.g.*, build a library of NEMS contacts with sufficient variety of materials); **(ii)** utilizing lifetimes of today's best NEMS, demonstrate reliable functions for niche-area applications (*e.g.*, devices that are power hungry and demand abrupt switching); **(iii)** start VLSI manufacturing of NEMS logic devices with improving uniformity and wider adoption.

VII. CONCLUSIONS AND PERSPECTIVES

We have briefly reviewed the rationales of studying NEMS for logic switches *without* off-state leakage and with promises for ultralow-power computing in high temperature (*e.g.*, highly densely packed systems with compromised heat removal) and other harsh environments. Particularly, we have demonstrated NEMS switches made of very small nanocantilevers/NWs in Si and SiC, with best lifetime among NEMS (all dimensions at ~100nm or sub-100nm levels, except length, device motional volume ~1μm^3 or smaller). This should stimulate large-scale manufacturing and refinement of such devices with uniformity at wafer scale. This also presents great opportunities towards 3D NEMS-nanoFETs co-integration.

ACKNOWLEDGMENT

The author thanks NEMS teams and colleagues (from both Caltech & Case) for discussions, and the support from DARPA MTO (NEMS Program, under Grant No. N66001-07-2039, NBCH1090007, D11AP00292), National Science Foundation (Grant No. CCF-1116102), DTRA Basic Scientific Research Program (Grant No. HDTRA1-15-1-0039), T. Keith Glennan Fellowship, and Case School of Engineering.

REFERENCES

[1] Y. Taur and T.H. Ning, *Fundamentals of Modern VLSI Devices* (2004).

[2] T.N. Theis and P.M. Solomon, *Proc. IEEE*, **98**, 2005-2014 (2010).

[3] X.M.H. Huang, *et al.*, *New J. Phys.* **7**, Art. No. 247 (2005).

[4] M.L. Roukes, *Tech. Digest of IEDM'04*, 539-542 (2004).

[5] V. Pott, *et al.*, *Proc. IEEE*, **98**, 2076-2094 (2010).

[6] X.L. Feng, *et al.*, *Nano Letters* **10**, 2891-2896 (2010).

[7] X.L. Feng, *et al.*, *Nano Letters* **7**, 1953-1959 (2007).

[8] R. He, *et al.*, *Nano Letters* **8**, 1756-1761 (2008).

[9] T. Ernst, *et al.*, *Tech. Digest of IEDM'08*, 745-748 (2008).

[10] J. Arcamone, *et al.*, *Tech. Digest of IEDM'11*, 669-672 (2011).

[11] P.X.-L. Feng, *et al.*, *US Patent*, No. US 8,115,344 B2 (2012).

[12] P.X.-L. Feng, *et al.*, *US Patent*, No. US 8,258,899 B2 (2012).

[13] R. Yang, *et al.*, *Proc. IEEE MEMS'13*, 229-232 (2013).

[14] R. Yang, *et al.*, *J. Micromech. & Microeng.* **25**, in press (2015).

[15] T. He, *et al.*, *Proc. IEEE MEMS'13*, 516-519 (2013).

[16] T. He, *et al.*, *Tech. Digest of IEDM'13*, Paper No. 4.6, 108-111 (2013).

[17] T. He, *et al.*, *Tech. Digest of Transducers'13*, 669-672 (2013).

[18] T. He, *et al.*, *Proc. IEEE MEMS'14*, 1079-1082 (2014).

[19] T. He, *et al.*, *Proc. IEEE NEMS'13*, 554-557 (2013).

[20] IBM, *'Airgap Microprocessor' Press Release*, May 3, (2007).

Wide Band Study of Silicon-on-Insulator Photodiodes on Suspended Micro-Hotplates Platforms

N. André[1], G. Li[1,2], P. Gérard[1], O. Poncelet[1], Y. Zeng[2], S. Z. Ali[3], F. Udrea[3,4], L. A. Francis[1] and D. Flandre[1]

[1]Institute of Information and Communication Technologies, Electronics and Applied Mathematics, Université catholique de Louvain, 3 Place du Levant, 1348, Louvain-la-Neuve, Belgium, nicolas.andre@uclouvain.be

[2]School of Physics and Microelectronics Science, Hunan University, 410082, Changsha, China, liguoli_lily@hnu.edu.cn

[3]Cambridge CMOS Sensors, St Andrew's House, CB2 3BZ, Cambridge, United Kingdom

[4]Department of Engineering, Electrical Engineering Division, University of Cambridge, 9 JJ Thomson Avenue, CB3 0FA, Cambridge, United Kingdom

Abstract—In this paper, the performances of a lateral thin-film PIN photodiode based on silicon-on-insulator technology are reported for applications from blue to red wavelengths. The platform consists of a micro-hotplate with a suspended heater and a photodiode. Responsivities of 0.01 to 0.05 A/W were obtained for 450-900 nm light range in reverse bias operation. Suspended photodiodes give up to 5x responsivity improvement with regard to the photodiodes on substrate. In addition to photodetection, the diode can monitor the temperature with a linear voltage decreasing by about 1.4 mV/K, under 50 µA constant forward current for a large range of temperature (measured from 25 to 300°C).

Keywords—Semiconductor photodiode, Lateral thin-film PIN diode, Temperature sensor, Silicon-on-Insulator, Harsh environment, CMOS technologies.

I. INTRODUCTION

Many photodiode implementations on diverse substrates (SiC, GaP, (Al)GaN, Si) have been presented over the years [1]. Thin-film lateral Silicon-on-Insulator (SOI) PIN photodiodes are often presented as a solution to specifically and efficiently absorb low wavelengths with silicon-based technologies [2], tuning the photodiode optical bandwidth by the layer thicknesses of the SOI CMOS process. Responsivities with values as high as 0.1 A/W have been reached for wavelengths below 400 nm with a Si film thickness of 80 nm and a specific anti reflective coating (ARC) [3], while the responsivity sharply drops by 10x when the wavelength is increased in the visible range. A SOI miniature light/temperature suspended sensor with fast response time is then of great interest for in-situ monitoring system from biomedical [4] to environmental applications. For example, flame indicators can control combustion in engines [5-6].

Several methods exist to improve the photodiode sensitivity: light scattering on gold nanoparticles [7], anodic aluminum oxide (AAO) layer [8], anti-reflective coatings [9-10]. However, few groups have studied the resonant cavity enhancement (RCE) effect with SOI photodiodes on membrane (theoretical work in [11], thick SOI for IR application in [12]), beside their great interest for combining the SOI advantages of high speed, low dark current and low capacitance. Furthermore, on the contrary to Ga- or SiC-based technologies, SOI CMOS technology enables the implementation of high-performance (low cost, low power and reliable) analogue/digital signal processing circuits towards a full system on a chip (SoC). We propose here to study interdigitated electrodes photodiode designs with various intrinsic lengths and to compare the performance of suspended and on-substrate diodes from blue to red wavelengths.

The novelty of our sensor can be summarized in the combination of following four points: (1) thin film SOI photodiode; (2) on suspended microhotplate; (3) temperature and light multisensing; (4) straightforward SoC integration in CMOS process.

II. FABRICATION

A. SOI micro-hotplate platform

The micro-hotplate platform is designed to measure humidity and gas levels when interdigitated electrodes and specific sensing layers are present at the surface [13], while having at the same time all the advantages provided by a commercial CMOS foundry. In addition to SOI technology, tungsten is used as interconnect metal and heater material which allows operations at high temperatures as a result of its extremely high melting point. Tungsten is also less sensitive to electro-migration compared with aluminum.

Fig. 1 shows the micro-hotplate typical cross-section, fabricated using standard SOI-CMOS process [14-15]. Table 1 lists the superposed materials and their thicknesses.

The thin buried oxide acts as a natural etch-stop during the back-etching, ensuring membranes high uniformity and, as a result, excellent reproducibility. The membranes micro-hotplates have a circular shape to minimize stress effect and to enhance their mechanical robustness. Their thickness is ~ 5 µm with membrane diameters of 600 µm and hotspot areas of 100 µm, respectively. The typical power consumption of fabricated micro-hotplates is 35 mW for heating at 600°C (efficiency of about 15°C/mW).

978-1-4799-7670-6/15 $31.00 © 2015 IEEE

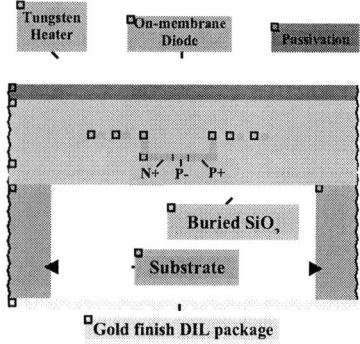

Fig. 1. Sensing platform cross-section with diode and heater.

TABLE I. MATERIALS THICKNESS LIST

Functions	Materials	Thickness
Passivation	Si_3N_4 PECVD	500 nm
Membrane	SiO_2 PECVD	2.7 μm
Heater	Tungsten	500 nm
Active thin film	Silicon	250 nm
BOX	SiO_2	1 μm
Substrate	Silicon	380 μm

Under constant current, the diode yields a forward voltage proportional to the membrane temperature, measured from room temperature (RT) up to 300°C (fig. 2). Similar designs were proved to operate in an extremely wide temperature range of -200°C to 700°C [16-17]. Moreover, the temperature diode accuracy and linearity will offer better accuracy for the light detection by allowing precise temperature effects cancellation.

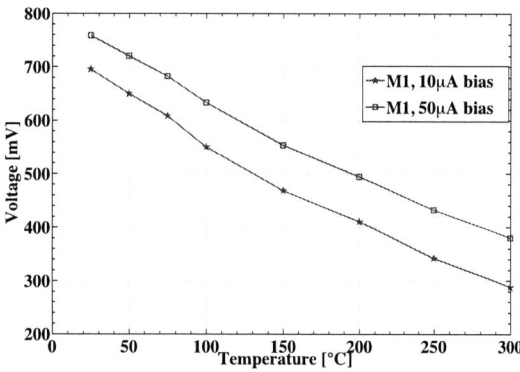

Fig. 2: Voltage drop for on-membrane photodiode M1 at 10 μA and 50 μA forward constant bias, from RT to 300°C.

The monolithic integration of these devices is fully feasible due to the use of an identical commercial SOI CMOS technology for hotplate and circuit, i.e. 1 μm-thick BOX oxide, 250 nm-thick active silicon and tungsten interconnection lines.

B. SOI $p^+/p^-/n^+$ photodiode

The photosensor consists of a number of p^+-p^--n^+ finger diodes interdigitated in parallel over a given sensing area.

The PIN diode design lies in a trade-off concerning the size of the intrinsic length, i.e. the lateral distance between the p^+ and n^+ areas. The responsivity can be increased when increasing the depletion length but will saturate when the intrinsic length becomes greater than the carrier diffusion length. A large intrinsic length could then increase the electrons-holes recombination and decrease the collection of photogenerated carriers [18], while it will reduce the number of fingers for the same total device area and thus reduce the dark current. A too large intrinsic zone decreases the number of interdigitated diodes for the same area, which is contrarily opposed to a maximal responsivity. Finally, for a too small intrinsic length, the increased number of p^+/n^+ regions with their metal interconnects will shadow much of the light.

Fig. 3 presents the SOI micro-hotplate devices on a $1.6x1.6$ mm^2 chip, which contains 6 photodiodes.

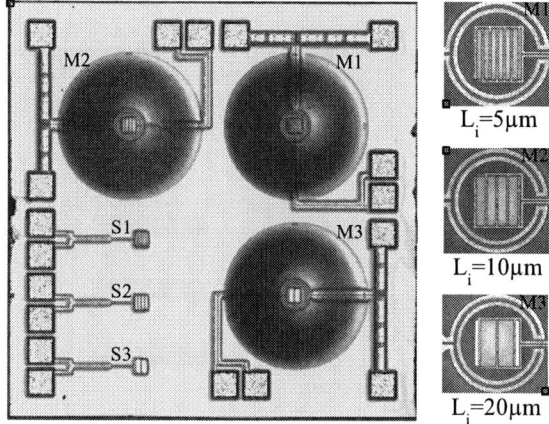

Fig. 3: Chip top view with 3 different intrinsic lengths (5/10/20 μm) and 2 different locations: M-series = on membrane with a surrounding heater and S-series = on substrate.

TABLE II. PHOTODIODE GEOMETRIES

#Device	$L_{electrode}$ [μm]	$W_{fingers}$ [μm]	L_i [μm]	#fingers	Effective area [μm^2]
M1, S1	4	60	5	6	30*60
M2, S2	4	60	10	4	40*60
M3, S3	4	60	20	2	40*60

The so-called photodiodes S1, S2, and S3 lie above the buried oxide and Si substrate of the standard SOI wafer with the geometries as specified in Table II. The photodiodes M1, M2 and M3 have identical geometries but are suspended on a membrane and surrounded by a circular tungsten heater. Three different intrinsic lengths are implemented: 5, 10 and 20 μm.

The active silicon film thickness of 250 nm corresponds to the penetration depth in Si of a light wavelength of 445 nm. Multiple light reflections will then occur from blue to red wavelengths in the multilayer structure formed by the Si_3N_4 / SiO_2 / Si / SiO_2 / 'cavity or substrate' stack. For the on-membrane structures, we need to consider that the chip is mounted in a DIL 16 package over a gold finish bottom layer which will act as a back mirror.

978-1-4799-7670-6/15 $31.00 © 2015 IEEE

III. RESULTS

A. Transfer matrix simulations

Based on the electromagnetic theory and considering the boundary conditions for the electric and magnetic fields of incident plane waves, we used the transfer matrix characterizing the interference in multilayer films and made the detailed calculation in Matlab to obtain the light absorption of active silicon layer. As shown in Fig. 4, the black curve is for the device structure with 380 μm silicon substrate, the red curve is for the membrane structure i.e. without substrate but considering the reflection from the bottom gold finish layer.

For specific wavelength ranges marked by A, B, C (around 510, 590 and 730 nm), we observe a strong improvement of silicon absorption due to the reflection from the bottom gold, by about a factor of 2 when compared to the on-substrate structure.

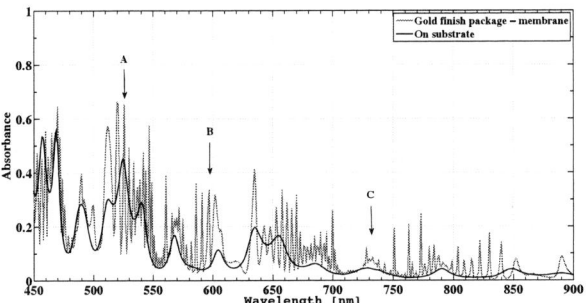

Fig. 4: Silicon Absorption with Transfer-matrix method, by step of 1 nm.

B. Wide band illumination measurements

The measurement set-up consists of four main parts : light supply, measurement PCB circuit, semiconductor analyzer and power supply. For the light supply, we use Muller LXH 100 as the light source and a monochromator to select a single wavelength (from 200 to 1600 nm). The micro-hotplates packaged in DIL-16 ceramic packages were embedded in three identical test printed circuit boards (PCBs). With the power source (K2400), we can heat the membrane device from the room temperature to 300°C. For the measurement, we use Semiconductor Parameter Analyzer (Agilent, HP4156) and Low Leakage Switch Mainframe (Agilent, E5250A) to obtain the current-voltage curves.

As depicted in Fig. 5, the responsivity is extracted for a reverse voltage applied to the anode equal to -2.0V.

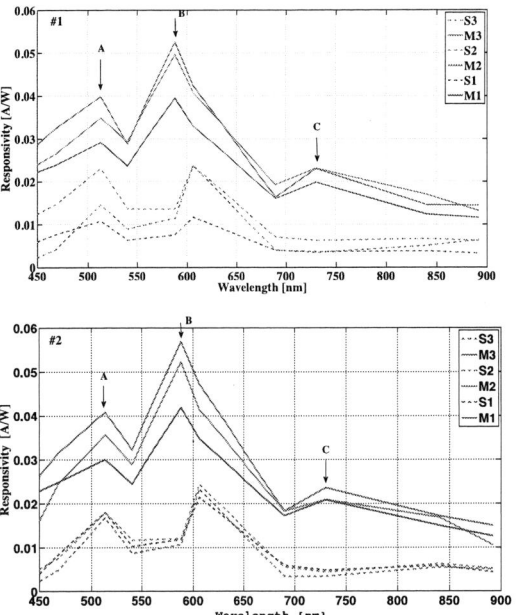

Fig. 5: Photodiodes responsivity at -2V under single wavelength illuminations for 2 different chips #1 and #2.

The incident power density for the 9 wavelengths considered in our measurements (468, 513, 540, 588, 606, 690, 730, 840 and 892 nm) is on the order of 10^{-5} W/cm^2. Each measurement was conducted three times. Responsivity results for the six photodiodes per chip were obtained during the same illumination session.

The occurrence and intensity of the three responsivity peaks at 513, 588 and 730 nm can be explained by the better absorption in the active Si layer, expected from Fig. 4 simulations, for suspended photodiode after reflection on the gold finish Dual-In-Line (DIL) package when compared to the on-substrate structures. The responsivity increase ranges from 2x to 5x over the blue to red bandwidth.

C. Wide band illumination measurements at HT

Fig. 6 presents the measurements of two photodiodes embedded on membranes (M1 and M3), exposed to white light illumination. The local tungsten microheater was used to increase the local temperature from 20 to 200°C. The temperature is estimated operating the same diode in forward mode and using the calibration curves of Fig. 2. Up to 200°C, the net photocurrent can be discriminated from the increasing dark current. The dark current increase follows the Si intrinsic carrier concentration dependence on temperature confirming that it is dominated by volume generation in the intrinsic region.

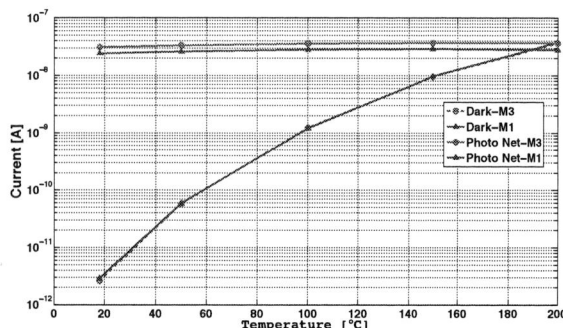

Fig. 6: *Photodiodes current for two on-membrane devices biased at -2V with local heating: (bottom curves) in the dark and (top curves) under white light illumination.*

If required, the extension to high temperature working range of a SoC photosensing system is straightforward by the use of CMOS circuitry built in the same SOI technology and so able to operate at ambient temperatures up to 225°C. More specifically, a light/temperature membrane sensor for methane-flame monitoring with fast response time can find a natural use into domestic boilers.

IV. CONCLUSIONS

The thin-film lateral PIN suspended photodiodes presented here have shown significanty better responsivity under blue to red light illumination than the same photodiode on substrate, thanks to a multilayer resonant cavity enhancement effect as predicted by simulations.

Moreover, the micro-hotplate platform offers multiple advantages in addition to the photodetection: (i) commercial technology with high yield , (ii) rapid drying in wet environment, (iii) possible thermal cleaning (>400°C) in dirty environment, (iv) temperature dependence cancellation, (v) accelerated aging tests, (vi) temperature sensing, (vii) no substrate loss for high-speed applications [19].

The addition of a top ARC layer is considered to increase absorption in the Si film for specific wavelengths, preferably made of silicon oxide (SiO_2) which can be deposited by ALD and whose refractive index is adapted to the membrane material stack.

ACKNOWLEDGMENT

The authors are very grateful to P. Simon for his help during opto-electrical characterizations.

REFERENCES

[1] M. Razeghi, « Short-Wavelength Solar-Blind Detectors—Status,
[2] A. Afzalian, D. Flandre, "Physical modeling and design of thin-film SOI lateral PIN photodiodes", IEEE Transactions on Electron Devices, Vol. 52, Issue 6, pp 1116-1122, 2005
[3] O. Bulteel, N. Van Overstraeten-Schlögel, A. Afzalian, P. Dupuis, S. Jeumont, L. Irenge et al., « Low-Wavelengths SOI CMOS Photosensors for Biomedical Applications, Biomedical Engineering »,

Trends in Electronics, Communications and Software, Mr Anthony Laskovski (Ed.), ISBN: 978-953-307-475-7, InTech, 2011.
[4] O. Bulteel, A. Afzalian, D. Flandre, « Fully integrated blue/UV SOI CMOS photosensor for biomedical and environmental applications », Analog Integrated Circuits and Signal Processing, Vol. 65, Issue 3, pp. 399-405, 2010
[5] D.M. Brown et al., in 3rd Int. High Temperature Electronics Conf., p. X-23, 1996
[6] J. Ballester and T. Garcia-Armingol Diagnostic techniques for the monitoring and control of practical flames, Progress in Energy and Combustion Science, 36 (2010) 375–411
[7] A. Ono, Y. Matsuo, H. Satoh, H. Inokawa, « Sensitivity improvement of silicon-on-insulator photodiode by gold nanoparticles with substrate bias control », Applied Physics Letters 99, 062105, 2011
[8] Y. Chen, T. Cheng, C. Cheng, C. Wang, C. Chen, C. Wei, Y. Chen, "Highly sensitive MOS photodetector with wide band responsivity assisted by nanoporous anodic aluminum oxide membrane", Optics Express 56, Vol. 18, No. 1, 2010.
[9] J. Chu, Z. Han, F. Meng and Z. Wang Spectral response of blue-sensitive Si photodetectors in SOI, Solid-State Electronics 55 (2011) 54–58
[10] H. Zhitao, C. Jinkui, M. Fantao, J. Rencheng," Design and simulation of blue/violet sensitive photodetectors in silicon-on-insulator", Journal of Semiconductors, Vol. 30, No. 104008, pp. 1-4, 2009
[11] A. Afzalian, D. Flandre, "Speed performances of thin-film lateral SOI PIN photodiodes up to tens of GHz", Proceedings of the IEEE SOI conf., pp. 83-84, Niagara Falls, New York, October 2006.
[12] B. Cheng, F. Yao, C. Xue, J. Zhang, R. Mao, C. Li, L. Luo, Y. Zuo, Q. Wang,"Si-based membrane resonant cavity enhanced photodetectors" Group IV Photonics, 2005. 2nd IEEE International Conference on, pp. 105-107, 2005
[13] N. André, G. Pollissard, N. Couniot, P. Gérard, Z. S. Ali, F. Udrea et al., "A Silicon-on-Insulator Platform Functionalized By Atomic Layer Deposition for Humidity Sensing." In International Meeting on New Sensing Technologies and Modelling for Air-Pollution Monitoring, Aveiro, Portugal, 15 October 2014
[14] Gas sensing semiconductor devices, by J. A. Covington, J. W. Gardner, F. Udrea. (2005, May 30). US Patent 7,495,300
[15] P.K. Guha, S.Z. Ali, C.C.C. Lee, F. Udrea, W.I. Milne, T. Iwaki, J.A. Covington, J.W. Gardner," Novel design and characterization of SOI CMOS micro-hotplates for high temperature gas sensors", Sensors and Actuators B 127, pp. 260–266, 2007
[16] A. De Luca, V. Pathirana, S.Z. Ali, F. Udrea, Silicon on Insulator thermodiode with extremely wide working temperature range, Proceedings of the 17th International Conf. on Transducers & Eurosensors XXVII, 2013, pp. 1911-1914, Barcelona, Spain, 16-20 June.
[17] S. Santra, P. K. Guha, S. Z. Ali, I. Haneef, F. Udrea, J. Gardner, "SOI diode temperature sensor operated at ultra high temperatures – a critical analysis", *Proceedings of IEEE Sensors Conf.*, 2008, pp. 78-81, Lecce, Italy, 26-29 October.
[18] O. Bulteel, Silicon-on-Insulator Optoelectronic Components for Micropower Solar Energy Harvesting and Bio-Environmental Instrumentation, PhD Thesis, 2011
[19] A. Afzalian, D. Flandre, "Design of Thin-Film Lateral SOI PIN Photodiodes with up to Tens of GHz Bandwidth", Advances in Photodiodes, Prof. Gian Franco Dalla Betta (Ed.) 2011, ISBN: 978-953-307-163-3, InTech

Evaluation of 32-Bit Carry-Look-Ahead Adder Circuit with Hybrid Tunneling FET and FinFET Devices

Tse-Ching Wu, Chien-Ju Chen, Yin-Nien Chen, Vita Pi-Ho Hu, Pin Su and Ching-Te Chuang

Department of Electronics Engineering & Institute of Electronics, National Chiao Tung University, Hsinchu, Taiwan
E-Mail: tseching.ee02g@g2.nctu.edu.tw ; ctchuang@mail.nctu.edu.tw

Abstract—In this paper, we investigate the hybrid TFET-FinFET 32-bit carry-look-ahead adder (CLA) circuit and compare the delay, power and power-delay product (PDP) with all FinFET and all TFET implementations in near-threshold region. We use atomistic 3D TCAD mixed-mode simulations for transistor characteristics and HSPICE circuit simulations with look-up table based Verilog-A models calibrated with TCAD simulation results. In the hybrid design, TFETs are used for the top critical path to reduce the longest path delay, and FinFETs are used for the rest of the circuit to reduce switching power and leakage power. The PDP of the hybrid TFET-FinFET CLA circuit is better than the circuits with all FinFET and all TFET implementations in the vicinity of V_{DD}=0.3V. However, as the operating voltage is further reduced, the lower-ranked critical paths (e.g. 2nd critical path) with some FinFET devices in the path stick out, and the delay and PDP become inferior to all TFET implementation.

I. INTRODUCTION

The subthreshold slopes of MOSFET and FinFET are limited by carrier thermionic emission. TFET device, which utilizes the band-to-band tunneling (BTBT) as the conduction mechanism, exhibits steep sub-threshold slope [1-3]. Recent research works on TFET based circuits have shown significant performance improvement and power reduction at low operating voltage [4-6]. However, due to the source-channel barrier in TFET device, the capacitance of TFET is dominated by C_{gd} (Miller capacitance) in weak and strong inversion region, which is larger than MOSFET and FinFET device. The enhanced Miller capacitance induces voltage spike during circuit switching, which may increase power consumption. Furthermore, the I_{on} of TFET at moderate voltage is lower than that of FinFET. Hybrid TFET-FinFET (or MOSFET) design approach enables power-delay trade-off to achieve the best energy efficiency for some special conditions. Several studies on hybrid TFET-CMOS circuits have been reported [7-8], but a physics-based performance and power assessment for large scale logic circuits is lacking. In this work, we provide an in-depth physics-based assessment on the power-delay trade-off of hybrid TFET-FinFET design for 32-bit CLA circuit.

II. DEVICE STRUCTURES AND CHARACTERISTICS

In this work, the basic TFET structure comprises a gated p-i-n tunnel diode under reverse bias with asymmetrical source/drain doping. TFET device utilizes the band-to-band tunneling as the major conduction mechanism. For n-type TFET, the source is p+ region with dominant electron conduction, the channel is gated intrinsic region, and the drain is n+ region. When N-TFET is "On" ($V_{GS}>0$), the conduction band edge of the channel is pulled down below the valence band edge of the source, and carriers can tunnel into available empty states of the channel region. When N-TFET is "Off"

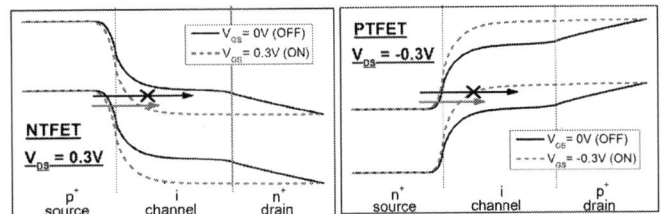

Fig. 1. Energy band diagrams of n-type (left) and p-type (right) TFET in ON/OFF state.

TABLE I. PARAMETERS OF TFET AND FINFET DEVICES

TFET and FinFET			
L_{eff} = 25nm	W_{fin} = 7nm	H_{fin} = 20nm	EOT = 0.65nm
	nTFET	*pTFET*	*FinFET*
Nch (cm⁻³)	undoped	undoped	1×10^{17}
Ns (cm⁻³)	4.5×10^{19} (p-type)	1×10^{20} (n-type)	1×10^{20}
Nd (cm⁻³)	2.0×10^{17} (n-type)	2×10^{17} (p-type)	1×10^{20}

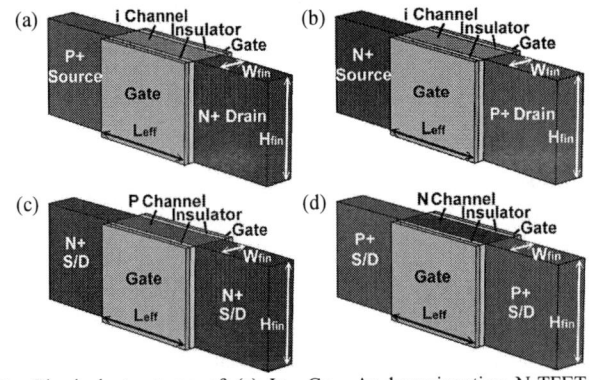

Fig. 2. Physical structures of (a) $In_{0.53}Ga_{0.47}As$ homojunction N-TFET, (b) $Ge_{0.925}Sn_{0.075}$ homojunction P-TFET, (c) $In_{0.53}Ga_{0.47}As$ N-FinFET and (d) Ge P-FinFET.

(V_{GS}=0), the valence band edge of the source is below the conduction band edge of the channel, and the band-to-band tunneling probability is low due to lack of available states in the channel region and wide barrier at source-channel junction. For P-TFET, the source is n+ region with dominant hole conduction, applying $V_{GS}<0$ turns P-TFET "ON". The band diagrams of TFET in ON/OFF states are shown in Fig. 1.

We consider the $In_{0.53}Ga_{0.47}As$ homojunction N-TFET and $Ge_{0.925}Sn_{0.075}$ homojunction P-TFET due to their high I_{on} and compatible I_{DS}-V_{GS} characteristics. The work function and I_{on} (I_{DS} at $V_{DS} = V_{GS} = 0.3V$) of N-TFET/P-TFET are 4.53/4.82 eV and 66.99/87.83 nA, respectively. $In_{0.53}Ga_{0.47}As$ N-FinFET and Ge P-FinFET are considered for comparison. The work function and I_{on} of N-FinFET/P-FinFET are 4.88/4.27 eV and 26.09/18.53 nA (at $V_{DS} = V_{GS} = 0.3V$), respectively. Fig. 2 shows the 3D TFET and FinFET device structures constructed for atomistic TCAD simulations. The device parameters and doping are shown in TABLE I. We use the non-local

978-1-4799-7670-6/15 $31.00 © 2015 IEEE

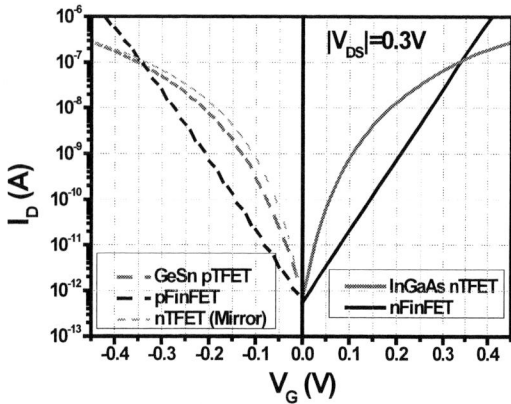

Fig. 3. I_{DS}-V_{GS} characteristics at $|V_{DS}|$=0.3V of $In_{0.53}Ga_{0.47}As$ N-TFET, $Ge_{0.925}Sn_{0.075}$ P-TFET, $In_{0.53}Ga_{0.47}As$ N-FinFET and Ge P-FinFET.

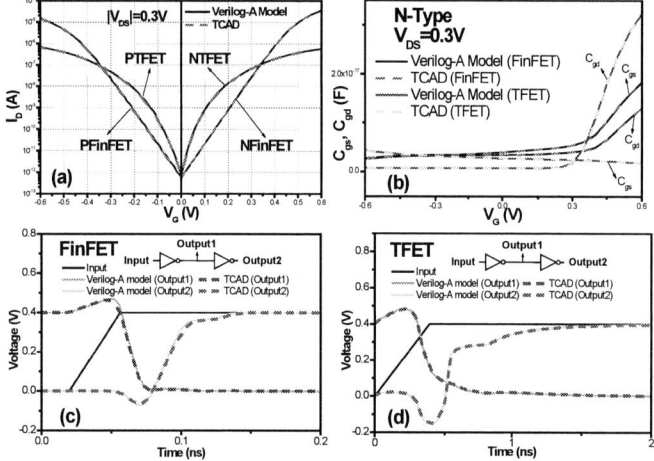

Fig. 4. Flowchart for HSPICE with look-up table based Verilog-A model generated from atomistic 3D TCAD simulation results [4, 6].

Fig. 5. The Verilog-A model is calibrated with TCAD results on (a) I_{ds}-V_{gs} characteristics, (b) C_{gd}/C_{gs}-V_{gs} characteristics, and (c) (d) transient switching waveforms for basic buffer (FO = 1) of FinFET and TFET respectively.

band-to-band tunneling model which is applicable to arbitrary tunneling barrier with non-uniform electric field for TFET simulations [9], and the parameters used in the model are calibrated with [10-11]. Fig. 3 shows the I_{DS}-V_{GS} characteristics of TFETs and FinFETs with comparable I_{off} at V_{DS} = 0.3V. The I_{off} (I_{DS} at V_{DS} = 0.3 V and V_{GS} = 0V) of N-TFET/ P-TFET / N-FinFET / P-FinFET is 0.596/0.761/0.546/ 0.425 pA.

III. SIMULATION AND CALIBRATION METHODOLOGY

We use the atomistic 3D TCAD device simulations to investigate the characteristics of TFET and FinFET. However, for complex circuits with large transistor counts, TCAD mixed-mode simulations require prohibitively long simulation time and face the convergence problems. To overcome these obstacles, look-up table based Verilog-A model has been employed for TFET circuit simulations in some studies [4, 6]. Nevertheless, previous approach is not physics-based, and the Verilog-A model does not accurately reflect the device characteristics. In this work, in order to simulate 32-bit CLA circuits and accurately capture the TFET and FinFET device characteristics, we extract the transfer characteristics of TFET and FinFET devices from atomistic 3D TCAD device simulations to build the two-dimensional Verilog-A look-up tables with $I_{ds}(V_{gs}, V_{ds})$, $C_{gs}(V_{gs}, V_{ds})$ and $C_{gd}(V_{gs}, V_{ds})$ characteristics across the voltage range of interest. The flow chart for small signal Verilog-A model generation is shown in Fig. 4. The Verilog-A models of devices are employed in HSPICE circuit simulations. We calibrate the Verilog-A model with TCAD results on I-V, C-V characteristics, and transient waveforms for basic buffer of FinFET and TFET respectively. Accurate I_{ds}-V_{gs}, C_{gd}/C_{gs}-V_{gs}, and transient characteristics for both FinFET and TFET using Verilog-A model with respect to TCAD results are shown in Fig. 5. In Fig. 5(c)(d), the accuracies of delays of Verilog-A model are within 0.90/1.05% of FinFET TCAD results, and within 0.02/1.30% of TFET TCAD results for one- and two-stage inverters at V_{DD} = 0.4V, respectively.

IV. HYBRID TFET-FINFET 32-BIT CLA CIRCUIT

At low operating voltage, TFET shows outstanding on-current compared with FinFET device because of its superior sub-threshold slope. However, TFET-based circuit design needs to consider the structure asymmetry. In FinFET device, the gate-to-source capacitance (C_{gs}) dominates the total gate capacitance as its drain and source are "linked" by the conducting inversion channel at on-state. On the contrary, the gate capacitance of TFET device, which is larger than that in FinFET, is dominated by gate-to-drain capacitance (C_{gd}, Miller capacitance) because of the source-channel barrier in the weak and strong inversion region. During circuit switching, the larger Miller capacitance of TFET leads to significant voltage "spike", resulting in larger power consumption.

In this work, we consider the tradeoff between power consumption and circuit delay for 32-bit CLA circuit as shown in Fig. 6. In terms of circuit delay, TFET exhibits higher performance than FinFET even with TFET's larger Miller capacitance which degrades the delay. On the other hand, FinFET with smaller Miller capacitance shows better (less) switching power than TFET device. To improve the energy

978-1-4799-7670-6/15 $31.00 © 2015 IEEE

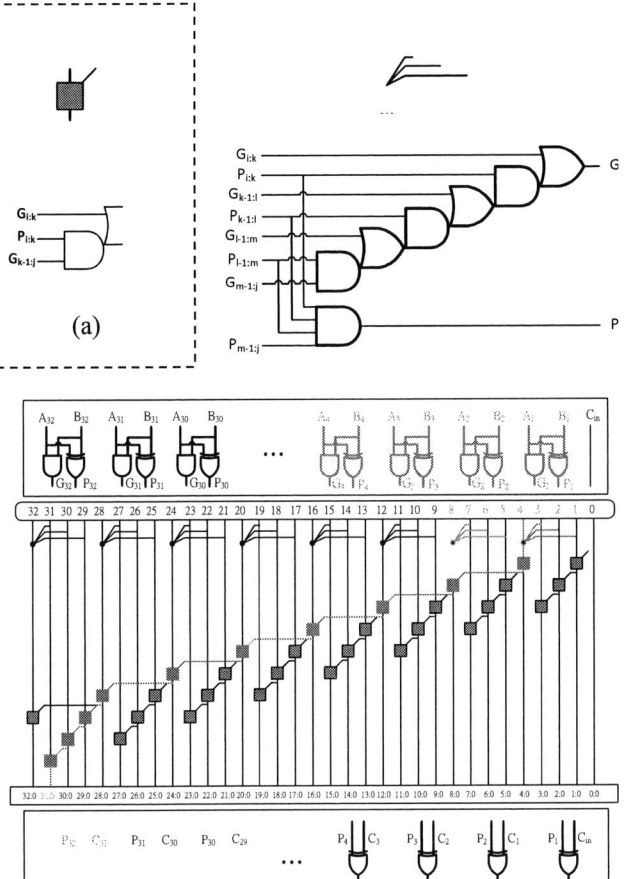

Fig. 6. (a) An AND-OR gate. (b) Valency-4. (c) The hybrid TFET FinFET 32-bit CLA circuit. The top critical path (red path) is implemented with TFETs, while FinFETs are used for the rest of the circuit. The secon critical path starts from the green path.

replace the devices on the top critical path with TFETs. In this hybrid TFET-FinFET implementation, the large Miller capacitances of TFETs and their impacts on switching power are minimized, and the delay of the top critical path is reduced. The delay and PDP improvement, however, existed only for certain operating voltage range (in the vicinity of $V_{DD}=0.3V$ in our case). This is because as the operating voltage is further reduced, the lower-ranked critical paths (e.g. 2nd critical path) with some FinFET devices in the path would stick out, and the delay and PDP become inferior to all TFET implementation. This phenomenon is discussed in detail in the next section.

V. COMPARISON OF 32-BIT CLA IMPLEMENTATIONS

In this section, the hybrid TFET-FinFET implementation, all FinFET implementation and all TFET implementation are comprehensively analyzed and compared using HSPICE simulations with Verilog-A model calibrated with 3D mixed-mode TCAD simulations.

A. Circuit Delay

We consider the top critical paths for all FinFET implementation and all TFET implementation. For hybrid

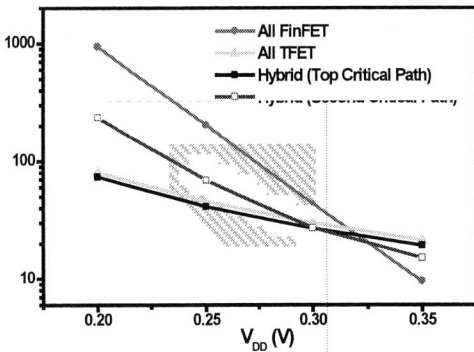

Fig. 7. Delay of 32-bit CLA circuit versus V_{DD} from 0.2V to 0.35V for the hybrid TFET-FinFET implementation including top and second critical path and the all FinFET and all TFET implementation.

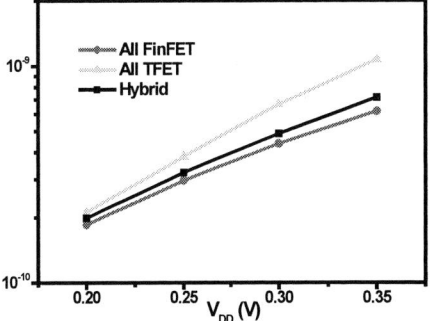

Fig. 8. Switching power (P_{sw}) of 32-bit CLA circuit versus V_{DD} from 0.2V to 0.35V for the hybrid TFET-FinFET implementation and the all FinFET and all TFET implementation.

TFET-FinFET implementation, the devices in the top critical path for the "base" FinFET implementation are replaced with TFETs. As such, the lower-ranked critical paths in the hybrid design would have some FinFETs in the paths. Fig. 7 shows the delay of 32-bit CLA circuit versus V_{DD} from 0.2V to 0.35V for the top two paths of the hybrid TFET-FinFET implementation, and the top critical paths of all FinFET and all TFET implementation.

The switching delay is commonly calculated as $\tau = (C_g V_{DD})/I_{on}$. At low operating voltage (<0.3V), TFET offers significantly larger current, which dominates over the impact of Miller capacitance, to speed up the circuit operation. Therefore, all TFET implementation exhibits lower delay time than all FinFET implementation. In the hybrid TFET-FinFET implementation, the delay of top critical path is close to that of all TFET implementation. However, for $V_{DD} < 0.3V$, the second critical path (which has some FinFETs in the path) sticks out and its delay becomes longer than that of the "top" critical path. Thus, the "real" critical path delay for the hybrid implementation comprises two parts: one part dominated by the top critical path for $V_{DD} \geqq 0.3V$, and the other part dominated by the second critical path for $V_{DD} < 0.3V$. Notice that the delay of the hybrid implementation is smaller than the all FinFET implementation for $V_{DD} < 0.32V$.

B. Switching Power and Leakage Power

The switching power (P_{sw}) of the 32-bit CLA circuit versus V_{DD} from 0.2V to 0.35V for the hybrid TFET-FinFET implementation and all FinFET and all TFET implementation

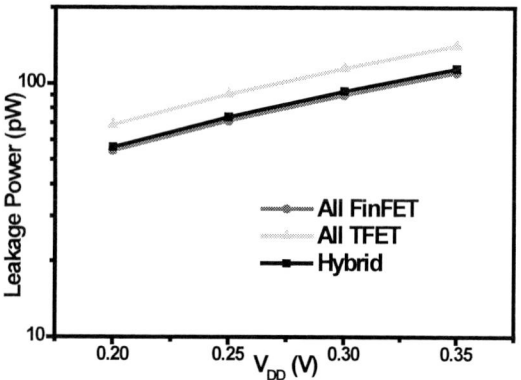

Fig. 9. Leakage power ($P_{leakage}$) of 32-bit CLA circuit versus V_{DD} from 0.2V to 0.35V for the hybrid TFET-FinFET implementation and the all FinFET and all TFET implementation.

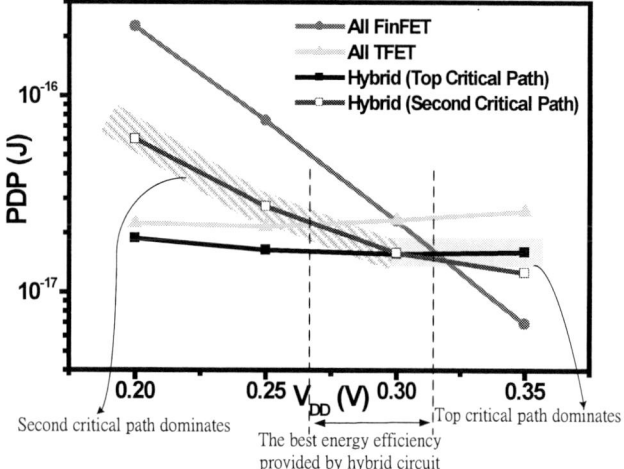

Fig. 10. PDP of 32-bit CLA circuit versus V_{DD} from 0.2V to 0.35V for the hybrid TFET-FinFET implementation and the all FinFET and all TFET implementation.

are shown in Fig. 8. Due to the larger Miller capacitance of TFET, the switching power of all TFET implementation is larger than that of all FinFET implementation. The P_{sw} for the hybrid implementation is slightly higher than that of all FinFET implementation.

Fig. 9 shows the leakage power ($P_{leakage}$) of the 32-bit CLA circuit versus V_{DD} from 0.2V to 0.35V for the hybrid TFET-FinFET implementation, all FinFET and all TFET implementation. As V_{DD} is reduced, the leakage power decreases. Notice that the $P_{leakage}$ of the hybrid implementation is almost the same as that of the all FinFET implementation, and both are lower than that of the all TFET implementation.

C. Power-Delay Product

Power and delay are often in conflict for circuit operation. PDP is a figure of merit to evaluate the energy efficiency of logic circuits. It represents the energy dissipated during a switching event, and can be calculated as: $PDP = C_g V_{DD}^2$.

The PDP of 32-bit CLA circuit versus V_{DD} from 0.2V to 0.35V for the hybrid TFET-FinFET implementation and all FinFET and all TFET implementation are shown in Fig. 10. Compared with all FinFET implementation, all TFET implementation exhibits better PDP at low operating voltage (<0.3V), but worse PDP in high operating voltage (>0.3V). For

the hybrid implementation, the PDP curve comprises two parts (similar to delay), one part dominated by the top critical path for $V_{DD} \geq 0.3$V, and the other part dominated by the second critical path for $V_{DD} < 0.3$V. Notice that the hybrid implementation offers the best PDP compared with the other implementations in the vicinity of V_{DD}=0.3V. The voltage range, in which the hybrid implementation provides PDP advantage, can be extended by replacing FinFETs in the second critical path with TFETs.

VI. CONCLUSION

We evaluate the hybrid TFET-FinFET implementation of 32-bit CLA circuit with HSPICE circuit simulations using Verilog-A device models extracted from atomistic 3D TCAD simulations with calibrated model and device parameters. The circuit delay, dynamic/leakage power and PDP of all TFET-based, all FinFET-based and mixed TFET-FinFET 32-bit CLA are analyzed and compared. For circuit delay, the results indicate that the hybrid implementation offers delay close to all TFET implementation for $V_{DD} \geq 0.3$V, but the second critical path (with some FinFETs in the path) delay would stick out for $V_{DD} < 0.3$V. The power of the hybrid implementation is close to that of all FinFET implementation. The hybrid implementation exhibits the best PDP in the vicinity of V_{DD}=0.3V.

ACKNOWLEDGMENT

This work was supported by the Ministry of Science and Technology in Taiwan under contract MOST 103-2221-E-009-196-MY2. The authors thank the National Center for High-Performance Computing in Taiwan for the software and facilities.

REFERENCES

[1] K. Boucart and A. M. Ionescu, "Double-gate tunnel FET with high-k gate dielectric," IEEE Trans. Electron Devices, vol. 54, no. 7, pp. 1725–1733, Jul. 2007.

[2] Q. Zhang, et al., "Low-subthreshold-swing tunnel transistors," IEEE Electron Devices Lett. , vol. 27, no. 4, pp. 297–300, 2006.

[3] W. Y. Choi, et al., "Tunneling field-effect transistors (TFETs) with subthreshold swing (SS) less than 60 mV/dec," IEEE Trans. Electron Devices, vol. 28, no. 8, pp. 743–745, 2007.

[4] V. Saripalli, et al., "Variation-tolerant ultra low-power heterojunction tunnel FET SRAM design," in Proc. IEEE/ACM Int. Symp. Nanoscale Architectures (NANOARCH), pp. 45–52, Jun. 2011.

[5] M. Cotter, et al., "Evaluation of Tunnel FET-Based Flip-Flop Designs for Low Power, High Performance Applications," IEEE Int. Symp. on Quality Electronic Design (ISQED), pp. 430 - 437, 2013.

[6] S. Datta, et al., "Tunnel transistors for energy efficient computing," IEEE Int. Reliability Physics Symposium (IRPS), pp. 6A.3.1-6A.3.7, 2013.

[7] V. Saripalli, et al., "An energy-efficient heterogeneous CMP based on hybrid TFET-CMOS cores," IEEE Int. Symp. on Design Automation Conference (DAC), pp.729-734, June 2011.

[8] Zhi Li, et al., "Hybrid CMOS-TFET based Register Files for Energy-Efficient GPGPUs," IEEE Int. Symp. on Quality Electronic Design (ISQED), pp. 112–119, March 2013.

[9] "Sentaurus TCAD Manual," Sentaurus Device, 2011.

[10] L. Liu, et al., "Scaling Length Theory of Double-Gate Interband Tunnel Field-Effect Transistors," IEEE Trans. Electron Devices, vol. 59, pp.902-908, 2012.

[11] R. Kotlyar, et al., "Bandgap engineering of group IV materials for complementary n and p tunneling field effect transistors," Appl. Phys. Lett., vol. 102, no. 11, pp. 113106-1–113106-4, Mar. 2013.

Area and Routing Efficiency of SWD Circuits Compared to Advanced CMOS

Odysseas Zografos*†, Praveen Raghavan, Yasser Sherazi*, Adrien Vaysset*, Florin Ciubatoru*,
Bart Sorée*, Rudy Lauwereins*, Iuliana Radu, Aaron Thean

imec, Leuven, Belgium

*also with Department of Electrical Engineering, KU Leuven, Belgium

†Email: zogra@imec.be

Abstract—In this paper, we present a standard cell design methodology for Spin Wave Device (SWD) circuits. We perform Place and Route (P&R) experiments against a 10nm FinFET CMOS technology and compare the area, the routing and metal distribution of several arithmetic benchmarks. We show that SWD circuits although they require more metal layers than CMOS designs and although they contain double the number of nets, their pin density and net length distribution makes them easier (2× shorter nets) and cheaper (13% less wiring required) to route than CMOS, without impact the area of the designs.

I. INTRODUCTION

Expanding the future of nano-electronics by enhancing the functionality while overcoming the power consumption limitations has been actively pursued by studying novel beyond-CMOS devices and circuits. Spin-based logic is among the popular emerging state variable technologies that will help evolve the roadmap both on the device and architecture level [1]. Spin Wave Devices (SWD) are components that use as state variable propagating magnetization oscillation in ferromagnetic materials and were introduced in [2]. The first approach of circuit design was presented in [3] and benchmarked in [4].

In SWD circuits computation is achieved due to interference of spin waves in ferromagnetic wave buses. The information is encoded in one (or more) of the spin wave properties, such as phase, amplitude and frequency. This wave computation scheme, radically different than the boolean computation of CMOS circuits, brings advantages of intrinsic majority gate and data parallelism.

Area, power and performance estimations for SWD circuits were presented in [5] and [6] and compared with nanoscale CMOS reference designs. Purpose of this work is to investigate the crucial aspect of wiring and interconnecting these circuits.

The remainder of the paper is organized as follows. In Section II, we present the basic components, principles and advantages of the spin wave technology. The methodology to design SWD circuits and evaluate their routing efficiency is described in Section III. Section IV presents the P&R experiments performed and the benchmarking results, comparing SWD circuits with 10nm CMOS reference designs. Section V concludes the paper.

II. BACKGROUND & MOTIVATION

This section summarizes the basic technology material and operation assumptions for SWDs.

A. SWD Technology

Spin wave generation, detection and propagation have been studied both experimentally [7], [8] and by simulations and modeling [9]–[11]. In [12], the authors use SWDs to compose circuits that largely outperform their CMOS refernce designs. Finally, [4], [5] introduce a first order benchmarking of SWDs as circuits.

The operation of SWD circuits as described in [3], is based on the Magneto-electric cell (ME cell) component [2]. In Fig. 1, we show a schematic view of such component. The ferromagnetic spin wave bus (NiFe) has a magnetostrictive layer embedded (Ni), which in turn has a piezoelectric layer (PE) on top.

Fig. 1. Schematic view of ME cell stack connected to SW ferromagnetic bus.

When an input voltage is applied on the metal cap of the piezoelectric layer, it causes deformation which applies strain on the Ni layer. In turn the magnetization of the Ni layer is perturbed and a spin wave is created that propagates in the spin wave bus (NiFe). The inverse process takes place for the read-out of a spin wave.

B. Canted Magnetization of the Magneto-Electric Cell

The critical aspect of the ME cell operation is that the magnetization of the Ni layer is canted by an angle θ_{ME} off the magnetization saturation axis of the spin wave bus (assumed to be the out-of-plane axis). Fig. 2, depicts a schematic view of the canted magnetization of the magnetostrictive layer.

This canting can be created by using materials who have perpendicular anisotropies [3]. There are two available stable canted states for the magnetization of the Ni layer in such

978-1-4799-7670-6/15 $31.00 © 2015 IEEE

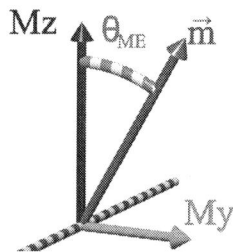

Fig. 2. Schematic representation of canted magnetization state in the magnetostrictive layer of an ME cell.

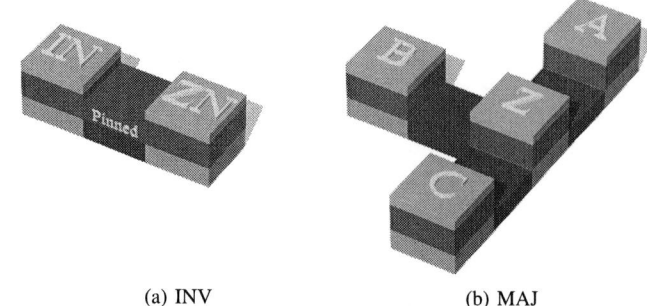

(a) INV   (b) MAJ

Fig. 3. Gate primitives used for SWD circuits.

configuration, one canted at an angle $+\theta_{ME}$ and one at an angle $-\theta_{ME}$ off the z-axis. Due to this bi-stability, the ME cell magnetization can store the information encoded in the spin wave's phase, by switching from one state to the other.

Motivation of this work stems from the non-volatile potential of the ME cell structure and the wave computation capabilities of SWDs. The non-volatile aspect of SWD circuits that operate with low voltages [3] can lead in ultra-low power designs. Wave computation can be employed to implement logically enhanced circuits with more binate gates (like Majority gates), a capability not efficiently available by CMOS devices. In this paper we explore what's the impact of placing and interconnecting such SWD gates, compared to state-of-the-art CMOS.

III. SWD CIRCUIT DESIGN METHODOLOGY

In this section we introduce the main components of a SWD circuit and their standard cell designs.

A. Gate Primitives

SWD technology is based on a wave computation scheme, it provides the capability of implementing simple and compact majority gates (MAJ), that can be produced by merging three wave buses. Majority gates enhance logic power of a design because they can emulate both AND and OR operation and is one of the basis for basic operation of binary arithmetic [13]. In order to fully utilize them we used a Majority synthesis methodology to synthesize the SWD designs, Majority-Inverter Graph (MIG) [14].

MIG has proven to be an efficient synthesis methodology for CMOS design optimization [14] and can be further exploited for SWD technology. The MIG is a logic representation structure consisting of three-input majority nodes and regular/complemented edges. This means that only two logic components are required for this representation, a MAJ gate and and inverter (INV). Figure 3 presents the two primitive gates we have considered to be implemented in SWD technology.

In Fig. 3a we present the INV component which is a simple wave bus, with a magnetically pinned layer on top, that inverts the phase of the propagating signal. The MAJ gate (Fig. 3b) is the merging of three wave buses. For the gates presented in Fig. 3, we assume minimum propagation length equal to one wavelength of the spin wave which in our study is assumed at 48nm, since the wavelength is defined/confined by the width of the spin wave bus.

B. Standard Cell Design

In order to evaluate the routability of SWD designs synthesized by MIG we created two standard cells depicted in Fig. 4. These cells are created to have the same cell height and spacing regions between the edge of the layers to the boundary of the cells to enable regular placement and routing.

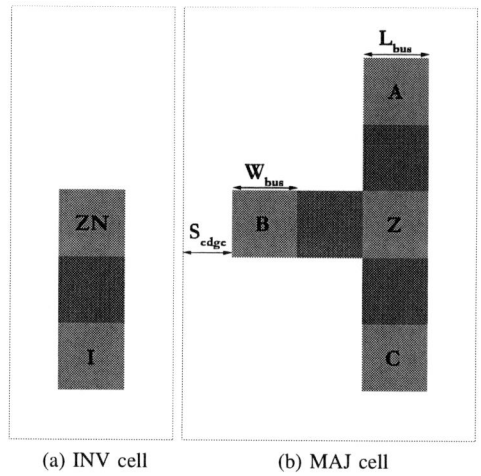

(a) INV cell   (b) MAJ cell

Fig. 4. Layout views of the two standard cells used to P&R the SWD circuits.

In the cells shown in Fig. 4a and 4b, metal1 layer is used to emulate the ferromagnetic spin wave bus layers and metal2 is used to emulate the contacts of the input and output ME cells of the gates. The dimensions and area of the SWD cells are shown in Table I.

TABLE I
SWD CELL DIMENSIONS

Dimension	Value (nm)
W_{bus}	48
L_{bus}	48
S_{edge}	36
INV	120×312
MAJ	216×312

Given that the metal1 and metal2 width of the 10nm CMOS reference technology is 24nm [15]. The widths and lengths of the cells, shown in Table I, are chosen to be double that size. The spacing of the cells is chosen to be 1.5× the metal1/metal2 width of the CMOS layers. This way there is enough space to route horizontally and vertically around the cell blockage.

978-1-4799-7670-6/15 $31.00 © 2015 IEEE

We assume that a complete SWD circuit is routed as any standard CMOS design, meaning that it consists of standard cells (Fig. 4) that are interconnected with metal wires. Using the same routing methodology, we will be able to obtain a fair comparison between the two technologies. However, the SWD circuits can lend themselves to another routing methodology and cascading, which can be achieved by used extra wave buses to interconnect input and output ME cells. This ferromagnetic routing would impose a restriction to the fabrication of the metal layers which is not analyzed in this paper.

IV. PLACE AND ROUTE EXPERIMENTS

This section presents the benchmarks we used and the evaluation methodology we followed, along with the main results of the P&R experiments.

A. Benchmarks

The gains of SWD circuits are quantified using relatively large combinational benchmarks. The four benchmarks selected are shown and described in Table II. All benchmarks were generated using the *Arithmetic module generator* [16].

TABLE II
DESCRIPTIONS OF ARITHMETIC BENCHMARKS

Name	Description
CSA464	4-operand 64-bit Carry-Skip Adder
MAC32	3-operand 32-bit (7,3) counter tree MAC with a Brent-Kung adder
MUL32	2-operand 32-bit Wallace tree Multiplier with Booth recording and a CLA adder
GFMUL	Mastrovito multiplier for irreducible polynomial: $x^{17} + x^8 + x^3 + x + 1$

In order to evaluate the impact in area and routing of these designs we used commercial P&R tools to place and route the netlists with both technologies. In the results presented in part IV-B we used different configurations of the SWD technology, where we varied the number of metal layers available to the router, from 4 layers (SWD-M4) to 7 layers (SWD-M7). The CMOS reference designs were P&R-ed only with 7 metal layers available (CMOS-M7). All designs where P&R-ed with the maximum density that didn't cause DRC errors for all four designs which is 75% in a square aspect ratio core.

B. Results

An overview of the P&R results is given in Table III. We observe that although the SWD designs consist of a much larger number of cells (2.2× more on average), the total P&R-ed area is roughly the same. Also, we can observe that the number of nets in the SWD designs is much higher (2.1× more on average) than in CMOS designs. However, the total wire length is 13.4% less with SWD technology. This shows that SWD circuits can be more cost-effective to route due to shorter wires.

Another metric to evaluate the routing of SWD circuits is the average metal distribution. In Fig. 5, we present the average metal distributions over all four benchmarks for all four technology configurations. We can observe that the SWD designs have higher usage of metal1, metal4, metal5 due

to large and dense metal2 pins and larger number of nets. Also, Fig. 5 shows that all four benchmarks could be P&R-ed successfully with only four metal layers available. This shows that SWD circuits can be as efficiently routed as CMOS circuits.

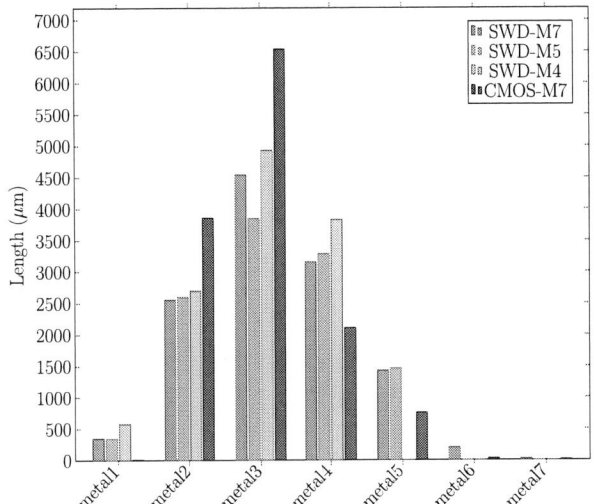

Fig. 5. Metal layer distribution in P&R-ed designs with SWD cells providing varying number of metal layers (SWD-M7 to SWD-M4) and 10nm CMOS cells providing seven metal layers (CMOS-M7).

To verify that indeed the nets in SWD designs are shorter in Fig. 6 we calculated the net length distribution of the MAC32 benchmark which has the larger number of nets for both SWD and CMOS technologies. This histogram clearly shows that in the SWD design there are many more shorter net compared to the CMOS design. More specifically the number of short nets with SWD-M7 is 5× larger.

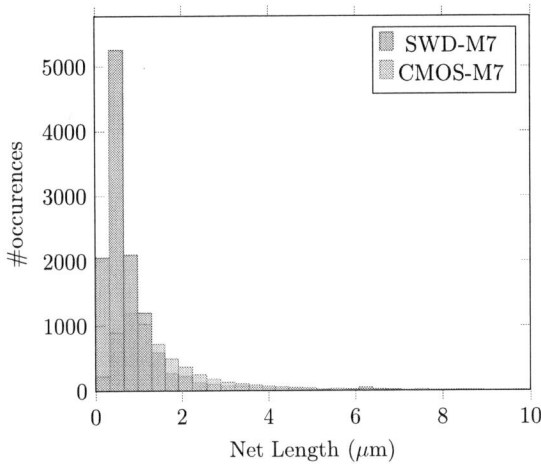

Fig. 6. Net length distribution in P&R-ed designs with SWD cells and 10nm CMOS cells for the MAC32 benchmark.

To better quantify the difference in net distribution and routing between the two technologies we calculated the empirical cumulative distribution function of the net length in the MAC32 benhcmark, shown in Fig. 7.

978-1-4799-7670-6/15 $31.00 © 2015 IEEE

TABLE III
SUMMARY OF P&R RESULTS

Name	CMOS - M7				SWD - M7			
	Area (μm^2)	#Cells	#Nets	Wire length (μm)	Area (μm^2)	#Cells	#Nets	Wire length (μm)
CSA464	137.17	200	646	9921.48	158.10	2171	2427	4520.40
MAC32	979.40	6117	6442	20846.97	896.03	12327	12423	20101.08
MUL32	874.23	5247	5390	19408.00	807.95	11141	11205	18367.67
GFMUL	117.68	894	928	3045.54	144.90	2081	2115	3104.64
Average	527.12	3114	3351	13305.50	501.74	6930	7042	11523.45

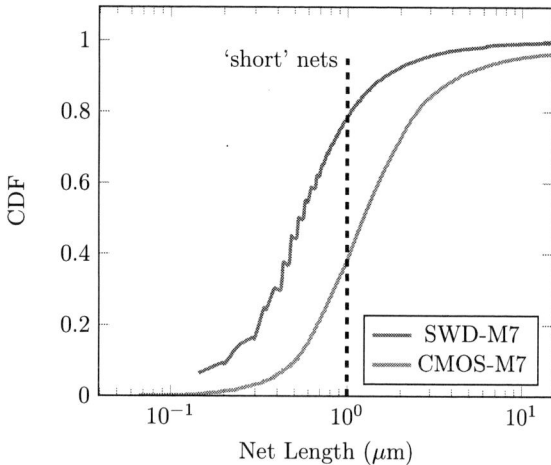

Fig. 7. Empirical cumulative distribution function of net lengths in P&R-ed designs with SWD cells and 10nm CMOS cells for the MAC32 benchmark.

We set an arbitrary threshold for net length at $1\mu m$ (vertical dashed line), which corresponds to a net spanning vertically over 3 MAJ gates in the SWD design or equivalently 2 standard cells in the CMOS design. Defining as a short all the nets with length equal or less than $1\mu m$, we get that 80% of SWD-M7 nets are short compared to only 40% of the CMOS-M7 nets. The results shown in Figs 6 and 7 are extracted from the experiments for the MAC32 but the same trend holds for the rest the benchmarks.

The SWD circuits can be routed efficiently with less wiring and shorter nets, so effectively the routing is easier than CMOS circuits without sacrificing any area. Also, the fact that SWD designs were successfully P&R-ed without great use of metal1 shows that there is the capability of a layer restriction in order to introduce ferromagnetic (spin wave bus) interconnection additional to the metal routing.

V. CONCLUSIONS

In this paper, we show that SWD technology and the use of Majority gate logic leads to circuits that can be routed more efficiently compared to 10nm CMOS circuits. More specifically the designs have on average 13% less wiring required and contain 2× more short nets, meaning that the routing is simplified. These results show the potential integrability of SWD technology and the fact that ferromagnetic interconnection can be efficiently combined with metal interconnection.

REFERENCES

[1] K. Bernstein, R. Cavin, W. Porod et al., "Device and architecture outlook for beyond cmos switches," Proceedings of the IEEE, vol. 98, no. 12, pp. 2169–2184, Dec 2010.

[2] A. Khitun and K. L. Wang, "Nano scale computational architectures with spin wave bus," Superlattices and Microstructures, vol. 38, no. 3, pp. 184 – 200, 2005. [Online]. Available: http://www.sciencedirect.com/science/article/pii/S0749603605000716

[3] ——, "Non-volatile magnonic logic circuits engineering," Journal of Applied Physics, vol. 110, no. 3, pp. –, 2011. [Online]. Available: http://scitation.aip.org/content/aip/journal/jap/110/3/10.1063/1.3609062

[4] D. Nikonov and I. Young, "Overview of beyond-cmos devices and a uniform methodology for their benchmarking," Proc. of the IEEE, vol. 101, no. 12, pp. 2498–2533, Dec 2013.

[5] O. Zografos, P. Raghavan, L. Amaru et al., "System-level assessment and area evaluation of spin wave logic circuits," in Nanoscale Architectures (NANOARCH), 2014 IEEE/ACM International Symp. on, July 2014, pp. 25–30.

[6] P. Shabadi, A. Khitun, K. Wong, P. Amiri, K. Wang, and C. Moritz, "Spin wave functions nanofabric update," in Nanoscale Architectures (NANOARCH), 2011 IEEE/ACM International Symposium on, June 2011, pp. 107–113.

[7] M. Madami, S. Bonetti, G. Consolo et al., "Direct observation of a propagating spin wave induced by spin-transfer torque," Nature nanotechnology, vol. 6, no. 10, pp. 635–638, 2011.

[8] A. Chumak, P. Pirro, A. Serga et al., "Spin-wave propagation in a microstructured magnonic crystal," Applied Physics Letters, vol. 95, no. 26, p. 262508, 2009.

[9] S.-K. Kim, "Micromagnetic computer simulations of spin waves in nanometre-scale patterned magnetic elements," Journal of Physics D: Applied Physics, vol. 43, no. 26, p. 264004, 2010. [Online]. Available: http://stacks.iop.org/0022-3727/43/i=26/a=264004

[10] S. Dutta, D. Nikonov, S. Manipatruni et al., "Spice circuit modeling of pma spin wave bus excited using magnetoelectric effect," Magnetics, IEEE Trans. on, vol. 50, no. 9, pp. 1–11, Sept 2014.

[11] J. Alzate, P. Upadhyaya, M. Lewis et al., "Spin wave nanofabric update," in Nanoscale Architectures (NANOARCH), 2012 IEEE/ACM International Symp. on. IEEE, 2012, pp. 196–202.

[12] P. Shabadi, S. N. Rajapandian, S. Khasanvis et al., "Design of spin wave functions-based logic circuits," in SPIN, vol. 2, no. 03. World Scientific, 2012.

[13] J. Von Neumann, "Non-linear capacitance or inductance switching, amplifying, and memory organs," Dec. 3 1957, uS Patent 2,815,488.

[14] L. Amarú, P.-E. Gaillardon, and G. De Micheli, "Majority-inverter graph: A novel data-structure and algorithms for efficient logic optimization," in Proc. of the 51st Annual Design Automation Conf., ser. DAC '14, 2014, pp. 1–6.

[15] J. Ryckaert, P. Raghavan, R. Baert et al., "Design technology co-optimization for n10," in Custom Integrated Circuits Conf. (CICC), 2014 IEEE Proc. of the, Sept 2014, pp. 1–8.

[16] T. U. Aoki laboratory, "Arithmetic module generator," http://www.aoki.ecei.tohoku.ac.jp/arith/.

Low-Phase Noise Variation VCO Implementing Resistorless Digitally Controlled Varactor

Mohammed Aqeeli, Abdullah Alburaikan, Xianjun Huang, and Zhirun Hu

School of Electrical and Electronic Engineering

The University of Manchester

Manchester, UK

mohammed.aqeeli@manchester.ac.uk, Abdullah Alburaikan@manchester.ac.uk, xianjun.huang@manchester.ac.uk, and z.hu@manchester.ac.uk

Abstract—A novel resistorless digital capacitor switching array (DCSA) has been implemented into a wideband CMOS VCO for 5-GHz WiMAX/WLAN applications. The proposed DCSA is added both in series and parallel to nMOS varactors. Based on this, a wideband VCO is achieved, which not only exhibits lower phase noise in comparison with reported state-of-the- art wideband VCOs, but also has low phase noise variation of less than 5 dBc/Hz. In addition, it has demonstrated low power consumption, improved linearity of the f-V curve and extended tuning range. The proposed VCO has been designed using UMC 130 nm CMOS technology. It operates from 3.65 GHz to 6.34 GHz, with a phase noise of -132.70 dBc/Hz at 1 MHz offset, a figure of merit (FoM) of -202.9 dBc/Hz, less than -41 dBm spurious harmonics and total VCO core power consumption of 2.88 mW from a 3.2 V supply voltage.

Keywords— *CMOS; phase noise variation, VCO gain, tuning range, figure of merit.*

I. INTRODUCTION

Phase noise can vary dramatically over the tuning range of a VCO, but the published literature often overshadow this problem by measuring or simulating phase noise at the center frequency only[1]. It is difficult to obtain simultaneously a large tuning range and small phase noise variation, particularly while accommodating MOS or diode varactors. Using a MOS transistor as a varactor increases the probability of converting AM noise to FM noise[2]. This will grow in line with the difference between the maximum and the minimum capacitances and may become indistinguishable from phase noise. Consequently, when the tuning range is more than 30 percent, switching capacitors are mandatory to maintain minimum phase noise variations, tuning linearity and low VCO sensitivity[3]. Studies have been conducted which consider switching capacitors inside the LC tank in order to extend the tuning range and improve VCO phase noise. However, these studies used switches that were implemented using stacked devices and large resistors to gain high impedance for RF signals, resulting in occupying a larger die area as well as exhibiting higher power dissipation and phase noise. This article presents a novel VCO which consists of two sets of high-performance 4-bit digital capacitor switching arrays (DCSA) added both in series and parallel to nMOS varactors. Such an approach enables the VCO to have an extended tuning bandwidth and minimizes the phase noise

variation throughout the entire tuning range, while maintaining minimum die size. Previous resistors employed[4][5][6] have been replaced with much smaller nMOS transistors, which perform as resistors. These in turn occupy a much smaller area, resulting in less parasitic capacitance affecting the RF nodes of the VCO. As a result, the proposed VCO has many advantages, such as providing a solution to the problem of high phase noise variation associated with wideband tuning, and minimizing the sensitivity of the VCO by optimizing gain. Finally, it has an output frequency which varies linearly with the control voltage. The proposed VCO's bandwidth is increased by more than 6 percent, phase noise has decreased by more than 12.7 dBc/Hz, and phase noise variation is less than 4.9 dBc/Hz in comparison with most recently reported works. This paper is organized as follows: Section II explains the VCO's core design and the implementation process, Section III presents the resonator tank design, Section IV illustrates the varactor characterization and simulation results, followed by the conclusion in Section V.

II. CIRCUIT DESIGN AND IMPLEMENTATION

A complementary cross-coupled VCO based on a binary-weighted switched capacitor bank and nMOS varactors is depicted in Fig. 1(a). This topology is chosen due its ability to achieve lower phase noise when compared to other configurations. The LC tank consists of a differentially-tuned varactor group and a very small, symmetrical, center-tapped inductor. A cross-connected differential pair provides negative resistance to neutralize the tank losses; this results in less current consumption and is thus more power-efficient. The wider the tuning bandwidth, the higher the phase noise, this means an RF VCO with sub-bands is the best choice for broadband implementation. Equation (1) shows clearly that the phase noise of the VCO is proportional to the square of the VCO gain[10].

$$S_{\phi n} = \left(\frac{K_{VCO}}{S} \left| Z(\omega_0 + \Delta\omega_0) \right| \right)^2 \cdot \overline{i_n}^2 \cdot \alpha \frac{1}{Q^2} \left(\frac{K_{VCO}}{2\Delta\omega_0} \right)^2 \quad (1)$$

where K_{VCO} is the gain of the VCO, ω_0 is the carrier frequency, $\Delta\omega_0$ is the offset frequency, i_n is the noise current Q is the quality factor of the LC tank. In order to obtain low phase noise, low K_{VCO} is necessary. Furthermore, to minimize phase noise variation, the K_{VCO} must be minimized and remain

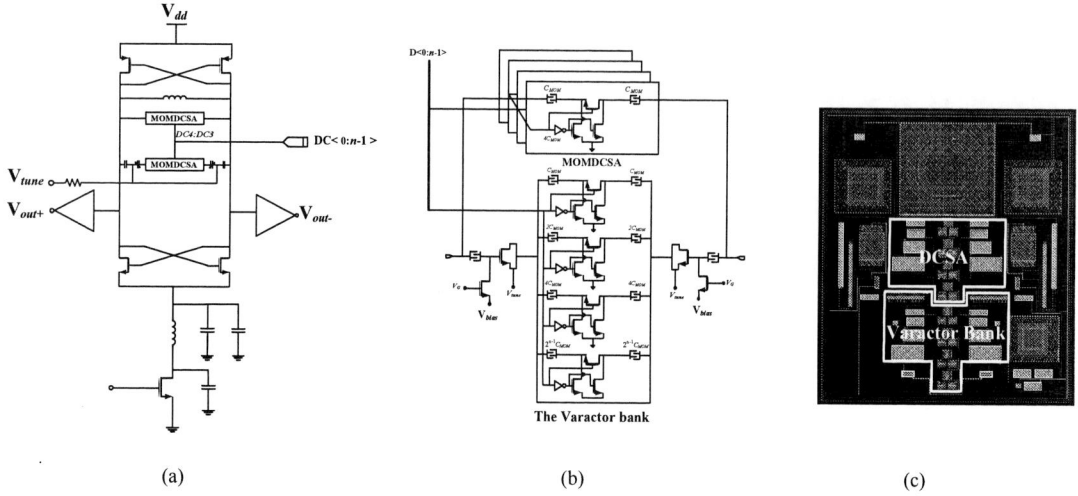

Fig. 1. (a) Schematic diagram of the proposed 5.0 GHz 130 nm CMOS VCO. (b) DCSA and its logic control with varactor bank. (c) Layout of the VCO with a die area of 720× 586 μm2.

constant across the tuning range. This means that the oscillation frequency will vary linearly with the tuning voltage.

III. NOVEL VARACTOR FOR LOW PHASE NOISE VARIATION

Apart from wideband tuning range performance, it is highly desirable for a VCO to provide constant phase noise across its operating frequency range, or at least to maintain an acceptable phase noise variation for all frequencies[11]. To achieve a wide tuning range for lower phase noise and minimum variation, two sets of DCSA are used in the proposed VCO. The first one is in series with the nMOS varactors, and the second is in parallel with the inductor as depicted in Fig. 1(b), to divide the tuning range into multiple frequency bands. The nMOS transistors occupy a smaller die area compared to pMOS transistors, and thus fewer parasitic capacitances are added to the VCO core. Such an arrangement will extend the linear tuning range of the varactors and importantly reduce the K_{VCO}, phase noise and phase noise variation[11]. It was found that this topology provides a relatively constant output swing frequency and minimal phase noise over the operating frequency range. Consequently, phase noise variation challenges are optimized. The operation of the DCSA was implemented using symmetric-type MOM capacitors, because of their large capacitance, low parasitic capacitance, symmetrical plate design, superior RF characteristics and their requirement of no additional masks or process steps[10][11]. In order to reduce phase noise variations we had to find the VCO design parameters which could compensate for the K_{VCO} variations. The DCSA capacitance value can be calculated as:

$$C = (2^{N-1})\left(\frac{1}{C_{MOM}} + \frac{1}{C_{DF}}\right)^{-1} \quad (2)$$

where C_{MOM} is the capacitance of the DCSA, and CDF is the drain fringe capacitance of the switch devices. Total capacitance Cα of the series MOMDCSA, including fixed

parasitic capacitance of the active devices, C_{par} can be calculated as:

$$C_\alpha = (2^{N-1})\left(\frac{1}{C_{MOM}} + \frac{1}{C_{DF}}\right)^{-1} + C_{PAR} \quad (3)$$

where C_{DF} is the drain fringe capacitance of the series MOMDCSA. Total capacitance C_β of the parallel MOMDCSA to the inductor and C_α is:

$$C_\beta = (2^{N-1})\left(\frac{1}{C_{MOM}} + \frac{1}{C_{DF\beta}}\right)^{-1} \quad (4)$$

where $C_{DF\beta}$ is the drain fringe capacitance of the parallel MOMDCSA. Total capacitance C_γ of the nMOS varactor and the series MOMDCSA can be calculated as:

$$C_\gamma = \frac{C_\alpha . C_{Var}}{C_\alpha + C_{Var}} \quad (5)$$

where C_{var} is the capacitance of the nMOS varactor

$$C_{\tan k} = \frac{C_\alpha . C_{Var}}{C_\alpha + C_{Var}} + C_\beta \quad (6)$$

The oscillation frequency (f_{osc}) of the VCO can be written as follows:

$$f_{Osc} = \frac{1}{2\pi\sqrt{L[(C_\alpha . C_{Var}/C_\alpha + C_{Var}) + C_\beta]}} \quad (7)$$

K_{VCO} can be derived as follows:

Assuming,

$$C_\delta = \frac{C_{Var}(C_\alpha + C_\beta) + C_\alpha . C_\beta}{C_\alpha + C_{Var}} \quad (8)$$

978-1-4799-7670-6/15 $31.00 © 2015 IEEE

44

then,

$$K_{VCO} = \frac{\partial f_{Osc}}{\partial V_{Cont}} = -\frac{C_\alpha{}^2}{4\pi\sqrt{L}.C_\delta{}^{3/2}} \cdot \frac{1}{(C_\alpha + C_{Var})^2} \cdot \frac{\partial C_{Var}}{\partial V_{Cont}} \quad (9)$$

where V_{cont} is the controlled voltage. From (7), we have,

$$C_\delta = \frac{1}{4\pi^2.f_{Osc}{}^2.L} \quad (10)$$

Substituting (10) into (9) we obtain,

$$\left|K_{VCO}\right| = 2\pi^2.f_{Osc}{}^3.L.\frac{C_\alpha{}^2}{(C_\alpha + C_{Var})^2} \cdot \frac{\partial C_{Var}}{\partial V_{Cont}} \quad (11)$$

Equation (11) shows the third-order exponential relationship between the K_{VCO} and the frequency of oscillation. It demonstrates that the gain of the proposed VCO is a function of the LC tank.

IV. Characterisation and Simulation

The proposed VCO was implemented using a commercially available UMC 130 nm, 6-metal, mixed-mode CMOS process and simulated using Cadence Virtuoso analogue design environment tools (CVADET). Fig. 1(c) shows the layout of the proposed VCO which has a die area of 720× 586 µm2, not including the bond pad. To verify the proposed theory, two VCOs were designed, the first one with DCSA connected in parallel to the varactor bank while the second one has DCSA connected in both series and parallel to the varactor bank, the phase noise variation and VCO gain of the proposed VCO are simulated as a function of the control code of the switchable capacitor bank, while the varactor tuning voltage is fixed at 1.0 V. As can be seen, the first VCO shows a gain sensitivity of 23.5MHz/V and phase noise variation of 12.7 dBc/Hz as shown in Fig. 3. The proposed VCO resulting in better simulated performance in terms of K_{VCO}, which reduced to 5.7MHz/V and phase noise variation which was less than 4.9 dBc/Hz, as shown in Fig.4. As a result, the new varactor achieves steep transition for tuning voltages varying between 0 and 2.5 V. This steep transition occurs similarly at all tuning curves. After completing the physical design (layout), both design rule check (DRC) and layout versus schematic (LVS) check was performed using the Cadence tools. Parasitic capacitances and resistances in the layout were strongly affecting the performance of the proposed VCO. To evaluate the effects of parasitic and to gain a higher degree of confidence, Cadence quantus extraction tool (QRC) was used because it provides a convenient and accurate methodology to predict the performance in critical building blocks such the proposed LC-tank VCO. Simulating the design with parasitic capacitances and resistances accounted for, the phase noise was increased from -136.9 dBc/Hz to -132.7 dBc/Hz at 1MHz offset frequency. The simulation results indicate that the VCO exhibits a wider tuning range between 3.65 GHz and 6.34 GHz. Phase noise variations range from -136.34 dBc/Hz to -132.56 dBc/Hz and FoM varies from -202.1 dBc/Hz to -204.8 dBc/Hz at 1 MHz offset frequency. The new varactor achieves steep transition for tuning voltages, as depicted in Fig. 2.

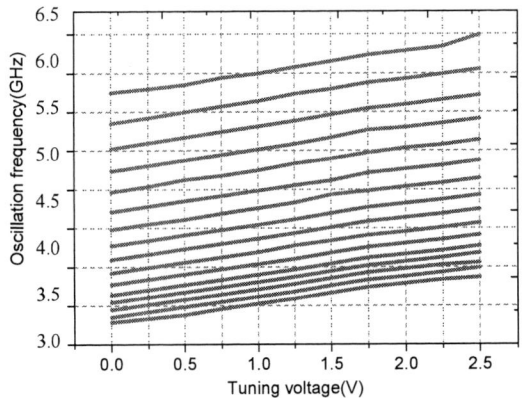

Fig. 2. Oscillation frequency characteristics versus tuning voltage.

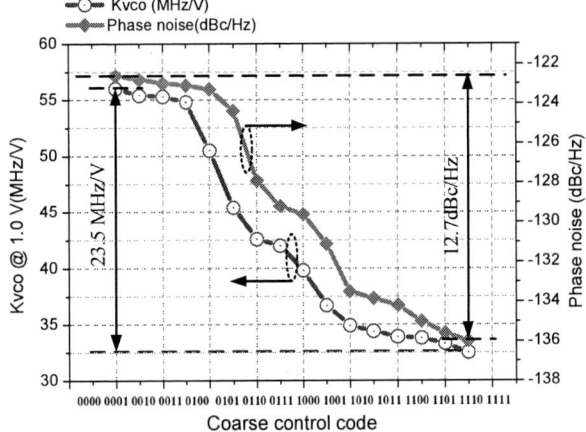

Fig. 3. VCO gain and phase noise of the proposed VCO with DCSA connected in parallel to the varactor bank.

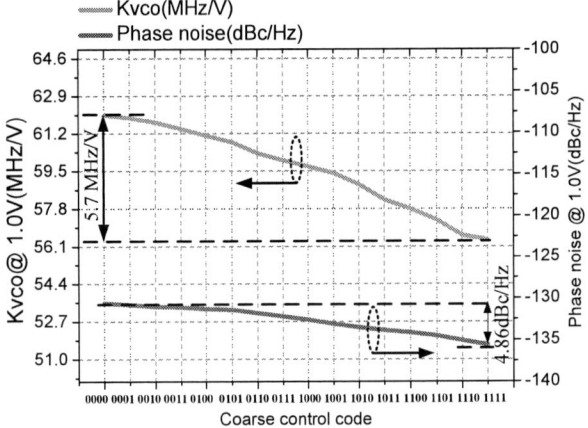

Fig. 4. VCO gain and phase noise of the proposed VCO with DCSA connected in series and in parallel to the varactor bank.

The proposed VCO has better performance parameters, due mainly to the MOMDCSA and the characteristics of the new topology. Importantly, phase noise variations remain relatively small for all tuning ranges, due to the advantages of the

978-1-4799-7670-6/15 $31.00 © 2015 IEEE

Table I: Performance comparison of CMOS VCOs

Process	F (GHz)	P (mW)	PN (dBc/Hz)	Offset	TR (GHz)	FoM (dBc/Hz)	Ref.
0.18μmCMOS	5.00	7.20	-90.2	1.0	4.10-5.00	-154.80	[4]
0.18μmCMOS	4.20	4.50	-119.0	1.0	4.10-4.80	-184.47	[6][S]
0.18μmCMOS	5.00	13.00	-121.5	1.0	5.10-5.36	-192.10	[9][S]
0.18μmCMOS	3.50	8.80	-125.9	1.0	3.05-3.92	-191.70	[10]
0.18μmCMOS	5.80	10.08	-117.0	1.0	5.27-6.41	-184.00	[11]
0.09μmCMOS	3.95	6.60	-147.0	10.0*	3.40-4.50	-191.00	[12]
0.13μmCMOS	5.00	2.88	-132.7	1.0	3.65-6.34	-202.90	This work

F: Centre frequency, P: power, PN: Phase noise at 1 MHz, *at 10 MHz offset TR: Tuning range, FoM: Figure of Merit, S: Simulation

discrete frequency ranges. The topology of this VCO can be used for the implementation of local oscillator (LO) synthesizers in WLAN radio applications. Circuit performance, which was simulated using CVADET to show the transient analysis of the VCO, demonstrates clearly that steady-state oscillation starts at approximately 6.27 ns. The circuit generates stable periodic signals with a harmonic index and measured output power of approximately 3.73 dBm at the resonant frequency, with tuning voltage varying from 0 to 2.5 V. The FoM model adopted by Ham and Hajimiri[8] normalizes measured phase noise with respect to center frequency and power consumption, as shown by equation (12):

$$FOM = L\{\Delta f\} - 20\log\left\{\frac{f_0}{\Delta f}\right\} + 10\log\left\{\frac{P_d}{1mW}\right\} \quad (12)$$

where $L\{\Delta f\}$ is the phase noise at offset frequency Δf, f_0 is the oscillating frequency, and P_d (W) is the power dissipation of the VCO. The performance of the novel VCO in this study is compared in Table I to other state-of-the-art designs in 0.09 μm, 0.13 μm and 0.18 μm CMOS technologies. It is evident that the proposed VCO has better characteristics with regard to phase noise, power consumption and FoM.

V. CONCLUSION

Using a novel design, a wideband LC-VCO has been designed and simulated to meet Wi-MAX/WLAN specification and the most stringent phase noise requirements. An ideal VCO has a broadband tuning range of approximately 49.2 per cent which varies linearly with the control voltage. The high performance VCO specifications include simulated phase noise of -132.7 dBc/Hz at an offset frequency of 1 MHz away from the 5.0 GHz FoM of -202.9 dBc/Hz, while the VCO core's total power consumption is 2.88 mW away from 3.2 V supply voltage. The proposed design is expected to be fabricated using multi-projects wafer fabrication in the near future.

REFERENCES

[1] D. B. Lesson, "A simple model of feedback oscillator noise spectrum," Proceedings of the IEEE, vol. 54, no. 2, Feb. 1966.

[2] M. Tieboud, "Low power VCO design in CMOS," Springer, Berlin, Heidelberg, Netherlands, 2006.

[3] A. Akatas and M. Ismail, "CMOS PLLs and VCOs for 4 G Wireless," Springer Science + Business Media Inc, USA, 2004.

[4] J. Jin, X. Yu, X. Liu, W. Lim and J. Zhou, "A Wideband Voltage-Controlled Oscillator With Gain Linearized Varactor Bank," IEEE Transactions on Components, Packaging, and Manufacturing Technology, pp.1-6, 2014.

[5] A. Thomas, W. Bakalski1, T. Ussmuller and R. Weigel, "A MIM-cap free digitally tunable nMOS capacitor," the 8th European Microwave Integrated Circuits Conference (EuMIC), pp.21-24, 2013.

[6] X. Tang, F. Huang, M. Shao, Y. Zhang, "A wideband 0.13 μm CMOS LC-VCO for IMT-Advanced and UWB Applications," IEEE MTT-S International Microwave Workshop Series on Millimeter Wave Wireless Technology and Applications, pp. 1-4, 2012.

[7] M. Aqeeli, A. Alburaikan, C. Muvianto, X. Huang, Z. Hu "Wideband Gain Linearized Microwave Voltage Controlled Oscillator with Low Phase Noise Variation in Nanometer CMOS Technology" Journal of circuits, systems and computers, Vol. 24, No 3,pp. 1-14, 2015.

[8] M. Aqeeli, A. Alburaikan, C. Muvianto, X. Huang, Z. Hu "An Ultra-Wideband and Gain Linearized CMOS VCO with Minor Phase Noise Variation " European Microwave Association, Asia Pacific Microwave Conference (APMC), pp. 965-967, 2014..

[9] K. Kuo, C. Wu, "A low phase noise VCO for 5 GHz WiMAX/WLAN frequency synthesizer," IEEE International Conference of Electron Devices and Solid-State Circuits (EDSSC)", pp. 1-2, 2013.

[10] M. Wei, S. Chang, Y. Zhang, Y. Yang and R. Negra "2.4 GHz / 3.5 GHz Dual-band wide-tuning-range Quadrature VCO using Harmonic-Injection Coupling Technique," The 14th IEEE Topical Meeting on Silicon Monolithic Integrated Circuits in RF Systems (SiRF), pp. 107-109, 2014.

[11] P. Ruippo, T. A. Lehtonen, and N. T. Tchamov, "An UMTS and GSM Low Phase Noise Inductively Tuned LC VCO," Microwave and Wireless Components Letters, IEEE, vol. 20, pp. 163-165, 2010.

[12] L. Fanori and P. Andreani, "Highly Efficient Class-C CMOS VCOs, Including a Comparison With Class-B VCOs," IEEE Journal of Solid-State Circuits, vol. 48, pp. 1730-1740, 2013.

978-1-4799-7670-6/15 $31.00 © 2015 IEEE

Dimensioning for power and performance under 10nm: the limits of FinFETs scaling

M. Garcia Bardon, P. Schuddinck, P. Raghavan, D. Jang, D.Yakimets, A. Mercha, D. Verkest, A. Thean

imec, Kapeldreef 75, 3001 Leuven, Belgium

Marie.GarciaBardon@imec.be

Invited paper

Abstract— **In this paper, we review the conditions at which FinFETs could meet system requirements at the 7nm node. The device parasitics appear as most important performance limiters. Following a top-down approach, we find the design space that allows to meet speed and power targets, then explore the optimization of the geometry in combination with disruptive solutions such as air gap spacers and wrapped contacts, the benefits and drawbacks of increased fin height, and a design level solution consisting in fin depopulation. The efficiency of each solution depends on the balance between interconnect and device parasitics.**

Keywords— FinFETs, scaling, design technology co-optimization (DTCO).

I. INTRODUCTION

Thanks to their improved channel control, FinFETs have enabled to continue scaling below 20 nm dimensions, with two generations of FinFETs already demonstrated after the 28nm planar technology [1,2]. However, scaling down under 10nm raises numerous challenges to continue performance, power, area and cost (PPAC) improvement. To push performance, different disruptive solutions are considered at device level, such as the transition to lateral nanowires (NW), or to high mobility channels (SiGe PFETs) [3]. However, an important limitation of performance at these scaled dimensions relies in the parasitic capacitances and resistances, both at device and at the interconnect levels. In this paper, we investigate how FinFETs could meet the 7nm targets by optimizing the geometry and addressing the device parasitics. We first find what device drive current is needed to meet system targets depending on the parasitic resistance and capacitances. To meet the obtained values of parasitics, geometry optimization is not sufficient and innovations on the spacers and contacts modules are needed. In term of geometry optimization, an increase in fin height seems appealing to increase performance and area efficiency [1], but it is shown that the advantage at 7nm node is limited by device parasitics. On the other hand, fin depopulation (reduction of the number of fins) appears as a powerful knob to decrease dynamic power while preserving speed.

II. TECHNOLOGY SCALING AND PERFORMANCE TARGETS

At the 7nm node, Contacted Gate pitch (CGP) and Metal Pitch (MP) are expected to be scaled to less than 45 nm and 35 nm respectively to continue the *x0.5* area scaling node to node (Fig. 1). Area scaling could be pushed more aggressively at design level by reducing the height of the standard cells (e.g., from 9 tracks cells to 7.5 tracks or lower), resulting in a reduction of

fins per active device. A node to node speed improvement of 20% is assumed and used as system target for High Performance (HP) devices, from 14nm to 10nm nodes and from 10nm to 7nm nodes, at constant leakage current and dynamic power (Fig. 1). The dimensions assumed for each FinFET generation are summarized in Table I.

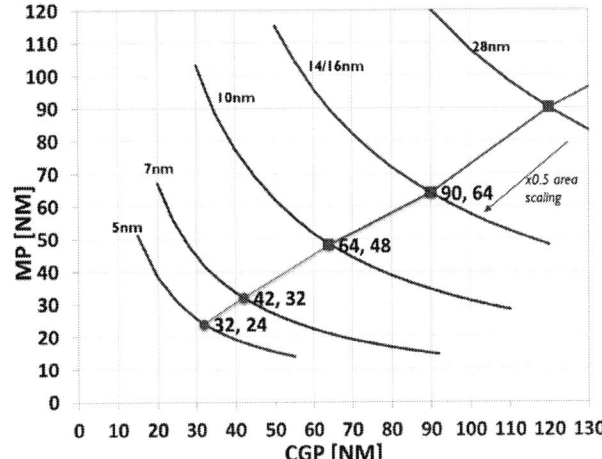

Fig. 1. CGP and MP to enable area scaling from 28nm node to 5nm node. MP scaling has been increased compared to CGP scaling. The area scaling can be pushed further by reducing the height of standard cells from node to node and hence the number of fins per device.

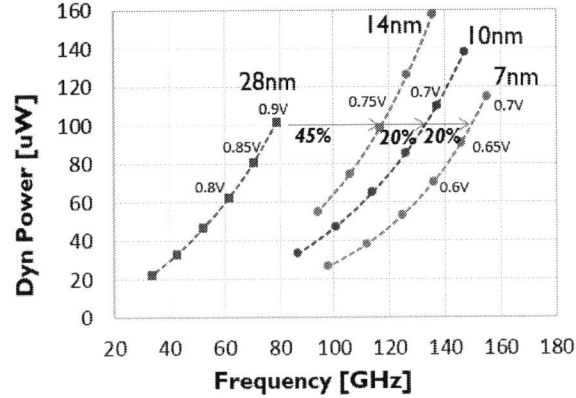

Fig. 2. Expected speed improvement from 28nm planar to FinFETs 14nm, 10nm, 7nm, at constant dynamic power and leakage power for High performance devices (Ioff≈15nA per device) on Ring oscillators of inverters with Fan-Out 3, 9Tracks library, no BEOL load.

TABLE I. PROCESS ASSUMPTIONS ACCROSS NODES

Technology		14nm	10nm	7nm
CGP	nm	90	64	42
MP	nm	64	48	32
Fin Pitch	nm	48	36	24
Gate length	nm	30	24	18
Fin width TFIN	nm	10	7	5
Fin height HFIN	nm	30	30	35
Tspacer	nm	14	8	6
Rbeol	Ω/um	25	60	135
Cbeol	F/um	0.195	0.175	0.16
Vdd	V	0.75	0.7	0.65

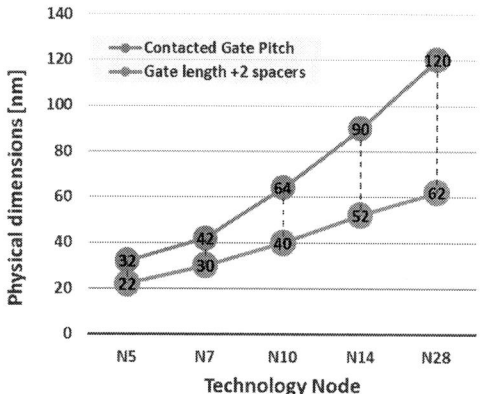

Fig. 3. Node to node CGP reduction results in a trade off for gate length, spacer width and contact size reductions.

The aggressive scaling of CGP translates into gate length, gate spacer, and contact size scaling (Fig. 3), leading to a trade-off at the 7nm node: gate length reduction degrades electrostatics, spacer reduction increases capacitance and contact reduction increases series resistance. In FinFETs, the subthreshold slope (SS) can be maintained while scaling the gate length by reducing accordingly the fin width TFIN (Fig. 4). SS=70mV/dec at nominal Lg=18 nm is expected from TCAD simulations for TFIN=5nm with DIBL of 50 mV/dec. Further fin width reduction does not seem feasible from processing and variability point of views, and further gate length scaling would need the transition from FinFETs to lateral nanowires to maintain SS to maximum 70mV/dec.

The choice of a maximum of 70mV/dec relies on both speed and power considerations: SS larger than 70 mV/dec could still meet speed targets for high performance devices (Ioff around 30 nA/um), but would have too degraded performance for low power devices at this scaled Vdd (0.65V), leading to an unbalance in the SoC design, where the relative speed of LP and HP devices matter to minimize the total system power [4]. With CGP=42nm and gate length 18nm, and assuming a spacer thickness of 5 nm, the contact size is 12 nm only, placing a high constraint on contact resistivity as will be seen in the next section.

With these dimensions (Table I) and electrostatics defined by TCAD (SS=70mV/dec, DIBL=50mV/dec), the target drive current per fin needed to reach the target speed for the 7nm node

can be obtained for various parasitic capacitance values of the device and series resistance as shown in Fig.5. For reference, the current is also translated in drive current normalized to the effective width of the transistor (2HFIN+TFIN). The target current has a high sensitivity to the device parasitic capacitance. The series resistance has almost no impact on the target current, but the higher R_{series}, the more supply voltage is lost on this resistance and the higher the requirement on the channel mobility and velocity saturation (Fig.5c).

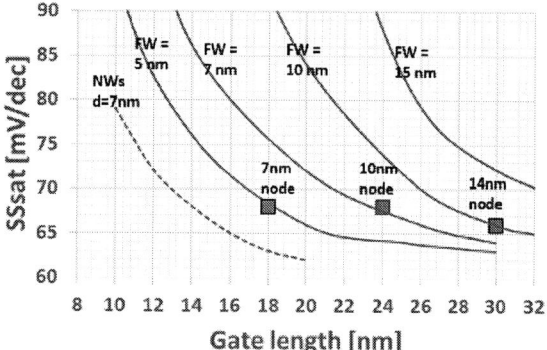

Fig. 4. Electrostatics in FinFETs is preserved accross nodes by fin width scaling.

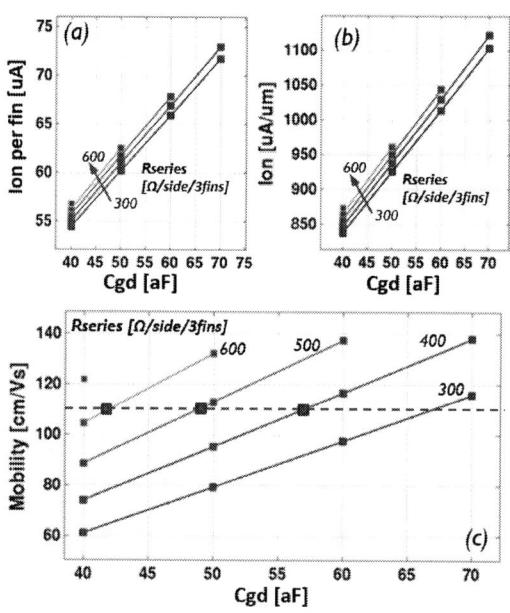

Fig. 5. (a) Target drive current per fin to reach speed targets depending on the device parasitics and series resistance, per side (S/D), for 3 fins (b) Target current normalized per Weff (=2*HFIN+TFIN) (c) Target mobility, needed to reach these target currents at each parasitics conditions.

At 18nm gate length, the effective mobility and velocity saturation depend on the stress in the channel, that will vary from 0.5 GPa to 1.5 GPa depending on the performance boosters (S/D stressors to 0.75 GPa, SRB for higher stress [6]). At these stress values, mobility in Si FinFETs is expected to be between 100 and 130 cm^2/Vs, and velocity saturation varies

978-1-4799-7670-6/15 $31.00 © 2015 IEEE

accordingly between 1x10⁷ cm/s and 1.6x10⁷ cm/s [5,6]. An average nfet/pfet mobility around 110 cm²/Vs could be achieved with Si channels for n- and p-channels assuming a stress of 1 GPa, or for 0.75 GPa assuming SiGe p-channels. The corresponding possible combinations of (R_{series}, C_{gd}) are shown in Fig.5(c), showing that the needed relative improvements, i.e. 600 Ω for 41 aF, 500 Ω for 49 aF, 400 Ω for 56 aF .

In the next section, we analyze what are the values for R_{series} and C_{gd} expected from analytical modelling at nominal dimensions, the sensitivity of these parasitics and the corresponding performance to geometrical parameters, and at which conditions target values can be reached.

III. FEOL PARASITICS OPTIMIZATION

A. FEOL parasitic capacitances and spacer optimization

The main contributor to parasitic capacitances in the front end is the parallel plate capacitor from gate to S/D areas through the gate spacer C_{gd}. The spacer thickness reduction from node to node increases this contribution but the resulting increase in C_{gd} is compensated by the fin pitch reduction that reduces the parallel plate area between gate and S/D contacts (Fig. 6). The total capacitance is thus almost constant between 55 aF and 60 aF per side (S/D) for a 3 fins device across 14nm, 10nm and 7nm technology. The targeting of previous section showed that these values are too high for the 7nm node, mainly because of the increased importance of this capacitance relative to BEOL capacitances. Also, as seen in the targets, an effort on C_{gd} allows to reduce the constraint on the series resistance (and on mobility). C_{gd} can be reduced by increasing T_{spacer} (spacer thickness) or reducing the spacer dielectric permittivity. The other geometrical changes to decrease C_{gd} are limited: further fin pitch reduction also increases R_{series}, fin height reduction reduces the drive current per footprint.

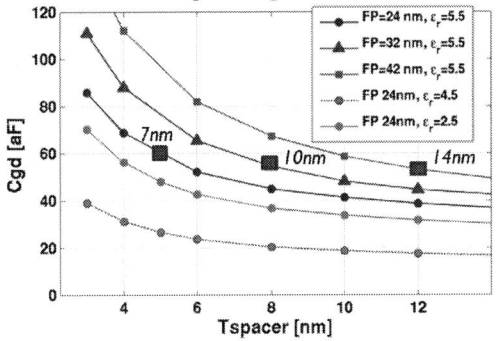

Fig. 6. Parasitic capacitance per side (S/D) accross nodes for a device with 4 fins, maintained at around 50aF from node to node despite the decrease in Tspacer, due to the simultaneous reduction of fin pitch. The 7nm node would benefit from further reduction, which requires lower spacer permittivity ε_r .

T_{spacer} has a very limited range of optimization. The optimum T_{spacer} for speed is between 5 and 6 nm depending on the BEOL load (Fig. 7), larger thickness degrading performance by increasing R_{series}, while smaller thickness degrades the speed due to C_{gd} increase, and is anyway not acceptable for reliability reasons. However, Fig.7b shows that a change in the spacer

material towards a lower dielectric constant ε_r is a powerful knob to reduce C_{gd} and increase speed. The transition from a SiON spacer (ε_r =5.5) to SiCN (ε_r =4.5) brings a speed improvement of 10% for unloaded case to 5% for 300-CGPs. If the spacers can be replaced by air gaps, supposing ε_r drops to 2.5 (ε_r =1 probably not realistic), the benefit is a reduction of C_{gd} to 25 aF per side, leading to an improvement in frequency by 16% (300-CGP load) to 32% (unloaded) or relaxing much more the constraint on R_{series} to reach same speed. This also shows that performance is much more sensitive to spacer permittivity (ε_r) than to spacer thickness (T_{spacer}).

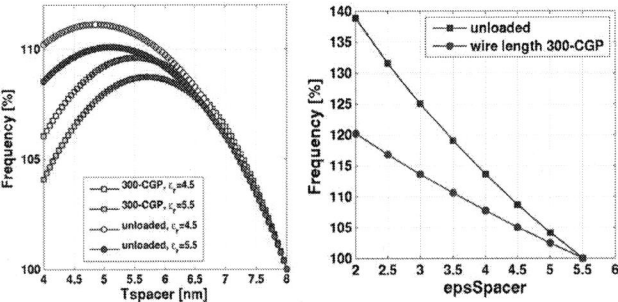

Fig. 7. Sensitivity of RO frequency to the spacer thickness and spacer permittivity for different loading expressed in wire length

B. Parasitic series resistance and contact optimization

The main contributors to series resistance are the contact and the extension. The contact resistance increases rapidly with the reduction of the contact width $T_{contact}$ (Fig. 8), assumed around 12nm for the 7nm node seen the constraints on gate length and minimum spacer width. To preserve the total series resistance around 600 Ω per side for a 3 fins device, the contact resistivity has to decrease to ρ=5x10⁻⁹ Ω-cm² if an epitaxial growth with diamond shape on S/D is used as in previous nodes (ρ is assumed 10x10⁻⁹ Ω-cm² at 10nm node). To go below this resistance value, alternative contact schemes might be needed.

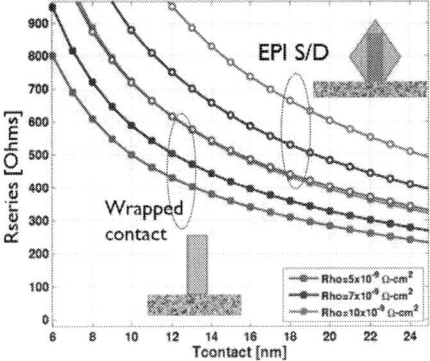

Fig. 8. Rseries per side versus Tcontact for EPI S/D and for wrapped contact (direct contact on the fin), for different contact resistivities, 4 fins per device.

An alternative is to use wrapped around contact, meaning contacting directly on fin, with the contact area using the whole fin height. In that case, 600 Ω could be reached at $T_{contact}$=12nm with ρ=10x10⁻⁹ Ω-cm². The choice between both schemes for

978-1-4799-7670-6/15 $31.00 © 2015 IEEE

the 7nm thus depends also on the effort on contact resistivity. In both cases, reduction lower than 600 Ω per side appears as high processing effort, showing the optimization of C_{gd} and thus spacer dielectric constant as easiest knob. The next sections consider what geometry optimization could bring as advantage from there.

IV. FIN HEIGHT OPTIMIZATION

Fin height increase is often seen as a performance element since increasing fin height increases the drive current per fin and therefore per footprint, thus theoretically improving the area efficiency. In practice, the benefit of HFIN increase will be limited by parasitics, leading to limited or no benefit on speed (Fig.9a) and simultaneous increase in dynamic power. The first reason for this double disadvantage is that fin height also increases C_{gate} and C_{gd}. The second limitation is, in the case of an EPI S/D contact, the series resistance does not decrease with HFIN increase (Fig.9b), and the higher current over the same resistance leads to a higher voltage drop (V_{gs} and V_{ds}), with consequence to unbias the device and limit the effective increase of current per fin compared to what could be expected. As a consequence of the combined effect of reduced drive current increase and high C_{gd}, in the case of an EPI contact, the frequency will actually be degraded for HFIN increase higher than 35nm. This effect was not yet present in previous nodes. For wrapped contact, HFIN increase up to 40 to 45nm results in a 5% improvement in frequency. However in both cases, the increase in C_{gd} and gate capacitance results in higher dynamic power consumption.

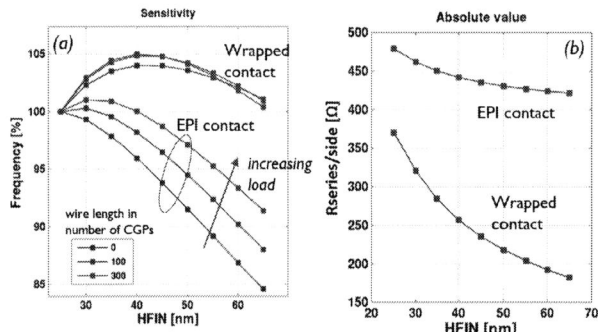

Fig. 9. (a) Frequency versus HFIN for the two contacting schemes and for different loading (in wire length). (b) Series resistance variation versus fin height HFIN for EPI contact and wrapped contact for same contact resitivity.

V. FIN DEPOPULATION

Fin depopulation is the reduction of the number of fins per device, meaning that the height of the standard cells is also reduced. For the standard cells heights as expressed in number of metal tracks, 9 tracks allows 4 fins per device, 7.5 tracks 3 fins, 6 tracks 2 fins. Fig. 10 shows that reducing the number of fins per active device allows to reduce drastically the dynamic power while preserving speed, especially for ROs with low or no BEOL loading. In the totally unloaded case, the delay is proportional to CV/I, and both C and I are reduced proportionally when the number of fins varies, leaving the delay unchanged. As the loading increases, the speed starts to degrade

more when removing fins since the BEOL capacitances start to dominate over the FEOL capacitances, and does not change with the number of fins, and the reduction of current becomes a penalty. Still, because of the high ratios of FEOL compared to BEOL capacitances, reducing to 3 fins per device brings a significant benefit in active power while being not of too much penalty to frequency in all the cases presented. Below 3 fins, the frequency drops too significantly dues to the drastic higher in series resistance, especially for one fin case.

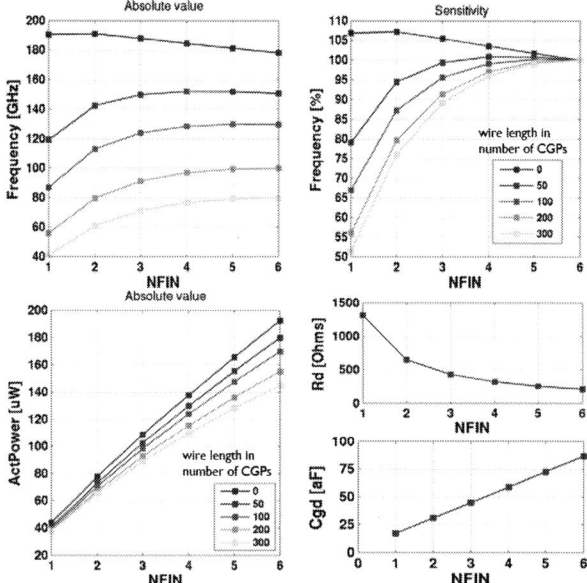

Fig. 10. Variation of frequency and Dynamic power in absolute value and normalized to the performance for one fin, for different loading expressed in wire length as number of CGPs.

VI. CONCLUSIONS

FinFETs could still provide enough electrostatic control at the 7nm node to achieve 20% speed improvement for HP devices, both for nSi-pSi channels and nSi-pSiGe, if breakthroughs are made to lower the device parasitic capacitances and resistances. Since there is no more room for geometry improvement in term of fin height, spacer width, contact size, this could be realized by changing the contacting mode from EPI S/D to direct contact wrapped on the fin and using spacers with lower permittivity constant, with this latter appearing as a powerful knob for speed increas. Fin depopulation down to 2 or 3 fins per active device provides drastic reduction in active power consumption while preserving speed. Beyond FinFETs, the transition to nanowires will have interest to improve low power devices or to push further gate length scaling while relieving the requirements on contact.

REFERENCES

[1] S. Natarajan, et al., *IEDM Tech. Dig.*, p. 71, 2014.

[2] S.Y Wu, et al., *IEDM Tech. Dig.*, p. 9.1.1, 2013.

[3] K. J. Kuhn, *IEEE Trans. Electron Devices*, vol. 59, no. 7, Jul. 2012

[4] M.G. Bardon, et al., VLSI Symposium, p. 1-2 2014

[5] V. Moroz, et al., *IEDM Tech. Dig.*, p. 180, 2014.

[6] G. Eneman, et al., *IEDM Tech. Dig.*, p. 131, 2012.

Impact of Fin Shape Variability on Device Performance towards 10nm Node

Kazuyuki Tomida, Keizo Hiraga, *Morin Dehan, *Geert Hellings, *Doyoung Jang, *Kenichi Miyaguhi, *Thomas Chiarella,
*Minsoo Kim,*Anda Mocuta, *Naoto Horiguchi, *Abdelkarim Mercha, *Diederik Verkest, *Aaron Thean

SONY corporation, Asahi-cho 4-14-1, Atsugi, 243-0014 Kanagawa, Japan / *imec, Kapeldreef 75, 3001 Leuven, Belgium
E-mail: Kazuyuki.Tomida@jp.sony.com, Phone: +32-16-28-7875

Abstract—**A transition from planar to FinFET brings additional variability sources from 3D channel structure. In this study, the impact of fin shape variability on device performance, especially from the view point of short channel effect control, is investigated with using Si-validated TCAD. This reveals that the width, height and taper angle of fin have significant impact on the electrostatics of the device. In addition, through the statistical Monte-Carlo simulations with compact model, the impact of fin shape variability is visualized in comparison with conventional device variability sources, i.e., gate length, work function, and equivalent oxide thickness. As a result, fin width and fin angle are found to be major variability source in addition to gate length. This indicates that the suppression of the process variability in fin width and fin angle is key to control device variability, especially in advanced node.**

Keywords—FinFET, variability, 14nm, 10nm, short channel effect, TCAD, Compact Model, Monte-Carlo

I. INTRODUCTION

One of the important benefits to introduce 3D channel FinFET is to improve the electrostatics for short channel effect (SCE) control, which enables low power operation [1-4]. Generally speaking, this benefit is emphasized when the fin gets narrower [5]. However, the width of the fin is just one parameter to define the fin shape, and the other parameters potentially influence the electrostatics, which could be a critical issue in terms of variability control. In other words, introducing 3D channel brings additional variability sources compared to the conventional planar devices. In this paper, the fin shape is divided into 6 components as shown in **Fig. 1**, and the effect of small variation in each component on the electrostatics is discussed based on Si-validated TCAD simulations, both with 14nm and 10nm nodes ground rule. In addition, the TCAD analysis results are implemented into Compact Model (CM), and the key factor to induce variability is discussed by using statistical Monte-Carlo (MC) simulation.

II. FINFET SHAPE VARIABILITY

Fig. 2 shows the basic process to fabricate bulk Si FinFET. For the device fabrication in this work, Self-Aligned Double Patterning (SADP), Replacement Metal Gate (RMG), and Raised Si Source/Drain (S/D) with highly doped drain implantation are employed. The ground rule of the device is mostly comparable to that of nominal 14nm generation FinFET (**Table 1**) [1-4, 6]. An exception is gate pitch, which is 110nm in the hardware of this study. The shape of the channel fin is mostly determined by fin module, and an additional impact could be given by the high-k / metal gate stack process. By tuning the etching process condition in fin module, the shape of fin can be varied from center to edge within a wafer, which influences the electrostatic characteristics as shown in **Fig. 3**. Here, two different etching processes were tested (indicated as process_A and process_B). Both processes give a wide variation in fin width and fin height as shown in **Fig. 3(b)** and **3(e)**. The electrostatic impact from different fin shape is summarized in **Fig. 3(c)** and **3(f)**, where Drain Induced Bias Lowering (DIBL) is taken as a benchmark index because DIBL is one of the most sensitive parameters to indicate device SCE. Fig. 3(c) and 3(f) indicate that both processes result in a lower DIBL at the edge of the wafer. In comparison with the trend of fin width and height (Fig. 3(b) and Fig. 3(e)), it is clear that a lower DIBL, meaning better SCE, is obtained when the fin shape of the channel gets narrower and higher.

Parameters	(a) Fin_Height	(b) Fin_Angle	(c) Footing	(d) Fin_Width	(e) Rounding	(f) Trench_Height
Definition in this paper	Distance from fin top to STI surface	Taper angle of sidewall of fin	Elevation angle of STI footing from middle of two fins	Fin width at STI surface level	Curvature radius of fin top corners	Distance from STI surface to bottom of trench

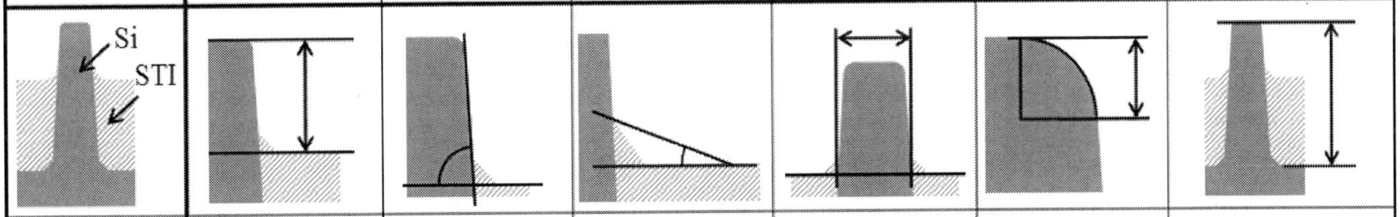

Fig. 1 Schematics of 6 parameters to describe fin shape and definition of them in this paper. Dark and light parts indicate Si (substrate / fin) and Shallow Trench Isolation (STI), respectively.

978-1-4799-7670-6/15 $31.00 © 2015 IEEE

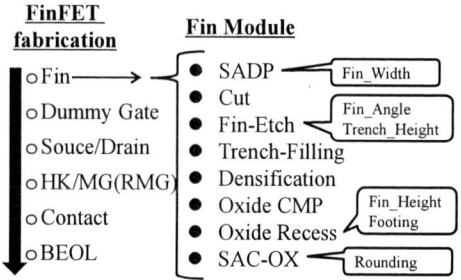

Fig. 2 Schematic flow of FinFET device fabrication and detail processes of the fin module. Main process to determine fin shape as shown in Fig. 1 is also indicated.

Table 1 Target ground rules of the hardware used in this study, and those of 14nm/10nm nodes to be implemented in TCAD and CM study.

	Hardware in this study	14nm	10nm
Fin Pitch [nm]	45	48	42
Gate Pitch [nm]	110	80	64
Fin Height [nm]	30	30	30
Fin Width [nm]	10	10	7
Fin Angle [nm]	86	86	86
Step Height [nm]	75	75	75
Lg [nm]	28	24	20
Vdd [V]	0.9	0.8	0.7

To understand the dominant factor influencing on DIBL, TCAD set up was established with 6 parameters of fin shape as shown in Fig. 1. The definition of each parameter in TCAD is summarized also in Fig. 1 (bottom row). First, a set of simulation was performed to confirm that the TCAD set up can reproduce the same DIBL trend as what Si device shows. The results are shown in Fig. 3(c) and 3(f), where the general trends are well matched between Si and TCAD. With using this TCAD set up, the effect of each parameter on DIBL was investigated. In this simulation, the center condition was set according to nominal 14nm ground rule (Table 1). **Fig. 4** shows the TCAD results of DIBL trend as function of gate length (L_g). As expected, shorter L_g gives higher DIBL both in NFET and in PFET. **Fig. 5** shows DIBL trend as functions of fin channel shapes. It is clearly shown that the impact of fin height, width and angle variation is much larger than the other 3 factors, and that there is no clear difference between NFET and PFET. According to this result, the further study was done mostly on PFET with using 3 key parameters, fin height, width and angle. **Fig. 6** shows the TCAD results of PFET on/off state current density as functions of 3 impacting parameters. Compared to the nominal case (in the left column), higher, narrower and more tapered fins show less off current. This is consistent with the results that lower DIBL is given in those conditions in Fig. 5.

Next, the TCAD simulation was performed with the ground

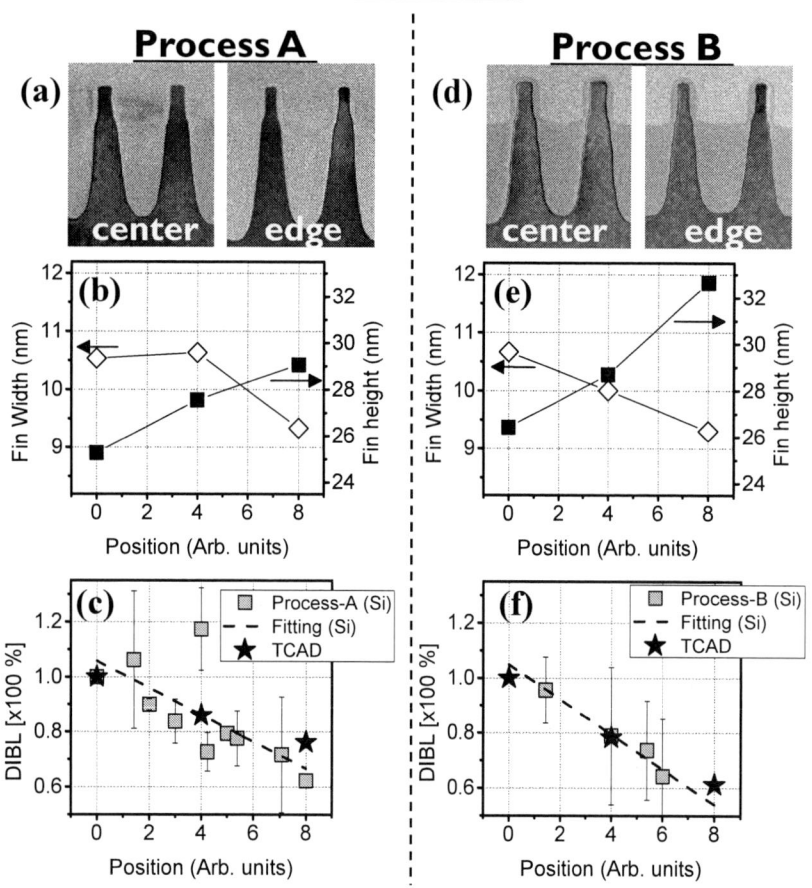

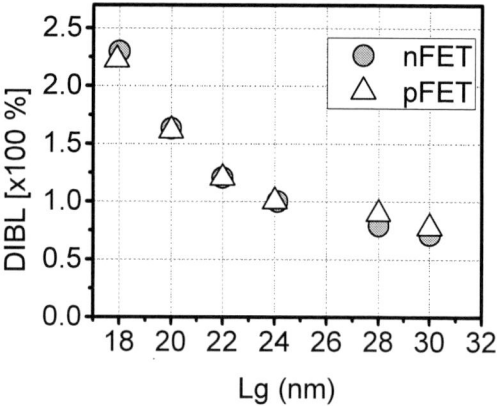

Fig. 4 TCAD results for DIBL trend in 14nm node as function of L_g.

Fig. 3 (a), (d) TEMs of Bulk Si FinFET, **(b), (e)** calibrated in-line data of fin height and width, and **(c), (f)** correlation between Si-data and TCAD used in this study. Position 0 and 8 corresponds to center and edge of 300mm wafer. Process A and B differ in the fin etching condition.

978-1-4799-7670-6/15 $31.00 © 2015 IEEE

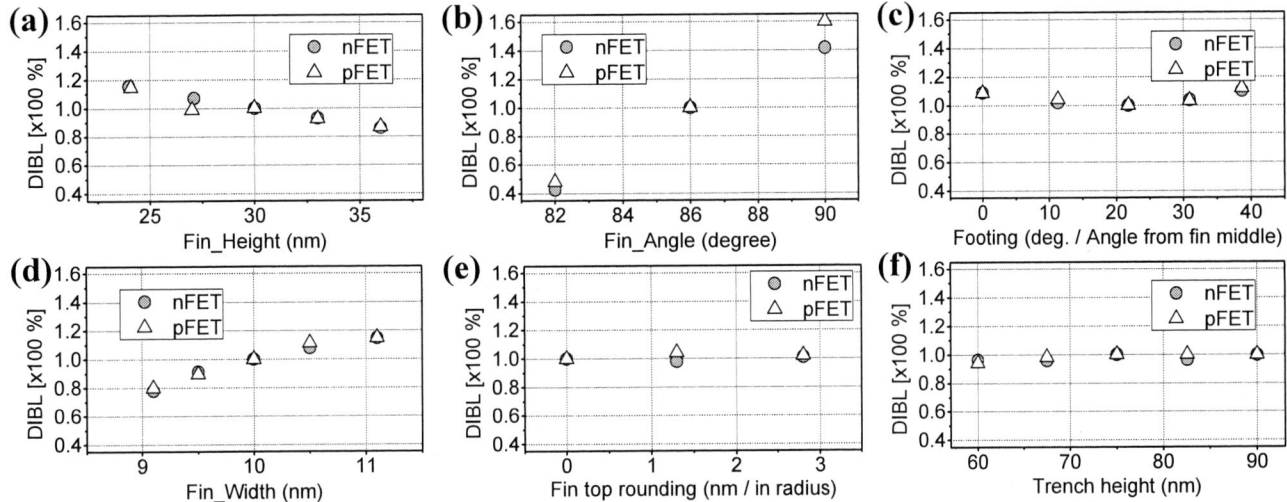

Fig. 5 TCAD results for DIBL trend in 14nm node as functions of **(a)** Fin height, **(b)** Fin angle, **(c)** footing, **(d)** Fin width, **(e)** Rounding and **(f)** Trench height. Strong impacts are found in (a), (b) and (d), as expected from Fig. 5.

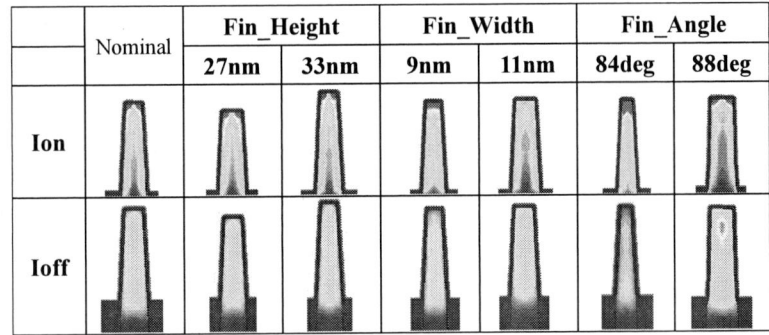

Fig. 6 I_{on} (top row) and I_{off} (bottom row) current density at the middle of channel as functions of fin shape.

rule assumption of 10nm node (Table 1). The DIBL trends of fin height, angle, width and L_g are shown in **Fig. 7**, where the x-axis is set to be 3σ of each parameter that is defined by process controllability assumption (**Table 2**). The relative impact on DIBL is summarized also in Table 2. In addition to L_g, the impact from fin width and fin angle drastically increases through the node transition. The reason for the large impact from fin width and angle in 10nm node can be explained as below. Since the absolute value of fin shape variability in 10nm node is same as that in 14nm node at best, the relative impact is enlarged due to a narrower fin width ground rule assumption in 10nm node. Regarding fin angle, the same explanation is adaptable because more tapered fin has narrower part at the top of fin. This analysis indicates that the fin width variability needs to be carefully suppressed in 10nm and further advanced nodes.

III. STATISTICAL ANALYSIS

To capture statistical distributions with CM, the 14nm and 10nm CM were tuned with TCAD results (**Fig. 8**). **Fig. 9** shows MC simulation results for (a), (c) I_{on}-I_{off} and (b), (d) DIBL distribution, respectively. In this simulation, the effect from nominal 3 parameters (L_g, metal gate Work Function (WF), and Equivalent Oxide Thickness (EOT)) are also simulated. The results clearly show that the fin shape variation brings additional variation in device property. It is also shown

that the effect is emphasized in 10nm node compared to 14nm node, which is as expected from TCAD results (Fig. 7 and Table 2). **Figure 10** shows the impact of each component on DIBL distribution, where the values are normalized with 14nm DIBL 3σ, and the height of the bar is given by a simple summation of each component. The major contributors are indicated to be L_g, fin width and fin angle. Especially in 10nm, L_g is the dominant factor to give the electrical variability. In Fig. 10, a case of better process control in L_g (3σ is assumed to be 5%) is also shown. Even with this optimistic assumption, the variability is much larger than 14nm case. In other words, an improvement of the fin width/angle variation is seriously required in advanced nodes.

IV. CONCLUSION

The impact of fin shape on electrostatics is studied by using Si-validated TCAD and CM. As a result, it is revealed that the height, width and angle of the fin have the dominant impact on electrostatics. The TCAD results are implemented into 14nm and 10nm CM to see the statistical distribution of device parameters. This exercise clarifies that the highly impacting parameters on DIBL variation are L_g, fin width and fin angle. Especially in advanced node, it is suggested that the suppression of the fin width and angle variability, on top of the L_g fluctuation, would be more essential to guarantee the total device variability.

978-1-4799-7670-6/15 $31.00 © 2015 IEEE

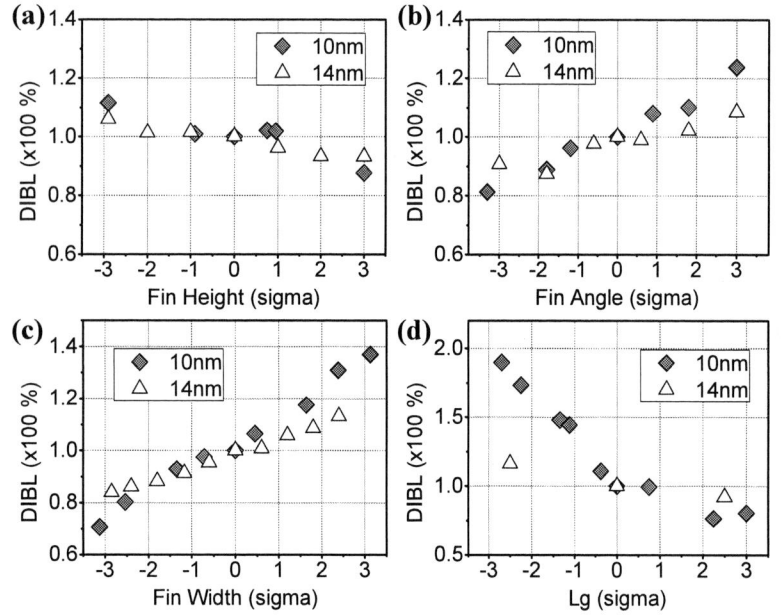

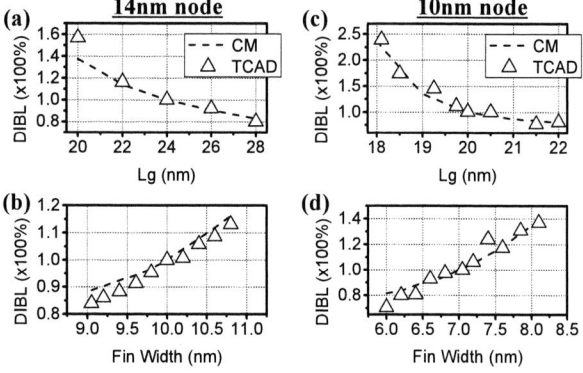

Table 2 Ideal 3σ of 3 key parameters + L_g and DIBL shift at when each factor takes 3σ variation.

	3-sigma in process	Impact on DIBL for 14nm	Impact on DIBL for 10nm
Fin Height	± 3.0nm	+5.0% / -7.8%	+11.4% / -14.4%
Fin Width	± 1.0nm	+15.9% / -17.6%	+36.0% / -28.5%
Fin Angle	± 0.5 deg	+9.0% / -10.9%	+22.3% / -16.9%
Lg	± 10 %	+19.5% / -9.4%	+99.4% / -24.9%

Fig. 7 (left) PFET DIBL trend in 14/10nm nodes from TCAD results as functions of 3 key parameters and L_g. X-axis in all plots are set to be +/- 3σ (table 2).

Fig. 8 CM tuning to align TCAD DIBL trend in 14nm **(a)**, **(b)** and 10nm **(c)**, **(d)**. Tuning was done with 4 parameters discussed in Fig. 7.

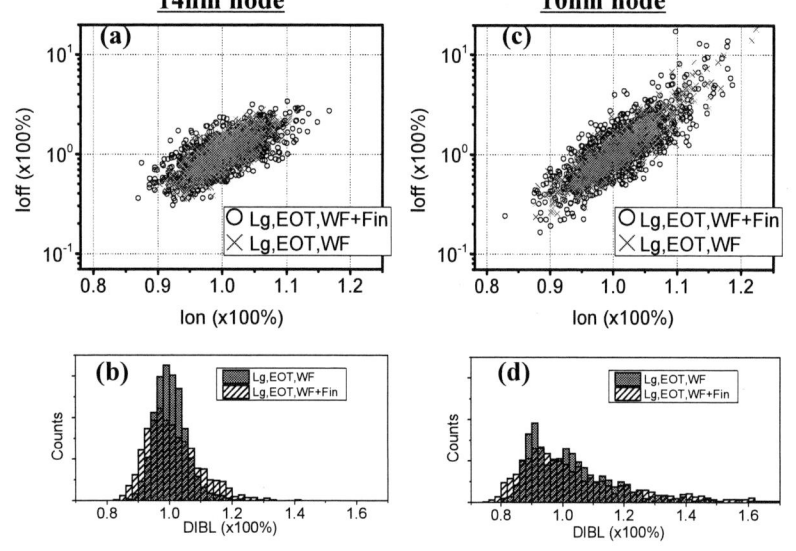

Fig. 10 (up) variability components in 14/10nm nodes. Values are normalized with 14nm.

Fig. 9 (left) MC statistical simulation results for **(a)**, **(b)** I_on-I_off and **(c)**, **(d)** DIBL distribution.

REFERENCES

[1] C. Auth *et al.*, "A 22nm High Performance and Low-Power CMOS Technology Featuring Fully-Depleted Tri-Gate Transistors", Symp VLSI Tech. Dig., pp. 131-132 (2012).

[2] S. Y. Wu *et al.*, "A 16nm FinFET Technology for Mobile SoC and Computing Applications" IEDM Tech. Dig., pp. 224-227 (2014).

[3] S. Y. Wu *et al.*, "An Enhanced 16nm CMOS Technology Featuring 2nd Generation FinFET Transistors and Advanced Cu/low-k Interconnect for Low Power and High Performance Applications" IEDM Tech. Dig., pp. 48-51 (2014)..

[4] S. Natarajan *et al.*, "A 14nm Logic Technology Featuring 2nd-Generation FinFET, Air-Gapped Interconnects, Self-Aligned Double Patterning and a 0.0588 m² SRAM cell size", IEDM Tech. Dig., pp. 131-132 (2014).

[5] A. R. Brown *et al.*, "Comparative Simulation Analysis of Process-Induced Variability in Nanoscale SOI and Bulk Trigate FinFETs" IEEE Trans. Electron Devices, vol.**60**, pp. 3611-3617 (2013).

[6] P. Schuddinck *et al.*, "Impact of measured process variations on performance and power consumption of loaded inverter cells in N14 / N10 / N7 Bulk FinFET technology nodes", Symp VLSI, submitted (2015).

ACKNOWLEDGMENT

This work is performed as Imec Industrial Affiliation Program (IIAP).

Modeling FinFET Metal Gate Stack Resistance for 14nm Node and Beyond

Kenichi Miyaguchi, Bertrand Parvais, Lars-Åke Ragnarsson, Piet Wambacq, Praveen Raghavan, Abdelkarim Mercha, Anda Mocuta, Diederik Verkest and Aaron Thean

Imec, Kapeldreef 75, B-3001 Leuven, Belgium
E-mail: Kenichi.Miyaguchi@imec.be

Abstract—A FinFET high-k replacement metal gate stack resistance model is proposed. Introduction of non-negligible contact resistance existing in boundaries between metal layers achieves a good model accuracy which is validated by FEM-based simulation results in 14nm and 10nm technology nodes. Impact of the contact resistance on digital and analog circuit is investigated, resulting in 20% degradation of analog speed by 5 $\Omega \cdot \mu m^2$ contact resistance. The derived gate resistance model is applicable to further downscaled FinFET technology.

Keywords—*Modeling, FinFET, gate resistance, high-k replacement metal gate, contact resistance.*

I. INTRODUCTION

CMOS technology scaling brought a change of gate structure from polysilicon gate to high-k metal gate. Migrating from planar to FinFET in addition to introduction of replacement metal gate (RMG) scheme employing multi work function metals to control V_t value, gate stack structure becomes complicated resulting in non-straight signal path on the FinFET gate stack due to its 3-D structure. Gate resistance model plays a significantly important role in analog/RF circuit design [1][2]. However, existence of multiple metal layers and high-k dielectric in the RMG makes it difficult to estimate its gate resistance.

The goal of this paper is to address how the gate resistance of the complicated FinFET gate stack structure is formulated. FinFET device assumption for the gate resistance study is described in section II, then we demonstrate that introduction of contact resistance which exists in boundaries between the metals is a key factor for accurate prediction of the gate resistance. Before conclusions, impact of the gate resistance on digital and analog circuits is presented in section IV.

II. FINFET DEVICE ASSUMPTION

We assume that 14nm and 10nm FinFET technology nodes (N14 and N10) feature bulk Si substrate, Cu back-end-of-line (BEOL), high-k replacement metal gate (RMG) and source/drain (S/D) SiGe epitaxial stressor for PFET (Fig. 1). Key design parameters of N14 and N10 devices are given in Table 1.

TABLE I. KEY DEVICE PARAMETERS FOR N14 AND N10

	N14	N10
Gate length, L_G (nm)	30	24
Contacted gate pitch, *CGP* (nm)	90	64
Gate spacer width, T_{SP} (nm)	14	8
Gate height, *dH* (nm)	38	30
Fin width, *TFIN* (nm)	10	7
Fin height, *HFIN* (nm)	30	30
Fin pitch, *PFIN* (nm)	48	36
Supply voltage, V_{DD} (V)	0.75	0.7

In this study, a commercially available finite-element-method (FEM) based simulation software, COMSOL [3], is used to build an accurate three-dimensional model of the FinFET gate stack structure for analysis of gate resistance. Fig. 2 shows gate stack model built in COMSOL where gate extensions are also drawn as an access part into FinFET core part. The length of the gate extension is fixed to 30nm in the simulations.

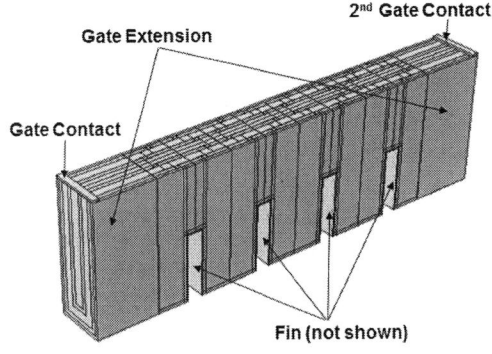

Fig. 2. FinFET gate stack model built in COMSOL (4-fins). Gate extensions are also drawn as an access part into FinFET core part.

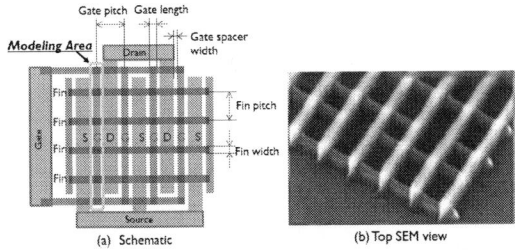

Fig.1. FinFET device (a) Schematic, (b) top SEM view

978-1-4799-7670-6/15 $31.00 © 2015 IEEE

To provide multiple options of threshold voltage depending on applications, gate work-function metals (WFM) are deposited with barrier metals and a conductive filling metal [4]. Assuming the high-k last PMOS 1st gate formation process for N14 and N10, dielectric and metal layer composition of the gate stack is dependent on the device dimension. Fig. 3 represents cross-sectional schematics of NMOS and PMOS in N14 and N10. A 1st TiN layer above the high-k dielectric layer of HfO$_2$ is followed by deposition of TiAl/TiN, resulting in a TiN/TiAl/TiN WFM for NMOS. The PMOS gate stack is processed with extra WFM and the barrier layers, resulting in TiN/TaN/TiN/TiAl/TiN. Deposition of Tungsten (W) as a filling metal finalizes the gate stack formation process. There is no space left for Tungsten in N10 PMOS. Thickness of the metals is given in the figure. The resistivity of the metals applied to the FinFET gate stack model is listed in Table II.

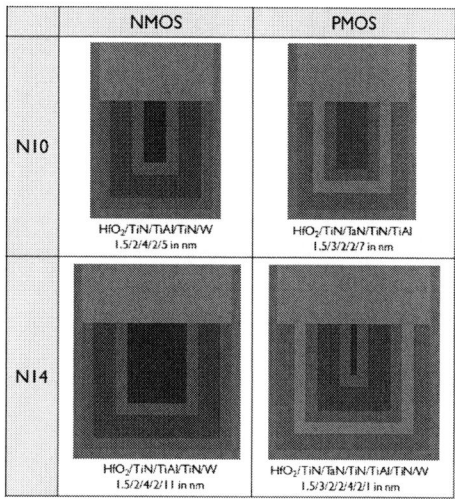

Fig. 3. Cross-sectional schematic of gate stack on top of fin. The gate of N10 PMOS has no space left for Tungsten as a conductive filling metal.

TABLE II.   RESISTIVITY OF METALS ASSUMED IN GATE STACK

	ρ ($\mu\Omega\cdot$cm)
TiN	200
TiAl	300
TaN	3000
W	30

III. FINFET GATE STACK RESISTANCE MODEL

A. Analytical Formula

The FinFET 3-D structural gate stack makes a signal path in not only parallel direction along the gate but also vertical one. Fig. 4 illustrates a network schematic of the gate stack resistance studied in the paper. In addition to parallel (R_{para}) and vertical (R_{ver}) components of the gate resistance, we introduce the contact resistance (R_c) which exists in boundaries between the metals. An analytical formula of the gate stack used in this study is expressed in Eq. (1).

$$R_g = R_{para} + R_{ver} + R_c + R_{ext}, \qquad (1)$$

where R_{ext} is a resistance of the gate extension part. Each components of the gate resistance are simply scalable to *NFIN* and *NF* as listed in Eq. (2)-(6):

$$R_{para} = \rho_{para} \cdot 1/3 \cdot PFIN/(L_g \cdot dH) \cdot NFIN/NF/NGCON^2, \qquad (2)$$

$$R_{ver} = \rho_{ver} \cdot (dH/2 + HFIN/3)/(L_g(PFIN - TFIN))/NFIN/NF, \qquad (3)$$

$$R_c = R_{c,eff}/(L_g \cdot (W_{eff} + \alpha)), \qquad (4)$$

$$R_{ext} = r_{ext}/NGCON/NF, \qquad (5)$$

$$W_{eff} = NF \cdot NFIN(TFIN + 2HFIN). \qquad (6)$$

where W_{eff} is effective width of the device. ρ_{para} and ρ_{ver} represent effective resistivity for parallel and vertical direction along the gate, respectively. Those resistivity depend on composition of high-k and metal materials forming the gate stack. *NFIN* is number of fin, *NF* is number of gate finger and *NGCON* is number of gate contact (1 or 2). $R_{c,eff}$ has to represent an effective contact resistance per unit size of the device since each metal boundary shows different contact resistance value. α is introduced into the term of the contact resistance as a fitting parameter to achieve a good accuracy.

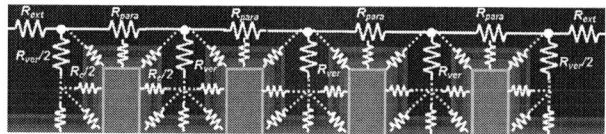

Fig. 4. A network schematic representing gate stack resistance. The contact resistance exists in boundaries between metals.

B. Parameter Extraction

COMSOL simulations to calculate total gate stack resistance of N14 and N10 FinFET devices were performed with sweeping *NFIN* in both cases of single and double gate contact (*NGCON*=1 and 2). Since we lack experimental data of the contact resistance of each metal layer boundaries (TiN/TiAl, TiN/W and TiN/TaN), a certain value of the contact resistance is assumed in the COMSOL gate stack model. The assumed contact resistance value was also swept in the simulations to investigate impact of the contact resistance at following section.

First, r_{ext} is extracted through simulation for only gate extension part model built in COMSOL. Second, at given r_{ext}, 2 sets of (R_{para}, R_{ver}) values are obtained by simulations with parameters of 2 different *NFIN* and setting of zero contact resistance (R_c=0), and then ρ_{para} and ρ_{ver} are calculated through Eq. (2) and (3). Lastly $R_{c,eff}$ and α have to be determined. Subtracting simulated data in case of zero contact resistance from one with non-zero contact resistance leaves only the term expressed by Eq. (3), resulting in:

$$\Delta R_g = R_{g,non\text{-}zero\text{-}contR} - R_{g,zero\text{-}contR} = R_{c,eff}/(L_g \cdot (W_{eff} + \alpha)). \qquad (7)$$

Rewriting Eq. (7) leads to Eq. (8).

$$L_g \cdot W_{eff} = R_{c,eff}/\Delta R_g - L_g \cdot \alpha \qquad (8)$$

A linear trend line between $L_g \cdot W_{eff}$ and $1/\Delta R_g$ gives slope ($R_{c,eff}$) and intercept point ($L_g \cdot \alpha$).

Table III summarizes the gate stack resistance parameters extracted from the simulation results for NMOS and PMOS in N10 and N14. In the table, $R_{c,eff}$ is maximum value which is validated in this study. PMOS devices show higher resistivity than NMOS due to no or much less space filled by W as the conductive filling metal.

TABLE III.   GATE RESISTANCE PARAMETERS EXTRACTED FROM SIMULATION DATA

	N10 NMOS	N10 PMOS	N14 NMOS	N14 PMOS
ρ_{para} ($\mu\Omega\cdot$cm)	132	312	88	250
ρ_{ver} ($\mu\Omega\cdot$cm)	64	111	75	120
$R_{c,eff}$ ($\Omega\cdot\mu m^2$)	$\leq$4.2	$\leq$2.8	$\leq$3.6	$\leq$2.6
r_{ext} (Ω)	89	274	48	156
α @$NGCON$=1 (μm)	0.31	0.50	0.19	0.28
α @$NGCON$=2 (μm)	0.44	0.70	0.27	0.40

C. Comparison between Model and Simulation Results

Fig. 5 shows the total gate stack resistance as a function of number of fin in N10 NMOS and PMOS for both single and double gate contact. Good agreement between model and simulation is achieved with less than 10% error. The vertical components are dominating at small number of fin as the signal transverses mainly in vertical direction through the contact resistance in boundaries between the metal layers. The parallel component becomes visible at large number of fin with decrease of the vertical and contact resistances which are inversely NFIN scaled.

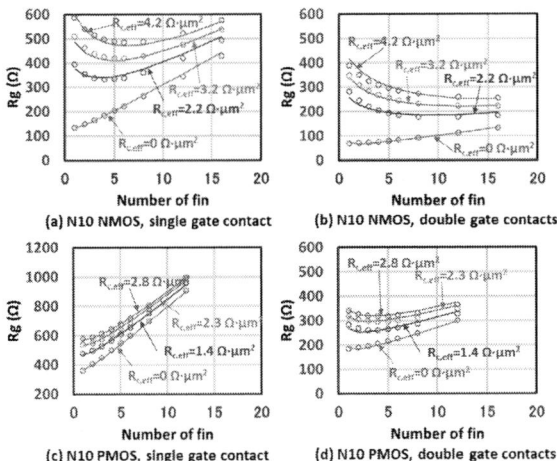

Fig. 5.   Comparison between model and simulation results in N10 NMOS and PMOS for both single and double gate contact (line: model, circle: simulation). The effective contact resistance values are extracted from the simulation data.

N14 NMOS and PMOS fitting results of model against simulation results are also shown in Fig. 6, validating that the analytical formula proposed in this study can be scalable to different gate length.

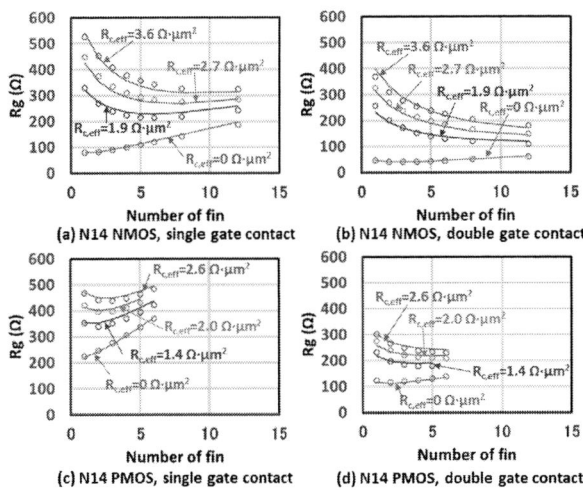

Fig. 6.   Comparison between model and simulation results in N14 NMOS and PMOS for both single and double gate contact (line: model, circle: simulation). The effective contact resistance values are extracted from the simulation data.

IV. CIRCUIT ASSESSMENT

In this section, impact of the contact resistance on digital/analog circuits is investigated. We use N10 and N14 predictive FinFET compact models for circuit simulations [5].

A. Digital

A 15-stages inverter-chained ring oscillator circuit is employed as a test bench circuit for digital use. We run the ring oscillator simulations using 4-fins FinFET device with 2 different load conditions: (i) fan-out 3 without BEOL wire load, (ii) fan-out 1 with 100CGP-length BEOL wire load. As shown in Fig. 7, a degradation of the ring oscillator delay is limited within 2% compared to zero $R_{c,eff}$ if the contact resistance is able to be assumed up to 5 $\Omega\cdot\mu m^2$ even in both the test benches for N10 and N14.

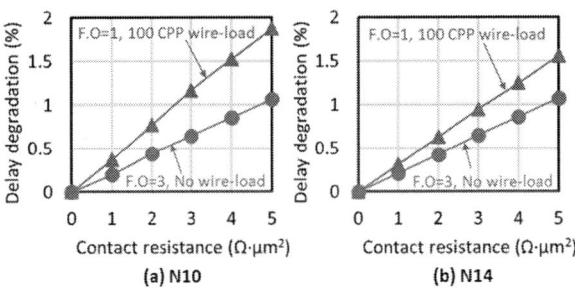

Fig. 7.   Impact of the contact resistance on delay of 15-stages inverter-chained ring oscillator in N10 and N14. Two different conditions are tested: (i) Fan-out 1 plus wire-load with length of 100 gate pitch, (ii) Fan-out 3 without wire-load.

978-1-4799-7670-6/15 $31.00 © 2015 IEEE

B. Analog

Investigating impact of the gate stack resistance on analog performances starts with device figure-of-merit (FOM). Since analog circuits, especially input and output stage of amplifiers, typically employs large size devices from mismatch and maximizing output power point of view, a maximum oscillation frequency (f_{MAX}) of device having a constant product of *NF* and *NFIN* with double gate contact is calculated at $V_{ds}=V_{dd}$. A peak f_{MAX} as a function of *NFIN* for NMOS and PMOS in N10 and N14 are plotted in Fig. 8. The larger *NFIN* is, the f_{MAX} decreases more due to increase of total gate resistance. with sweeping the contact resistance. Compared to the curve of zero constant resistance which is not realistic (dotted line in the figure), the introduction of 5 $\Omega\cdot\mu m^2$ contact resistance (solid line) shifts the f_{MAX} curve down by 20% at maximum. Predicting the contact resistance precisely by experimental data or simulation is a key factor to achieve a good accuracy of the analog circuit design because the contact resistance is not negligible for high frequency circuit design unlike logic circuit.

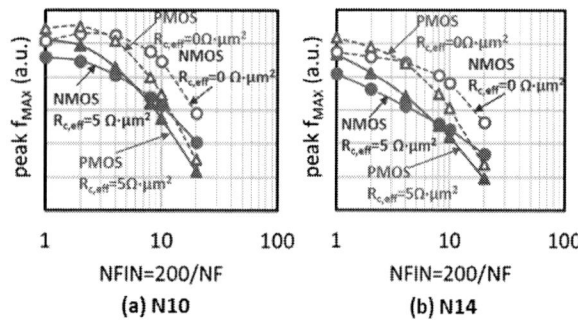

Fig. 8. Maximum oscillation frequency, f_{MAX}, as a function of number of fin for NMOS and PMOS in (a) N10 and (b) N14. Device size is constant (NF*NFIN=200). The larger NFIN is, the f_{MAX} decreases more due to increase of total gate resistance.

To make more clear the impact of the contact resistance, Fig. 9 shows the peak f_{MAX} of the device with 4-fins and 50-gate fingers as a function of the contact resistance at $V_{ds}=V_{dd}$. We see a linear relationship between degradation of the peak f_{MAX} and the contact resistance in N10 and N14.

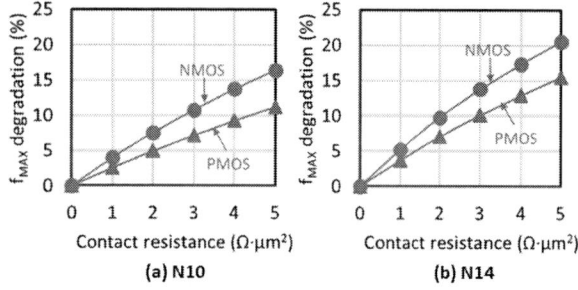

Fig. 9. Impact of the contact resistance on maximum oscillation frequency, f_{MAX}, of NMOS and PMOS with 4-fins and 50-gate fingers in (a) N10 and (b) N14. f_{MAX} degrades by around 20% in case of contact resistance of 5 $\Omega\cdot\mu m^2$.

Extending analog performance assessment from device to circuit level, we discuss an LC-VCO operating in 60GHz range shown in Fig. 10. At a fixed NMOS size of the cross-coupling pair, simulation results with sweeping *NFIN* reveal that phase noise performance degrades by less than 2dB when the contact resistance increases to 4.2 $\Omega\cdot\mu m^2$. Also, 3dB variation of the phase noise can be seen if *NFIN* changes between 1 and 20 if the contact resistance is assumed to be zero, but the variation becomes larger (4dB) n case of 4.2 $\Omega\cdot\mu m^2$ contact resistance.

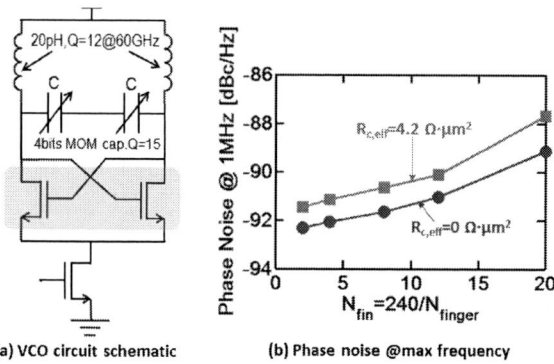

(a) VCO circuit schematic   (b) Phase noise @max frequency

Fig. 10. Impact of the contact resistance on LC-VCO. The phase noise degradation is estimated less than 2dB if contact resistance is assumed to be 4.2 $\Omega\cdot\mu m^2$.

V. CONCLUSIONS

We have demonstrated the contact resistance in boundaries between metal layers deposited on the high-k metal gate stack in FinFET becomes a dominant factor of total gate resistance, and play an important role in estimation of analog circuit performances. The verified analytical model can be applicable for further downscaled FinFET technology node in which the gate resistance increases.

ACKNOWLEDGEMENT

The imec logic insite program and core CMOS device development members are acknowledged for their support and discussion.

REFERENCES

[1] W. Wu, et al., "Gate Resistance Modeling of Multifin MOS Devices," IEEE Electron Device Letters, vol. 27, No. 1, pp. 68-70, January 2006.

[2] R. A. Wachnik, et al., "Gate Stack Resistance and Limits to CMOS Logic Performance," IEEE CICC, vol. pp. 1-4, September 2013.

[3] http://www.comsol.com/

[4] L.-Å. Ragnarsson, et al., " Highly Scalable Bulk FinFET Devices with Multi-V_T Options by Conductive Metal Gate Stack Tuning for the 10-nm Node and Beyond," VLSI Technology Symposium Digest, p. 56, June 2014.

[5] J. Ryckaert, et al., "Design Technology Co-optimization for N10," IEEE CICC, vol. pp. 1-8, September 2014.

Static and Dynamic Power Management in 14nm FDSOI Technology

O. Weber, E. Josse, J. Mazurier, and M. Haond

Abstract— **This work presents a 14nm technology designed for high speed and energy efficient applications using FDSOI transistors. -34% speed delay at same static power with -100mV V_{dd} supply voltage operation vs 28nm FDSOI is demonstrated.**

The specific FDSOI features for adjusting the threshold voltage and managing power are highlighted in this paper. It is shown that a light channel doping and reverse back bias are effective to reduce the static leakage and, on the other hand, forward back bias (FBB) can provide dynamic power saving at same speed. All this process & design techniques, in addition to the poly bias capability, makes FDSOI a highly flexible technology to maximize the speed/leakage/power compromise for each application product.

Index Terms— **Fully Depleted silicon-on-insulator (FDSOI) technology, Threshold voltage, Static power, Dynamic Power, Mismatch, Back bias.**

I. INTRODUCTION

As CMOS technology scales down, two paths are pursued by the industry to overcome the fundamental limits of traditional planar bulk transistors. One is the introduction of a Tri-Gate or FinFET transistor at the 22 and 16 nm nodes [1, 2]. These architectures provide impressive drive currents per footprint at low supply voltages because of the 3-D conduction channel and excellent electrostatic control. Conversely, they have high gate and parasitic capacitances, proportional to the 3-D effective W increase, which negatively impacts both the speed and active power consumption. Alternatively FDSOI provides an evolutionary path. FDSOI includes excellent mismatch properties, a simplified planar manufacturing process vs 3-D finFET technology and capitalization of existing design techniques. First introduced at the 28nm node [3] and demonstrated using ARM-based chips operating up to a record frequency of 3 GHz [4], the scaling of the technology to the 14nm node using strain-engineered transistor has been demonstrated in 2014 [5]. The FDSOI technology also extends the possibility of back biasing and therefore offers unique "smart" solutions for dynamic power optimization [6]. This paper presents the process & design solutions to tune the

O. Weber, J. Mazurier, are with CEA-Leti, MINATEC Campus, 17 rue des Martyrs, F38054 GRENOBLE, Cedex 9, France, and are assignees in STMicroelectronics, 850 rue Jean Monnet 38926 Crolles, France (e-mail: olivier.weber@st.com).

E. Josse, M. Haond are with STMicroelectronics, 850 rue Jean Monnet 38926 Crolles, France.

This work was partly supported by the Catrene Dynamic-ULP, Places2be KETs and Nano2017 projects, through the French ministry of Industry.

transistors threshold voltage (Vt) in 14nm FDSOI technology for managing static and dynamic power in a large range.

II. TECHNOLOGY PERFORMANCE

A 0.55x area scaling with respect to the 28nm FDSOI technology was achieved with the introduction of local interconnect and the adoption of fixed layout shapes (or "constructs"). Compared to the 28nm technology, new Front-End process elements include a dual SOI/SiGeOI N/P channel, a dual workfunction gate-first HKMG integration scheme and a dual in-situ doped Si:CP/SiGeB N/P raised source-drain [5]. TEM pictures for NMOS and PMOS are presented in Fig.1.

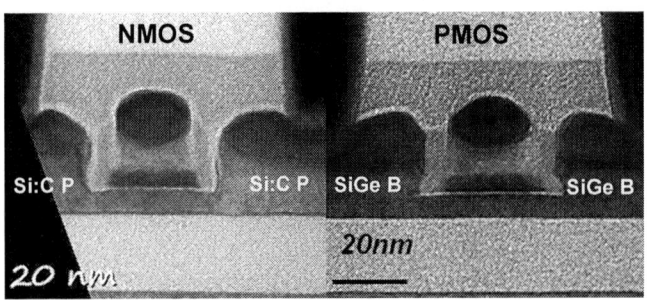

Fig.1: NMOS and PMOS transistor TEM cross-sections in 14nm FDSOI technology.

I_d-V_g plots at V_{dd}=0.8V supply voltage for L=20nm and L=30nm are presented in Fig.2. Transistors show a DIBL of 85mV and a sub-threshold slope of 85mV/dec. for both NMOS and PMOS at L_{nom}=20nm. The gate length increase from 20nm up to 30nm in the same 90nm contacted poly pitch (CPP) allows reducing the off-leakage by more than 1 decade. Poly length biasing is the first knob to tune the transistor Vts and the static leakage in FDSOI. Due to improved electrostatic compared to bulk technology the minimum gate length allowed is shorter and the poly bias range is larger than in bulk.

As a result of low C_{gd} and large PMOS drive current, 14nm FDSOI technology demonstrated in [5] a -20% delay gain with the Fan-Out 3 (FO3) RO inverters at the same static leakage and a 100mV V_{dd} reduction (0.8V vs 0.9V) over the 28nm FDSOI technology. From this previous work, the transistor performance has further progressed and the delay boost is now established at -34% with -100mV V_{dd} operation, as shown in Fig.3. It means >50% speed frequency in 14FDSOI at 0.8V V_{dd} vs 28FDSOI at 0.9V V_{dd}. It means also that 14nm FDSOI can run as fast as 28nm FDSOI with supply voltage much lower than V_{dd}=0.8V, and thus with much lower dynamic power consumption.

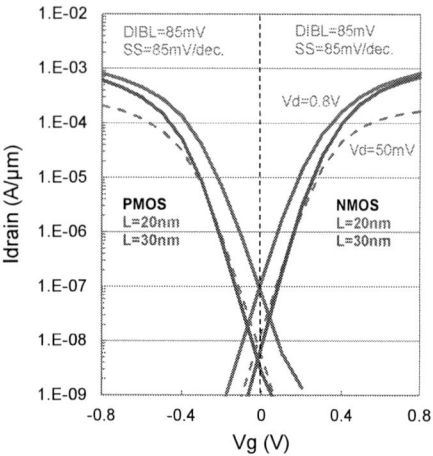

Fig.2: I_d-V_g plots for L_{nom}=20nm and L=30nm for both NMOS and PMOS transistors.

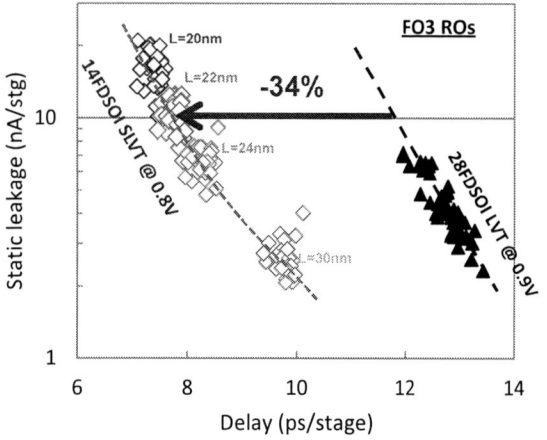

Fig.3: Delay/I_{stat} in FO3 ring oscillators in 14nm FDSOI technology at V_{dd}=0.8V compared to the 28nm FDSOI technology at V_{dd}=0.9V.

III. TRANSISTOR WELL ARCHITECTURE: FROM BULK TO FDSOI TECHNOLOGY

Two kinds of well architecture can be used in FDSOI as shown in Fig.4. The "regular well" architecture is the same as in bulk technology, i.e. a Pwell under NMOS and a Nwell under PMOS, such as FDSOI can be blind ported from a bulk design. The well straps have the same size as a conventional bulk technology, and the "Hybrid" CAD layer used to define the opening in the buried oxide does not induce surface penalty (Fig.5). Hybrid bulk areas also provide a space for passive devices and ESD FETs to be built [7]. A perfectly flat bulk to SOI transition is achieved between these areas (Fig.5).

The second well architecture, named "flipwell", is obtained by reversing the wells from the previous regular well architecture, i.e. a Nwell under the NMOS and a Pwell under the PMOS. This is a FDSOI specific feature, since active areas of SOI MOSFETs are naturally isolated from wells by the buried oxide (BOX). To avoid the diode conduction in direct

mode between Pwell and Nwell in the flipwell configuration, the Pwell under the PMOS has to be grounded, which changes the typical well biasing of the bulk technology.

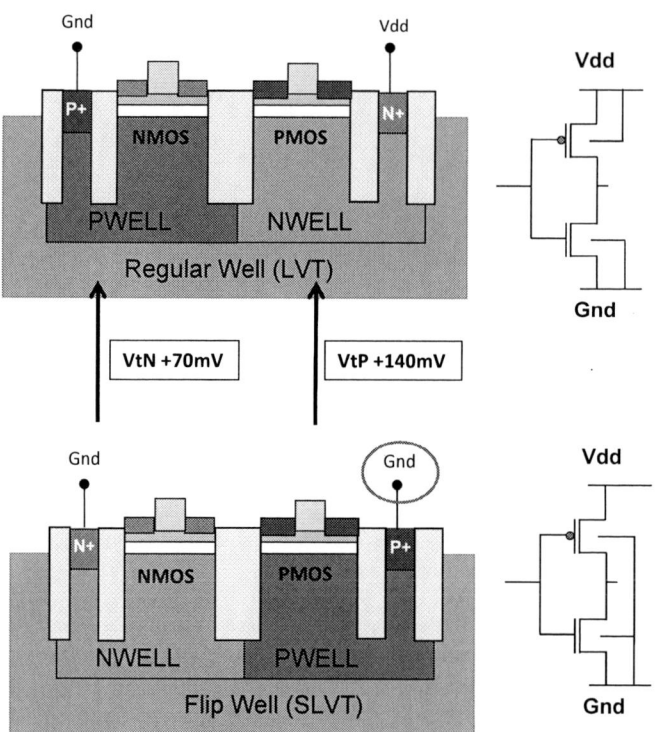

Fig.4: NMOS & PMOS schematics for both regular well and flipwell device architectures.

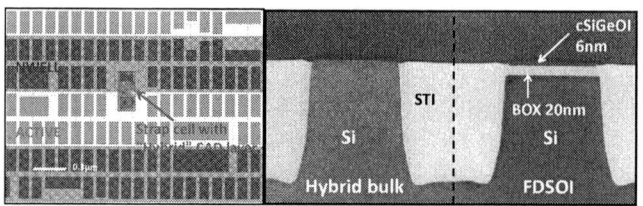

Fig.5: Example of 14FDSOI std cell layout, showing the "Hybrid" CAD layer (in red) on a well strap (left) and TEM picture showing adjacent SOI/Bulk active areas (right).

The well swap between regular and flipwell architectures in NMOS transistors induces a back gate worfunction change of ~1eV, which in turn induces a Vt shift of 70mV for FDSOI transistors with a 20nm BOX and a body factor of 70mV/V. In PMOS, in addition to the well swap, the well bias change has to be considered and the Vt shift is thus about 140mV between regular and flip well architectures (Fig.5). As a result, Vts for NMOS and PMOS cannot be equilibrated at the same time for both regular well and flipwell architectures. To overcome this issue, it may be optimal to re-adjust the PMOS Vt in the regular well architecture by introducing a second metal gate workfunction (separated by 70meV from the former one) or to ground the PMOS Nwell, as shown in Fig.6.

At the end, Regular and flipwell architectures allow to offer two Vt flavors separated by ~70mV (LVT and SLVT, respectively) while keeping the channel undoped and without adding any mask. One drawback is that these two Vt flavors

978-1-4799-7670-6/15 $31.00 © 2015 IEEE

are not easily mixable since LVT and SLVT transistors are not sharing the same wells. A process solution for mix-Vt inside these two well architecture families is then presented later in section V. One major advantage is that these two well architectures provide a large extend of the back biasing capability, as described in the next section.

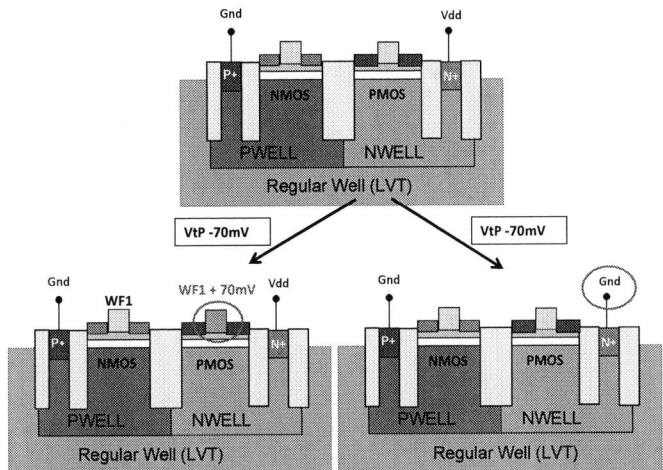

Fig.6: PMOS optimization that can be done in regular well architecture to recover the N/P Vt equilibrium.

IV. REVERSE AND FORWARD BACK BIASING (RBB AND FBB)

Well types are adapted to use either a large reverse back bias (RBB) on regular well LVT architecture minimizing leakage, or a large forward back bias (FBB) on flipwell SLVT architecture maximizing speed. Fig.7 illustrates the large delay/Istat modulation provided by 1.2V RBB on regular well LVT and by 1.2V FBB on flipwell SLVT devices. No additional well straps are required to apply back bias, it is only needed to have bias generators, same as for bulk technology when body bias is used. Since transistor channels are isolated from wells by the BOX, there is no RBB or FBB limitation coming from source-drain to well junction leakage. RBB and FBB ranges in LVT and SLVT can easily be extended to the breakdown voltage of the N/P well diode in reverse mode or to about 5V.

By tuning the Vt between the slow and fast process corners, back bias can be used for process compensation, but also as a solution to maximize the power efficiency of a product while maintaining the same speed. Indeed, Fig.8a shows that devices running at $V_{dd} \sim 0.63V$ with a 2V FBB are as fast as devices running at $V_{dd}=0.8V$ with no back bias, reflecting a 40% dynamic power saving at same speed. Of course, because FBB is associated with Vt lowering, static leakage strongly increases at the same time, as show in Fig.8b. But, with this flexibility offered by the large back bias capability, FDSOI appears as a powerful technology to find the right performance at the right power and at the right leakage for each application product.

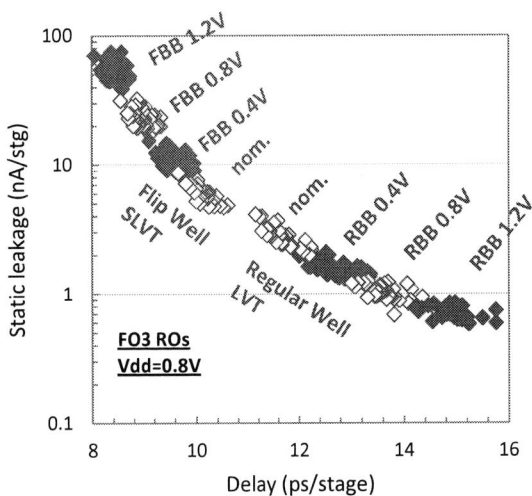

Fig.7: Delay vs I_{stat} for regular well LVT and flipwell SLVT nominal devices with back bias application.

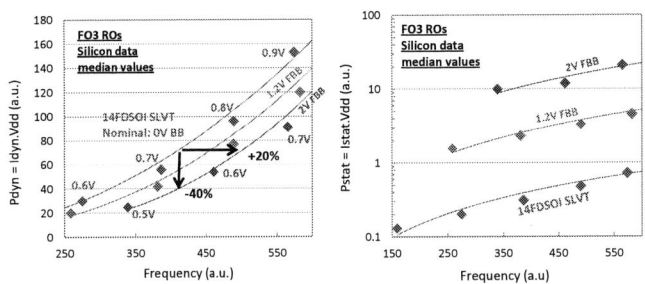

Fig.8a (left): RO Frequency vs P_{dyn} for various V_{dd} and various FBB up to 2V. Fig.8b (right): RO Frequency vs P_{stat} for various V_{dd} and various FBB up to 2V.

V. CHANNEL LIGHT DOPING FOR VT MIXABILITY

The introduction of a light channel doping has been evaluated in order to add a second Vt flavor in each device family (regular well and flipwell), as shown in Fig.9.

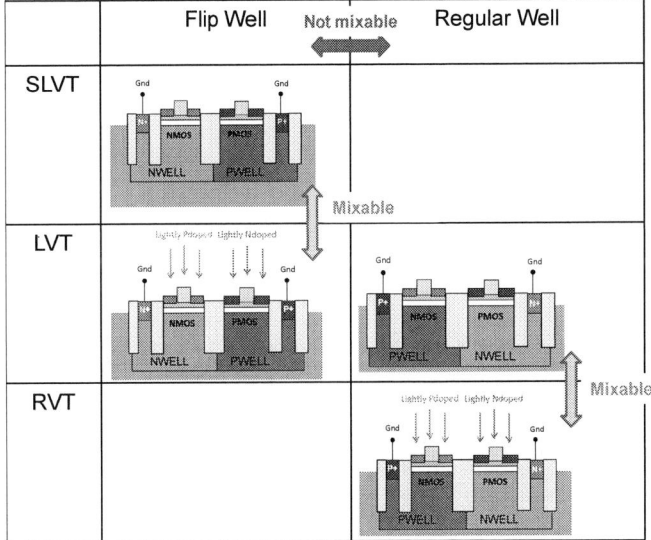

Fig.9: Table illustrating the Vt flavor and the Vt mix capability that can be allowed in FDSOI.

978-1-4799-7670-6/15 $31.00 © 2015 IEEE   61

As a result, we created a new LVT device in the flipwell family which is mixable with the SLVT one, and a new RVT device in the regular well family which is mixable with the previous LVT device described in section III (Fig.9). Combined with an adequate poly bias mapping, this Vt mixability strongly extends the leakage/speed optimization of the FDSOI critical paths.

Fig.10 presents the I_{eff}/I_{off} tuning for both NMOS and PMOS nominal devices after light channel boron and phosphorous implants, respectively. More than 1 decade leakage reduction can be covered by adjusting implant conditions.

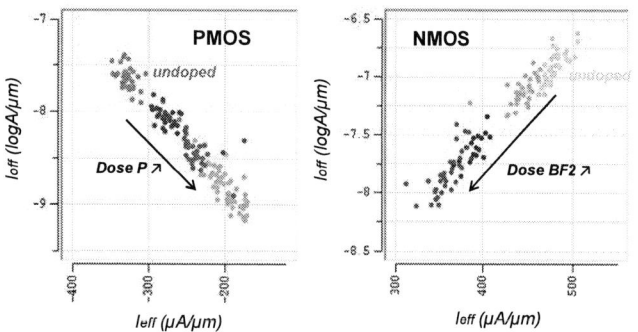

Fig.10: Ieff/Ioff for nominal NMOS and PMOS transistors at V_{dd}=0.8V after applying a light channel doping implantation.

Vt mismatch for devices with the highest dose in Fig.10 has been compared to undoped channel transistors in Fig.11. A moderate degradation from AVt=1.1mV.µm up to AVt=1.4mV.µm is observed after channel implantation but these AVt values remain highly competitive compared to bulk technologies.

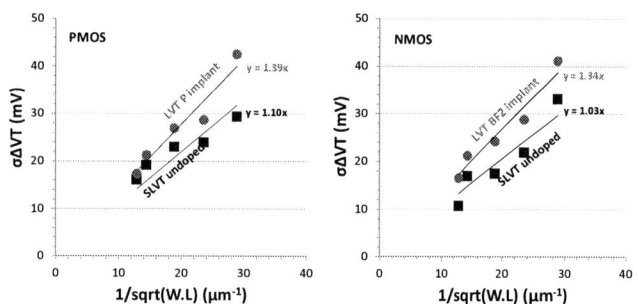

Fig.11: PMOS (left) and NMOS (right) Vt mismatch pelgrom plots for doped channel FDSOI devices compared to undoped transistors.

VI. CONCLUSION

The leading edge FDSOI technology presented in this paper illustrates the scalability from the 28nm to the 14nm node, with -34% speed delay enhancement demonstrated at same static leakage and with -100mV V_{dd} operation. Process & design solutions to tune the transistors threshold voltage (Vt) in 14nm FDSOI technology for managing static and dynamic power in a large range has been presented. This work demonstrates the suitability of FDSOI technology for optimizing the speed/leakage/power for each application product.

REFERENCES

[1] C.-H. Jan et al., "A 22nm SoC Platform Technology Featuring 3-D Tri-gate and High-k/Metal Gate, Optimized for Ultra Low Power, High Performance and High Density SoC Applications", IEDM Tech. Dig., p.44, (2012).

[2] S.-Y. Wu et al., "A 16nm FinFET CMOS Technology for Mobile SoC and Computing Applications", IEDM Tech. Dig., p.633, (2013).

[3] N. Planes et al., "28nm FDSOI Technology Platform for High-Speed Low-Voltage Digital Applications", Symposium on VLSI Technology, 2012, pp. 133-134.

[4] Press release, EETimes, Feb 2013.

[5] O. Weber et al., "14nm FDSOI Technology for High Speed and Energy Efficient Applications", VLSI Symposium Tech. Dig., (2014),

[6] F. Arnaud et al., "Switching Energy Efficiency Optimization for Advanced CPU thanks to UTBB Technology", IEDM Tech. Dig., p.3.2.1, (2012).

[7] D. Golanski et al., First demonstration of a Full 28nm high-k/metal gate circuit transfer from Bulk to UTBB FDSOI technology through hybrid integration", VLSI Symposium Tech. Dig., p.124, (2013).

Lateral NWFET Optimization for Beyond 7nm Nodes

D. Yakimets[1], D. Jang, P. Raghavan, G. Eneman, H. Mertens, P. Schuddinck, A. Mallik,

M. Garcia Bardon, N. Collaert, A. Mercha, D. Verkest, A. Thean, K. De Meyer[1]

imec, Kapeldreef 75, 3001 Leuven, Belgium   [1] also with KU Leuven, Dept. ESAT, Belgium

dmitry.yakimets@imec.be

Abstract—**In this study, different S/D contacting options for lateral NWFET devices are benchmarked at 7nm node dimensions and beyond. Comparison is done at both DC and ring oscillator levels. It is demonstrated that implementing a direct contact to a fin made of Si/SiGe super-lattice results in 13% performance improvement. Also, we conclude that the integration of internal spacers between the NWs is a must for lateral NWFETs in order to reduce device parasitic capacitance.**

Keywords—**Lateral gate-all-around FET, nanowire, scaling, design technology co-optimization (DTCO).**

I. INTRODUCTION

Although FinFETs have been enablers for CMOS scaling at the last technology nodes [1], scaling down to the 7nm node will be a challenge. Gate-all-around or nanowire (NW) FETs are the candidates for the ultimate CMOS scaling due to their excellent electrostatics integrity [2]. The channel in the NWFETs may be oriented both horizontally and vertically. Although the latter configuration has certain advantages, it is considered as very disruptive, as major changes should happen at both technology and design sides [3], [4]. Therefore, it is desired to try to optimize lateral devices. They may be made of several stacked NWs to boost the performance. However, such approach causes issues discussed in the paper. One is related to the formation of the spacer between stacked NWs, while another concern is related to the S/D contacting efficiency.

Different technological options for lateral NWFETs are compared in this study by running DC and RO simulations. The focus of the paper is the 5nm technology node. However, as NWFETs may be introduced already at 7nm technology, some of the results are discussed across these two nodes.

II. TECHNOLOGY DESCRIPTION

We assume that NWFETs channels are made of Si for NMOS and $Si_{0.5}Ge_{0.5}$ for PMOS. Both devices use a common strain relaxed buffer (SRB) of $Si_{0.75}Ge_{0.25}$. Such a Ge content in the SRB results in about 1.5 GPa strain level for both NMOS and PMOS [5]. The simplified process flow is shown in Fig. 1.

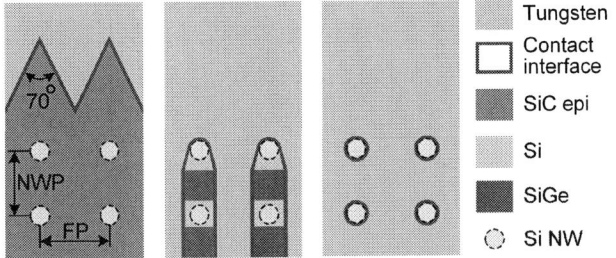

Fig. 2. Explored contacting options. Left to right: contact to an epitaxially grown S/D structures ("Epi" in text), contact to a "fin" made from the Si/SiGe superlattice ("Lattice" in text), and direct contact to NWs ("NW" in text). The illustration is for NMOS, but the same principles apply to PMOS as well.

The focus of this study is on two particular process steps: S/D stressor epitaxy and the internal spacer formation. It is common to use epitaxially grown S/D structures as they introduce strain in the channel and increase the contacting area. However, the direct (wrap) contact may be a better option at tight fin pitches (FP) and tall fins (or several stacked NWs). Several contacting options are compared in this study (Fig. 2).

In case the process flow is followed without internal spacer formation, the gate stack appears in between the stacked NWs instead of spacer (Fig. 3), which significantly increases the parasitic capacitance. The necessity of having a spacer between the NWs is discussed in this paper.

The key design parameters for the 7nm and 5nm technology nodes are given in Table I. The contacts to S/D are made of tungsten. Its resistivity increases exponentially with decreasing the cross section (Fig. 4a). For the back end of line (BEOL), the low-k ($\varepsilon_r = 2.1$) copper dual damascene process is assumed with 1 nm metal barrier. The BEOL load is considered at the ring-oscillator (RO) level by loading it with 100CGP-long (Contacted Gate Pitch) wire (Fig. 4b).

- Bulk Si wafer with SRB on top
- Well implantations
- SiGe/Si epitaxy
- Fin patterning
- Low-temperature STI fill
- Dummy gate patterning
- Extension doping
- Spacer deposition
- Spacer etch and fin recess
- Etch stop liner epitaxy
- SiGe (SiC) S/D epitaxy
- ILD0 deposition and CMP
- Poly removal
- Sacrificial layer etch
- Internal spacer fabrication
- HK/MG fill
- Silicidation + Contacts
- BEOL

Fig. 1. Process flow for NWFETs.

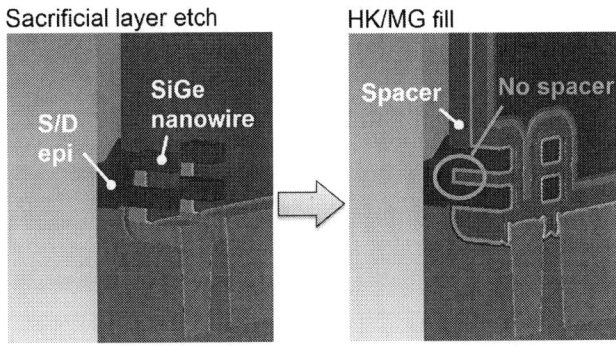

Fig. 3. Process flow emulation indicating an issue of internal spacer absence.

TABLE I. PROCESS ASSUMPTIONS FOR 5NM AND 7NM NODES

	7nm	5nm
Gate length, L_G (nm)	14-18	14
NW diameter (nm)	7	7
Metal pitch (MP) (nm)	32	24
Contacted gate pitch (CGP) (nm)	42	32
Spacer thickness (Tsp) (nm)	4-7	4-5
V_{DD} (V)	0.65	0.6
R_{BE} (Ω/μm)	136	317
C_{BE} (aF/μm)	161	153

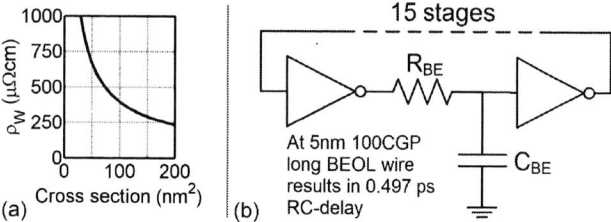

Fig. 4. (a) Tungsten resistivity dramatically increases with conductor cross section reduction. (b) Schematic representation of the INV-based 15 stages RO loaded with 100CGP-long BEOL wire used for benchmarking.

Cell heights are assumed to be nine or ten tracks (a single track corresponds to one metal pitch (MP)). Fin pitch, the number of fins per devices and MP choices are tightened together as they should be on the grid. The possible FP and MP combinations considered in this paper are shown in Table II.

III. MODELLING METHODOLOGY

All the simulations were performed in SPICE with the BSIM-CMG compact model used for device modelling [6]. The compact model was fitted to the TCAD data for electrostatics and drive. For all the simulations, the off-current was fixed to 10 nA by adjusting the work function. The device RC-parasitics were assessed with the 3D finite element modeling software.

A conventional drift-diffusion (DD) model has described the carrier transport well until the 14nm technology node. However, beyond 7nm node, the quasi-ballistic (QB) transport cannot be neglected. To take into account the impact of QB transport, Synopsys Sentaurus Band Structure [7] simulations were performed to determine the fully ballistic current as a function of applied voltage and stress in the channel. Then, these currents were multiplied by the ballistic ratio from [8] to calculate the actual current for different stress levels and channel lengths.

The ballistic ratio was assumed to be independent of the channel materials (Si, SiGe) due to small differences between them [5]. The stress levels were varied in the range from 1 GPa to 2 GPa (tensile for NMOS and compressive for PMOS) since in addition to the 1.5 GPa stress from the SRB, we can expect some extra stress to come from embedded S/D epi or Si/SiGe super-lattice depending on the contact options.

TABLE II. FIN PITCH – NUMBER OF FINS PAIRS FOR 5NM AND 7NM NODES

	7nm		5nm			
Fin pitch (nm)	24	32	18	20	24	27
9 tracks	4 fins		4 fins			2 fins
10 tracks		3 fins		4 fins	3 fins	

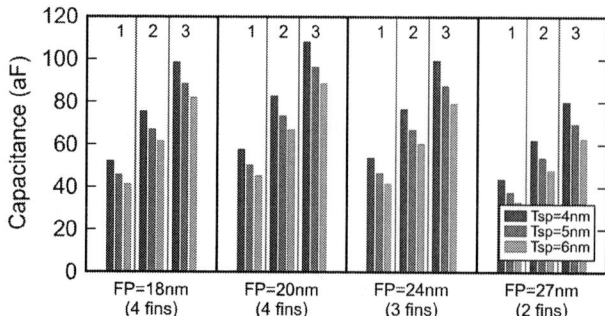

Fig. 5. Parasitic drain capacitances for different devices at the 5nm node. Digits on the top indicate the number of stacked NWs. The efficiency of spacer thickness relaxation drops with increasing number of NWs. NW pitch is 18 nm. Spacer relative permittivity $\varepsilon_r = 5.5$.

The extracted data were used to calibrate the compact model by proper adjustment of the low-field mobility and saturation velocity. Instead of conventional long-channel mobility, the low-field mobility was replaced by the apparent mobility due to ballistic mobility reduction and additional scattering mechanisms [9]. It can be computed with Eq. 1.

$$\frac{1}{\mu_{app}} = \frac{1}{\mu_{long}} + \frac{\alpha_\mu}{L_G},\qquad(1)$$

where μ_{long} is a low field mobility in the long-channel and α_μ is the mobility degradation factor. The μ_{long} was assumed to be 300 cm^2/Vs and 400 cm^2/Vs for n-Si and p-SiGe NW-FETs under the maximum stress conditions (i.e. +2 GPa for NMOS and –2 GPa for PMOS) and the α_μ was assumed to be 0.07 nmVs/cm^2 for a relaxed channel and 0.05 nmVs/cm^2 for a stressed channel ($\pm$2 GPa), as α_μ decreases with stress [10-12].

IV. PARASITICS ANALYSIS

Parasitic capacitances are independent from the choice of contacting option, yet parameters like spacer thickness (Tsp) or number of NWs are of importance. As the FP relaxation comes together with the reduction of number of fins (Table II), the parasitic capacitances remain similar for all the devices (Fig. 5). Capacitance increases linearly with every extra stacked NW.

In case there is no internal spacer in the device (Fig. 3), parasitic capacitances increase by 40 to 60 percent, depending on the geometry. This penalty increases with the number of stacked NWs (Fig. 6). There would be no difference between 5nm and 7nm technologies, as, most likely, the NW pitch (NWP) would vary together with the fin pitch.

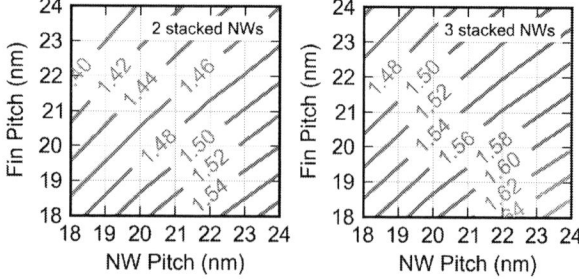

Fig. 6. Ratio of parasitic capacitances in case there is no internal spacer between the NWs, and in case there is a spacer between them.

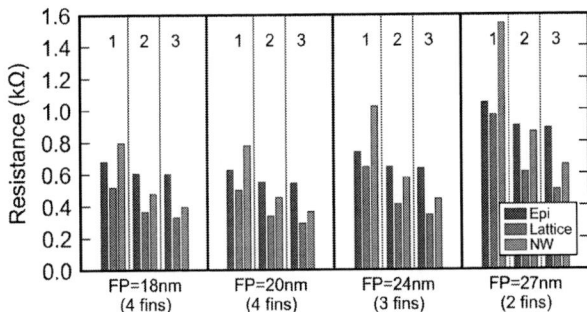

Fig. 7. Parasitic drain access resistance for different NMOS devices at the 5nm node. Digits on the top indicate the number of stacked NWs. Direct contact to a "fin" made from the Si/SiGe superlattice ("Lattice") is the most efficient option. Same holds true for PMOS. Direct contact to the NWs ("NW") brings benefits only in case of several stacked NWs. NW pitch is 18 nm, spacer thickness is 4 nm.

Comparison of access resistance for different contacting options is shown in Fig. 7. For all the cases, we assumed a specific contact resistivity of $5 \cdot 10^{-9}$ Ωcm^2 [13]. The assumed active doping concentration in S/D regions was 3E20 cm^{-3} for both NMOS and PMOS. PMOS data are similar to NMOS, but the actual numbers are slightly different due to various S/D resistivities. First trend to notice is that resistance wise it is better to have more tightly spaced fins rather than fewer of them at relaxed pitch. Second, direct contacting to the epi limits the number of stacked NWs: after two stacked NWs access resistance is limited by contact resistance. Similar effect happens for the wrapped contacts, but the saturation happens later and it is fin pitch dependent. The reason behind is that the tungsten resistance between the "fins" is very high for a tight fin pitch and for several stacked NWs it becomes comparable to the contact resistance. Third, contact to a "fin" made of a Si/SiGe super-lattice is the best option for all the devices and it outperforms the direct contact to NWs because of the increased contacting area.

However, direct contact to a "fin" means that the super-lattice in the S/D regions remain intact and thus, due to the alternating Si/SiGe layers it results in undesired stress. Similar, in the case of embedded epi structures in the S/D regions additional stress boosting is expected. Stress levels from either the super-lattice or the S/D epi should be at least three times lower than the stress from SRB, due to their small volume [10].

V. IMPACT OF CONTACT ON DEVICE PERFORMANCE

The impact of different contacting schemes on the drain saturation current is shown in Fig. 8. Not only the access resistance is the highest for the loosely placed fins, but also the current, because of the smaller number of channels.

The best performance is for a 10-tracks cell made with four fins per device at a fin pitch of 20 nm. For all the cases except the case of the most relaxed fin pitch (27 nm), a device with three stacked NWs and a contact landing on the S/D epi are the worst, even in case the most optimistic stress boosting (+500 MPa) is assumed. In case of two stacked NWs contact to the epi is still the worst. However, in case embedded epi can add 500 MPa of stress to the channel and Si/SiGe superlattice degrades the stress by 500 MPa, all the contact options yield roughly the same saturation current.

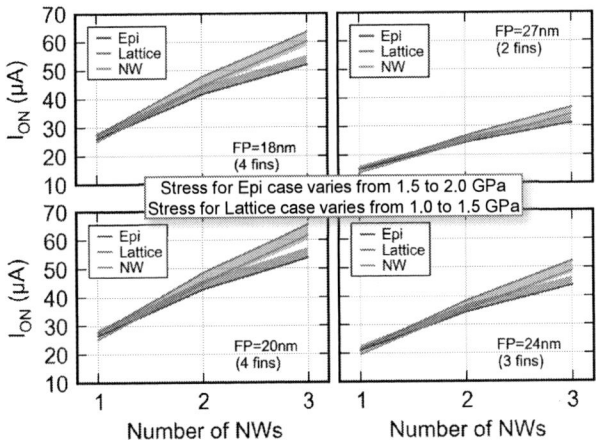

Fig. 8. Impact of different contact options on saturation current for different NMOS 5nm node devices. Bold lines correspond to 1.5 GPa stress (from SRB only). Vertical NW pitch is 18 nm, spacer thickness is 4 nm.

Fig. 9a shows the impact of vertical NW pitch on device performance. In case of contacting to a super-lattice based "fin", the current does not change, as improvements in contact resistance (due to increased contacting surface) are balanced with the degradation in metal resistance. In case of relaxed FP (27 nm) the saturation current for the "Lattice" contact option increases (not shown here). The use of "Epi" is rather insensitive to the NW pitch, because the increase in epi resistance is not really visible with respect to a contact resistance.

VI. RING OSCILLATOR PERFORMANCE

Absence of the internal spacer increases the parasitic capacitances by 50% (Fig. 6). This directly reflects in the RO performance being degraded by 13 to 23% depending on the geometrical configuration (Fig. 10). On top of that, power consumption increases. Thus we conclude that having an internal spacer is a must for NWFETs. The rest of the results will be presented for the case with an internal spacer between the NWs. The RO performance degradation is less than the parasitic capacitance difference. This may be explained by the extra load from the BEOL wire, which somewhat screens the FEOL issues.

NW pitch has almost no impact on the device saturation current, however, the RO is quite sensitive to it. The relative performance of devices with different contacting options matches the DC data (Fig. 9b).

Though in section V we demonstrated that the current constantly increases with every new stacked NW it is not the case for the RO (Fig. 11). Its performance drops for three stacked

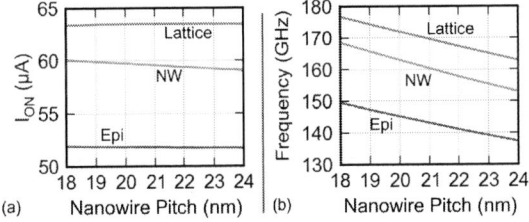

Fig. 9. Impact of vertical NW pitch on (a) saturation current (from NMOS); (b) RO frequency. Spacer thickness is 4 nm, stress is 1.5 GPa, fin pitch is 18 nm, three NWs stacked.

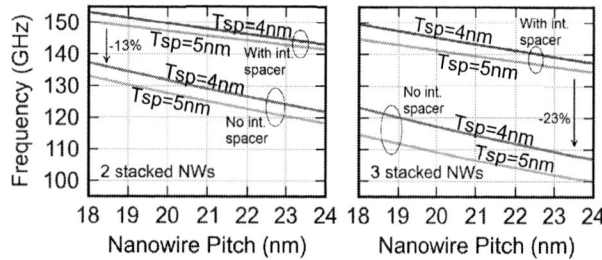

Fig. 10. Impact of vertical NW pitch on the performance of the RO made with devices having and not having internal spacer between the NWs. Stress is 1.5 GPa, fin pitch is 18 nm, epi-contact is used.

NWs if the contacting to epi is used. This may be explained by analyzing the data in section IV. Resistance improvements with every new stacked NW are minor and saturate, especially at tight FP, while the capacitance keeps increasing linearly. The tendency for saturation is also visible for the wrap contacts. Best RO results are achieved for the 10-tracks cells, 4 fins configuration (FP = 20 nm) and a direct contact to a super-lattice. However, performance drop for going from 10-tracks to 9-tracks is very small and area-wise it would probably be better to use a 9-tracks library. Also, in case the super-lattice generates "bad" stress in the channel (500 MPa), a direct contact to NWs may be used: the results are very similar.

Although for the three stacked NWs and "Epi" contact case, the RO performance gets penalized because of high FEOL capacitance, increasing the spacer thickness from 4 nm to 5 nm does not result in performance improvement, meaning that the bottleneck is still the access resistance (Fig. 10).

VII. DISCUSSION ABOUT 7NM TECHNOLOGY NODE

Gate pitch at the 7nm technology node is relaxed which leaves some room for spacer thickness and gate length optimization. If we assume a 5nm node gate length (14 nm, as it yields subthreshold slope right below 70 mV/dec), we can better trade off the FEOL RC-parasitics. Relaxed FP favours the "Epi" contact. At the same time, because of the relaxed NWP, "NW" contact is always worse than the "Lattice" contact (assumed NW diameter remains same, 7 nm). Therefore, for the 7nm node we

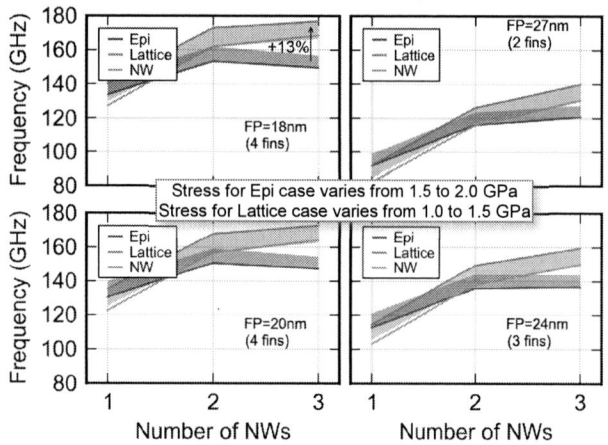

Fig. 11. Impact of different contact options on RO performance for different 5nm node devices. Bold lines correspond to 1.5 GPa stress (from SRB only). NW pitch is 18 nm, spacer thickness is 4 nm.

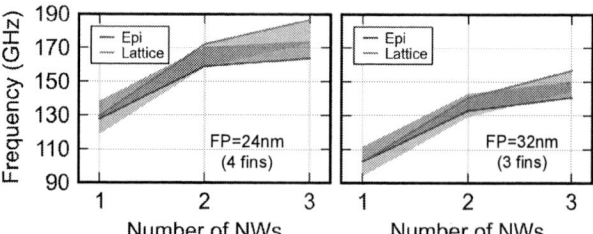

Fig. 12. Impact of different contact options on RO performance for different 7nm node devices. Bold lines correspond to 1.5 GPa stress (SRB only). The NW pitch is 24 nm, spacer thickness is 4 nm. Stress for the "Epi" case varies from 1.5 to 2.0 GPa and for the "Lattice" case it varies from 1.0 to 1.5 GPa.

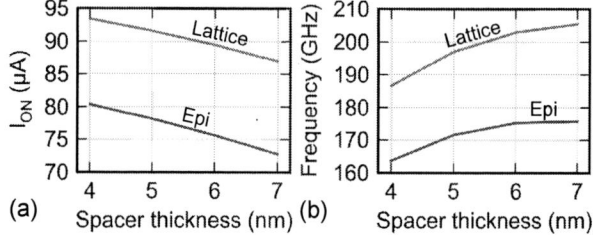

Fig. 13. Tsp impact on (a) the I_{ON} current from the 7nm NMOS devices and (b) the RO performance. Stress is 1.5 GPa, fin pitch is 24 nm, three NWs stacked.

demonstrate results only for "Lattice" and "Epi" contacts. The RO performance as a function of the number of stacked NWs is shown in Fig. 12. The impact of spacer thickness is shown in Fig. 13: at 7nm node the co-optimization of FEOL RC-parasitics is possible with the spacer thickness relaxation.

VIII. CONCLUSION

We compared device designs at the 5nm technology featuring different contacting options. The standard approach relies on epitaxially grown S/D structures. However, at tight fin pitches better performance is achieved in case the direct contacting to a Si/SiGe super-lattice is used, even in case there is some undesired stress from the super-lattice in the channel. For the relaxed fin pitch the difference between the contacting options gets smaller, but the fin pitch relaxation means fin depopulation, which has a negative impact on performance.

Also, certain integration methods of internal spacers between the stacked NWs should be developed, because the presence of gate stack between the NWs penalizes the performance by 20%.

REFERENCES

[1] S. Natarajan, et al., *IEDM Tech. Dig.*, p. 71, 2014.

[2] K. J. Kuhn, *IEEE Trans. Electron Devices*, vol. 59, no. 7, Jul. 2012.

[3] D. Yakimets, et al., *Device Research Conference (DRC)*, p.133, 2014.

[4] T. H. Bao, et al., *ESSDERC*, p. 102, 2014.

[5] W. Guo, et al., *IEDM Tech. Dig.*, p. 168, 2014.

[6] http://www-device.eecs.berkeley.edu/bsim/?page=BSIMCMG

[7] Synopsys, Sentaurus Band Structure User Guide, I-2013.12.

[8] V. Moroz, et al., *IEDM Tech. Dig.*, p. 180, 2014.

[9] G. Bidal, et al., *Silicon Nanoelectronics Workshop (SNW)*, p. 25, 2009.

[10] G. Eneman, et al., *IEDM Tech. Dig.*, p. 131, 2012.

[11] J. L. Huguenin, et al., *Solid. State. Electron.*, vol. 54, no. 9, p. 883, 2010.

[12] K. Huet, et al., *ESSDERC*, p. 382, 2007.

[13] ITRS 2013. http://www.itrs.net/

Nonparabolicity and confinement effects of IIIV materials in novel transistors

M. Ali Pourghaderi, Anda Mocuta and Aaron Thean

IMEC, B-3001, Kapeldreef 75, Heverlee, Belgium.
E-mail – pourgham@imec.be, Ph: +32-16-28-8779, Fax: +32-16-28-94-00.

Abstract

Employing a 8 band k.p solver, the self-consistent band structure of the rectangular IIIV nanowires (NW) has been calculated. It is shown that the strong confinement combined with the band nonparabolicity will considerably change the effective masses and the band gap. The mass tensor elements get heavier than the bulk values and improve the density of state (DOS) and centroid capacitance accordingly, while in return the mobility will be degraded. The band widening has also been calculated for different width and height combinations. It is shown that oxide thickness scaling cannot compensate the poor DOS of IIIV, where the silicon device exhibits a continuous performance boost by thinning the oxide layer. Possible improvements of DOS through the width and the mole fraction modulation have been investigated.

Introduction

To meet the performance-power need of upcoming technologies (7nm and 5nm), high-mobility materials like IIIV and Ge are being evaluated as transistor channel materials. Besides transport enhancement, devices targeted for these technologies will require good electrostatics, which imply the need for strong confinement across the channel. However, the 2-D quantization effects relevant to 7nm/5nm device structures (FinFET/Nanowire) are not well understood for IIIV channel material. Key concerns are the geometrical scaling influences on band gap, DOS, the transport mass and the quantum dark-space, which will directly influence the device performance and possibly change the way to scale gate capacitance of these devices.

Tight-binding calculation has already confirmed that effective mass drastically deviate from bulk values[1]. In essence, the initial assessment based on bulk properties needs to be readjusted to account for strong confinement effects. This simulation study has exploited the 8-band k.p to address the confinement issue of IIIV NW .

8-band k.p solver

Within the operational range of transistors, a 8-bands k.p solver can capture the band mixing and nonparabolicity effects of IIIV materials. As it comes to the calculation, we are dealing with a coupled partial differential equation with closed and open boundary conditions in the confinement and the transport direction, respectively[2]. A finite element solver has been developed for an arbitrary 2D cross section. This solver calls the Arnoldi subroutines from ARPACK library to handle the corresponding generalized-eigen-value problem[3]. To avoid the spurious solutions, the momentum matrix element has been calibrated to recover the desired elliptic form of the equations [4] and rather than enforcing the symmetrization procedure, the correct ordering of the operators has been adopted[5]. The electron density is calculated assuming the Fermi-Dirac statistics and coupled to the Poisson solver through the Gummel-Iteration Scheme[6].

Band structure of the confined systems

The k.p parameters used in this study has been listed in table.1. As it is shown in Fig.1, the correct setting for momentum matrix parameter and the order of the operators result in a spurious-free solution. Depicted band structure in Fig 1 predicts a considerable band gap widening of 150 meV for 7x20 nm $In_{0.53}Ga_{0.47}As$ wire. To study the impact of the dimension on band gap widening, we run a set of simulations for various pairs of width and height listed in table.2. The simulation results in Fig.2 and 3 suggest that the band gap widening is asymmetric on conduction and valance band side and it is more pronounced for materials with lighter mass. Fig.4 shows the effective mass in the transport direction for the first four conduction bands of InGaAs NW. Due to the band nonparabolicity the effective mass linearly increases as a function of energy. The parabolic approximation of the conduction band is just applicable to a small region at the vicinity of the conduction band minima. The aggressive scaling of the wires will shift up the energy of states away from the parabolic region, therefor the nonparabolicity effects will be well pronounced. Fig 5 and 6 illustrate the contrasts of the parabolic and nonparabolic bands. For parabolic band solver, a fixed effective mass

approximation (EMA) is used. Due to nonparabolicity the confinement energy decreases, the transport mass gets heavier and DOS is significantly boosted. In essence, nonparabolicity of the bands will add to the gate capacitance but in return the heavier mass degrades the mobility. Fig 7 illustrates the effective mass of the first conduction band for different materials. The corresponding nonparabolicity factor is reported in table 3. As predicted by the theory, the nonparabolicity factor is inversely proportional to the band-gap[7].

To study the band mixing effect we simulate the $In_{0.53}Ga_{0.47}As$-InP HEMT structure in Fig.8. Fig9-11 show that for high energies the thin quantum well is not effective and perhaps we should increase the barrier height. As it is shown in Fig.12 for the narrow channels the effective mass increases due to a larger wave function penetration in InP layer. The overall capacitance for different channel thicknesses is illustrated in Fig. 13 concluding higher capacitance for thinner channels.

Quantum confinement and nonparabolicity effects on CV

The gate capacitance is the series combinations of oxide and DOS capacitance. The bottom of conduction band is coupled to the gate terminal through the C_{OX} , while C_{DOS} expresses the rate of stored charge with respect to Fermi level increments. Fig 14 and 15 show the capacitive components of Silicon and InGaAs NW. In the case of Silicon, C_{DOS} is much larger than C_{OX}, therefor the gate capacitance is dominated by C_{OX}. This means that we can directly improve the transconductance by thinning the oxide layer. Contrary to Silicon, for InGaAs device C_{OX} is larger than C_{DOS} and the gate capacitance is limited by the latter one. This indicates a weak impact of the oxide scaling on the transconductance, where anyhow the poor DOS will degrade the overall capacitance. Fig 16 compares the gate capacitance of Silicon and InGaAs NW together. Although the nonparabolicity largely enhances the InGaAs capacitance, almost 3 times, still the gate capacitance is 4 times smaller than Silicon. This means that for an order of magnitude boost in the performance the mobility of InGaAs should be at least 40-30 times higher than Silicon. Fig. 17 compares the typical electron density profile of Silicon and InGaAs for 7x20 nm NW. The volume inversion pattern is recognized for InGaAs, while Silicon still performs in surface inversion mode. Perhaps the surface roughness wont hamper the InGaAs device as it may do in Silicon case.

Fig.18 shows that changing the mole fraction can slightly affect the CV. The heavier mass of GaAs gives slightly higher DOS, where on the other end the large nonparabolicity factor of InAs increases the mass at high bias points and shrinks the gap with the reference case of InGaAs.

The width of the wire is a key parameter for both the DOS and electrostatic control. The conduction bands get closer to each other in energy domain for wider wires. Fig.19 shows the DOS capacitance for various widths. For the narrowest wire the capacitance is slightly higher at low surface potentials, which is the consequence of the strong quantization and nonparabolicity. Since the conduction bands of the wider wire are closely spaced, the wider wires engage more bands at higher surface potentials and their DOS will be increased. The overall capacitance is depicted in Fig. 20. The wider wire gives a higher capacitance and a lighter mass in the transport direction but in the same time the electrostatic will be aggravated.

Conclusion

Considering the band nonparabolicity of InGaAs, the gate capacitance is estimated to be quarter of Si NW (7x20nm). These results lead to a quantitative criteria on mobility boost for a given performance boost.

References:

[1] S. H. Park *et al.*IEEE trans. Electron Devices **59** (8), 2107-2114 (2012).

[2] O. Stier and D. Bimberg, Phys. Rev. B **55** (12), 7726-7732 (1997).

[3] http://www.caam.rice.edu/software/ARPACK/

[4] R. G. Veprek *et al* Phys. Rev. B **76** (16), 165320(1-9)(2007).

[5] B. A. Foreman, Phys. Rev. B **56, R12748**(1997).

[6] H. K. Gummel, IEEE Trans. Electron Devices, **ED-11**, 455-456(1964)

[7] V. A. Altschul *et al*, J. Appl. Phys. **71**(9)4382-4384 (1992).

	GaAs	In$_{0.53}$Ga$_{0.47}$As	InAs	InP
γ_1	6.98	11.01	20	4.95
γ_2	2.09	4.18	8.5	1.65
γ_3	2.93	4.84	9.2	2.35
E_g(meV)	1520	750	420	1420
Δ(meV)	340	356	380	110
m_{el}/m_0	0.067	0041	0.026	0.079
E_p(meV)	20	18.9	18	16

Table 1- The k.p parameters have been used in this study.

Width(nm)	Height(nm)
5	20
6	20
7	20
8	23
9	25
10	27
11	30

Table 2. Height and width matrix for NW at Fig.2 and 3. The EOT for these devices is set to 0.8 nm and rounding radius is fixed as 2nm.

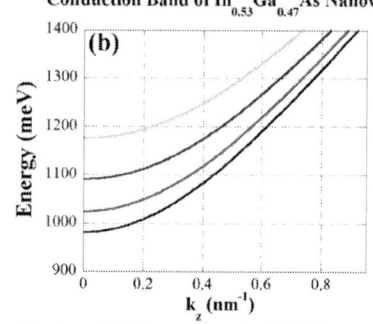

Figure 1. The valance and conduction bands of undoped-7x20 nm nanowire. EOT is 0.7 nm and Fermi level is set at the mid gap. The band gap of 900 meV is deduced.

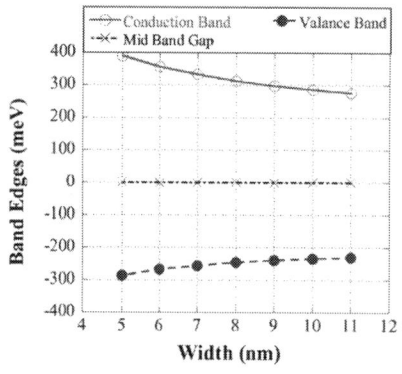

Figure 2. The band gap widening for InAs NW. since the electrons are lighter than holes the confinement effect is more pronounced on the conduction band.

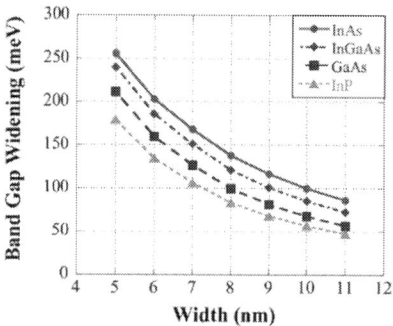

Figure 3. The increase of the band gap as a function of NW width for different IIIV materials.

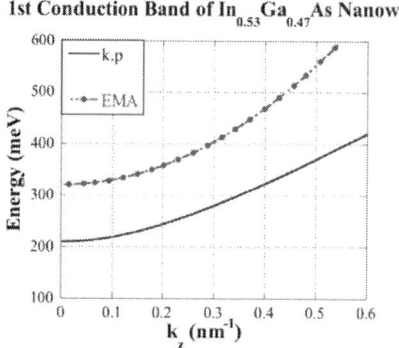

Figure 4. The effective mass for the first 4 conduction bands of 7x20 nm In$_{0.53}$Ga$_{0.47}$As NW. The electron density is fixed at 10^{18} cm^{-3}.

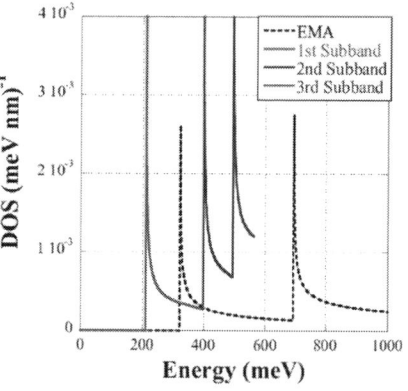

Figure 5. The first conduction band of 7x9 nm NW. The nonparabolicity effect will increase both the confinement and the transport mass.

Nano-Wire DOS: k.p vs. EMA

Figure 6. The DOS comparison of parabolic and nonparabolic bands for 7x9 nm In$_{0.53}$Ga$_{0.47}$As NW. Nonparabolicity will add considerably to DOS of IIIV.

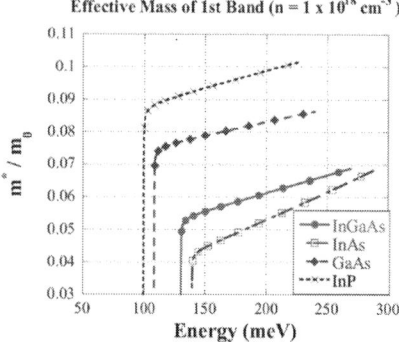

Figure 7. The effective mass for 7x20 nm NW. The electron density is set as 10^{18} cm^{-3} and EOT is 0.7 nm.

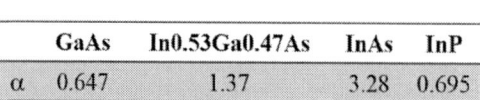

	GaAs	In0.53Ga0.47As	InAs	InP
α	0.647	1.37	3.28	0.695

Table 3. the nonparabolicity factor for IIIV materials. The mass has been fitted as :

$$m(E) = m_{el}(1 + 2\alpha E)$$

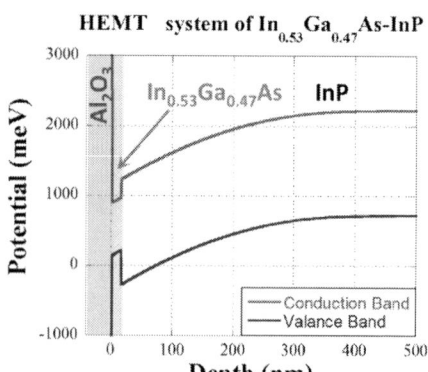

Figure 8. The cross section of HEMT system. InP layer is lightly p-doped with concentration of 10^{16} cm^{-3} and its band offset is 260 meV.

978-1-4799-7670-6/15 $31.00 © 2015 IEEE

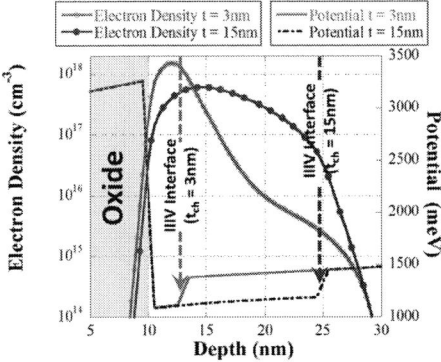

Figure 9. Electron density and potential profile for 3 and 15nm channel thickness. The charge integral is fixed as 5×10^{11} cm^{-2} and EOT = 5.2nm. carriers largely penetrate in InP layer for 3nm channel.

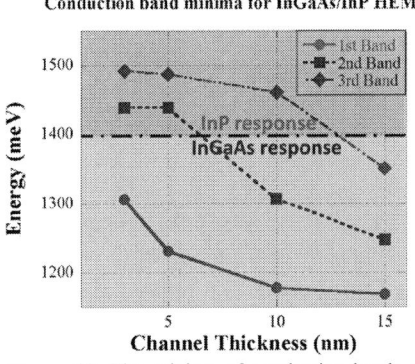

Figure 10. The minima of conduction bands for various channel thicknesses. In the case of high energy states, carriers will respond to InP system and do not see the quantum well at all. As the result their sensitivity to InGaAs thickness is faded away.

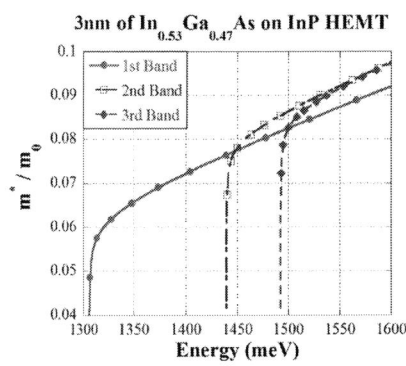

Figure 11. The effective mass for 3nm InGaAs channel thickness. The carriers of second and third bands are not effectively confined in quantum well and mainly interact with InP lattice.

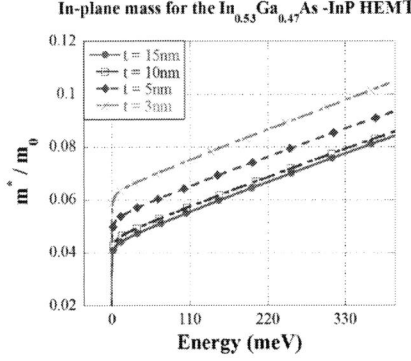

Figure 12. The transport effective mass at off-state for different channel thicknesses. For thin channels the wave function largely penetrates in InP and the band mixing is effective, while for the thick channels the in-plane mass is just similar to the bulk.

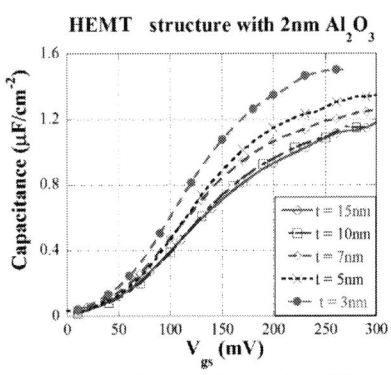

Figure 13. The capacitance for different channel thicknesses. The high density of state for thin channel devices will enhance the gate capacitance.

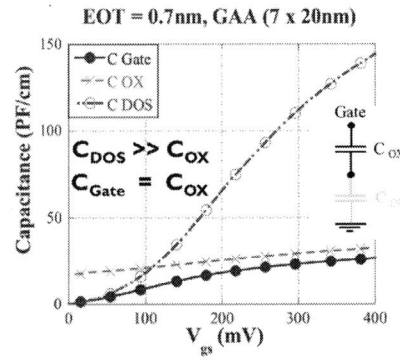

Figure 14. Capacitance decomposition for Silicon NW. The overall capacitance is limited by oxide capacitance.

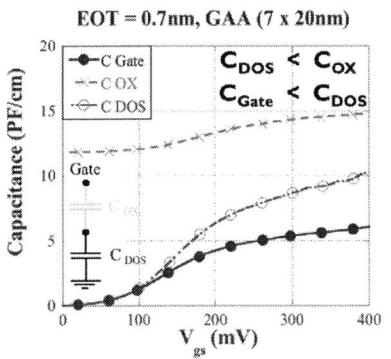

Figure 15. Capacitance decomposition for InGaAs NW. The overall capacitance is limited by DOS capacitance.

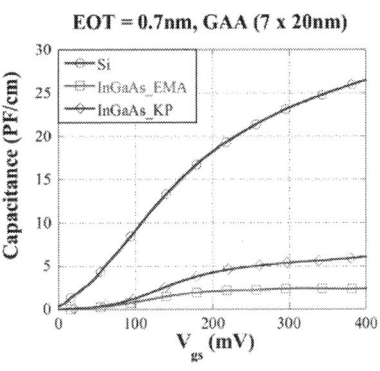

Figure 16. The comparison of Si and InGaAs capacitance. Despite the sizeable raise due to nonparabolicity still InGaAs suffers from poor DOS.

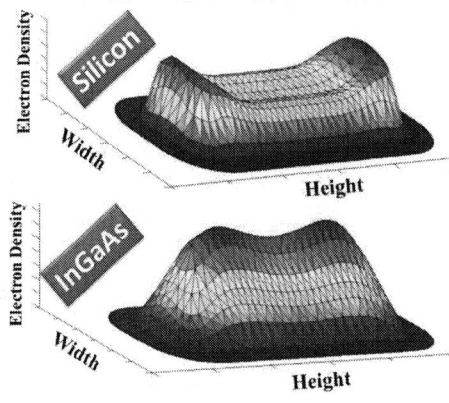

Figure 17. The typical inversion pattern for Si and InGaAs 7x20 nm NW. For InGaAs the peak of carrier density is located at the core of the wire and CET is higher

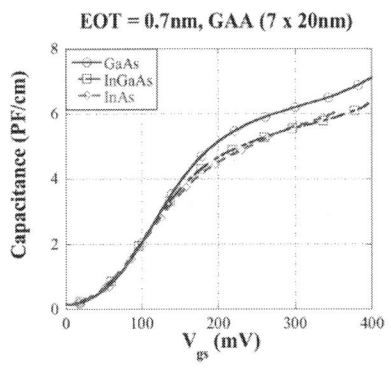

Figure 18. The capacitance for GaAs, InGaAs and InAs NW. The sensitive to the mole fraction is rather small.

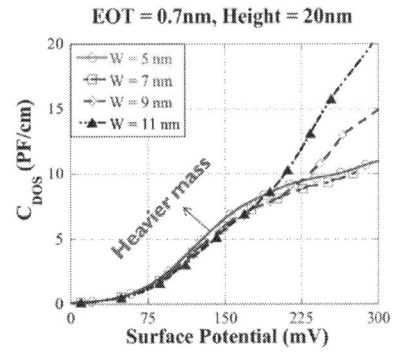

Figure 19. DOS capacitance of InGaAs NW for various width. The wider wires provides higher DOS.

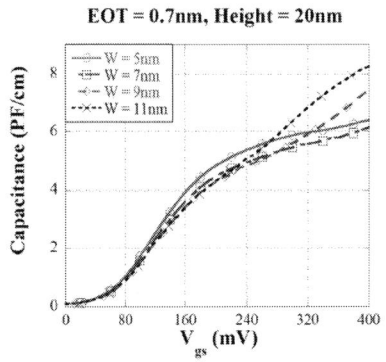

Figure 20. Gate capacitance of InGaAs NW for various width. The 11nm -wire results in the larger C_G at high V_G.

978-1-4799-7670-6/15 $31.00 © 2015 IEEE   69

PBTI for N-type Tunnel FinFETs

W. Mizubayashi, T. Mori, K. Fukuda, Y. X. Liu, T. Matsukawa, Y. Ishikawa, K. Endo, S. O'uchi,
J. Tsukada, H. Yamauchi, Y. Morita, S. Migita, H. Ota, and M. Masahara

National Institute of Advanced Industrial Science and Technology (AIST)
1-1-1 Umezono, Tsukuba, Ibaraki 305-8568, Japan
E-mail: w.mizubayashi@aist.go.jp

Abstract—**This paper reports the positive bias temperature instability (PBTI) characteristics for n-type fin-channel tunnel field-effect transistors (TFETs) with high-k gate stacks. The subthreshold slope (SS) is not degraded at all while the threshold voltage (V_{th}) shifts in the positive direction by the PBTI stress. The activation energy of ΔV_{th} for TFETs is almost the same as FinFETs, indicating that the PBTI mechanism for TFETs is almost the same as FinFETs. It was found that by applying a positive bias to the n$^+$-drain, the PBTI lifetime is dramatically improved as compared with that in the conventional stress test. This is because carrier injection from the n$^+$-drain is the main cause of the PBTI, especially for n-type TFETs. Thus, to accurately predict the PBTI lifetime of n-type TFETs, it is necessary to apply a drain bias for the reliability test.**

Keywords—n-Type tunnel FinFETs, threshold voltage, PBTI, lifetime prediction, activation energy

I. INTRODUCTION

The increasing power consumption of complementary metal-oxide-silicon (CMOS) devices is a critical issue. To reduce the power consumption, it is necessary to lower the operation voltage. To reduce the operation voltage, tunnel field-effect transistors (TFETs) are considered as a promising candidate because of their steeper SS than the limitation for conventional metal-oxide-silicon field-effect transistors (MOSFETs) (60 mV/dec) [1-4]. There are main two different points between TFETs and MOSFETs. One is the operation mechanism and the other is the polarity of the source/drain. TFETs operate by band-to-band tunneling from the source to the channel, and the polarity of the source/drain for TFETs is asymmetric (Fig. 1).

Since TFETs are assumed to operate at an ultralow voltage of below 0.5 V, the allowable BTI lifetime becomes more severe. By asymmetric polarity of the source/drain, the electric field is concentrated at the source/gate edge under operation conditions (Fig. 1). Furthermore, the asymmetric source-drain doping for TFETs makes it difficult to understand the PBTI degradation as shown in Fig. 1. Nevertheless, the PBTI for TFETs has scarcely been studied [5, 6] and the mechanism of PBTI degradation for TFETs has not been comprehensively understood yet.

In this work, we systematically investigated the PBTI characteristics of n-type TFETs.

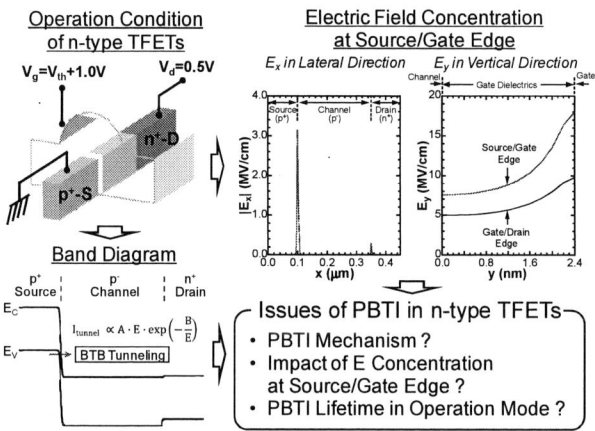

Fig. 1. Schematic illustration of issues of PBTI in n-type TFETs.

II. EXPERIMENTAL METHODS

N-type fin-channel TFETs and n-type FinFETs were fabricated on the same wafer. A (110) fin channel was formed on a (100) SOI substrate. n$^+$poly-Si/HfAlO$_x$/SiO$_2$ (equivalent oxide thickness (EOT) = 2.2 and 2.4 nm) gate stacks were formed. BF$_2^+$ was implanted at 5 keV with doses of 1.0×10^{15} cm^{-2} in the source region, while As$^+$ was implanted at 5 keV with doses of 1.0×10^{15} cm^{-2} in the drain region [7]. To suppress the spread of the implanted dopants by activation annealing, we performed flash lamp annealing (1200 °C, 3 ms) [2]. Finally, the back-end process was carried out.

In this study, V_{th} was defined as the gate voltage (V_g) at a drain current (I_d) of 10^{-11} A/μm.

III. RESULTS AND DISCUSSION

A. PBTI Mechanism for n-type TFETs

Figure 2 shows ΔV_{th} as a function of the stress time for n-type tunnel FinFETs under various stress gate voltages. V_{th} shifts in the positive direction by the PBTI stress. Also, ΔV_{th} follows a power law of the stress time ($\Delta V_{th} = \alpha t^n$) regardless of the measurement temperature. Figure 3 shows the exponent n in the power law as a function of the stress voltage for the TFETs. The exponent n is 0.1–0.2 regardless of the EOT, the

978-1-4799-7670-6/15 $31.00 © 2015 IEEE

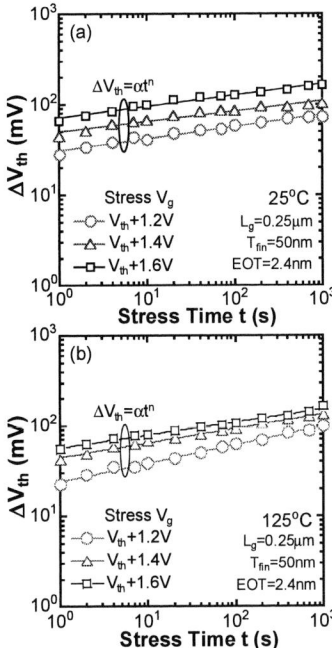

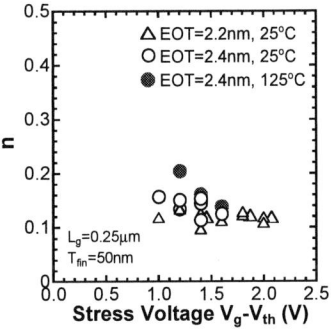

Fig. 2. ΔV_{th} as a function of stress time in n-type TFETs. The measurement temperatures are (a) 25 °C and (b) 125 °C. The PBTI stress conditions are as follows: the stress bias was only applied at the gate, and the source, drain, and substrate were grounded. ΔV_{th} follows a power law of stress time regardless of the measurement temperature.

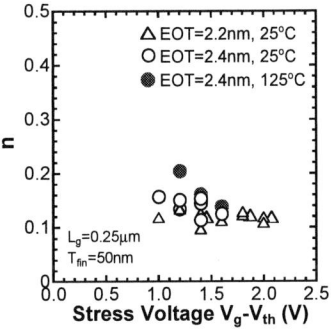

Fig. 3. Exponent n as a function of stress voltage for n-type TFETs. n is about 0.1–0.2 regardless of the EOT and measurement temperature.

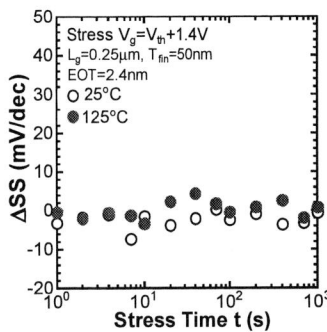

Fig. 4. ΔSS as a function of stress time for n-type TFETs. ΔSS hardly changes with the stress time. This means that the SS is not degraded by the PBTI stress.

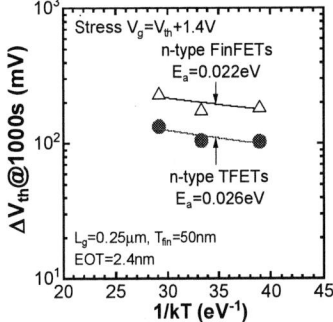

Fig. 5. Arrhenius plots of ΔV_{th} at 1000 s at stress $V_g = V_{th}+1.4V$ for n-type TFETs and n-type FinFETs. The activation energy of ΔV_{th} in the n-type TFETs is almost the same as that in the n-type FinFETs.

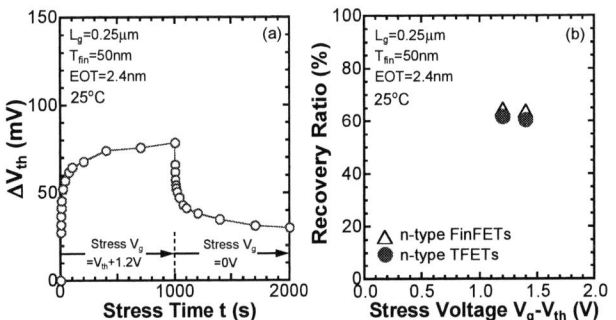

Fig. 6. (a) ΔV_{th} as a function of stress time for n-type TFETs before and after the PBTI stress. (b) Recovery ratio as a function of stress voltage for n-type TFETs and n-type FinFETs. (a) The recovery phenomenon is observed in the n-type TFETs. (b) The recovery ratio in the n-type TFETs is almost the same as that in the n-type FinFETs regardless of the stress voltage.

stress voltage, and the measurement temperature. On the other hand, the SS is hardly degraded by the PBTI stress (Fig. 4).

To understand the PBTI mechanism for n-type TFETs, we investigated the activation energy (E_a) of ΔV_{th} by the PBTI degradation. Figure 5 shows Arrhenius plots of ΔV_{th} at 1000 s at stress $V_g = V_{th}+1.4V$ for n-type TFETs and n-type FinFETs. E_a of ΔV_{th} for the TFETs is estimated to be 0.026 eV, which is almost the same as that for the FinFETs ($E_a = 0.022$ eV). This result clearly indicates that the PBTI mechanism for n-type TFETs is almost the same as that for n-type FinFETs. Since the SS of the n-type TFETs is hardly changed by the PBTI stress,

the PBTI degradation of n-type TFETs is caused by electron trapping in high-k gate dielectrics.

We investigated the recovery characteristics of the n-type TFETs (Fig. 6). The recovery test was performed at 25 °C for suppressing the impact of trap-assisted tunneling (TAT). The following PBTI tests were also carried out at 25 °C. Figure 6(a) shows ΔV_{th} as a function of the stress time before and after the

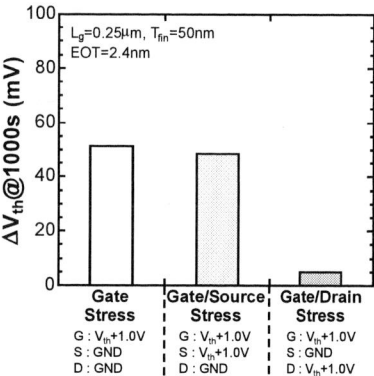

Fig. 7. ΔV_{th} at 1000 s for n-type TFETs under various stress conditions.

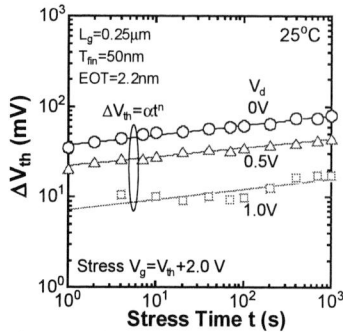

Fig. 8. ΔV_{th} as a function of stress time at stress $V_g = V_{th}+2.0$ V and $V_d = 0$, 0.5, and 1.0 V for n-type TFETs.

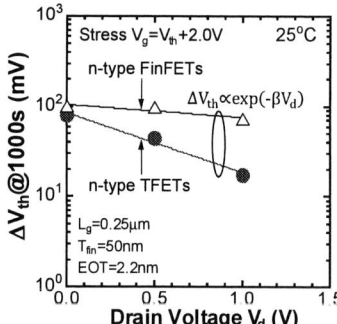

Fig. 9. ΔV_{th} at 1000 s as a function of drain voltage for n-type TFETs and n-type FinFETs. ΔV_{th} at 1000 s decreases with increasing V_d for n-type TFETs.

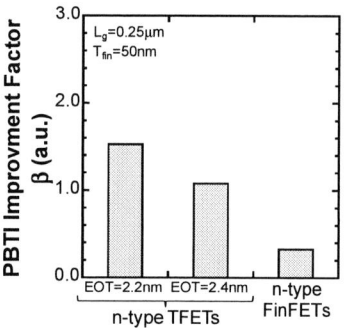

Fig. 10. PBTI improvement factor β for n-type TFETs and n-type FinFETs. β for n-type TFETs increases with thinning EOT and is about three times that for n-type FinFETs.

PBTI stress. The recovery phenomenon is clearly observed in the n-type TFETs. The recovery ratios are about 60% for both the n-type TFETs and n-type FinFETs (Fig. 6(b)).

B. Impact of Electric Field Concentration at Source/Gate Edge

In the case of n-type TFETs, since they have a p^+-source, the electric field (lateral E_x and vertical E_y) is concentrated at the source/gate edge (Fig. 1). We investigated the impact of the E_y concentration at the source/gate edge on PBTI degradation (Fig. 7). To separate the effects on the source/gate and the gate/drain edge, the PBTI test was performed under gate/source or gate/drain stress. The gate/source stress mainly corresponds to carrier injection from the drain, whereas the gate/drain stress corresponds to the electric field concentration at the gate/source edge. ΔV_{th} at 1000 s is almost the same as the conventional gate stress and the gate/source stress. On the other hand, ΔV_{th} at 1000 s for the gate/drain stress is very small as compared with the other stress conditions. This result indicates that the E_y concentration at the source/gate edge has no impact on PBTI degradation and that the carrier injection from the n^+-drain is the main cause of PBTI for n-type TFETs.

C. Impact of Drain Bias on PBTI Lifetime for n-type TFETs

To accurately predict the realistic PBTI lifetime for n-type TFETs, we investigated the impact of the drain bias on the PBTI characteristics (Fig. 8). ΔV_{th} decreases with increasing the drain bias and also follows a power law of stress time under all drain bias conditions (Fig. 8). The exponent n in the power law is about 0.1 regardless of the drain bias (Fig. 8). Figure 9 shows ΔV_{th} at 1000 s as a function of the drain voltage for n-type TFETs and n-type FinFETs under a stress gate voltage of 2 V. ΔV_{th} at 1000 s for the n-type TFETs decreases with increasing V_d. By applying a positive bias to the n^+-drain, the electric field in the gate dielectric at the gate/drain edge decreases, and thus carrier injection from the drain is suppressed [8]. This is the main reason for the PBTI improvement by the drain bias. Furthermore, the measured ΔV_{th} is fitted to

$\exp(-\beta V_d)$, where β is defined as the PBTI improvement factor. Since the carrier injection from the drain to the gate follows the tunneling mechanism through the gate dielectrics, the relational equation can be described as $\exp(-\beta V_d)$. β for the n-type TFETs is about three times that for the n-type FinFETs and becomes large with thinning EOT (Fig. 10).

Figure 11(a) shows the PBTI lifetime as a function of the stress voltage for n-type TFETs. The PBTI lifetime was defined as the stress time at $\Delta V_{th} = 50$ mV. The PBTI stress test was performed under no drain bias or $V_d = 0.5$ V. The PBTI

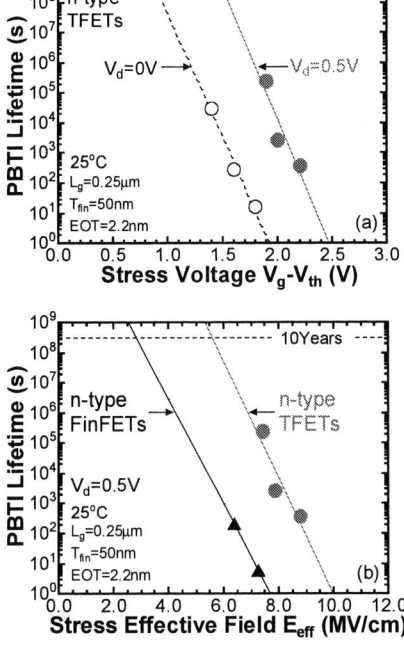

Fig. 11. (a) PBTI lifetime at ΔV_{th} = 50 mV as a function of stress voltage for n-type TFETs. The PBTI stress test was performed under V_d = 0 V or 0.5 V. (b) PBTI lifetime at ΔV_{th} = 50 mV as a function of stress effective field for n-type TFETs and n-type FinFETs. The stress effective field was estimated from the C-V_g characteristics. The drain bias applied in the PBTI stress test was 0.5V to assume operation conditions. The lifetime longer than 1000 s was estimated from the extrapolation of the measured data.

lifetime for V_d = 0.5 V shifts 0.5–0.6 V in the positive direction as compared with that in the case of no drain bias, meaning that the PBTI lifetime is improved by applying drain bias. Thus, since the PBTI lifetime changes by the drain bias, to accurately predict the PBTI lifetime, it is necessary to apply a drain bias during the PBTI stress test. Next, we compared with the PBTI lifetime for the n-type TFETs and n-type FinFETs (Fig. 11(b)). V_{th} for the n-type TFETs is different from that for the n-type FinFETs [8]. On the other hand, since V_{fb} for n-type TFETs is almost the same as that for n-type FinFETs [8], the lateral axis in Fig. 11(b) was normalized the effective field estimated from the C-V_g characteristics. In this case, the drain bias in the PBTI stress test was applied 0.5V to assume operation conditions. It is clearly shown that the PBTI lifetime for the TFETs is dramatically improved as compared with that for FinFETs. In the TFETs, the carrier injection from the drain is dramatically suppressed by applying the drain bias. This is the main reason for the dramatic improvement of PBTI in the n-type TFETs.

IV. CONCLUSIONS

We systematically investigated the PBTI characteristics of n-type TFETs. The PBTI mechanism for n-type TFETs is almost the same as n-type FinFETs. We clarified that the PBTI degradation of TFETs is mainly caused by carrier injection from the drain, while the electric field concentration at the source/gate edge has no impact on PBTI degradation. Since the PBTI lifetime of n-type TFETs changes by the drain bias, to accurately predict the PBTI lifetime, it is necessary to apply a drain bias during the PBTI stress test.

ACKNOWLEDGMENT

This research was granted by the Japan Society for the Promotion of Science (JSPS) through the "Funding Program for World-Leading Innovative R&D on Science and Technology (First Program)," initiated by the Council for Science and Technology Policy (CSTP).

REFERENCES

[1] A. Villalon, C. Le Royer, M. Cassé, D. Cooper, B. Prévitali, C. Tabone, J.-M. Hartmann, P. Perreau, P. Rivallin, J.-F. Damlencourt, F. Allain, F. Andrieu, O. Weber, O. Faynot and T. Poiroux, "Strained tunnel FETs with record I_{ON}: first demonstration of ETSOI TFETs with SiGe channel and RSD," in VLSI Symp. Tech. Dig., 2012, pp. 49-50.

[2] T. Mori, K. Fukuda, T. Yasuda, A. Tanabe, T. Maeda, S. O'uchi, Y. X. Liu, W. Mizubayashi, M. Masahara, and H. Ota, "EOT scaling in tunnel field-effect transistors: trade-off between subthreshold steepness and gate leakage," in Ext. Abst. of the 2012 SSDM, 2012, pp. 74-75.

[3] Y. Morita, T. Mori, S, Migita, W. Mizubayashi, A. Tanabe, K. Fukuda, T. Matsukawa, K. Endo, S. O'uchi, Y.X. Liu, M. Masahara, and H. Ota, "Synthetic electric field tunnel FETs: drain current multiplication demonstrated by wrapped gate electrode around ultrathin epitaxial channel," in VLSI Symp. Tech. Dig., 2013, pp. T236-T237.

[4] R. Rooyackers, A. Vandooren, A.S. Verhulst, A. Walke*, K. Devriendt, S. Locorotondo, M. Demand, G. Bryce, R. Loo, A. Hikavyy, T. Vandeweyer, C. Huyghebaert, N. Collaert, and A. Thean., "A new complementary hetero-junction vertical tunnel-FET integration scheme," in IEDM Tech. Dig., 2013, pp. 92-95.

[5] G. F. Jiao, Z. X. Chen, H. Y. Yu, X. Y. Huang, D. M. Huang, N. Singh, G. Q. Lo, D.-L. Kwong, and M.-F. Li, "New degradation mechanisms and reliability performance in tunneling field effect transistors," in IEDM Tech. Dig., 2009, pp. 741-744.

[6] G. Han, Y. Yang, P. Guo, C. Zhan, K. L. Low, K. H. Goh, B. Liu, E.-H. Toh, and Y.-C.Yeo, "PBTI characteristics of n-channel tunneling field effect transistor with HfO_2 gate dielectric: new insights and physical model," in 2012 Symp. on VLSI-TSA Proc., 2012, p. T82.

[7] T. Mori, T. Yasuda, T. Maeda, W. Mizubayashi, S. O'uchi, Y. X. Liu, K. Sakamoto, M. Masahara, and H. Ota, "Tunnel field-effect transistors with extremely low off-current using shadowing effect in drain implantation," Jpn. J. Appl. Phys., vol. 50, pp. 06GF14-1-3, 2011.

[8] W. Mizubayashi, T. Mori, K. Fukuda, Y. X. Liu, T. Matsukawa, Y. Ishikawa, K. Endo, S. O'uchi, J. Tsukada, H. Yamauchi, Y. Morita, S. Migita, H. Ota, and M. Masahara, "Accurate prediction of PBTI lifetime for n-type fin-channel tunnel FETs," in IEDM Tech. Dig., 2014, pp. 824-827.

Reliability impact of Advanced Doping techniques for DRAM peripheral MOSFETs

Alessio Spessot[1,*], Romain Ritzenthaler[2], Tom Schram[2], Marc Aoulaiche[1], Moonju Cho[2], Maria Toledano Luque[2], Naoto Horiguchi[2], Pierre Fazan[1]

[1]Micron Technology Belgium, imec Campus, Kapeldreef 75, 3001 Leuven Belgium
[2]imec, Kapeldreef 75, 3001 Leuven Belgium

(* e-mail: aspessot@micron.com)

Abstract—**We have evaluated the impact on the reliability of an innovative process flow, specifically designed for peripheral MOSFETs of DRAM memories. Al and MgO layers are deposited, diffused into the gate stacks of NMOS and PMOS and finally removed. We have demonstrated an anomalous yet predictable PBTI behavior, coupled with a more standard NBTI one. Decent lifetime is achieved for both gate stacks, demonstrating the feasibility of the proposed process flow concerning BTI.**

Keywords—NBTI; PBTI; MOSFET; HKMG

I. INTRODUCTION

The stringent requirements of low leakage and high performance in the DRAM specifications are forcing the adoption in the related periphery transistors of characteristics already used by high performance logic devices. High-k dielectrics combined with Metal Gate (HKMG) represent the only way to meet the specifications for the next generation nodes in terms of speed and bandwidth, enabling the reduction of operating voltage down to 1.0V and below without losing performance. In addition to that, significant area scaling can be achieved, obtaining further cost reduction. [1] It has been proved that to fully exploit the HKMG benefits in terms of low threshold voltage (V_{TH}) and improved matching, it is necessary to include a Work Function (WF) shifter material directly in the gate stack, such as La, Al, Y [2] for NMOS and Al or Al_2O_3 for the PMOS. On top of that, when a standard process flow for transistor needs to be embedded into a Stand-alone Memory chip, it is subjected to additional process steps required by the memory element, which involves in the case of DRAM long thermal treatments. When such thermal treatments are taken into account, the solutions already proposed by the industry for logic applications are not working anymore, and different process integration flows need to be followed.

A possible working solution to obtain the desired Work Function shift, needed by the peripheral transistors, and sustain the strong thermal treatments required by the memory elements has been recently proposed [3]. Such an approach presents some limitations, due to the gate stack asymmetry and the dependence of the final threshold voltage shift to the entire thermal treatment steps. Another approach has been

recently investigated, which implies a) the selective deposition of MgO and Al_2O_3 to include the V_{TH} shifter dipoles into the gate stacks of N-and P-MOS devices, b) the diffusion of the dipoles with a single Diffusion Anneal (DA) and c) removing the layers and finally depositing a fresh Metal Gate [4]. This approach has not been used before in industrial applications to the knowledge of the authors, and the first results of a possible integration flow have been recently proposed [5]. In the present paper, we deeply investigated the reliability impact of such integration flow.

II. EXPERIMENTAL RESULTS AND DISCUSSION

The proposed HKMG gate stack is fabricated by using an HKMG/Metal Inserted Poly Si gate (MIPS) process [2], using a 1.2 nm Interfacial Layer of Silicon dioxide (IL) followed by 2 nm of HfO_2 deposited by ALD. The NMOS solution is based on a MgO layer, source of V_{TH} shifter, embedded in a sandwich of TiN layers (TiN/MgO/TiN). The PMOS V_{TH} shifter is obtained by a layer of Al_2O_3 deposited on top of the high-k and covered by another TiN layer. Both layers are deposited, annealed by the same DA and then removed, guaranteeing the diffusion of the dipoles to adjust the effective Work Functions at the desired values for NMOS and PMOS. Example of the process flow is shown in Fig 1.

The impact on Negative and Positive Bias Temperature Instability (N- and P-BTI) have been characterized to understand the impact of Al_2O_3 and MgO, respectively, with this innovative process scheme. The process parameters

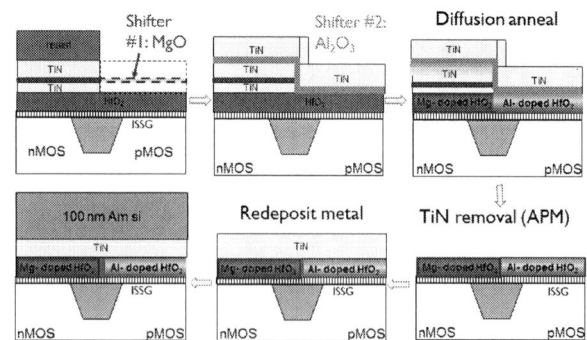

Fig. 1: Example of the proposed integration flow. A single Metal Gate is used for both N- and P-MOS.

978-1-4799-7670-6/15 $31.00 © 2015 IEEE

investigated are the diffusion anneal (DA) conditions and the dopant dose. For the NMOS, the interaction with an additional, complementary V_{TH} shifter (As ions implanted in the gate stack) has been investigated also. The measurements are done accordingly to the eMSM method, presented elsewhere [6].

A. NBTI

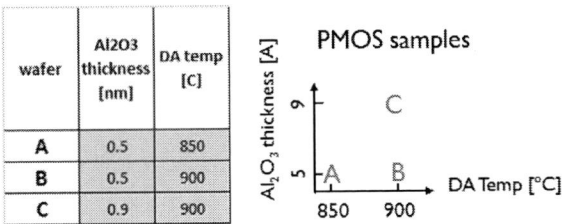

Fig. 2: schematic view of the analyzed samples. Two conditions of DA and dopant thickness have been explored

For NBTI, different samples with DA ranging between 850C and 900C (t<60 sec) have been explored. The considered dose of Al_2O_3 is equivalent to a thickness of deposited layer between 0.5 nm and 0.9nm. A schematic graph of the presented sample is shown in Fig. 2. An example of the measured ΔV_{TH} shift (T=125°C) vs stress time is shown in Fig 3, for one of the sample (C). It is visible that in general the time exponent of the various stresses is similar, however for higher field acceleration a different ΔV_{TH} kinetics can be observed, likely related to a different level of charge trapping due to existing trap into the bulk of the high-k layer.

Concerning relaxation, all the presented samples show a standard relaxation behavior (ΔV_{TH} reduction in longer relaxation time), as shown in Fig 4. Considering a 10 years lifetime criterion based on 30mV of V_{TH} shift at temp=125°C, all the samples achieved the lifetime. Depending on the diffusion anneal conditions and the Al_2O_3 deposition dose (layer of 0.5 nm vs 0.9 nm), respectable values (between 0.78 V and 0.81 V) of maximum applicable gate voltage overdrive can be applied to the analyzed gate stacks. Considering that EOT ranges from 1.3nm to 1.4 nm, the maximum applicable

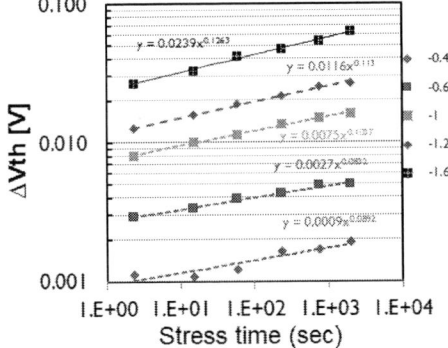

Fig. 3: V_{TH} shift vs stress time, for one of the analysed sample (wafer C). The time exponent of the various curves is similar to 0.1 for all the considered overdrive conditions (between -0.4V and -1.6V)

V_{DD} is well above 1.0V for all the measured applicable overdrive. A summary of the results is shown in Fig 5 (left), which shows the lifetime plot for the three considered samples. It is clearly visible in the table of Fig 5(right) that all

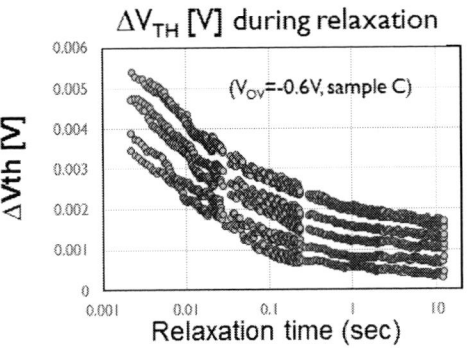

Fig. 4: Relaxation kinetics vs stress time, for one of the analyzed sample (wafer C), under stress bias close to operating conditions (V_{OV}=-0.6V). Each curves is linked to progressively increasing stress. An almost completely recovery is observed for the earliest stress time.

the samples outperform the specification of V_{DD}=-1.0V, and the higher achievable V_{DD} seems more linked to the thicker oxide rather than the DA conditions.

B. PBTI

As mentioned before, the NMOS are fabricated by using a sandwich of MgO embedded in a TiN layer. Concerning PBTI, an abnormal relaxation behavior has been found, similar to samples obtained with La and a different integration scheme [7]. Accordingly, both positive and negative V_{TH} shifts can be obtained, depending on the stress conditions (stress time, stress bias, stress/relaxation durations). We have characterized four splits, as shown in Fig 6.

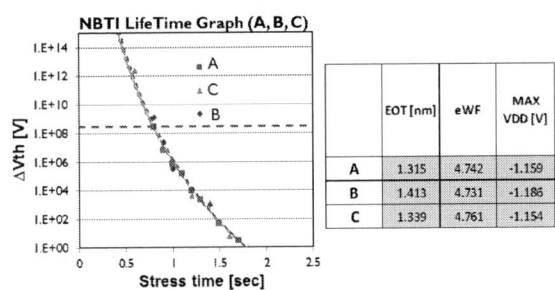

Fig. 5: (*left*) lifetime plot for the three considered sample. (*right*) summary table of the Equivalent Oxide Thickness (EOT), measured effective Work Function (eWF), and max applicable V_{DD}. All the samples outperform the stringent criteria of max 30mV of V_{TH} shift for V_{DD}=1.0V at 10 years, t=125°C.

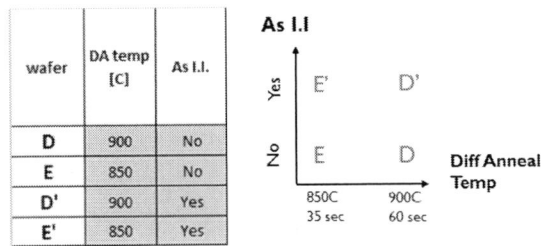

Fig. 6: schematic view of the analyzed samples. Two conditions of DA have been analyzed, as well as the selective inclusion of As ion implant into the gate stack.

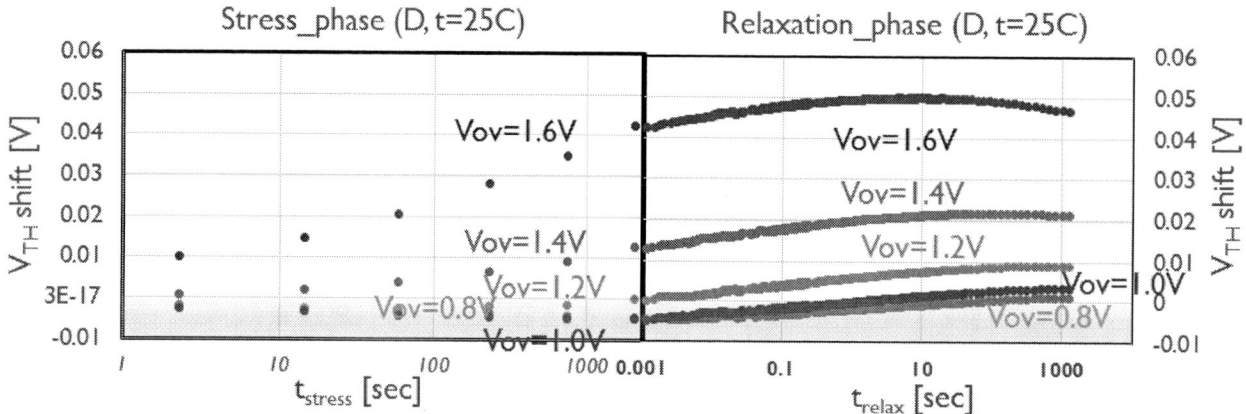

Fig. 7: (*left*) Example of V_{TH} shift measured at t=25°C, for different gate voltage overdrive stresses. For low stress bias the V_{TH} shift in the observed range is negative, while for higher stress conditions a purely positive trend is observed. (*right*) Relaxation kinetics for the various overdrives considered. All the conditions show a further initial increase of V_{TH}, followed by a decrease.

An example of the experimentally observed trends is shown in Fig 7 for constant stress (*left*) and the subsequent relaxation phase (*right*). It is visible that for lower stress bias, close to operating condition (V_{OV}=1.0V) the V_{TH} shift during the constant stress phase is negative, while for higher stress conditions a purely positive trend is observed. Fig 7 (*right*) shows the relaxation phase. Despite the fact that all the conditions show a further initial increase of V_{TH}, all the trends are followed by a final decrease. Such a behavior could be explained by the interaction of two distinct components, an

electron trapping weakly thermally activated [8] and a holes trapping, which is more sensitive to the temperature [9]. By using a proper modelling, as presented in [9], the two components responsible for the PBTI degradation can be separated and independently evaluated:

$$\Delta V_{TH} = \frac{\Delta V^+_{TH,0}}{1 + B^+ \left(t_{relax}/t_{stress} \right)^{\beta^+}} + \frac{\Delta V^-_{TH,0} \cdot F(V_{stress}, t_{stress})}{1 + B^- \left(t_{relax}/t_{stress} \right)^{\beta^-}} + \Delta V_{TH,P} = R^+ + R^- + P$$

As shown in the above equations, three terms need to be taken into account to properly explain the experimental trends. On

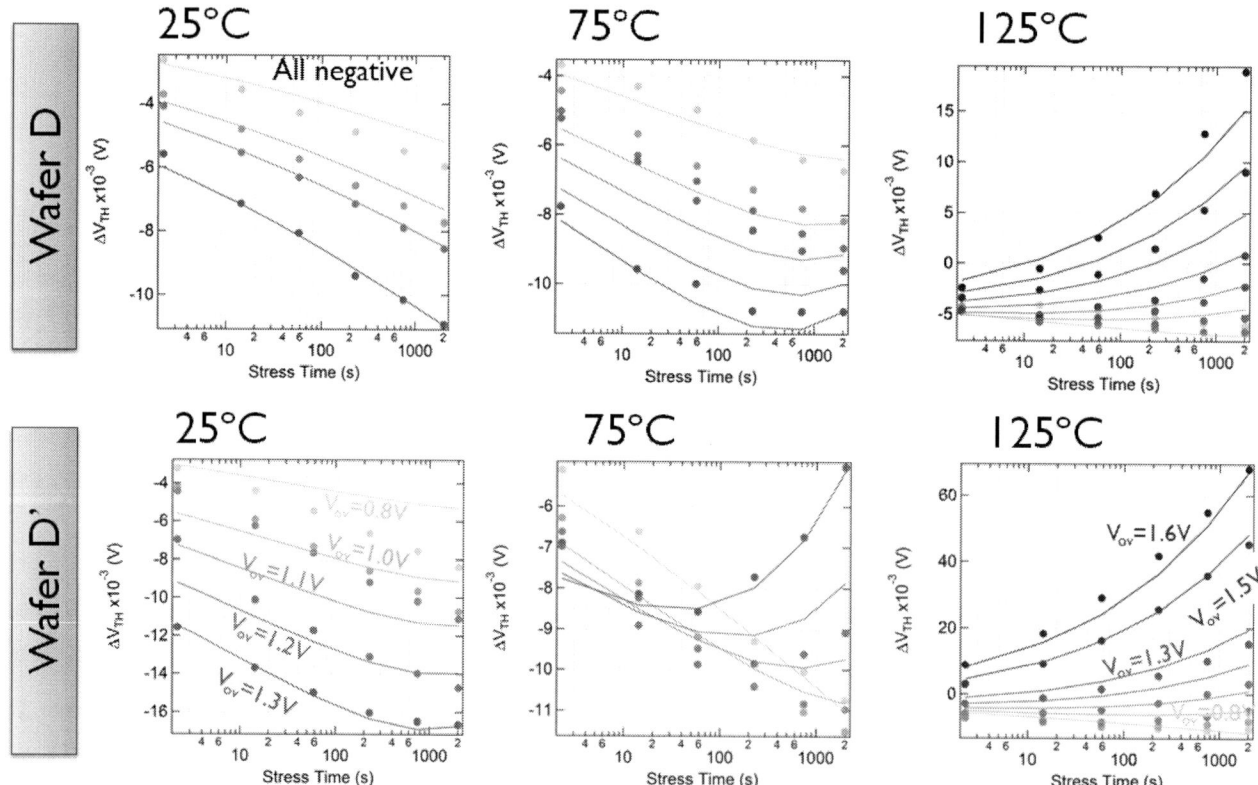

Fig. 8: (*left*) Example of V_{TH} shift measured at t=25°C, for different overdrive stress. Consistently with Fig 7, for low stress bias the V_{TH} shift in the observed range is negative, while for higher stress conditions a purely positive trend is observed. (*right*) Relaxation kinetics for the various overdrives considered. All the conditions show a further initial increase of V_{TH}, followed by a decrease. (V_{OV} between 0.8 and 1.6V)

	Temp (C)	25	75	125
Negative Component	A^-	6.07E-03	8.96E-03	1.02E-02
	n^-	0.14	0.14	0.16
	m^-	2.83	1.40	1.22
Positive Component	A^+	9.26E-04	2.51E-03	5.81E-03
	n^+	0.28	0.23	0.22
	m^+	3.38	3.18	3.10

Wafer D

	Temp (C)	25	75	125
Negative Component	A	3.68E-03	5.57E-03	9.28E-03
	n	0.10	0.12	0.11
	m	1.61	1.57	1.16
Positive Component	A	2.98E-05	4.54E-04	4.36E-03
	n	0.38	0.33	0.17
	m	2.60	2.25	2.63

Wafer D'

$$R^+ = A^+ \cdot t_{stress}^{n^+} \cdot V_{stress}^{m^+} \qquad\qquad R^- = A^- \cdot t_{stress}^{n^-} \cdot V_{stress}^{m^-}$$

Fig. 9: Resulting parameters of the fitting procedure, for the three measured temperatures, on wafer D and D'. The equations used are shown in the bottom part of the. It can be see that both the pre-factors A^+ and A^- tend to increase with the temperature.

the basis of the fitting, meaningful extrapolations can be achieved.

An example of the good fitting that can be obtained is shown in Fig 8, depicting the V_{TH} shift measured at three different temperatures (25°C, 75°C and 125°C) for two samples (DA=900°C t<60sec, with and without As I.I). The corresponding fitting lines are superimposed to the experimental points. The summary tables with the obtained parameters after the fitting are shown in Fig 9, where the resulting parameters after the fitting procedure have been obtained.

t=25, D'	Vov [V]								
Time [s]	0.8	0.9	1	1.1	1.2	1.3	1.4	1.5	1.6
1.E+00	-0.003	-0.004	-0.004	-0.004	-0.005	-0.006	-0.006	-0.007	-0.008
1.E+01	-0.003	-0.005	-0.005	-0.005	-0.006	-0.007	-0.008	-0.009	-0.010
3.E+01	-0.004	-0.005	-0.005	-0.006	-0.007	-0.008	-0.009	-0.010	-0.011
1.E+02	-0.004	-0.006	-0.006	-0.007	-0.008	-0.009	-0.010	-0.011	-0.012
3.E+02	-0.004	-0.006	-0.006	-0.007	-0.008	-0.009	-0.011	-0.012	-0.013
1.E+03	-0.005	-0.007	-0.007	-0.008	-0.009	-0.010	-0.012	-0.013	-0.014
3.E+03	-0.005	-0.008	-0.008	-0.009	-0.010	-0.011	-0.013	-0.014	-0.015
1.E+04	-0.006	-0.008	-0.008	-0.009	-0.011	-0.012	-0.013	-0.015	-0.016
3.E+04	-0.006	-0.009	-0.009	-0.010	-0.011	-0.013	-0.014	-0.015	-0.017
1.E+05	-0.007	-0.009	-0.009	-0.011	-0.012	-0.013	-0.014	-0.016	-0.017
3.E+05	-0.007	-0.009	-0.009	-0.011	-0.012	-0.013	-0.014	-0.015	-0.016
1.E+06	-0.007	-0.009	-0.009	-0.010	-0.011	-0.011	-0.012	-0.012	-0.012
3.E+06	-0.007	-0.008	-0.008	-0.008	-0.008	-0.008	-0.008	-0.007	-0.006
1.E+07	-0.005	-0.005	-0.005	-0.004	-0.003	-0.002	0.000	0.003	0.006
3.E+07	-0.003	0.000	0.000	0.002	0.005	0.008	0.013	0.018	0.024
1.E+08	0.002	0.008	0.008	0.013	0.020	0.027	0.036	0.046	0.057
3.E+08	0.009	0.022	0.022	0.031	0.042	0.055	0.070	0.087	0.107
1.E+09	0.022	0.046	0.046	0.062	0.082	0.104	0.130	0.159	0.192

t=125, D'	Vov [V]								
Time [s]	0.8	0.9	1	1.1	1.2	1.3	1.4	1.5	1.6
1.E+00	-0.005	-0.005	-0.005	-0.005	-0.004	-0.004	-0.003	-0.002	-0.001
1.E+01	-0.006	-0.006	-0.005	-0.005	-0.004	-0.003	-0.002	0.000	0.002
3.E+01	-0.006	-0.006	-0.006	-0.005	-0.004	-0.003	-0.001	0.001	0.004
1.E+02	-0.007	-0.006	-0.006	-0.005	-0.004	-0.002	0.001	0.003	0.007
3.E+02	-0.007	-0.007	-0.006	-0.005	-0.003	0.000	0.003	0.006	0.010
1.E+03	-0.007	-0.007	-0.006	-0.004	-0.001	0.002	0.005	0.010	0.015
3.E+03	-0.008	-0.007	-0.005	-0.003	0.000	0.004	0.009	0.015	0.021
1.E+04	-0.008	-0.007	-0.004	-0.001	0.003	0.008	0.014	0.021	0.030
3.E+04	-0.008	-0.006	-0.003	0.001	0.006	0.012	0.020	0.029	0.040
1.E+05	-0.008	-0.005	-0.001	0.004	0.011	0.019	0.029	0.040	0.053
3.E+05	-0.008	-0.004	0.001	0.008	0.016	0.027	0.039	0.053	0.069
1.E+06	-0.007	-0.002	0.005	0.013	0.024	0.037	0.053	0.071	0.091
3.E+06	-0.005	0.001	0.009	0.020	0.034	0.050	0.069	0.091	0.117
1.E+07	-0.003	0.005	0.016	0.030	0.047	0.068	0.092	0.120	0.152
3.E+07	-0.001	0.010	0.024	0.041	0.063	0.088	0.118	0.152	0.191
1.E+08	0.004	0.017	0.035	0.057	0.084	0.116	0.154	0.197	0.246
3.E+08	0.009	0.026	0.048	0.076	0.109	0.149	0.195	0.248	0.308
1.E+09	0.017	0.039	0.067	0.102	0.144	0.194	0.251	0.313	0.353

Fig. 10: Lookup tables with the expected V_{TH} shift under constant stress bias for t=25°C (*top*) and t=125°C (*bottom*), in the case of As I.I. When the overdrive is lower than 0.8V, the maximum V_{TH} shift extrapolated is lower than 30mV for 10years

Based on the reconstructed stress and model, the exact V_{TH} shift obtained under different operating conditions can be predicted. Fig 10 shows the reconstructed lookup tables obtained by fitting and extrapolating the data, under the assumption of constant stress. It is shown that for a constant stress of V_{OV}=0.8V, corresponding to V_{DD}>1.1V, a lifetime well above 10y is reached for both t=25C (*top*) and high temp (*bottom*). It is worth noting that the introduction of the As in the gate stack reduces the maximum lifetime, while higher diffusion anneal improves it. In the case of wafer E', a lower diffusion anneal temperature gives an higher relative contribution of the positive component. This results in an improved lifetime for lower overdrive, but in a worsening for higher applied bias (not shown). On the basis of the extrapolated lifetime, the circuit designers could optimize their schematics.

III. CONCLUSIONS

The lifetime of an innovative process flow for DRAM peripheral transistors have been deeply investigated, characterizing the stress and relaxation behavior of the proposed NMOS and PMOS solutions, and demonstrating their compatibility with the needs of the nowadays industry standards for memory.

Acknowledgment

This work was performed as part of imec's Core Partner Program. The authors would like to thank imec AMSIMEC p-line, and A. Thean of imec for the managerial support.

References

[1] S. Y. Cha, "DRAM Technology - History & Challenges", IEDM 2011 Short Course

[2] M. M. Frank, "High-K/Metal Gate Innovations Enabling Continued CMOS scaling", Proc ESSDERC, p. 25 (2011).

[3] Ritzenthaler et al, "A Low Power HKMG CMOS platform compatible with DRAM node 2X and beyond", TED Vol 61, P 2935 (2014)

[4] Patents EP2717308(pub), JP2014 078708 (pub), US 2014 0106556 (pub)

[5] Ritzenthaler et al, "A new high-k/metal gate CMOS integration scheme (Diffusion and Gate Replacement) suppressing gate height asymmetry and compatible with high-thermal budget memory technologies", Proc IEDM 2014

[6] B. Kaczer et al., Proceedings of International Reliability Physics Symposium (IRPS), Phoenix, AZ, USA, 2008. pp. 20–27.

[7] B. Kaczer, A. Veloso, Ph.J. Roussel, T. Grasser, G. Groeseneken, J. Vac. Sci. Technol. B 27 (2009) 459–462

[8] F. Crupi et al., Microelectron. Eng. 80 (2005) 130–133

[9] M. Toledano-Luque, B. Kaczer, M. Aoulaiche, A. Spessot, Ph. J. Roussel, R. Ritzenthaler, T. Schram, A. Thean, G. Groeseneken "Analytical model for anomalous Positive Bias Temperature Instability in La-based HfO2 nFETs based on independent characterization of charging components", Microelectronic Engineering 109 (2013) 314–317.

| Student Paper |

Impact of Random Telegraph Noise on Ring Oscillators Evaluated by Circuit-level Simulations

Azusa Oshima[1], Pieter Weckx[2], Ben Kaczer[2], Kazutoshi Kobayashi[1], Takashi Matsumoto[3]

Kyoto Institute of Technology[1], imec[2], VDEC[3]
Department of Electronics[1]
Kyoto, Japan[1], Leuven, Belgium[2], Tokyo, Japan[3]
aoshima@vlsi.es.kit.ac.jp[1]

Abstract—**Random Telegraph Noise (RTN) has become dominant with transistor rapid scaling in recent years. We simulate RTN-induced frequency fluctuation of Ring Oscillators (ROs) using a circuit-level simulator to replicate measurement results from previous works. Consequently, we can predict dependences of frequency fluctuation on operating voltages, number of ROs stages, gate widths, and body biases.**

Keywords—Reliability, Variability, Random Telegraph Noise, Ring Oscillator

I. INTRODUCTION

The variability issues in MOSFET devices can be classified into static (as-fabricated or time-zero) and dynamic (time-dependent). The latter is due to degradation effects such as HCI (Hot Carrier Injection), TDDB (Time Dependent Dielectric Breakdown), BTI (Bias Temperature Instability) and RTN (Random Telegraph Noise). Both static and dynamic variations have random and systematic (process-induced) components [1]. A major source of random variation, both static and dynamic, is random dopant fluctuation (RDF) [2]. With the rapid downscaling of the MOSFET devices, dynamic variation is becoming serious. In this paper, we deal with dynamic variations caused by transistor gate oxide defects. When a voltage is applied to the transistor gate, threshold voltage and delay increase with time due to BTI [3], which is accelerated under increased temperature. Even at constant bias, threshold voltage increase randomly by RTN [4]. Recently, it has been argued that the same mechanism is responsible for RTN and the recoverable component of BTI (Fig. 1) [5, 6]. When gate oxide defects capture carrier between source and drain, the channel current decrease and the threshold voltage changes. The degradation recovers when oxide defects emit carriers. In BTI, traps continue to capture carriers while the stress is applied. Soon after the stress removed, traps emit carriers. During RTN, capture and emission occur alternately while the operating voltage is applied, with the time constants defined as τ_c and τ_e [4, 5].

Since the RTN magnitude is sensitive for device size [7], it is becoming a new threat not only to analog circuits but also to digital circuits, for instance, SRAMs [8], flash memories [9] and image sensors [10]. Some works show threshold voltage variation of RTN is due to RDF [7, 11].

In this paper, we focus on RTN. To predict the impact of RTN, we simulate RTN-induced frequency fluctuation of ROs and replicate measurement results in a previous work [4].

II. MEASUEMENT RESULTS OF RTN IN PREVIOUS WORK

Our first goal is to replicate the experimental results of RTN on ROs in [4]. Frequencies of 7 stage ROs are measured for 80 s. Fig. 2 shows frequency time dependence of one of 12,600 ROs. F_{max} and ΔF of each RO are defined as the maximum frequency and the maximum frequency fluctuation respectively. Fig. 3 shows the measured time-zero, F_{max} distribution [4]. The distribution of measured (normalized) frequency fluctuation ΔF is given by in Fig. 4. Both quantities are distributed due to static and dynamic (RTN) variations. In the following, we simulate these distributions, and discuss the impact of operating voltage, number of ROs stages, device gate widths and body bias.

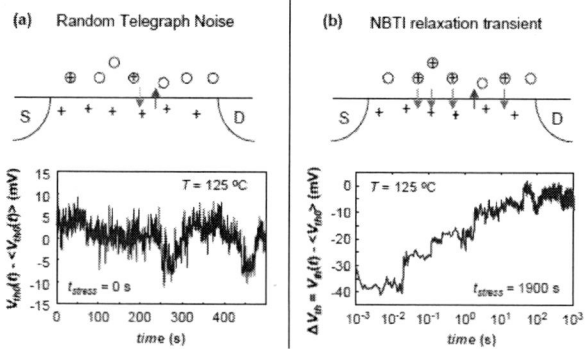

Fig. 1. Mechanism of RTN and BTI [5].

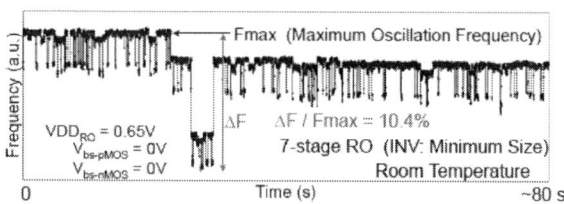

Fig. 2. Time dependency of frequency [4].

978-1-4799-7670-6/15 $31.00 © 2015 IEEE

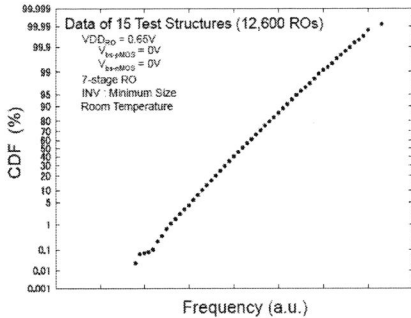

Fig. 3. Normal CDF plot of measured $F_{\max}$ caused by time-zero variations [4].

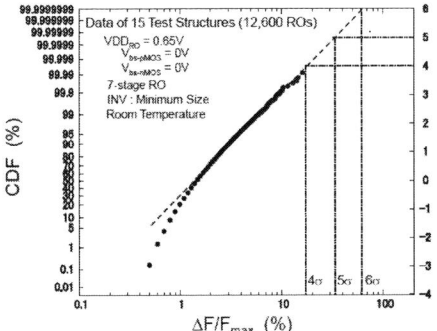

Fig. 4. Log-normal CDF plot of measured frequency fluctuation ($\Delta F/F_{\max}$) [4].

III. SIMULATION METHOD TO EVALUATE IMPACTS OF RTN

To evaluate the impact of RTN on circuit performance, we simulate ROs in a circuit-level simulator with static and dynamic (RTN) variations. We will show how to calculate RTN-induced ΔV_{th}.

A. RTN-induced single trap ΔV_{th} distribution

RTN-induced ΔV_{th} distribution has two parameters : the number of captured traps N in the MOSFET gate oxides and the mean ΔV_{th} per single trap η. In this section, we first assume the captured trap on a device is equal to one ($N = 1$) and discuss the distribution of η. Several works show that ΔV_{th} per trap follows an exponential distribution [5, 12]. The Probability Density Function (PDF) of ΔV_{th} is described as

$$f_\eta(\Delta V_{\text{th}}, \eta) = \frac{1}{\eta} e^{\frac{\Delta V_{\text{th}}}{\eta}} \quad (1)$$

The Cumulative Distribution Function (CDF) of ΔV_{th} is then

$$F_\eta(\Delta V_{\text{th}}, \eta) = 1 - e^{\frac{\Delta V_{\text{th}}}{\eta}} \quad (2)$$

η scales inversely with the device channel area [13].

$$\eta \propto \frac{1}{W\sqrt{L}} \quad (3)$$

where L and W are the device gate length and width, respectively.

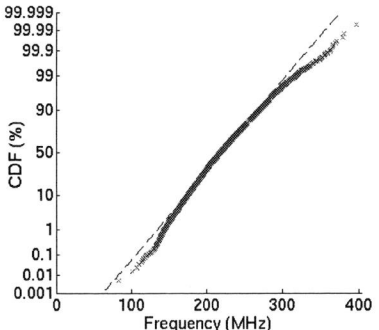

Fig. 5. Normal CDF plot of simulated time-zero frequency.

B. RTN-induced N traps ΔV_{th} distribution on a device

We now assume the number of captured traps on a device is the constant N. The ΔV_{th} distribution can be expressed as a convolution of N individual-trap exponential distributions (see the previous section) [5], and the PDF g_η and CDF G_η of ΔV_{th} are respectively

$$g_\eta(\Delta V_{\text{th}}, \eta) = \frac{e^{\frac{\Delta V_{th}}{\eta}}}{(n-1)!} \frac{\Delta V_{\text{th}}^{n-1}}{\eta^n} \quad (4)$$

$$G_\eta(\Delta V_{\text{th}}, \eta) = 1 - \frac{\Gamma\left(n, \frac{\Delta V_{th}}{\eta^n}\right)}{(n-1)!} \quad (5)$$

C. RTN-induced total ΔV_{th} distribution on a chip

An actual device will contain a Poisson-distributed number of defects n, with mean value N [5, 6, 14]

$$P_N(n) = \frac{e^{-N}N^n}{n!} \quad (6)$$

The average number of captured traps N is related to the oxide trap density N_{ot} as $N = WLN_{ot}$. The total ΔV_{th} CDF is given by the convolution of Eqs. 5 and 6.

$$H_{\eta,N}(\Delta V_{\text{th}}, \eta) = \sum_{n=0}^{\infty} P_N(n)\, G_\eta(\Delta V_{\text{th}}, \eta) \quad (7)$$

D. Simplified model of RTN-induced ΔV_{th} distribution

In the actual simulation, a Monte Carlo method is used combining Eqs. 2 and 6. Each $\Delta V_{\text{th,k}}$ sample is consisting of the sum of a Poisson distributed number of single defect $\Delta V_{\text{th},i}$, which are exponentially distributed.

$$\Delta V_{\text{th},k} = \sum_{i=1}^{N_k} \Delta V_{\text{th},i} \quad (8)$$

Using Eq. 8 a ΔV_{th} table is created for $k = 1 \dots m$, where m is the number of Monte Carlo runs, N_k is Poisson-distributed around the mean value N (Eq. 6) and $\Delta V_{\text{th},i}$ follows an exponential distribution with mean η (Eq. 2). The final ΔV_{th} distribution is then entirely characterized by the mean number of occupied traps N and the mean impact per trap η, described by Eq. 7.

978-1-4799-7670-6/15 $31.00 © 2015 IEEE

E. Time-zero ΔV_{th} distribution

Since time-zero variations have also a serious impact on circuits, they have to be considered together with RTN. Assuming time-zero variation induced-ΔV_{th} in both of NMOS and PMOS is normally distributed [16], we can generate the time-zero distribution of frequencies in the simulated ROs. The simulated distribution in Fig. 5 has the same shape as the measured distribution in Fig. 3.

F. Comparision with Measurement Results

Measurement results of F_{max} (Fig. 3) and $\Delta F/F_{max}$ (Fig. 4) are calculated from the frequency time dependence shown in Fig. 2 [3]. In our simulations we generate the impact of time-zero variation and RTN on frequency of 10,000 ROs at a fixed moment in time. ROs are simulated using 45 nm BSIM Predictive Technology Model (PTM). Each RO instance is simulated with (i) time-zero variation only and with (ii) the same time-zero variation and RTN, resulting in F_0 and F_{RTN+0}, respectively. Specifically,

(i). PMOS and NMOS in all RO stages have time-zero-induced normal-distributed ΔV_{th} and this distribution has the mean μ=52 mV and the standard deviation σ =31 mV.

(ii). PMOS and NMOS in all RO stage have same process variation as (i) and RTN-induced ΔV_{th}. RTN-induced parameters η and N, as well as the devices dimensions.

Finally, we calculate the main metrics of [3] as

$$\frac{\Delta F}{F_{max}} = \frac{F_0 - F_{RTN+0}}{F_0} \qquad (9)$$

IV. SIMULATION RESULTS OF RTN ON LOGIC CIRCUIT

We now discuss the results of our simulations.

A. RTN-induced Frequency Fluctuation

Fig. 6 shows the distribution of $\Delta F/F_{max}$ calculated for 10,000 ROs at V_{dd} =0.65 V. Both of simulation and measurement results is in the log-normal distribution. Note the simulation predicts a concave-up tail of distribution at *low percentiles*. This is because in the simulation, there is always a (small) probability that no defect is present in the RO with strong impact on ΔV_{th} and the frequency. The measurement results in Fig. 4, however, show *low-percentile* tail to converge to a constant minimum value. We speculate that this minimum value of $\Delta F/F_{max}$ is due to a finite resolution of the frequency measurement. The *high-percentile* tail of the $\Delta F/F_{max}$ is, however, crucial for the reliability predictions of real CMOS applications.

B. Discussion of parameter dependences

Figs. 7-10 show the dependences of frequency fluctuation on operating voltages, number of ROs stages, gate widths and body biases calculated for 840 ROs. As V_{dd} increases, the mean of $\Delta F/F_{max}$ decreases in Fig. 7. This tendency, due to increase of F_{max} with V_{dd} is the same as in the experimental results [3]. Fig. 8 then shows the impact of RTN decreases with the number of stages, also corresponding to measurement

results [4]. Specifically, the mean of $\Delta F/F_{max}$ is fixed but their distribution widths become small. Fig. 9 also shows the distribution widths of $\Delta F/F_{max}$ become smaller when gate widths increase corresponding to measurement results. This is because the number of captured traps N increases and ΔV_{th} per trap η decreases with increased device channel area following to Eq. 3.

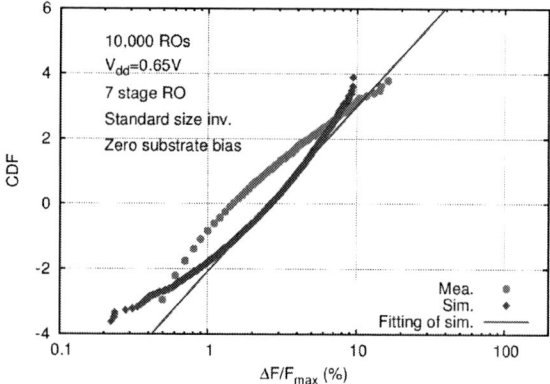

Fig. 6. Log-normal CDF plot of simulated RTN-induced frequency fluctuation ($\Delta F/F_{max}$).

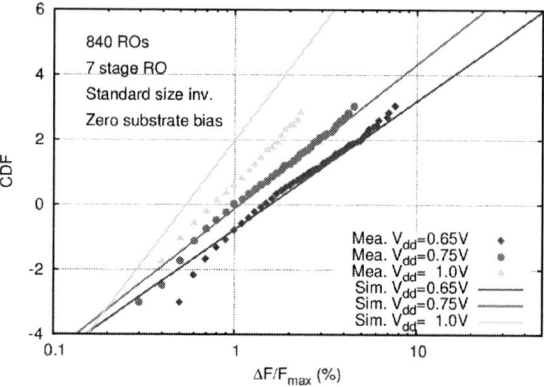

Fig. 7. Simulated V_{dd} dependence of RTN-induced frequency fluctuation.

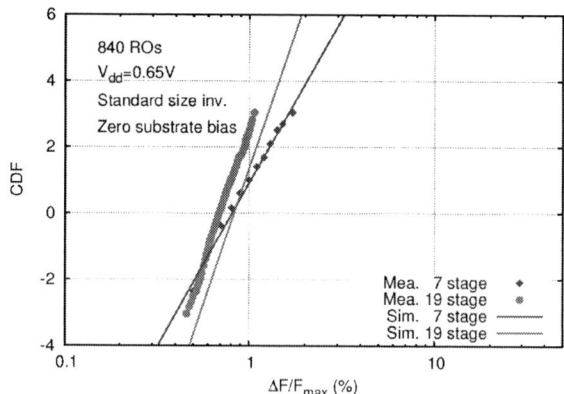

Fig. 8. Simulated the number of RO stages dependence of RTN-induced frequency fluctuation.

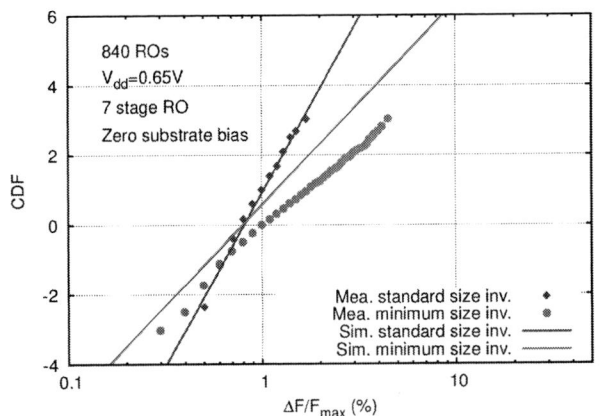

Fig. 9. Simulated gate width dependence of RTN-induced frequency fluctuation.

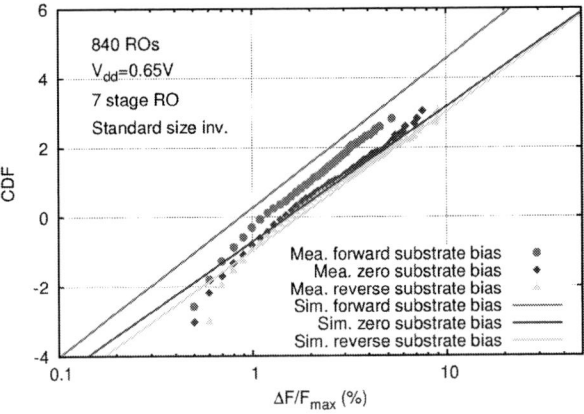

Fig. 10. Simulated body bias dependence of RTN-induced frequency fluctuation.

The ratios of PMOS and NMOS gate areas (W×L) of the minimum size inverters to the standard size inverters are 0.21 and 0.30 respectively. Fig. 10 shows forward and reverse bias dependence of $\Delta F/F_{max}$. Forward bias means the body bias for PMOS $V_{bs\_pmos} = +0.2$ V, the body bias for NMOS $V_{bs\_nmos} = +0.2$ V. Reverse bias means $V_{bs\_pmos} = -0.2$ V, $V_{bs\_nmos} = 0$ V. When forward bias is applied to PMOS and NMOS, it has been previously observed that η decreases and impact of RTN decreases [17]. On the other hand, reverse bias increases the mean of $\Delta F/F_{max}$ since η increases with reverse bias [16]. The body bias dependency of RTN is corresponding to the measurement results.

V. CONCLUSION

We have simulated the distributions of RTN-induced frequency fluctuations of ROs to replicate trends in on-chip previous measurements. We assume RTN-induced ΔV_{th} to be described as a convolution of η (ΔV_{th} of per trap) and N (number of captured traps). Simulation results have the same dependence of frequency fluctuation $\Delta F/F_{max}$ on the operating voltage, the number of RO stages, gate widths and body bias as measurement results. Because of the gate width and body bias dependence of η, the impact of RTN becomes smaller with increased gate widths and forward biasing. Impact of RTN can be reduced by increasing the operating voltage or increasing the number of RO stages.

REFERENCES

[1] B. Kaczer, J. Franco, P. Roussel, G. Groeseneken, T. Chiarella, N. Horiguchi, T. Grasser, "Extraction of The Random Component of Time-Dependent Variability Using Matched Pairs," Electron Device Letters, IEEE, 2015

[2] S.S. Chung, "The Variability Issues in Small Scale Trigate CMOS Devices : Random Dopant and Trap Induced Fluctuations", IPFA, 2013

[3] M. Yabuuchi, R. Kishida, K. Kobayashi, "Correlation between BTI-induced degradations and process variations by measuring frequency of ROs," , IMFEDK, 2014

[4] T. Matsumoto, K. Kobayashi, H. Onodera, "Impact of random telegraph noise on CMOS logic circuit reliability," CICC, 2014

[5] B. Kaczer, T. Grasser, P.J. Roussel, J. Franco, R. Degraeve, L.-A. Ragnarsson, E. Simoen, G. Groeseneken, H. Reisinger, "Origin of NBTI variability in deeply scaled pFETs," IRPS, 2010

[6] B. Kaczer, T. Grasser, J. Martin-Martinez, E. Simoen, M. Aoulaiche, P.J. Roussel, G. Groeseneken, "NBTI from the perspective of defect states with widely distributed time scales," Reliability Physics Symposium, 2009

[7] N. Tega, H. Miki, R. Zhibin, C.P. D'Emic, Z. Yu, D.J. Frank, M. A. Guillorn, D-G. Park, W. Haensch, K. Torii, "Impact of HK / MG stacks and future device scaling on RTN," IRPS, 2011

[8] M. Tanizawa, S. Ohbayashi, T. Okagaki, K. Sonoda, K. Eikyu, Y. Hirano, K. Ishikawa, O. Tsuchiya, Y. Inoue, "Application of a statistical compact model for Random Telegraph Noise to scaled-SRAM Vmin analysis," VLSIT, 2010

[9] H. Kurata, K. Otsuga, A. Kotabe, S. Kajiyama, T. Osabe, Y. Sasago, S. Narumi, K.. Tokami, S. Kamohara, O. Tsuchiya, "Random Telegraph Signal in Flash Memory: Its Impact on Scaling of Multilevel Flash Memory Beyond the 90-nm Node," Solid-State Circuits, IEEE Journal of , 2007

[10] J.-M. Woo, H.-H. Park, H.S. Min, H. Shick, C.H. Park, S.-M. Hong, C. H. Park, "Statistical analysis of random telegraph noise in CMOS image sensors," SISPAD, 2008

[11] K. Takeuchi, T. Nagumo, S. Yokogawa, K. Imai, Y. Hayashi, "Single-charge-based modeling of transistor characteristics fluctuations based on statistical measurement of RTN amplitude," VLSIT 2009

[12] B. Kaczer, P.J. Roussel, T. Grasser, G. Groeseneken,"Statistics of Multiple Trapped Charges in the Gate Oxide of Deeply Scaled MOSFET Devices—Application to NBTI," Electron Device Letters, IEEE , 2010

[13] J. Franco, B. Kaczer, M. Toledano-Luque, P.J. Roussel, J. Mitard, L.-A. Ragnarsson, L. Witters, T. Chiarella, M. Togo, N. Horiguchi, G. Groeseneken, M.F. Bukhori, T. Grasser, A. Asenov, " Impact of single charged gate oxide defects on the performance and scaling of nanoscaled FETs", IRPS, 2012

[14] M. Toledano-Luque, B. Kaczer, J. Franco, P.J. Roussel, T. Grasser, T.Y. Hoffmann, G. Groeseneken, "From mean values to distributions of BTI lifetime of deeply scaled FETs through atomistic understanding of the degradation," VLSIT, 2011

[15] T. Mizutani, A. Kumar, T. Hiramoto, "Measuring threshold voltage variability of 10G transistors," IEDM, 2011

[16] A. Ghetti, C.C. Monzio, A.S. Spinelli, A. Visconti, "Comprehensive Analysis of Random Telegraph Noise Instability and Its Scaling in Deca–Nanometer Flash Memories," Electron Devices, IEEE Transactions on , 2009

[17] J. Franco, B. Kaczer, M. Toledano-Luque, P.J. Roussel, G. Groeseneken, B. Schwarz, M. Bina, M. Waltl, P.-J. Wagner, T. Grasser, "Reduction of the BTI time-dependent variability in nanoscaled MOSFETs by body bias," IRPS, 2013

978-1-4799-7670-6/15 $31.00 © 2015 IEEE

Simple Technique for Prediction of Breakdown Voltage of Ultrathin Gate Insulator under ESD Testing

Yuichiro Mitani, Kazuya Matsuzawa

Advanced LSI Research Laboratory, Toshiba Corporation
1, Komukai-Toshiba-cho, Saiwai-ku, Kawasaki 212-8582
yuuichiro.mitani@toshiba.co.jp

Abstract— In this study, simple ramped voltage TLP (RV-TLP) measurement was utilized to predict breakdown voltage (BV_{OX}) and number of pulses to breakdown (N_{BD}) under ESD testing. The proposed prediction method does not require lengthy DC-TDDB measurements but instead utilizes quick Ramped Voltage (RV) stress measurements to calculate a voltage to breakdown (BV_{OX}) in the ESD timeframe. From voltage ramping rate dependence of Q_{BD} and breakdown current (J_{BD}), the power law between Q_{BD} and J_{BD} was obtained. By using this Q_{BD}-J_{BD} correlation, we succeeded the predictions of BV_{OX} and N_{BD} analytically, and these values correspond to that for conventional constant-voltage TLP measurement. Furthermore, according to the evaluation of Q_P, anode-hole-injection (AHI) model is still adaptable for the breakdown under nanosecond pulse ESD testing.

Keywords—Breakdown voltage; SiON; ESD; TLP; Anode-hole-injection model

I. INTRODUCTION

Immunity to the electrostatic discharge (ESD) is one of the serious problems for a reliability of ULSI devices [1-3]. In order to design ESD protection devices, guidelines for breakdown voltage (BV_{OX}) and number of pulses to breakdown (N_{BD}) of the gate insulators in internal circuits are important. However, as similarly observed in the case of TDDB under DC stress, it takes much time to directly estimate the N_{BD} from experiments using constant-voltage transmission line pulsing (CV-TLP) measurement. It has been reported that the nanosecond range breakdown can be predicted using experimental results of DC-TDDB [4, 5]. However, since constant-voltage (CV) TDDB testing requires lengthy measurement, the method which can estimate gate oxide breakdown immunity in short time is desperately needed. For the quick prediction of DC-TDDB lifetime, the transformation using Voltage Ramp Stress (VRS) data has been proposed [5-7]. In the same way, Voltage-Ramp TLP measurement with a fixed pulse width and increasing voltage level is applied to the estimation of gate oxide reliability under ESD condition [7-9]. In these methods, however, the measurement of DC-TDDB is indispensable in this method, because the power index of

voltage acceleration factor and effective stress voltage are indispensable to transform VRS data to CV-TDDB data. In addition, in Ref. 9, the authors have investigate the estimation of gate oxide breakdown voltage (BV_{OX}) systematically and they have pointed out the artifact from this methodology. That is, the rise time in the TLP pulse cannot be negligible and the real duration of the stress is less than the full pulse width. Therefore, in this paper, a simple methodology to quickly and accurately predict for both BV_{OX} and N_{BD} under ESD condition is investigated based on the method reported in Ref. 10. We focus on the step voltage (V_{STEP}) dependence in the ramped voltage TLP (RV-TLP) measurement, which is corresponds to the conventional TLP stress methodology commonly used for ESD characterizations.

II. EXPERIMENTAL

TLP system used in this study applies 100nsec pulses (10nsec rise time) to a device, while measuring the voltage across and current through it. In order to evaluate the gate insulator reliability, ramped voltage TLP (RV-TLP) measurement were utilized as depicted in Fig. 1(b), comparing to conventional constant-voltage TLP (CV-TLP) measurement.

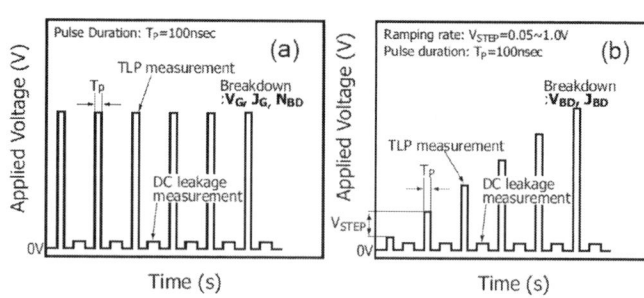

Fig. 1. Schematic measurement sequences for transmission line pulse (TLP) used in these experiments. (a) constant-voltage TLP (CV-TLP) stress, (b) ramped voltage TLP (RV-TLP) stress. The TLP system applies a 100ns pulse during which voltage and current are measured. Between TLP pulses, a DC leakage measurement is performed. When breakdown is occurred, breakdown voltages (V_{BD}), breakdown gate currents (J_G, J_{BD}) and number of pulse (N_{BD}) are determined.

978-1-4799-7670-6/15 $31.00 © 2015 IEEE

The breakdown points were determined through monitoring TLP voltage, TLP current and DC leakage current. The devices used in this study were N-channel MOSFET having ultrathin SiON as gate insulator.

III. EXPERIMENTAL RESULTS AND DISCUSSION

In DC-TDDB measurement, charge-to-breakdown (Q_{BD}) estimated from ramped voltage stress (RVS) and exponential ramped current stress (ERCS) measurements correspond to Q_{BD} under DC-TDDB measurement [10, 11]. By use of these experimental methods, Q_{BD} values are estimated as the sum of the amount of injected carriers through gate insulator during a ramping step. We applied this method to the TLP measurement and assess the relevance of this approach in predicting response to nanosecond pulse ESD events.

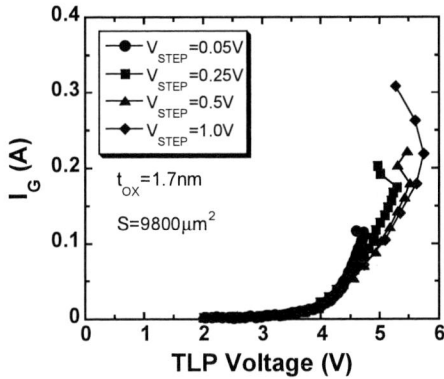

Fig. 2. Typical I-V characteristics of N-MOSFETs having 1.7 nm SiON. Voltage ramp rate (V_{STEP}) under RV-TLP stress varied from 0.02V to 1.0V. V_{BD} and J_{BD} increase with increasing V_{STEP}, and breakdown points were clearly detected, respectively.

Fig. 2 shows the typical I-V characteristics under several V_{STEP} conditions. From this result, the total fluence of carrier (Q_{BD}) can be estimated as shown in Fig. 3. Here, Q_{BD} values in the case of RV-TLP stress were approximately expressed as

$$Q_{BD} = \sum_{n=1}^{N_{BD}} J_G(n) \cdot T_P = \frac{10^{\frac{1}{s}}}{10^{\frac{1}{s}} - 1} J_{BD} \cdot T_P$$

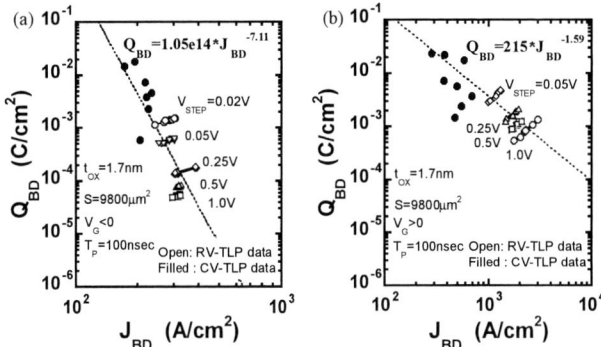

Fig. 3. Estimations of Q_{BD} values from (a) RV-TLP stress, comparing to (b) CV-TLP stress. The amount of injected carriers in one TLP pulse (Q_{INJ}) were obtained as $J_G \times T_P$. Our TLP system applies 100nsec pulses with 10nsec risetime to a device. However, in this work, T_P was simply assumed to be 100 nsec.

where s is the step numbers of TLP pulses, T_P is the total duration of each pulse, and J_{BD} is the current density of last step [10].

Figs. 4(a) and 4(b) show the Q_{BD}-J_{BD} correlation with various V_{STEP} under RV-TLP stress. It is found that Q_{BD} values increase with decreasing V_{STEP} and J_{BD}. In addition, as previously reported in the case of DC-RVS measurement [10, 12], Q_{BD} values under TLP stress also varied approximately following the power law,

$$Q_{BD} = \gamma \cdot J_{BD}^{\xi} .$$

These Q_{BD} values are compared with that for CV-TLP stress (filled symbols in these figures). It should be noted that Q_{BD} values in both cases run on the same line of the power law. This trend agrees with the reported results in DC measurement [10]. These experimental results suggest that we can predict Q_{BD} values and N_{BD} under CV-TLP stress through the simple RV-TLP measurement.

Fig. 4. Q_{BD} in RV-TLP stress, comparing to Q_{BD} in CV-TLP stress in N-MOSFETs. (a) TLP voltage is applied to the gate ($V_G>0$), (b) TLP voltage is applied to SDB ($V_G<0$). Open symbols correspond to the experimental data from RV-TLP measurements, and filled symbols correspond to that from CV-TLP measurement. It should be noted that Q_{BD} values in both cases run on the same line of the power law in both cases of gate polarities. This trend agrees with the experimental result under DC measurement in Ref. 10.

Next, in order to convert J_{BD} to BV_{OX}, J-V data are utilized as shown in Fig. 5. Measured J-V data are well fitted to the gate current estimated by

$$J_G = A \left(\frac{\phi_B}{V_{OX}}\right)\left(\frac{2\phi_B}{V_{OX}} - 1\right) E_{OX}^2$$
$$\cdot exp\left(-\frac{B\left[1 - \left(1 - \frac{V_{OX}}{\phi_B}\right)^{3/2}\right]}{E_{OX}}\right)$$

where ϕ_B is the barrier hight and V_{OX} is the oxide voltage [13, 14]. Figs. 6(a) and 6(b) show the predicted TLP voltage from the RV-TLP stress as shown in Fig. 4, comparing with measured Q_{BD} from CV-TLP measurement. Note that BV_{OX} can be accurately predicted through the simple RV-TLP stress. These results also suggest that N_{BD} at applied TLP voltage can be provided using Figs. 5 and 6 as

$$N_{BD} = \frac{Q_{BD}}{J_G \cdot T_P}$$

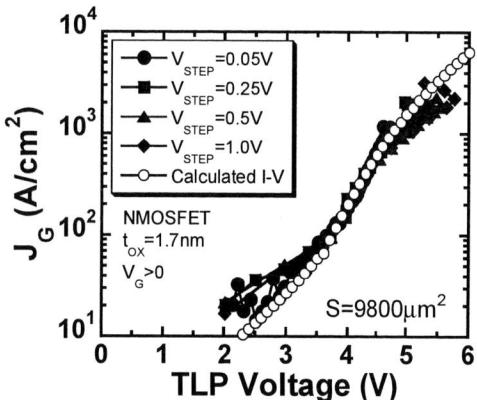

Fig. 5. I-V characteristics for RV-TLP stress. Calculated I-V data, which is estimated from F-N and DT expression in Ref. [13], was also plotted. From this result, breakdown voltages at J_{BD} can be estimated.

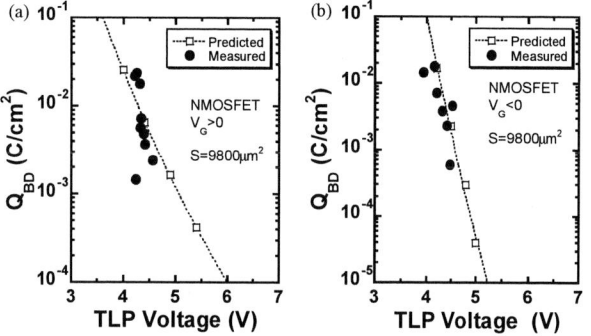

Fig. 6. Q_{BD} for CV-TLP stress as a function of TLP voltage (V_G) in N-MOSFET having 1.7nm SiON. (a) $V_G>0$, (b) $V_G<0$. Predicted Q_{BD} values from RV-TLP measurement (open squire) were estimated from Figs. 6 and 7. Note that Predicted Q_{BD} values correspond to the measured Q_{BD} values.

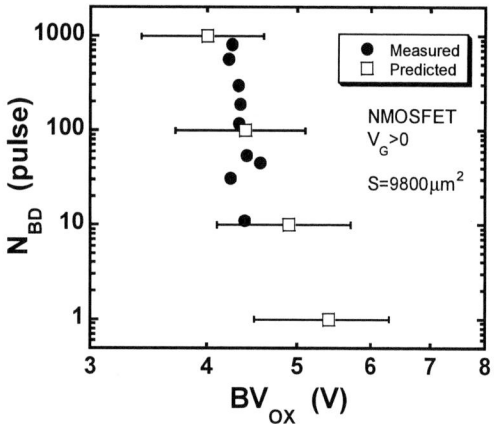

Fig. 7. Prediction of BV_{OX} - N_{BD} correlation for NMOSFETs with 1.7nm SiON. Error bar indicates BV_{OX} from 0.01% to 99.9% failure, which is estimated by considering Weibull distribution of Q_{BD} with Weibull slope $\beta=1.1$ [17]. Experimental data from CV-TLP measurement are also plotted.

Fig. 7 shows the correlation between BV_{OX} and N_{BD} in the case of N-MOSFETs having 1.7nm SiON. By taking into account the Weibull distribution of breakdown [16], predicted results correspond to experimental data well.

Fig. 4 also implies that the same breakdown mechanism is applicable to both RV-TLP and CV-TLP stresses. Proposed mechanisms for the dielectric breakdown include injection of anode hot holes (AHI) [18] or the release of hydrogen (AHR) [19] into the dielectric. In the case of ultra-thin gate dielectrics, power-low model for TDDB is accepted generally, where $t_{BD} = \alpha V_G^{-N}$ [20]. In this model, hydrogen is expected to play a central role in breakdown at low voltages, while there is insufficient energy of injected electron from cathode for impact ionization at the anode. On the contrary, it has been reported that anode-hole-injection (AHI) model can be adaptable to the breakdown mechanism in 3nm gate insulators [21,15]. In order to investigate the breakdown mechanism in the SiON used in this study, we estimate the total fluence of holes to breakdown, Q_P. Fig. 8 shows the J_G-V_G and J_{hole}-V_G characteristics for 1.7nm-SiON N-MOSFETs. Here, J_{hole} can be measured experimentally as the substrate current. Calculated J_{hole} is estimated by

$$J_{hole} = \alpha \cdot P_t \cdot J_G$$

where α is impact ionization rate at the anode, and P_t is tunneling probability of holes [13, 15].

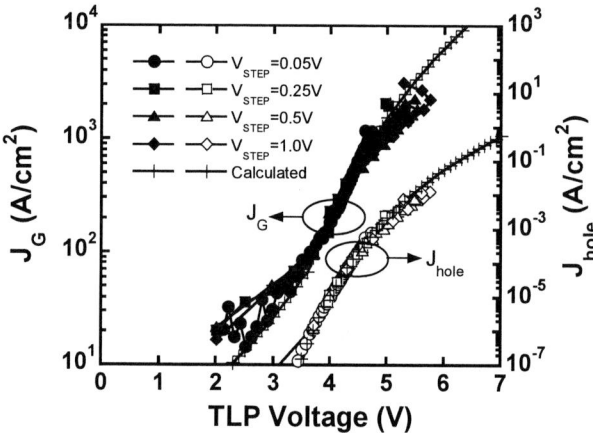

Fig. 8. J_G-V_G and J_{hole}-V_G characteristics for N-MOSFETs having 1.7nm SiON. J_{hole} were estimated using $J_{hole}=\alpha*P_t*J_G$. Here, α is an impact ionization rate and P_t is a tunnelling probability of holes [13, 15].

Fig. 9 shows Q_{BD} and Q_P estimated using experimental results in RV-TLP and CV-TLP measurements. Here, Q_P values are calculated as

$$Q_P = \sum_{n=0}^{N_{last}} (J_{hole}(n) \cdot T_P)$$

It is found that Q_P values are also constant, irrespective of stress condition, though Q_{BD} values monotonically increase with decreasing gate current. In addition, Q_P values under TLP measurement correspond to that under DC stress. This result suggests that the AHI model for breakdown is also applicable even in thinner gate insulators under very short ESD pulse stress conditions. This may be because higher voltage are applied to the samples during ESD testing rather than conventional DC-TDDB measurement.

978-1-4799-7670-6/15 $31.00 © 2015 IEEE

From these correlations, respective BV_{OX} can be estimated from only RV-TLP experimental data. Figure 10 shows the BV_{OX} under TLP measurement obtained in this study. It should be noted that, from the experimental result of V_{STEP} dependence in simple RV-TLP measurement, t_{OX} dependence of BV_{OX} under $1\sim1000$ N_{BD} can be obtained for ultrathin gate insulators. Furthermore, it is clearly shown that the predicted data almost coincides with the reference data which has been reported in Refs. 4 and 7.

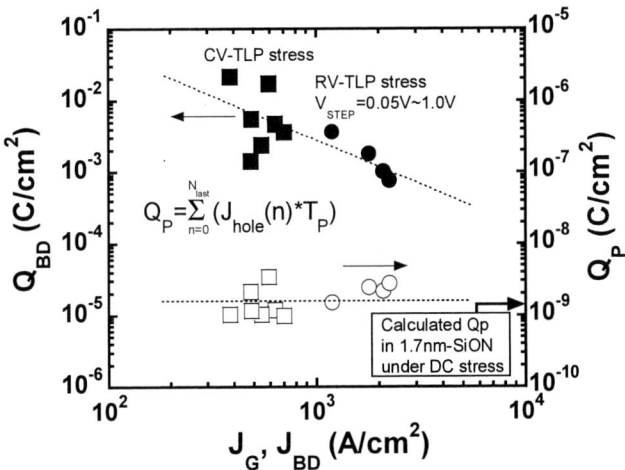

Fig. 9. Q_{BD} and Q_P in cases of CV-TLP and RV-TLP stress as a function of gate current (J_G, J_{BD}). Q_P values were estimated from $Q_P = \Sigma(J_{hole}(n){*}T_P)$. J_{hole} were determined from Fig. 9. It should be noted that Q_P values are constant, irrespective of gate current, applied stress type. Furthermore, these values correspond to the calculated Q_P values in 1.7nm SiON under DC stress.

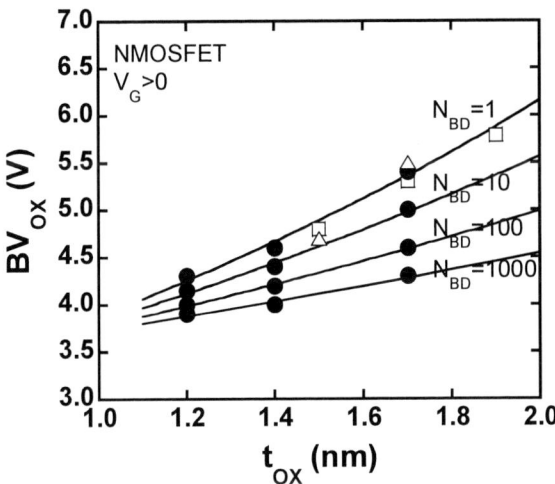

Fig. 10. Median breakdown voltage (BV_{OX}) in N-MOSFETs as a function of oxide thickness (t_{OX}). The breakdown voltage (BV_{OX}) is evaluated for normalized area of 13 mm^2 considering the Weibull scale. From RV-TLP measurement with various V_{STEP}, t_{OX} dependence of BV_{OX} under various N_{BD} can be obtained. Furthermore, the predicted data covers the reference data (open squires in [4] and open triangles in [7] in the case of $N_{BD}=1$).

IV. CONCLUSIONS

In this study, simple prediction technique for BV_{OX} and N_{BD} for ESD testing has been investigate base on the ramped voltage TLP measurement with various V_{STEP}. As a result, Q_{BD}

and BV_{OX} estimated from V_{STEP} dependence of RV-TLP stress correspond to those for CV-TLP stress even in MOSFETs with 1.4nm-SiON gate insulator. From the viewpoint of AHI model, constant Q_P is also observed in TLP measurement. In conclusion, this RV-TLP measurement with various V_{STEP} can be a powerful tool for the prediction for Q_{BD}, N_{BD} and BV_{OX} in nanosecond ESD testing even in ultrathin gate insulators.

ACKNOWLEDGMENT

The authors would like to thank Drs. A. Kinoshita, H. Kawashima, C. Sutou, J. Kurihara, and T. Hiraoka for their thoughtful discussions and comments.

REFERENCES

[1] S. Duvvuy and G. Boselli, "ESD and Latch-up Reliability for Nanometer CMOS Technologies," in IEEE IEDM Tech. Digest, pp. 933-936 (2004).

[2] H. Gieser and M. Haunschild, "Very fast transmission line pulsing of integrated structures and the charged device model," IEEE Trans, Compnents, Packaging, and Manufacturing Technology -Part C, Vol. 21, No. 4, pp. 278-285 (1998).

[3] J. Wu and E. Rosenbaum, "Gate Oxide Reliability Under ESD-Like Pulse Stress," IEEE Trans. Electron Devices, Vol. 51, No.8, pp. 1528-1532 (2004).

[4] B. E. Weir et al., "Gate dielectric breakdown in the time-scale of esd events," Microelectronics Reliability, Vol. 45, pp. 427–436 (2005).

[5] A. Kerber et al., "From Wafer-Level Gate-Oxide Reliability Towards ESD Failures in Advanced CMOS Technologies," IEEE Trans. Electron Devices, Vol. 53, No.4, pp.917-920 (2006).

[6] A. Kerber et al., "Impact of Failure Criteria on the Reliability Prediction of CMOS Devices With Ultrathin Gate Oxides Based on Voltage Ramp Stress," IEEE Electron Device Lett., Vol. 27, No. 7, pp. 609-611 (2006).

[7] A. Berman, "TIME-ZERO DIELECTRIC RELIABILITY TEST BY A RAMP METHOD," in 19th Annual International Reliability Physics Symposium, pp. 204–209 (1981).

[8] D. F. Ellis et al., "Prediction of Gate Dielectric Breakdown in the CDM Timescale Utilizing Very Fast Transmission Line Pulsing," in 47th Annual International Reliability Physics Symposium, pp. 585–593 (2009).

[9] A. Ille et al., "Ultra-thin gate oxide reliability in the esd time domain," in EOS/ESD Symposium, pp. 285-294 (2006).

[10] N. A. Dumin, "A New Algorithm For Transforming Exponential Current Ramp Breakdown Distributions Into Constant Current TDDB Space,And The Implications For Gate Oxide Q_{BD} Measurement Methods," in 36th Annual International Reliability Physics Symposium,, pp. 80-86 (1998).

[11] A. Martin et al., "Dielectric Reliability Measurement Methods: A Review," Microelectron. Reliab, Vol. 38, No.1, pp. 37-72 (1998).

[12] E. Rosenbaum et al., "Accelerated Testing of Si02 Reliability," IEEE Trans. ElectronDevices, Vol. 43, No. 1, pp. 70-80 (1996).

[13] K. F. Schuegraf and C. Hu, "Hole Injection SiO2 Breakdown Model for Very Low Voltage Lifetime Extrapolation," IEEE Trans. Electron Devices, Vol. 41, No. 5, pp. 761-767 (1994).

[14] W. -C. Lee and C. Hu, "Modeling CMOS Tunneling Currents Through Ultrathin Gate Oxide Due to Conduction- and Valence-Band Electron and Hole Tunneling," IEEE Trans Electron Devices, Vol. 48, NoO. 7, pp. 1366-1373 (2001).

[15] A. Kinoshita et al., "Breakdown Voltage Prediction of Ultra-Thin Gate Insulator in Electrostatic Discharge (ESD) Based on Anode Hole Injection Model," in 44th Annual International Reliability Physics Symposium,, pp. 623-624 (2006).

[16] E. Y. Wu et al., "Weibull slopes, critical defect density, and the validity of stress-induced-leakage current (SILC) measurements," in IEEE IEDM Tech. Digest, pp. 125-128 (2001).

[17] J. H. Stathis "Physical and predictive models of ultrathin oxide reliability in CMOS devices and circuits," IEEE Trans. Device Mater. Rel., Vol. 1, No. 1, pp.43-59 (2001).

[18] J.D. Bude et al., "Explanation of stress-induced damage in thin oxides," in IEEE IEDM Tech. Digest, pp. 179-182 (1998).

[19] D.J. DiMaria and J. Stasiak, "Trap creation in silicon dioxide produced by hot electrons," Journal of Applied Physics Vol. 65, 2342 (1989).

[20] E. Y. Wu et al., "Experimental Evidence of TBD Power-Law for Voltage Dependence of Oxide Breakdown in Ultrathin Gate Oxides," IEEE Trans. Electron Devices, Vol. 49, pp. 2244-2253 (2002).

[21] K. Matsuzawa et al., "Simulation of Number of Pulses to Breakdown during TLP for ESD Testing," in IEEE SISPAD, pp. 129-132 (2003).

Off-State Stress degradation mechanism on advanced *p*-MOSFETs

Moonju Cho, Alessio Spessot[*], Ben Kaczer, Marc Aoulaiche[*], Romain Ritzenthaler, Tom Schram,

Pierre Fazan[*], Naoto Horiguchi, Dimitri Linten

Imec, Kapeldreef 75, 3001 Leuven, Belgium,
[*]Micron Technology Belgium, imec Campus Kapeldreef 75,3001 Leuven Belgium
(e-mail: Moon.Ju.Cho@imec.be)

Abstract— Deep insights into the Off-State Stress (OSS) degradation mechanism on *p*-MOSFETs with High-K/Metal Gate technology are presented in this paper. Large subthreshold slope degradation, or positive V_{TH} shift is observed in high, or low V_{TH} devices, where both phenomena impact the off current degradation. The OSS degradation mechanism in *p*MOS is generated by (1) hot carrier generation close to the drain junction by impact ionization, then (2) hot electron injection into the oxide bulk defects, and (3) Si/oxide interface degradation. Both TCAD simulations and measurement with $V_{Gate-to-Drain}$ modulation demonstrate the mechanism. The V_{TH} shift in OSS is toward an opposite direction compared to CHC or BTI, which suggest a means to restore the V_{TH} to the initial value after the OSS degradation.

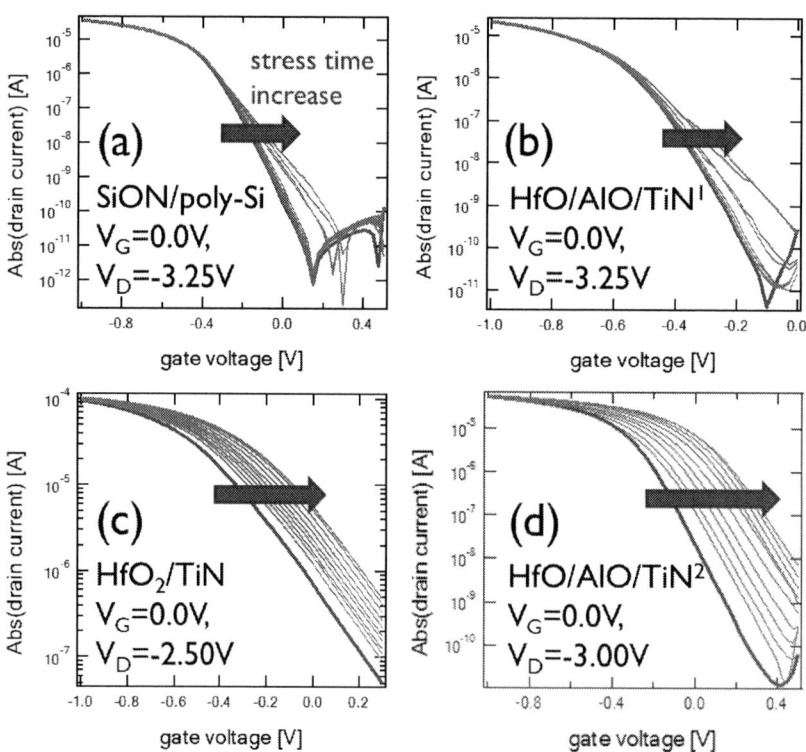

Fig. 1: The evolution of I_{Drain}-V_{Gate} current as a function of the stress time under OSS is shown for (a) Si/SiON/poly-Si, (b) Si/SiO$_2$/HfO/AlO/TiN, (c) Si/SiO$_2$/HfO$_2$/TiN and (d) Si/SiO$_2$/HfO/AlO/TiN devices. The devices in (c) and (d) show lower initial V_{TH} than the devices (a) and (b). The devices (b) and (d) include the same gate stack, however, initial V_{TH} is different due to different gate length and implants, linked to the use of two different mask-sets. Subthreshold slope degradation is mainly observed in the high V_{TH} devices in (a) and (b), positive V_{TH} shift is clearly seen in the low V_{TH} devices in (c) and (d).

978-1-4799-7670-6/15 $31.00 © 2015 IEEE

Keywords — MOSFETs, Semiconductor device reliability, high-k metal-gate, Off State Stress.

I. INTRODUCTION

Controlling the off current of MOSFETs becomes more challenging in scaled devices due to the short channel effect [1]. Additionally, V_{DD} is saturating at a level around 1.0V since the 65 nm node due to the non-scaling sub-threshold slopes of the MOSFET's [2]. This enhances the lateral field applied in the channel, and increases hot carrier generation by higher impact ionization, resulting in higher device degradation. Therefore, studying hot carrier induced off current degradation in short channel MOSFETs is a crucial topic for the low power device development toward the mobile applications.

While the channel hot carrier (CHC) degradation with 'on' state current stress has been extensively studied [3-5], 'off' state current stress (OSS) induced hot carrier degradation has not received much attention. Indeed, the SiO2/poly devices did not show a serious degradation by OSS [6]. However, we have previously demonstrated that the adoption of high-k layer in *n*MOSFET generates a serious concern for OSS, mainly due to the hot hole trapping into the high-k defects [7]. In this paper, we discuss the OSS mechanism in *p*MOSFET with and without the high-k dielectrics in an advanced technology node.

II. EXPERIMENTAL CONDITIONS

p-MOSFET devices have been fabricated on Si(100) wafers. Prior to the high-k deposition, a SiO2 layer was grown, followed by a 1.8 nm HfO2 layer or a 1.8 nm HfO2 /0.5nm AlO stack deposited by means of atomic layer deposition (ALD). As gate electrode, a 5nm TiN metal gate was deposited after the gate oxide formation. EOT was extracted as 1.01 nm for the SiO2/HfO2 device, and 1.3 nm for the SiO2/AlO/HfO2 devices. For the comparison, a SiON with poly-Si gate stack has been fabricated, giving an EOT of 2.0nm. During the Off-state stress (OSS), the Gate, Source and Bulk were grounded while the Drain was biased at constant stress voltage. Full I_{DRAIN}-V_{GATE} curves were measured after each Off-state stress, and the Threshold Voltage (V_{TH}) shift was monitored for the degradation. Short channel devices with 70nm of norminal L_{GATE} have been studied, and the OSS measurement was performed at room temperature.

III. OSS DEGRADATION MECHANISM

The I_D-V_G characteristics measured before and after OSS are shown in Figure 1, for the *p*MOS devices with (a) Si/SiON/poly-Si, (b) Si/SiO2/HfO/AlO/TiN, (c) Si/SiO2/HfO2/TiN and (d) Si/SiO2/HfO/AlO/TiN gate stacks. It is worth noting that among the Metal Gate devices, c) and d)

show lower initial V_{TH} than (b). The devices (b) and (d) include the same gate stack, however, initial V_{TH} is different due to different gate length and implants, linked to the use of two different mask-sets. The SiON/poly-Si device shows a clear sub-threshold slope and off current degradation after OSS, while the V_{TH} shift and G_m degradation are limited. This implies the degradation occurs mainly at the Si/SiON interface. The same characteristic is observed in the SiO2/HfO/AlO/TiN stack (Figure 1 (b)), regardless of adopting the high-k layer containing high density of defects.

In contrast to that, low V_{TH} devices with SiO2/HfO2/TiN and SiO2/HfO2/AlO/TiN stacks show mainly V_{TH} shift after OSS rather than the sub-threshold slope degradation (Figures 1 (c) and (d)), which also impacts on the off current degradation. The effect of the high or low initial V_{TH} is due to current characteristics at V_G=0.0V of OSS condition. For the high V_{TH} device in Figures 1 (b), the channel is not opened at V_G=0.0V, and the bulk to drain current dominates at the OSS condition, as shown in Figure 2. This explains the observation that the N_{it} degradation inside the drain is mainly observed. In the low V_{TH} devices, however, the device is already in the sub-threshold

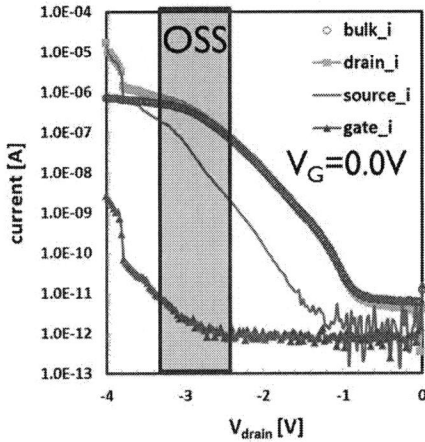

Fig. 2: Gate, Source, Drain, and Bulk currents vs Drain voltage is measured at V_G=0.0V for the high initial V_{TH} device shown in Figure 1 (b). The channel is not opened yet at V_G=0.0V, and the bulk to drain current is dominant at OSS condition.

regime at V_G=0.0V, and the channel current dominates at OSS condition. In this case, the carriers in the channel are controlled by the gate oxide field, and this enhances hot electron injection into the gate oxide. The V_{TH} shift in Figures 1 (c) and (d) supports this mechanism. Note that the low V_{TH} devices are degraded at lower OSS condition (V_{D_stress}), due to the higher current level at the stress. The SiON/poly-Si device in Figure 1 (a) also shows channel opening at V_G=0.0V, however, only N_{it} degradation is observed due to negligible oxide bulk defect density in SiON layer.

To verify the mechanism, a calibrated TCAD deck in TSUPREMIV [8] was used for the high-k/metal-gate device simulation. The impact ionization current simulated by MEDICI is shown in Fig. 3 (a) for the OSS condition of $V_G=V_S=V_B=0V$, $V_D=-3.0V$. It is clearly observed that the peak of the impact ionization is located close to and inside the drain. In Fig. 3 (b), the electrostatic potential distribution shows that high potential difference is applied close to the drain junction. Fig. 3 (c) summarizes the OSS degradation mechanism according to the TCAD simulations, highlighting that (1) the hot carriers are generated close to and inside the drain junction, then (2) the hot electrons damage the Si/oxide interface and/or are injected into the oxide bulk defect by the high field applied between the gate and the drain (note that the bias is relatively positive at the gate than the drain).

Figure 4 demonstrates the mechanism by measurement on the Si/SiO$_2$/HfO$_2$/TiN stack. To observe the impact of the field for the hot carrier injection into the bulk oxide defects as

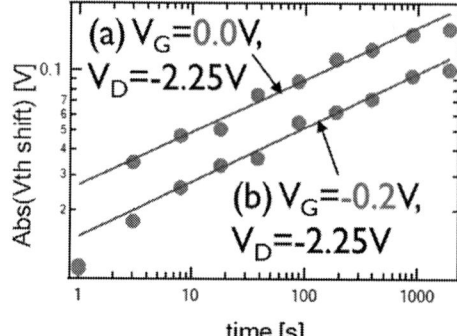

Fig. 4: V_{TH} shift vs stress time at (a) $V_{gate-drain}$=-2.25V, and (b) $V_{gate-drain}$=-2.05V. At lower gate-to-drain voltage, less V_{TH} shift is observed due to the reduced hot carrier injection by lower field applied between the gate and drain.

simulated in Figure 3 (c), the gate voltage during OSS is changed from 0.0V down to -0.2V. V_D is fixed at -2.25V, therefore the electric field between the gate and the drain is decreased at V_G=-0.2V. A clear V_{TH} shift reduction is seen at V_G=-0.2V in the Si/SiO$_2$/HfO$_2$/TiN stack, which shows less hot carrier injection during OSS.

IV. RESTORING THE VTH

The V_{TH} shift after OSS in pMOS is toward positive voltage, due to the hot electron injection into the oxide by the gate to drain applied field. This V_{TH} shift direction is opposite to the one shown by both channel hot carrier (CHC) degradation [9] and negative bias temperature instability (NBTI) [10]. CHC stress is applied at the maximum impact ionization rate, which is $V_G=V_D$ for short channel devices and $V_G\sim V_D/2$ for long channel devices [11]. Both hot electrons and hot holes are generated by the impact ionization, however, hot holes are injected into the gate oxide by the relatively higher V_G than V_D or V_B. This generates negative V_{TH} shift after CHC stress. In case of NBTI, the holes in the inversion layer are injected into the bulk oxide, and this also generates negative V_{TH} shift.

Figure 5 (a) and (b) show clearly the V_{TH} shift direction after OSS and CHC in the Si/SiO$_2$/HfO$_2$/TiN stack studied in this paper. This phenomenon is consistent also to what has been observed in the nMOS devices, with the remark that electrons and holes swap their roles at the opposite biases. To mitigate the effect of OSS, circuital solutions to restore the V_{TH} have been proposed for both nMOS and pMOS [12].

(a) Impact ionization (log scale)

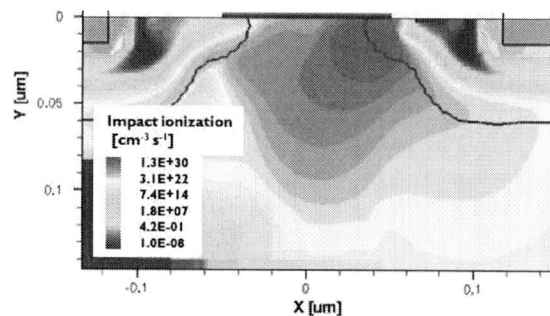

(b) Electrostatic potential

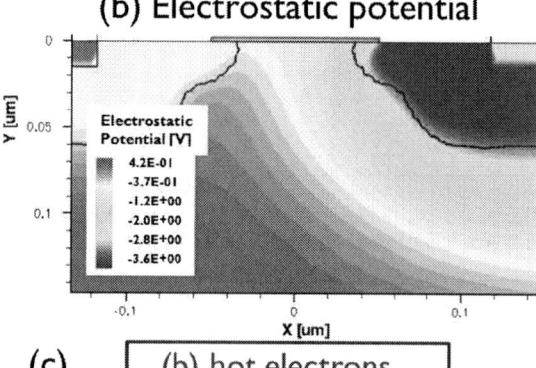

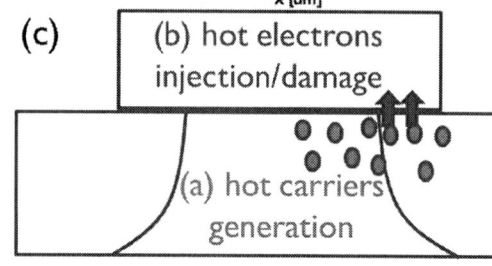

Fig. 3 : TCAD simulations during the OSS of the device for (a) impact ionization, and (b) electrostatic potential. The peak of impact ionization appears inside the drain junction, and the high field applied across the gate-drain overlap enables the hot electrons to damage the gate dielectric interface and to be filled into the oxide defects. (c) A schematic diagram summarizing the degradation mechanism is shown.

V. CONCLUSIONS

The mechanism of off-state stress in pMOSFET has been discussed. In high V_{TH} devices, interface degradation by hot carriers close to and inside the drain affects the sub-threshold and off-current degradation. In low V_{TH} devices, hot electron injection into the oxide bulk defect induces positive V_{TH} shift

It is also noted that the V_{TH} shift direction after OSS is opposite to the shift seen after CHC or BTI, which enables us to propose an algorithm to 'restore' the V_{TH} after OSS by applying BTI-like pulses.

ACKNOWLEDGMENT

The authors would like to thank the Amsimec, P-line, and the imec core partner program for the support.

REFERENCES

[1] S. M. Sze, *Physics of Semiconductor Devices*, John Wiley & Sons Inc., 2nd Ed., 1981.

[2] G. Groeseneken, R. Degraeve, B. Kaczer, K. Martens, "Trends and perspectives for electrical characterization and reliability assessment in advanced CMOS technologies", IEEE ESSDERC proceedings, pp.64-72, 2010.

[3] T. Grasser, *Hot Carrier Degradation in Semiconductor Devices*, Springer, 2014.

[4] A. Bravaix , C. Guerin, V. Huard, D. Roy, J.-M. Roux, E. Vincent, "Hot-Carrier Acceleration Factors for Low Power Management in DC-AC stressed 40nm NMOS node at High Temperature", IEEE International Reliability Physics Symposium (IRPS) Proc., pp. 531-548, 2009.

[5] S. E. Rauch, G. La Rosa, "The Energy-Driven Paradigm of NMOSFET Hot-Carrier Effects", IEEE Trans. on Device and Materials Reliab., Vol. 5, N°4, p. 701, 2005.

[6] V. Reddy, A. T. Krishnan, A. Marshall, J. Rodrigu, S. Natarajan, T. Rost, S. Krishnana " Impact of negative bias temperature instability on digital circuit reliability", Microelectronic Reliability, Vol 45, pp.31-38, 2005.

[7] A. Spessot, M. Aoulaiche, M. Cho, J. Franco, T. Schram, R. Ritzenthaler, B. Kaczer, "Impact of Off State Stress on advanced high-K metal gate NMOSFETs", IEEE ESSDERC proceedings, pp. 365-368, 2014.

[8] "Advanced Calibration User Guide", Version. A-2008.09, Synopsys, Mountain View, CA, 2008.

[9] Ben Kaczer, "Advanced Experimental Techniques for BTI Characterization", tutorial topic 221, IEEE International Reliability Physics Symposium (IRPS), 2012.

[10] E. Amat, T. Kauerauf, R. Degraeve, A. De Keersgieter, R. Rodríguez, M. Nafría, X. Aymerich, and G. Groeseneken, "Channel Hot-Carrier Degradation in Short-Channel Transistors With High-k/Metal Gate Stacks", IEEE Transactions on Device and Materials Reliability, Vol. 9, No. 3, pp. 425-430, 2009.

[11] M. Cho, P. Roussel, B. Kaczer, R. Degraeve, J. Franco, M, Aoulaiche, T. Chiarella, T. Kauerauf, N. Horiguchi, and G. Groeseneken, "Channel Hot Carrier Degradation Mechanism in Long/Short Channel n-FinFETs", IEEE Trans. on Electron Devices, vol. 60, no. 12, pp. 4002-4007, 2013.

[12] A. Spessot, M. Cho, "Restoring OFF-State Stress Degradation of Threshold Voltage", US patent application 14/570,592.

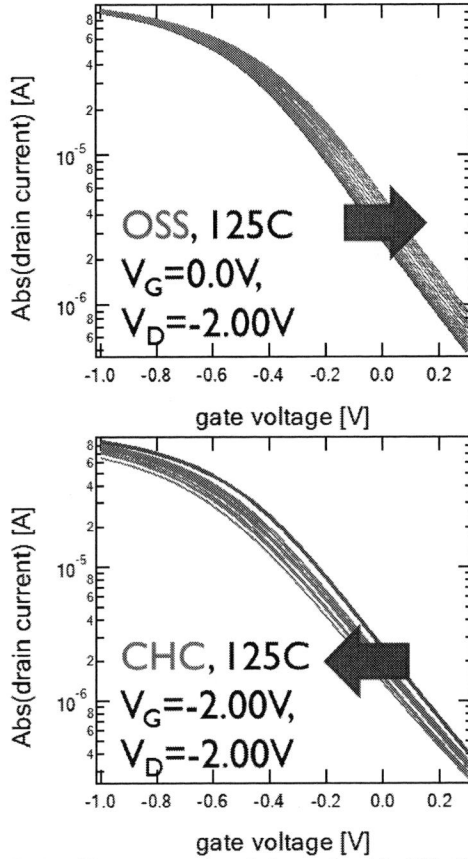

Fig. 5: I_{Drain}-V_{Gate} current degradation under (a) OSS (V_G=0.0V, V_D=V_{stress}) and (b) CHC stress (V_G=V_D=V_{stress}) shows opposite direction of V_{TH} shift. Depending on the bias applied between the gate and drain, either hot electrons are injected in (a), or hot holes are trapped in (b).

after OSS, and this results in the off-current degradation. It is also shown that the off-current degradation occurs in both SiON/poly and high-k/TiN devices, which can be a serious issue for further scaling of low power oriented circuits.

OSS degradation mechanism is confirmed by TCAD simulation. The hot carrier generation is observed close to and inside the drain junction, and the high electric potential difference close to the drain junction affect the hot electron injection into the bulk oxide defects and the hot carrier induced interface degradation. Reducing the gate-to-drain voltage in OSS shows lower V_{TH} shift, which additionally confirms the degradation mechanism, with hot carrier injection by the gate to drain field contributing to the OSS.

Deadspace-aware Power/Ground TSV Planning in 3D Floorplanning

Shengcheng Wang   Farshed Firouzi   Fabian Oboril   Mehdi B. Tahoori

Chair of Dependable and Nano Computing, Karlsruhe Institute of Technology, Karlsruhe, Germany

Email:{Shengcheng.wang, farshad.firouzi, fabian.oboril, mehdi.tahoori}@kit.edu

Abstract—The reliable Power Delivery Network (PDN) design is a challenging aspect in Three-Dimensional-Integrated-Circuits (3D-ICs). In order to ensure the robustness of the 3D PDN, the number and the locations of the Power/Ground (P/G) Through-Silicon-Vias (TSVs) should be carefully planned. Non-regular P/G TSV placement has superior performance compared to the regular one in terms of TSV count. However, the corresponding deadspace optimization is necessary, which complicates the traditional 3D floorplanning. In this work, we propose an efficient deadspace-aware P/G TSV planning combined with the 3D floorplanning to simultaneously place the 2D blocks and P/G TSVs to minimize the total wirelength and the number of inserted TSVs under IR-drop and fixed-outline constraints.

I. INTRODUCTION

Three-Dimensional-Integrated-Circuit (3D-IC) stacking has emerged as a promising technology to extend the 2D scaling trajectory predicted by Moore's Law. For a 3D-IC chip, multiple tiers are stacked together through vertical interconnects such as the *through-silicon-vias* (TSVs) [1]. The TSV-based 3D integration technology provides the possibility of arranging digital and analog functional blocks across multiple tiers at a very fine level of granularity. This results in a considerable routing wirelength reduction of the global interconnects, which naturally translates into less wire delay and power consumption [2]. For this reason, advances in TSV-based 3D-ICs are undoubtedly gaining momentum in semiconductor industry.

The design of robust power delivery is one of the most critical challenges in 3D-ICs [3]. Compared to a 2D chip, the higher integration density and smaller footprint result in a significantly increased power density in a 3D-IC, which leads to power integrity challenges and threatens the system reliability. An unreliable *power delivery network* (PDN) design may result in severe supply voltage variations, which can increase the gate delay and harm the functionality of the circuit. Particularly, since tier stacking has a higher impact on IR-drop in 3D-ICs [3], IR-drop is more important compared to Ldi/dt noise.

As an integral part of the PDN in a 3D-IC, P/G TSVs are used not only to deliver inter-tier power by vertically connecting the on-chip P/G network on different tiers, but also to alleviate the IR-drop challenge with optimal planning. However, it is not free to use P/G TSVs in 3D-ICs. Due to their larger sizes compared to signal TSVs [4], P/G TSVs occupy more area on the chip. Moreover, P/G TSVs prefer to be aligned throughout the 3D-IC [5, 6], which might result in routing congestion [7]. Consequently, the number of P/G TSVs should be reduced as much as possible, while IR-drop constraints must be satisfied as well. Compared to the regular placement, a non-regular P/G TSV placement algorithm [7] can insert less P/G TSV to satisfy the same IR-drop constraint.

To take the advantage of non-regular TSV topology, the P/G TSV placement should be performed based on an accurate power profile [7], obtained from a detailed floorplan. Therefore, the previous work [8–10] perform the P/G TSV planning after the floorplanning, in which the P/G TSVs are inserted in the deadspaces. However, due to the non-optimized deadspace distribution, such a "post-floorplan"

procedure results in suboptimal solutions as demonstrated in this paper. These require more TSVs or even may fail to satisfy the required IR-drop constrains. Moreover, all the optimized metrics and constraints (e.g., wirelength, packing area, and fixed-outline constraint) during the previous floorplanning stage might be degraded and/or violated by this "post-floorplan" P/G TSV planning.

In this work, we address the above issues with an integrated P/G TSV planning as a part of the 3D floorplanning step. During the floorplanning, we seek to co-optimize the locations of blocks and P/G TSVs to diminish the TSV count, while accounting for floorplan qualities (i.e., total wirelength, chip area and fixed-outline constraint) and IR-drop constraints simultaneously. To the best of our knowledge, this is the first work that combines 3D floorplanning and P/G TSV planning together. Our main contributions are as follows:

- We develop an efficient algorithm for P/G TSV planning based on a fast IR-drop estimation method and an effective approach for deadspace redistribution.
- We integrate this P/G TSV planning approach into a *simulated annealing* (SA)-based fixed-outline 3D floorplanner to simultaneously place the 2D blocks and P/G TSVs taking total wirelength, chip area, and IR-drop constraints into account.

The rest of the paper is organized as follows. The background and related work are presented in Section II, followed by a motivational example in Section III. Section IV describes our proposed methodology in detail. The experimental results are presented in Section V. Finally, conclusions are drawn in Section VI.

II. BACKGROUND & RELATED WORK

A. P/G TSV Impact on 3D-IC Layout

The target structure in this work is illustrated in Figure 1. We assume adjacent tiers are bonded in a face-to-back (F2B) fashion. For P/G TSVs, they are routed through stacked local vias in each tier, and hence affect all metal layers as well as device layers in a similar way as via-last TSVs. On the other hand, P/G TSVs are preferably aligned to limit electromigration and IR-drop [5, 6], which means that the TSVs have the same coordinates in the different tiers (as shown in Figure 1). This requirement cause more severe routing congestions compared to signal TSVs if many 3D connections are required [7]. Moreover, since the TSVs do not penetrate through the device layer of the top tier in the 3D-IC (Tier_3 in Figure 1), they can be placed in any location of this tier without deadspace consideration.

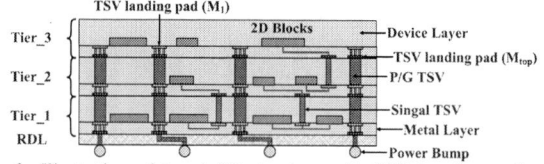

Fig. 1: Illustration of target 3D structure with F2B bonding style and via-first TSVs.

B. Power Delivery Network in 3D-IC

A schematic of a 3D PDN is illustrated in Figure 2. For each tier in a 3D-IC, there is an individual P/G network based on a mesh structure, which is the same as in 2D chips. To reach upper tiers, the supply power goes through P/G TSVs connecting different tiers. As shown in Figure 2, each P/G node, which is an intersection of a vertical and a horizontal power line, is considered as a possible candidate for a P/G TSV insertion.

In this work, the PDN is modeled as a series resistor chain along power line with current sources, which represents the power consumption of each block, based on a given floorplan and a power profile. Moreover, the P/G TSVs and power bumps are modeled as resistors, which are connected with adjacent tiers and ideal voltage sources, respectively.

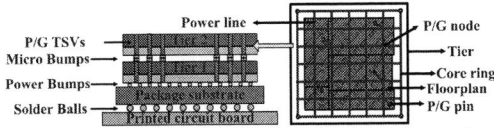

Fig. 2: Three-dimensional power distribution network.

C. Related work

In recent years, numerous works have explored P/G TSV planning. In [3, 5], a regular P/G TSV placement is considered, i.e., the locations of P/G TSVs are predetermined before the floorplanning. To reduce the number of inserted TSVs, several non-regular P/G TSV placement algorithms were proposed, and the majority of them tended to deal with the P/G TSV planning after floorplanning. Based on the generated floorplan, an accurate power profile can be obtained. And then the power delivery bottlenecks can be determined after IR-drop estimation. Nevertheless, the power delivery bottlenecks are usually occupied by 2D blocks and the IR-drop problem cannot easily be alleviated since P/G TSVs cannot be inserted directly into these regions [9]. Therefore, in [8] and [9], the P/G TSVs were inserted into nearby available deadspaces, which may require more TSVs or even may fail to satisfy the required IR-drop constrains. Therefore, the IR-drop noise could not be reduced to the desired threshold, unless the generated floorplan is modified to facilitate TSV insertion by optimizing the distribution of deadspaces. In [10], the deadspaces are redistributed after the generation of the floorplan for TSV insertion while accounting for multiple design optimization goals, which is implemented using the notion of spatial slacks [11]. In case the design goals are not met, the amount of deadspace was increased to ease TSV insertion. However, both of these manipulations might impact the qualities of the floorplan, which have been optimized during the floorplanning phase. On the one hand, the deadspace redistribution can cause degradation on total wirelength and overall chip area, and on the other hand, the deadspace insertion can violate the fixed-outline constraint.

III. MOTIVATIONAL EXAMPLE

In this section, we present an example to motivate the proposed methodology. Figure 3(a) shows the floorplan of a three-tier 3D-IC. From bottom to top in the 3D-IC the tiers are Tier 1, Tier 2 and Tier 3. For each tier, we define width W_{tier} and height H_{tier} as the fixed-outline constraints. As shown in Figure 3(a), the size of each TSV is 2, including the *keep-out-zone* (KOZ) [1].

As shown in Figure 3(a), the optimal candidate for TSV insertion is overlapped with the existing floorplan. Since the P/G TSV cannot

---

[1]The KOZ is the region around the TSV to prevent any logic gates from being impacted by TSV-induced stress.

---

be inserted through the block [2] [9], it cannot be placed in the optimal candidate. Therefore, deadspace redistribution is necessary to create enough space for the P/G TSV insertion.

However, there are cases in which the redistribution is not possible. For instance, in Figure 3(b), the amount of deadspaces in Tier 2 is insufficient with the requirement of TSV insertion. Since the "post-floorplan" deadspace redistribution can only shift the blocks inside each tier in limited ranges, its prospect entirely depends on the generated floorplan. In this case, we have to add extra deadspace for the TSV insertion, which might violate the fixed-outline constraint and degenerate the optimized wirelength and chip area during the floorplanning.

In fact, if we can perform the P/G TSV planning during the floorplanning, the deadspace can be adequately redistributed, which is beneficial for the TSV planning. For example, we can swap the two blocks G and I during the floorplanning, as shown in Figure 3(c). In this case, the deadspace in Tier 2 is sufficient for the TSV insertion. Moreover, since we consider the P/G TSVs planning during the floorplanning, the traditional floorplan metrics (such as wirelength, chip area and fixed-outline constraint) can be co-optimized along with the P/G TSV planning. In this work, we propose a deadspace-aware P/G TSV planning as part of the 3D floorplanning, which will be presented in the next section.

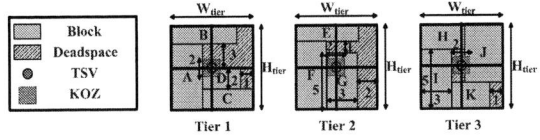

(a) Floorplan with sufficient deadspaces.

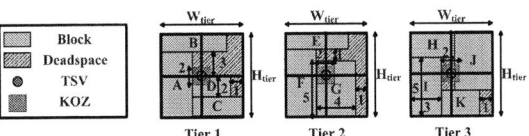

(b) Floorplan with insufficient deadspaces.

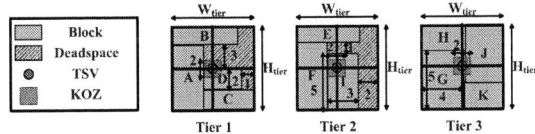

(c) deadspace redistribution during floorplanning.

Fig. 3: Illustration of motivational example.

IV. DEADSPACE-AWARE P/G TSV PLANNING IN 3D FLOORPLANNING

A. Problem Formulation

The deadspace-aware P/G TSV planning algorithm can be formulated as follows:

- **Input**: 1) A set of blocks B, where each block b_i in B has a fixed width and height [3], w_i and h_i; 2) A netlist E; 3) The number of tiers k; 4) The width and height of a tier, W_{tier} and H_{tier} [4]; 5) The width and height of a TSV, w_{TSV} and h_{TSV}.

---

[2]Here we consider a hard block, which means that the block has a fixed size and aspect ratio.

[3]We only consider hard blocks in this work. For a soft block, it can be sharped based on the guideline proposed in [12]. Note that, a TSV can be inserted through a soft block.

[4]In our experiments, we adopt the technique proposed by [13] to determine the width and height, given the allowable deadspace in the floorplan.

978-1-4799-7670-6/15 $31.00 © 2015 IEEE

- **Output**: The values of the coordinates (x_i, y_i, z_i) of each block b_i and the coordinates of each P/G TSV block.
- **Objective**: Minimize the wirelength, the number of inserted P/G TSVs and the number of P/G nodes which violate the given IR-drop constraint.
- **Constraints**: 1) $0 \leq x_i \leq W_{tier} - w_i$, $0 \leq y_i \leq H_{tier} - h_i$ and $1 \leq z_i \leq k$; 2) No two blocks overlap; 3) The P/G TSVs are inserted in the deadspaces between blocks; 4) The P/G TSVs in adjacent tiers are aligned.

B. Overview of the Proposed Algorithm

The overall flow of our approach is shown in Figure 4. As shown in this figure, the proposed methodology can be divided into two parts: 3D floorplanning and P/G TSV planning. Simulated Annealing (SA) is adopted to explore the solution space for 3D floorplanning. For each perturbed solution, the deadspace-aware P/G TSV planning will be performed iteratively. For each loop, one TSV will be inserted as follows: first, the optimal candidate for TSV insertion is determined by IR-drop estimation; In case that the optimal candidate is overlapped with blocks, the deadspace is redistributed to create the space for TSV insertion. After the P/G TSV planning, the related metrics (e.g., chip area, wirelength, aspect ratio, P/G TSV count, and the P/G nodes) are selected for the calculation of cost function in SA.

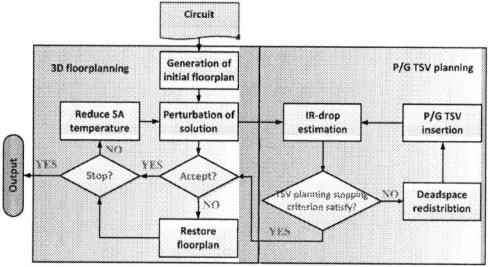

Fig. 4: Overall flow of proposed algorithm.

C. P/G TSV planning

Employing the IR-drop estimation and deadspace redistribution, the P/G TSV planning is performed as follows:

1) By an IR-drop estimation, the optimal candidate for the TSV insertion is identified as the P/G node which has the largest value of IR-drop; For each step, only one TSV is inserted;
2) For all tiers expect the top one, the blocks overlapped with this optimal candidate are shifted to expand enough deadspace capacity for the TSV insertion;
3) Due to the relocations of blocks and the insertion of TSV, the IR-drop distribution will be changed. Thus, the IR-drop value for each node is recalculated.

In case where the IR-drop constraint is not satisfied, we continue with the first step. This TSV planning will be finished once one of the following stopping criterions is met: i) all the nodes satisfy the IR-drop constraints; ii) The deadspace cannot be further redistributed under the fixed-outline constraint.

1) IR-drop estimation: In order to determine the IR-drop value, most previous work consider a modified nodal analysis (MNA) of a resistive equivalent circuit [14], which is very time-consuming. To avoid this issue, a simplified IR-drop estimation is adopted to guide the P/G TSV insertion, which is first proposed in [10]. Here we present some experimental results to verify the accuracy of the qualitative IR-drop analysis. Figure 5 shows an example which compares the initial largest IR-drop value predicted by MNA and qualitative IR-drop analysis. Both of them are performed on the GSRC benchmark [15] n100 with three-tier architecture and

50×50 PDN. We generate 15 floorplans randomly by the perturbation operation in SA. As shown in Figure 5, the qualitative IR-drop analysis matches the trend very well and underestimates the IR-drop obtained by MNA on average by only 6.5%. Moreover, the average runtime of this qualitative IR-drop analysis is just 0.42s, while MNA takes several seconds for each floorplan. Compared to the qualitative IR-drop analysis, MNA is infeasible for being carried out as part of SA.

With this simplified IR-drop estimation, the optimal candidate for TSV insertion is identified as the P/G node n which has the largest value of $IR'(n)$, and obtain the corresponding location. Note that, in this work we set the IR-drop constraint for each node n as:

$$IR'(n) \leq \lambda_{opt} \times IR'_{\max} \ (\lambda_{opt} \leq 1) \qquad (1)$$

Here, $IR'_{\max}$ is the maximal qualitative IR-drop of the initial floorplan, and λ_{opt} is a preset cost factor for the IR-drop optimization.

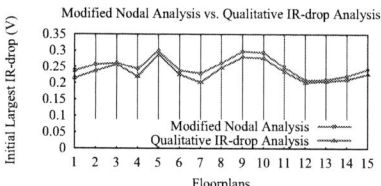

Fig. 5: Comparison between modified nodal analysis (accurate but slow) and qualitative IR-drop analysis (estimated but fast).

2) Deadspace redistribution: After the IR-drop estimation, the deadspace resources should be managed as needed by deadspace redistribution. In this work, a heuristic method based on BFT is used to tackle the deadspace redistribution problem efficiently. We divide the redistribution process into two parts: horizontal redistribution and vertical redistribution. As shown in Figure. 6, the basic idea is to shift the blocks horizontally (or vertically) for expanding the enough whitespace for P/G TSV insertion. In case that there is no enough for a single block, the neighbouring blocks should also be moved following the constraint graph (CG) to maintain a valid placement after the redistribution.

First, based on the initial floorplan, we construct the horizontal constraint graph (CG) for each tier separately. Second, we calculate the maximum moveable distance (MMD) for each block. The computation process is based on BFT, which starts from node T and ends at node S, as shown in Figure 6. The MMD of block i is given as:

$$\mathrm{MMD}[i] = \begin{cases} 0 & \text{if } i = T \\ \min_{j \in \mathrm{Radjacent}[i]} \mathrm{MMD}[j] + \mathrm{Weight}(i \to j) & \text{if } i \neq T \end{cases} \qquad (2)$$

Here $\mathrm{MMD}[i]$ is block i's MMD; Radjacent$[i]$ is right adjacent blocks of block i; Weight$(i \to j)$ is the distance from block i to block j. With Equation (2), we can calculate the MMD, and also the *left side space* (LSS) and *right side space* (RSS) for each block.

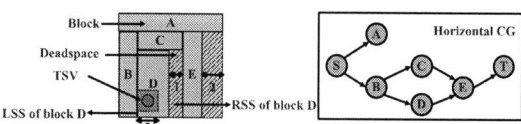

Fig. 6: Illustration of BFT-based deadspace redistribution.

After obtaining this information, the deadspace start to be redistributed for TSV insertion. The deadspace redistribution has two BFT searching processes: the first one travels from S to T, and the second one travels from T to S. In the first BFT process, if the capacity of the LSS of current travelling block is not enough, it will transfer the "expand to right" requirement to its RSS. However, if the RSS is also

978-1-4799-7670-6/15 $31.00 © 2015 IEEE

TABLE I: With-redistribution vs. Without-redistribution

Metric	Proposed approach			Without-redistribution[9]		
	n100	n200	n300	n100	n200	n300
Init. largest IR$'(V)$	0.2145	0.2677	0.4038	0.2145	0.2677	0.4038
violated nodes count	0	0	0	2761	3733	3869
Inserted TSVs count	97	104	125	46	58	71

TABLE II: Combined TSV planning and floorplanning vs. Disjointed TSV planning and floorplanning

Metric	Proposed approach			Disjoint-floorplanning[10]		
	n100	n200	n300	n100	n200	n300
Wirelength (mm)	137.51	336.03	412.11	145.42	367.57	424.55
Chip area (mm^2)	0.122	0.163	0.185	0.124	0.166	0.198
Inserted TSVs count	97	104	125	109	112	132
# of outline violated	0	0	0	2	1	4

not big enough, the block will send the "expand to right" sign to all its right side adjacent blocks.

After travelling the last node T, the second process is started, which lets each block inform its left side adjacent blocks the MMD, which is the horizontal space it can suppose after considering the requirement of TSV insertion. If all the related blocks have sufficient MMDs, the TSV insertion is feasible. The P/G TSV will be inserted in the desired candidate, and then the MMD, LSS, and RSS of each block will be updated. If the MMDs of the existing floorplan are not sufficient for the TSV insertion, the IR-drop estimation will be performed again to determine the new candidate for TSV insertion based on the generated floorplan after deadspace redistribution.

V. EXPERIMENTAL RESULTS

The proposed algorithm is implemented in C++ based on Corblivar [16] and run on a workstation with two Intel Xeon E5540 and 16 GB RAM. The representative GSRC benchmarks [15], n100, n200, n300, were used to test our algorithm. We consider F2B stacking and via-first P/G TSVs with a diameter of 8 μm and a KOZ of $12\mu m \times 12\mu m$ [10]. For the P/G network, we define a 100×100 power mesh for the simplified IR-drop estimation, and the cost factor of the IR-drop optimization λ_{opt} in Equation (1) is 30%. All the reported data in our experiments is collected from 10 independent runs. To demonstrate the effectiveness of the proposed algorithm, we will show the experimental results in the following three scenarios based on a three-tier stack configuration.

A. The necessity of deadspace redistribution

In [9], a region-aware P/G TSV planning was proposed, which does not consider the deadspace redistribution. In this scenario, we compare our algorithm against the one without redistribution. For a fair comparison, we re-implemented this algorithm, and used the same set of benchmarks for both cases.

Without the deadspace redistribution, the TSVs cannot be inserted in the optimal candidates. Moreover, since the P/G TSVs are preferably aligned, the deadspaces between adjacent tiers need to be aligned as well. This requirement further limits the possibility to insert the TSVs successfully, which brings the negative effect on the IR-drop reduction. Therefore, the proposed approach presents superior performance in terms of violated P/G nodes (i.e., the P/G nodes which are not satisfied with the IR-drop constraint.) compared to [9], as shown in Table I. Note that, the smaller TSV counts in column 4 to 7 are due to the insufficient deadspaces in that method, which result in the violation of IR-drop constraints.

B. Combined TSV planning and floorplanning vs. Disjointed TSV planning and floorplanning

In [10], a multi-objective TSV planning was proposed. Although the deadspace redistribution was considered, the TSV planning was performed independently after the 3D floorplanning. Therefore, the planning of TSVs might cause a degradation on the optimized floorplan metrics (such as wirelength and chip area) and can violate the fixed-outline constraint.

In this scenario, we compare our proposed approach against such "disjoint TSV planning and floorplanning" algorithm. Note that, since this algorithm considered a multi-objective optimization, including P/G, signal, clock, and thermal TSVs, we re-implement this algorithm only considering P/G TSV planning, for a fair comparison. As shown

in Table II, compared to our proposed method, the algorithm in [10] generates a suboptimal floorplan with larger chip area and longer wirelength, and an inferior TSV planning in terms of TSV count. For example, the overhead of wirelength is 6.1% and the overhead of TSV count is 8.6% in average for three benchmarks. Furthermore, due to the lack of consideration of fixed-outline constraint, the generated floorplan after TSV planning may violate this constraint. For example, in benchmark n300, there are four times from the ten independent runs that the fixed-outline constraint was violated in method of [10].

C. Runtime

Here, we present the runtime for the three benchmarks with different tiers. Our algorithm is very fast and only takes a few minutes for GSRC benchmarks with 100×100 PDN. The reason for the efficient planning is that: 1) a simplified IR-drop estimation is adopted to determine the optimal candidate for TSV insertion, without MNA of a huge resistive equivalent circuit, and 2) a BFT-based method is used to redistribute the deadspace.

TABLE III: Runtime of different tier stacking

Metric	2 Tiers			3 Tiers			4 Tiers		
	n100	n200	n300	n100	n200	n300	n100	n200	n300
Runtime (s)	156	379	613	177	451	813	204	582	976

VI. CONCLUSION

In this work, we considered the P/G TSV planning during the 3D floorplanning for co-optimizing wirelength and P/G TSV count while satisfying the fixed-outline and IR-drop constraints. We presented an efficient algorithm that simultaneously places the 2D blocks and the P/G TSVs in 3D-IC. Compared to the exiting P/G TSV planning approaches, we consistently produce better solutions in terms of wirelength, area, and TSV count.

REFERENCES

[1] M. Motoyoshi. Through-silicon via (tsv). *Proceedings of the IEEE*, 97(1):43–48, 2009.
[2] Y. Xie. Processor architecture design using 3d integration technology. In *VLSID*, 2010.
[3] N. H. Khan, et al. System-level comparison of power delivery design for 2d and 3d ics. In *3DIC*, 2009.
[4] Y.-J. Lee, et al. Co-design of reliable signal and power interconnects in 3d stacked ics. In *IITC*, 2009.
[5] M. B. Healy and S. K. Lim. Power delivery system architecture for many-tier 3d systems. In *ECTC*, 2010.
[6] M. B. Healy and S. K. Lim. Power-supply-network design in 3d integrated systems. In *Quality Electronic Design (ISQED), 2011 12th International Symposium on*, 2011.
[7] M. Jung and S. K. Lim. A study of ir-drop noise issues in 3d ics with through-silicon-vias. In *3DIC*, 2010.
[8] Z. Li, et al. Thermal-aware power network design for ir drop reduction in 3d ics. In *ASP-DAC*, 2012.
[9] S. Yao, et al. Efficient region-aware p/g tsv planning for 3d ics. In *ISQED*, 2014.
[10] J. Knechtel, et al. Multiobjective optimization of deadspace, a critical resource for 3d-ic integration. In *ICCAD*, 2012.
[11] S. N. Adya and I. L. Markov. Fixed-outline floorplanning: Enabling hierarchical design. *TVLSI*, 11(6):1120–1135, 2003.
[12] T.-C. Chen and Y.-W. Chang. Modern floorplanning based on b*-tree and fast simulated annealing. *Computer-Aided Design of Integrated Circuits and Systems, IEEE Transactions on*, 25(4):637–650, 2006.
[13] L. Xiao, et al. Fixed-outline thermal-aware 3d floorplanning. In *ASPDAC*, 2010.
[14] C.-W. Ho, et al. The modified nodal approach to network analysis. *TCS*, 22(6):504–509, 1975.
[15] *GSRC benchmarks*. http://vlsicad.eecs.umich.edu/BK/GSRCbench/.
[16] J. Knechtel, et al. Structural planning of 3d-ic interconnects by block alignment. In *ASPDAC*, 2014.

Impact of Device and Interconnect Process Variability on Clock Distribution

Nathalie Fiévet[*‡], Praveen Raghavan[‡], Rogier Baert[‡], Frédéric Robert[*],
Abdelkarim Mercha[‡], Diederik Verkest[‡], Aaron Thean[‡]
[*]Université Libre de Bruxelles, Brussels, Belgium
[‡]IMEC, Leuven, Belgium

Abstract—For sub-28nm, process variations became more important. Clock distribution networks are sensitive to those variations because they lead to increased clock skew, which translates to a deterioration of the performance. In this scope, it is the first time that different existing processes are compared. We consider self-aligned double patterning (SADP) and triple expose triple etch (LELELE). First we study the sensitivity of clock skew to interconnect capacitance and resistance. Next we present the influence of the geometry of the tree as the chip size and the clock tree depth. We also investigate the performance of adding air gaps between wires. The results show that the skew is more sensitive to the variation of resistance of the lower metal layers (Mx) and of capacitance of the upper metal layers (Mz). Thus we choose triple-expose triple-etching (LELELE) process for Mx and a relaxed metal pitch for Mz in order to optimize RC-variations. By increasing the depth of the tree the front-end of line (FEOL) influence on skew becomes more dominant with respect to the back-end of line (BEOL) as the number of drivers grows up exponentially with respect to the depth. In the end, we find a trade-off between power consumption and skew deviation with the introduction of air gaps between wires. For a reduction of 9% of the capacitance thanks to the air gaps, the power consumption decreases by the same percentage (6%) as the skew deviation.

Keywords—*clock network, clock skew, h-trees, process variations, self-aligned double patterning (SADP), multiple litho-etch*

I. Introduction

In synchronous logic designs, correct functioning depends on the synchronous arrival of clock signal at the signal storage elements. Differences in latencies in the clock distribution network reduce the maximum frequency at which the circuit can reliably operate. In order to reduce this clock skew, techniques like OCV aware routing [1], buffer and wire sizing [2], and others are employed. However process variations still causes skews in the network. For nodes beyond 28nm multiple patterning is introduced leading to more variability in front-end of line (FEOL) and back-end of line (BEOL) [3]. Process variability is characterized for BEOL through disparity on wire resistance and capacitance, and for FEOL through variation on mismatch parameters A_{Vt} and A_{beta}.

In contrast to other work [1][2], we compare different processes to minimize clock skew like self-aligned double patterning (SADP), triple expose triple etch (LELELE), and single patterning (SP). Furthermore to reduce skew, we investigate different metal pitch for single patterning. We also look at different constrains on the processes over etching, chemical

mechanical planarization (CMP), and critical dimensions (CD). In [6] and [7], they only studied the impact of variations.

This paper is structured as follows: in Section II a description of the setup used for the simulations with an explanation about the different BEOL processes is given. In Section III we describe the choice of the process to compare the influence of BEOL and FEOL on the clock distribution network. Here we also present the results of the influence of BEOL and FEOL on the network through geometrical parameters as the die size, the tree depth and the introduction of air gaps. Section IV concludes this paper.

II. Experimental Setup

In ASIC designs two types of clock networks are commonly used: H-tree for its simplicity [12] and mesh which tolerates better variations [8]. We modeled a symmetrical H-tree structure. Because of the symmetry, we can obtain zero skew when the network presents no mismatch.

The assignment of the clock network wire segments to interconnect layers is done by taking into account the average fan-out and gate density of an ASIC. Thus it can be calculated with an assumption on the ratio of the number of flip-flops (FF) in a design where GP is the gate pitch and MP is the metal pitch. We assume an insertion of two drivers on the first level of the tree to reinforce the clock signal as this level presents the longest wires as depicted in Figure 1.

$$\#Gates = \frac{die\_area}{4 * GP * 9 * MP} \quad (1)$$

$$\#FF = \frac{\#Gates * ratio\_FF}{4^{depth+1}} \quad (2)$$

In this section, the mapping strategy will be briefly explained. Clock networks present the longest wires among networks in ASIC design. Therefore, we investigate different processes for BEOL and their variability in addition to FEOL process variability.

A. Layer assignment and driver sizing

The aim of the layer mapping strategy is to have an acceptable slew rate at the leaves of the tree or mesh while minimizing the buffer sizes. For the simulations, we took a clock frequency of $1.5GHz$ with a slew rate of $10ps$ which is our targeted one at the termination of the tree. Indeed loosing

978-1-4799-7670-6/15 $31.00 © 2015 IEEE

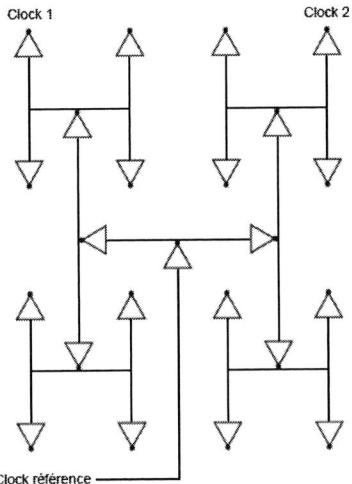

Fig. 1: Balanced H-tree clock distribution with symmetrical structure.

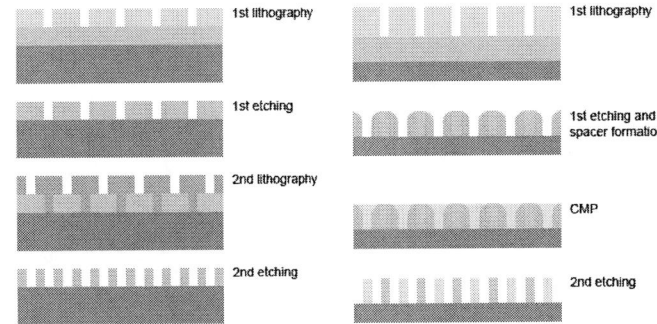

Fig. 2: Process flow for LELE (left) and self-aligned double patterning (right)

sharpness of clock signal implies reduction of its hold time. This means more time violations can occur [9]. Different combinations for driver sizes and metal layer mappings were simulated. To simplify computation, the solution chosen is the one where all drivers on the top-level of the tree have the same strength and the drivers on the lowest level are sized based on the number of flip-flops at the leaves. Furthermore all the clock signal routing is realized on Mz metal layer and only the last level before the leaves is done on Mx. Required driver strength (i.e. the number of finger per transistor) is determined based on the die area and tree depth which are combined into the wire segment length of the last level of the tree.

B. BEOL process assumptions and variability

The BEOL stack in the experimental setup consists of 9 metal layers: 5 layers of local and intermediate interconnect at minimum pitch (Mx), 2 layers of semi-global interconnect at double Mx pitch (My) and 2 global interconnect layer at 8 times Mx pitch (Mz). The BEOL stack process assumptions are summarized in Table I.

TABLE I: BEOL process assumptions

Layer Type	Mx	My	Mz
Minimum Pitch [nm]	48	96	384
Minimum CD [nm]	24	48	192
Aspect Ratio	2	2	2
Dielectric ϵ_r	2.2	2.5	3.1
Metal Barrier Thickness [nm]	3	3	3
Dielectric Barrier Thickness [nm]	10	15	30
Dielectric Barrier ϵ_r	5.5	5.5	5.5

For sub-28nm nodes, multiple patterning turns out to be the intermediate solution to EUV which is still under development [10]. Different lithography methods are considered for the Mx layers: self-aligned double patterning (SADP) and multiple litho-etch (LE) [13]. In the first process the building

of lines is manufactured by making primitive cores (mandrels) on the layer. Then spacers are deposited around them. After removing the cores we have multiple lines lying in the core and gap places as illustrated on Figure 2 on the right. Therefore we introduce several variability on the lines because of the variation on the core CD and the spacer CD, and others process parameters such as etching and CMP. In the other hand we have the triple expose triple etch which consists by printing lines with three different masks leading to overlay and line variability [13][14]. Those steps are illustrated for the LELE in Figure 2 on the left.

To estimate the electrical variability of an interconnect wire, we assume a normal distribution for the core CD, spacer CD, etch, CMP and overlay parameters, with standard deviations estimated based on empirical measurement data. Using a field solver for capacitance calculation and a semi-empirical resistivity model, the standard deviations and correlation coefficient for resistance and capacitance are estimated by running the model in a Monte Carlo loop.

C. FEOL process assumptions and variability

We assume a FinFET device targeted for the 10nm node with 20nm gate length. The devices used in clock signal repeaters have 4 fins per device, and a variable number of fingers to change the drive strength.

Process variability for FEOL is characterized by variation of threshold voltage and gain through the size-dependent mismatch parameters A_{Vt} and A_{beta}. These parameters capture the electrical variability of transistors due to, for example, dopant fluctuations, line edge roughness, and channel length modulation [4][5].

D. Skew estimation

The skew is computed as the maximum difference in clock arrival time between two flip-flops of two different branches (clock1-clock2 in Figure 1). Skew variability as function of FEOL and BEOL variation is obtained from Monte Carlo circuit level simulations with 500 runs.

For the nominal case in out benchmark we assume a square die area of $0.36mm^2$ and a clock tree depth of 6 levels. The fan-out of the leave nodes is 10 flip-flops.

978-1-4799-7670-6/15 $31.00 © 2015 IEEE

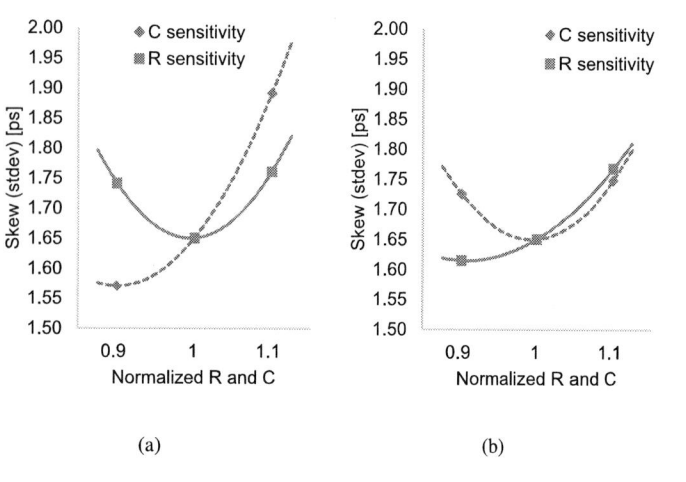
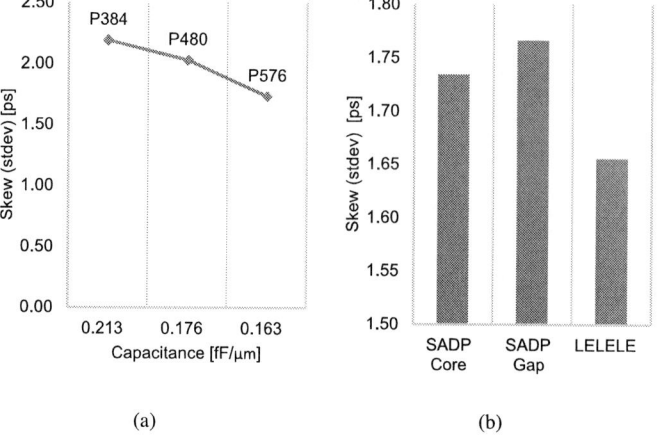

(a)   (b)   (a)   (b)

Fig. 3: Sensitivity of skew deviation on R and C values for (a) Mz layer, and (b) Mx layer on BEOL

Fig. 4: (a) Skew for different metal pitch for Mz and (b) skew deviation for different processes for Mx.

III. RESULTS

Process variability is investigated for different patterning options and different metal pitches. We only consider skew for different Mx and Mz layer options because the clock network is not mapped on My in our design.

A. Impact of Metal Layers Mx and Mz

We first try to determine the dominant parameter of BEOL, between resistance and capacitance. As can see from Figure 3(a), the clock network is more sensitive to capacitive variations for the upper levels Mz. On the other hand, on the lower levels, Mx, we can see in Figure 3(b) wires are resistive dominant as we have smaller length. Even if the design only includes the last level on Mx, the choice of the process has an influence on the mean of the skew and its deviation (Figure 4(b)). Thereby we increase metal pitch for Mz which reduces the deviation on the skew as shown in Figure 4(a) and choose LELELE process which has less variation on wire resistance. The assumption made for the simulations can be found in the Table II. The values of resistance and capacitance of each layer are listed as well as their process variability and correlation.

TABLE II: Skew with respect to near maximum relaxed constrains on different process parameters of BEOL

	$R[\Omega/\mu m]$	σ_R	$C[fF/\mu m]$	σ_C	MP $[nm]$
Mx	60.93	0.126	0.176	0.04	48
Mz	0.355	0.039	0.163	0.02	576

B. Power Consumption

Another parameter which is important is the impact is the power consumption of the clock distribution network (CDN). Process variability doesn't affect significantly CDN power as the deviation is less than 1% of the mean value. Therefore, we only compare the mean of the power consumption for different parameters and not its deviation.

C. Comparison FEOL and BEOL

In this Section, we will study the influence of the geometry, specifically the depth of the tree and the area of the die, on the skew of the clock network.

As expected the deviation on skew grows with respect to the increase of the area of the die. Indeed, the bigger the die is the longer the wires are, implying an increase of R and C values and their absolute variation. This also has an effect on the FEOL parameter due to dependence of the current on the load. As consequence current fluctuation leads to skew deviation. Furthermore, we can observe on Figure 5(a) that variability of FEOL becomes more dominant as the die size increases. Likewise the deviation on skew decreases with the depth of the tree as we have less wires and drivers (as shown on Figure 5(b)).

TABLE III: Comparison BEOL, FEOL, and both BEOL and FEOL variability

	BEOL	FEOL	BEOL and FEOL
Skew Deviation [ps]	1.66	2.37	3.28

In the Table III, we look at the influence between the BEOL, FEOL and both effect at the same time. We can notice FEOL variations have the highest impact compared to the BEOL on the clock network. The effect of both front-end and back-end don't cancel each other. Besides we can say they increase the deviation on the clock skew.

D. Air Gaps

We investigate the consequence of using air gaps (AG) instead of low-k dielectric in the BEOL stack. The main advantage is the power consumption decrease due to the lowering of the wire capacitance. Furthermore, the depth of the H-tree can at the same time be lowered in order to decrease the skew deviation on the BEOL (as depicted in Figure 6(a)) and even more the power consumption. For the FEOL, having one

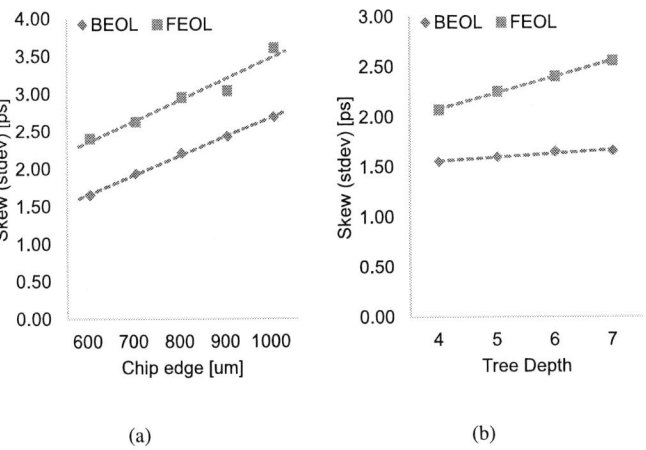

Fig. 5: (a) Skew deviation for different die size, (b) Skew deviation for different depth levels

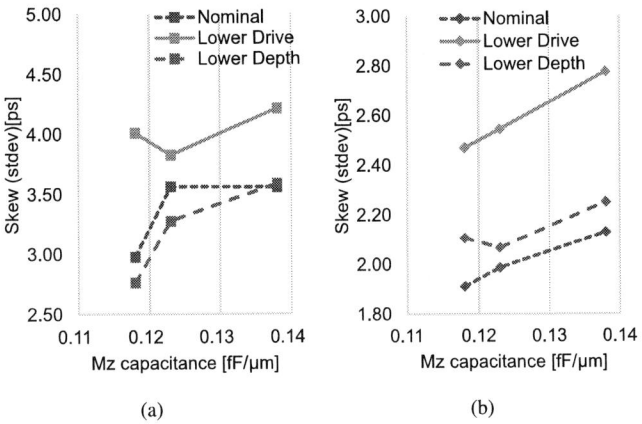

Fig. 6: (a) Skew due to BEOL and (b) to FEOL for network with air gaps on Mz for different configurations

TABLE IV: Global skew for Mz with and without air gaps

	Skew Deviation [ps]	Mean Power [mW]
Without AG (C_{Mz} = 0.164 fF)	3.28	63.97
With AG (C_{Mz} = 0.118 fF)	3.49	59.93
Gain of with over without AG	6.36 %	-6.31 %

sensitive to process variation in terms of resistance at Mx level. On the other hand Mz layer should present less process variation in terms of capacitance. Therefore this shows that low process variations leads to low skew deviation. Nevertheless improvements on the circuit level is done to reduce skew deviation. The two main modifications are on the depth of the tree and the chip size. As expected skew variability increases with the area as the wires length is proportional to the chip size. Besides with low tree depth the amount of drivers decreases and their required strength increases. This two phenomena leads to diminish skew deviation. Furthermore we note that the dominance of FEOL on BEOL is also decreasing with tree depth reduction. In the last section we investigate the effect of air gaps insertion. We reduce the power consumption by trading the clock skew deviation by the same percentage.

REFERENCES

[1] T.-B. Chan, K. Han, A. B. Kahng, J.-G. Lee and S. Nath, "OCV-Aware Top-Level Clock Tree Optimization," in *Proc. Great Lakes Symposium on VLSI*, 2014, pp. 33-38.

[2] G. Wilke and R. Reis "A new clock mesh buffer sizing methodology for skew and power reduction," in *Proc. IEEE Comput. Soc. Annu. Symp. VLSI*, IEEE, 2008, pp. 227-232.

[3] K. Jeong, A. Kahng, and R. Topaloglu, "Assessing Chip-Level Impact of Double Patterning Lithography," *IEEE Intl. Symp. Quality Electronic Design (ISQED)*, IEEE, March 2010, pp. 122-130.

[4] S. Borkar, *Designing reliable systems from unreliable components: The challenges of transistor variability and degradation*, IEEE Micro, vol. 25, no. 6,November/December 2005, pp. 10-16.

[5] G. Konstadinidis, "Chapter 8: Physical Design Considerations," in *Clocking in Modern VLSI Systems*, Springer, 2009, pp. 248-298.

[6] Y. Liu, S. R. Nassif, L. T. Pileggi, and A. J. Strojwas. "Impact of interconnect variations on the clock skew of a gigahertz microprocessor," in *Proc. DAC*, IEEE, 2000, pp. 168-171.

[7] A. Narasimhan and R. Sridhar "Impact of variability on clock skew in H-tree clock networks," in *Proc. ISQED*, IEEE, 2007, pp.458-466.

[8] P. Chakrabarti, V. Bhatt, D. Hill, and Aiqun Cao, "Clock mesh framework," in *Proc. ISQED*, IEEE, 2012, pp.424-431.

[9] C. Koh, J. Jzin, and S. Cauley, "Chapter 13: Synsthesis of Clock and Power/Ground Networks," in *Electronic Design Automation*, Elsevier Science & Technology, 2009, p.766.

[10] Lithography Chapter in ITRS 2013 Edition, http://www.itrs.net/Links/ 2013ITRS/2013Chapters/2013Litho_Summary.pdf.

[11] G. Konstadinidis, "Chapter 8: Physical Design Considerations," in *Clocking in Modern VLSI Systems*, Springer, 2009, pp.276-277.

[12] G. Konstadinidis, "Chapter 2: Modern Clock Distribution Systems," in *Clocking in Modern VLSI Systems*, Springer, 2009, pp.23-29.

[13] Lithography Chapter in ITRS 2011 Edition, http://www.itrs.net/Links/ 2011ITRS/2011Chapters/2011Lithography.pdf.

[14] M. Mirsaeedi, A.J. Torres, and M.H. Anis, "Litho-Friendly Decomposition Method for Self-Aligned Double Patterning," in *Very Large Scale Integration (VLSI) Systems*, IEEE, 2013, pp. 1469-1480.

[15] S.-C. Wong, K.-H. Pan, and D.-Y. Ma, "A CMOS Mismatch Model and Scaling Effects," in *IEEE Electron Device Letters*, Vol. 18, No. 6, June 1997, pp. 261-263.

level less means a higher current to supply as we have more flip-flops at the leaves. Thus there is more absolute variation on the current as we keep the same relative variation. As can be seen on Figure 6(b) decreasing the drive strength won't bring the diminution of the skew variation. Indeed, weakening the drivers will augment the current variability of each transistor as σv_t is proportional to the inverse of the number of fingers [15].

The drawback of having air gaps lies in the increase of the skew deviation as shown in Table IV. However in the case simulated, the loss of power consumption is translated in an increase of skew deviation.

IV. CONCLUSION

Clock skew variation tendency is to rise with scaling. In order to lessen this variability, modifications on the circuit level can be done. However in this article we show that we can also achieve skew deviation reduction by choosing LELELE process. Indeed we find skew variability is more

Impact of time-dependent variability on the yield and performance of 6T SRAM cells in an advanced HK/MG technology

P. Weckx[1,2,*], B. Kaczer[2], Ph. J. Roussel[2], F. Catthoor[1,2], G. Groeseneken[1,2]

[1]Katholieke Universiteit Leuven, ESAT-MICAS, Leuven, Belgium [2]imec vzw, Leuven, Belgium

[*]+3216281342, pieter.weckx@imec.be

(Invited Paper)

Abstract—Stochastic device degradation—due to individual oxide defects—like Random Telegraph Noise (RTN) and Bias Temperature Instability (BTI) causes a threshold voltage drift of transistors resulting in decreased SRAM yield and performance. BTI and RTN has been shown to follow an defect-centric behavior, which can be bimodal in nature for heterogeneous gate oxide stacks. Consequently the tail of the distribution can significantly deviate from a Gaussian distribution. In this paper we combine statistical silicon extracted from large transistor arrays (32k) designed and fabricated in an advanced 20nm High-k/Metal Gate process, with current state-of-the-art statistical assessment techniques in order to acquire a realistic impact of BTI degradation on the yield and performance of 6T SRAM cells.

Keywords-component; Bias temperature instability, time dependent variability, SRAM, high sigma

I. INTRODUCTION

Recent advances in characterizing and modeling degradation as time-dependent variability has led to a paradigm shift in the assessment of circuit reliability margins [1-3]. Accurate predictive reliability models, e.g. for ΔV_{TH} BTI, are imperative for a correct simulation of circuit Figure Of Merits (FOMs). Even more so when high yields are required, e.g. for SRAM arrays, high predictive power in the tail of the distribution is necessary. Consequently, coping with transistor parameter variability has become a significant design challenge for current State of the Art (SotA) VLSI design.

Contrary to the initial time-zero V_{TH} distribution, time-dependent ΔV_{TH} distributions, induced by Bias Temperature Instability (BTI) and Random Telegraph Noise (RTN), have been shown to deviate from a normal distribution far below 6σ [4,6,7]. Theoretical formulation of the ΔV_{TH} NBTI distribution has been described in [7] using a defect-centric picture, recently extended to a more generic model capable of handling PBTI distributions which can be bimodal in nature [8]. Besides accurate models, silicon data, taken from a large number of samples (test element group (TEG)) are complementary necessary to make accurate predictions on the main defect-centric parameters [8]. Finally, recent simulation methodologies allows evaluating the workload-dependent BTI impact on circuit performance for usage based scenarios [9]. In this paper we combine statistical silicon data extracted from TEG structures designed and fabricated in an advanced 20nm HK/MG process [8], with current state-of-the-art high-sigma statistical

assessment techniques [9] in order to acquire a realistic impact of BTI degradation on the yield and performance of advanced 6T SRAM cells.

II. SRAM SIMULATION SETUP

Fig. 1a shows the studied 6T SRAM cell. Transistor width sizing is chosen to be 1/1/2 and 1/1.5/2 for Pull-Up (PU), Pass Gate (PG) and Pull-Down (PD) transistors respectively. All transistors are of minimum gate length. The SRAM cell performance metrics are obtained by simulation using a 20nm Predictive Technology Model (PTM) in a SPICE like simulator.

A. Read margin

To characterize the read stability of the SRAM cells, the Static Noise Margin (SNM) is taken from the butterfly plot (Fig. 1b)[10] during read mode, i.e. when the Word Line (WL) and Bit Line (BL, BLbar) voltages equal the supply voltage VDD:

$$SNM = \min(SNM_1, SNM_2) \qquad for\ BLs = WL = VDD. \quad (1)$$

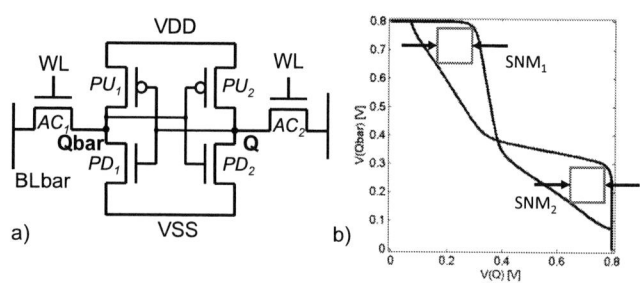

Figure 1. (a) The 6T-SRAM cell used. (b) SNM margins are obtained by analyzing the butterfly curve during read mode, where the Word and Bit Lines are are equal to VDD.

B. Write margin

To characterize the write stability and performance of the cell, both the Write Trip Point (WTP) (as a static margin) and the Write Margin (WM) (as a dynamic speed margin) is used. Fig. 2a shows the definition and extraction of the WTP: i.e. by lowering the BL and BLbar voltages until the cell is written. Fig. 2b shows the definition and extraction of the Write Time. The dynamic WM is then defined as the time the WL is high (set at 1ns) subtracted by the Write Time (Eq. 2).

This work has been partially supported by the Agency for Innovation by Science and Technology in Flanders (IWT)

978-1-4799-7670-6/15 $31.00 © 2015 IEEE

$$WM = 1ns - Write\ Time \qquad (2)$$

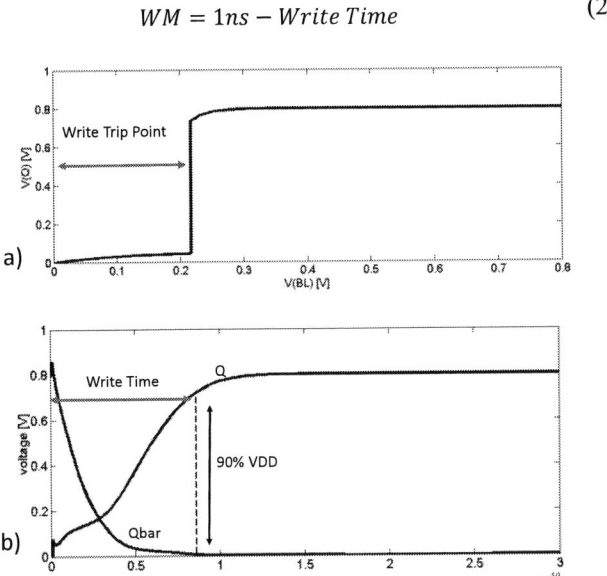

Figure 2. (a) Extraction of the WTP by lowering the BL voltage until the cell is written. (b) Extraction of the Write Time by flipping of the SRAM cell during a write operation when the WL is high. The Write Time is defined as the time needed for the node Q to reach 90% of the logic '1' level

C. Workload scenarios for SRAM cells

Shown in Fig.3 are various degradation scenarios that can be encountered for an SRAM cell. Due to the circuit topology and usage during read and write, highly correlated degradation occurs for the cell transistors. The degradation of the PU and PD transistors are dependent on the probability of storing a '0' or '1' which results in an average Duty Factor (DF) over the lifetime of the cell (frequency dependencies on the degradation are not taken into account in this work). Due to the cell design, the PU_{left} and PD_{righ} always degrade simultaneously, as does the PU_{right} and PD_{left}. Moreover the DF of the cells determines the level of degradation of all core transistors as given by

$$
\begin{aligned}
DF_{PUleft} &= DF_{PDright} \\
DF_{PUright} &= DF_{PDleft} \\
DF_{PUleft} &= 1 - DF_{PDleft}
\end{aligned}
\qquad (3)
$$

The degradation of the PG transistors depends on the activity of the cell for read and write operations. Nevertheless PG transistors will always have a lower degradation then the PU and PD transistors since it is not a realistic case of having a constantly activated WL in an SRAM array.

D. Workload dependence of BTI degradation

Workload dependence of BTI can be derived using capture/emission time (CET) maps [11,12,13] describing the probability density function of broadly distributed defect capture and emission times and their correlations acquired from experimental eMSM data. Evaluating the occupancy probability at each point in the CET map space for a given workload, allows capturing accurate workload dependence of the mean ΔV_{TH} for a variety of stress patterns. However eMSM data was not provided for the tested silicon in [8].

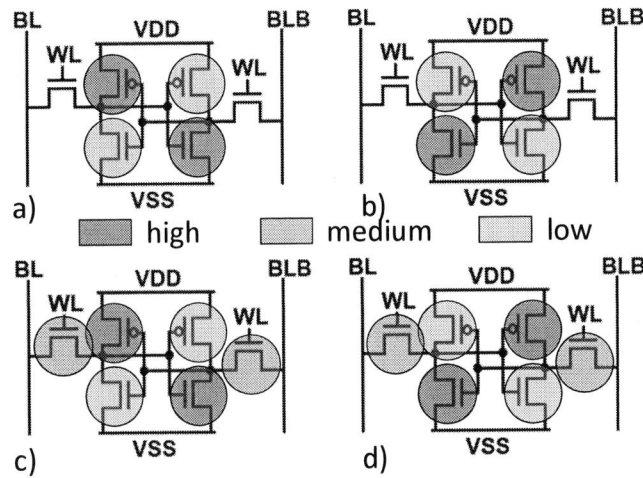

Figure 3. Different degradation scenarios for SRAM cells. Storing a '1' (a) or '0' (b) with minimum activity of the cell leading the PG transistor to be undegraded. High read/write activity can lead to the PG being partially degraded (c,d). Colors indicate the level of degradation.

Therefore a more empirical based model is used which fits the BTI-induced mean ΔV_{TH} by a simplified power law:

$$\Delta V_{TH} = A V_{OV}{}^{\gamma} t^{n}, \qquad (4)$$

where t is the time, V_{OV} the overdrive voltage, and A, n and γ are fitting coefficients, respectively. Since the mean number of occupied traps $N_T = \Delta V_{TH}/\eta$, it also can be fitted using a simplified power-law model [8].

E. Time-zero and time dependent variability

Time-dependent variability parameters were extracted for 20nm advanced HK/MG technology used TEG structures described in [8]. Time-zero variability is directly derived from the mismatch between two transistors using Pelgrom's mismatch parameter $A_{\Delta VTH}$ [14] and modeled using a Normal distribution with

$$\sigma_{V_{TH}} = \frac{A_{\Delta V_{TH}}}{\sqrt{2WL}} \qquad (5)$$

The BTI ΔV_{TH} distribution can be described from a defect-centric point of view through the statistics dependent on the average impact per defect η and number of active defects N_T, which has been shown to match with experimental data [7,8]. Measurements on large variability arrays allow to accurately assess both N_T and η. PBTI has been shown to follow bimodal defect-centric behavior, related to HK and IL/HK interface trapping described by [8]

$$F_{N_1,N_2,\eta_1,\eta_2}(\Delta V_{TH}) = \sum_{k_1=0}^{\infty} \sum_{k_2=0}^{\infty} \frac{e^{-N_1}N_1^{k_1}}{k_1!} \frac{e^{-N_2}N_2^{k_2}}{k_2!} F_{k_1,k_2,\eta_1,\eta_2}(\Delta V_{TH}). \qquad (6)$$

For NBTI an unimodal defect-centric distribution is used described by [7]

$$F_{N,\eta}(\Delta V_{TH}) = \sum_{k=0}^{\infty} \frac{e^{-N}N_2^{k}}{k!} F_{k,\eta}(\Delta V_{TH}) \qquad (7)$$

978-1-4799-7670-6/15 $31.00 © 2015 IEEE

Values for time-dependent variability parameters are shown in TABLE I originating from the work done in [8] together with Pelgrom's mismatch parameter $A_{\Delta VTH}$ taken as 1.8 mVμm. Fig. 4 then shows the resulting total ΔV_{TH} distribution for PU/PD/PG with and without 10 years BTI degradation under a fixed supply voltage V_{DD}.

Parameters\FETs	PU	PG	PD
type	PMOS	NMOS	NMOS
$A_{\Delta VTH}$ [mVμm]	1.8	1.8	1.8
η [mV]	7	3.6/0.77	3.6/0.77
γ	2.7	1.17/4.9	1.17/4.9
n	0.16	0.097/0.29	0.097/0.29

TABLE I Time-dependent variability parameters for 20nm advanced HK/MG technology published in [8]. Due to the bimodal behavior for PBTI the time-dependent variability parameters are given in pairs. Pelgrom's mismatch parameter $A_{\Delta VTH}$ is taken as 1.8 mVμm.

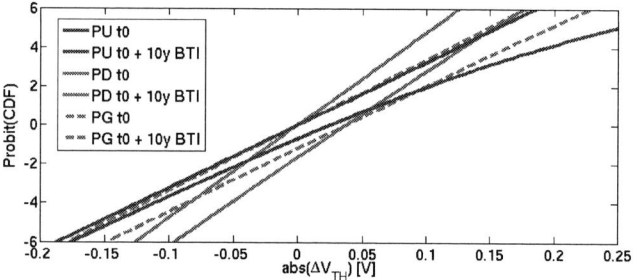

Figure 4. The total ΔV_{TH} distribution for PU/PD/PG with and without 10 years BTI degradation under a fixed supply voltage V_{DD}.

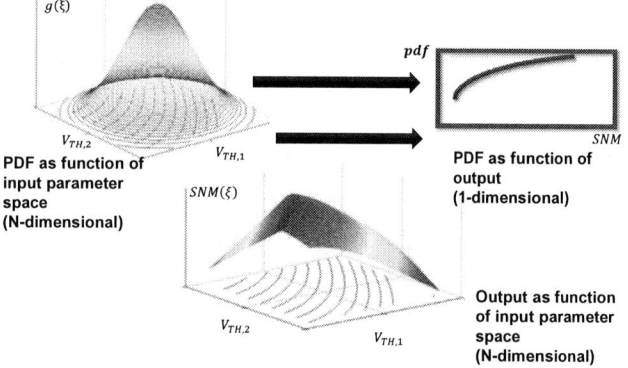

Figure 5. Illustration of the calculation of output statistics using the response surface for output parameter SNM(ξ) as function of input parameters ξ. The corresponding SNM Probability density function (PDF) can be evaluated by propagating the PDF g(ξ) of the input parameters via the output response surface SNM(ξ) [9].

F. High Sigma simulation of SRAM degradation

A numerical Non-Monte-Carlo methodology capable of reproducing circuit output distributions at very high quantiles is used where arbitrary input distributions can be handled [9]. Input distributions are propagated to output distributions using circuit response surfaces. For each FOM of the SRAM cell (SNM,WTP and WM) a response surfaces is calculated for an equidistant discrete input space (ΔV_{TH} for all PU, PG and PD transistors of

the SRAM cell). By proper definition of the input space boundaries, a resolution of -7 sigma is obtained which allows us to investigate the far end tail regions of the SRAM FOMs.

III. RESULTS AND DISCUSSIONS

A. Impact sizing on of time-zero variability

In Fig. 4 the impact of time-zero variability on the SNM, WTP and WM for 2 different cell sizing is shown.

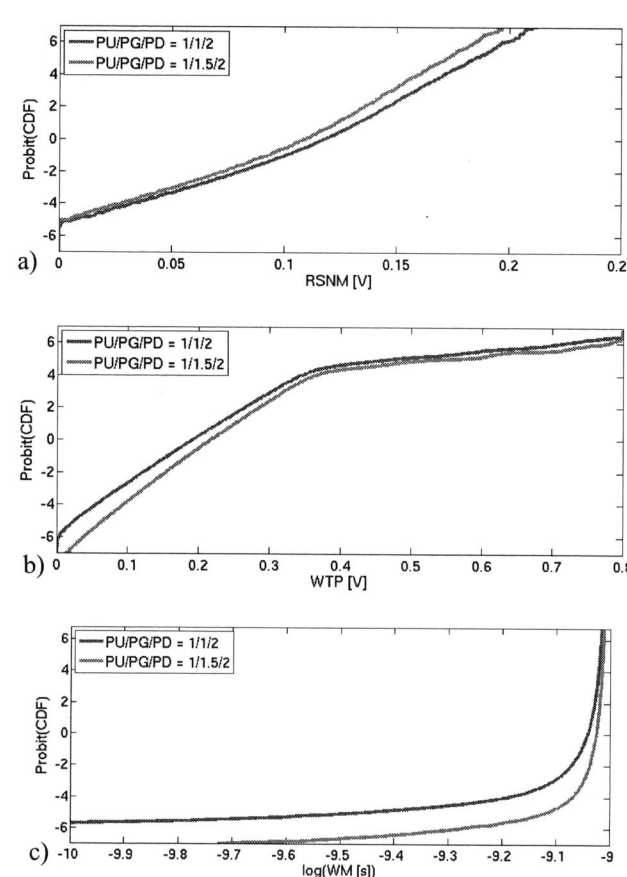

Figure 6. The impact of time-zero variability on the (a) SNM, (b) WTP and (c) WM for 2 different cell sizings 1/1/2 and 1/1.5/2 for the PU/PG/PD respectively.

Comparing the 1/1/2 and 1/1.5/2 sizing schemes, the latter provides better write ability, both in terms a stability (WTP) and speed (WM). This is not to be surprising since a wider sized PG allows for more writing current. However careful attention is needed to make sure that the read stability is not compromised. Fig. 5a shows that although the average SNM is reduced, the low tail region converges between the -5 and -6 sigma points. Hence the read stability for both sizing schemes is equivalent since the probability for a destructive read is at the same order of magnitude. The reason for this convergence is due to the reduced variability of the PG transistors which are 1.5x larger then minimum sized. For further reliability simulations, the 1/1.5/2 sized cell is used.

B. Impact of time-dependent variability

To simulate the impact of time dependent variability the different degradation scenarios for SRAM cells, shown in Fig.

978-1-4799-7670-6/15 $31.00 © 2015 IEEE

3, were investigated. In Fig. 7a the impact of variability on the SNM is shown for various degrees of degradation of the cell core (PU and PD) and PG transistors. The degradation calculated using the parameters from TABLE 1 is projected towards 10 year operating lifetime under a fixed V_{DD} supply voltage. Figs. 7a,b show a similar plot for WTP and WM respectively.

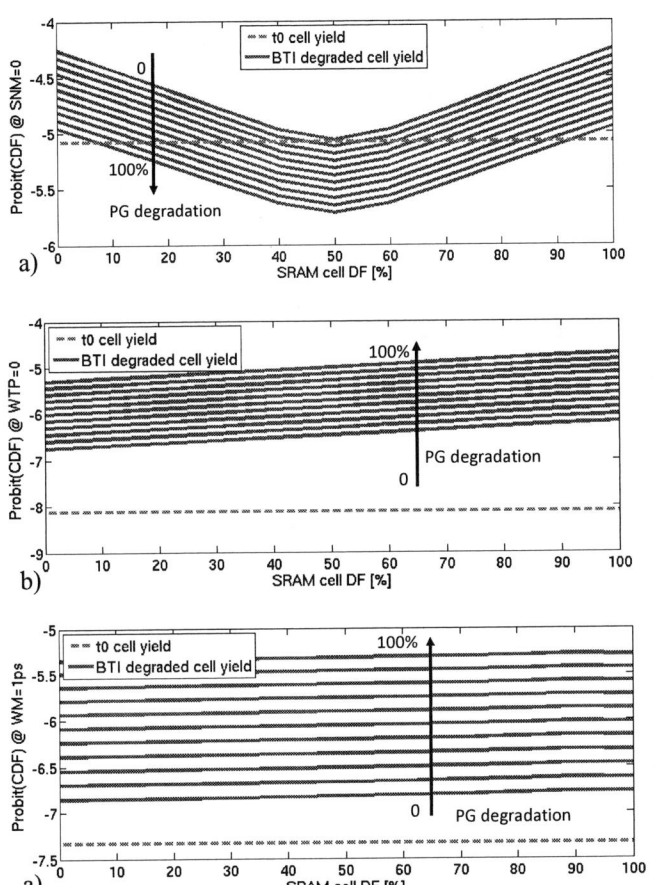

Figure 7. Cumulative probability of having (a) SNM=0, (b) WTP=0 and (c) WM<1ps as function of the cell DF and PG degradation. Time-zero cell yield is added for comparison (See also Fig. 6).

For SNM the probability of failure has a minimum at 50% DF and increases symmetrically for lower and higher DF, reaching a maximum for an always '0' and always '1' cell. Increased PG degradation allows for more margin since the weaker PG is less likely to cause a destructive read. For the WTP and WM the probability of failure is less sensitive to the cell DF as to the PG degradation. The probability of failure increases if the cell is more degraded on the side where the write occurs. Increased PG degradation increases the failure probability even more so.

IV. CONCLUSIONS

In this paper we have combined statistical silicon reliability data, extracted from TEGs designed and fabricated in an advanced 20nm HK/MG process, with current state-of-the-art statistical assessment techniques in order to acquire a realistic impact of BTI degradation on the yield and performance of advanced 6T

SRAM cells. It is shown that BTI reduces the write stability and speed where the PG transistors play a vital role and the core DF is of less importance looking at the far end of tail. On the other hand, for read stability, PG transistor degradation helps to increase the SNM. Here the core DF does play an important role where a maximum read stability occurs at 50% DF.

REFERENCES

[1] B. Kaczer, C. Chen, P. Weckx, P.J. Roussel, M. Toledano-Luque, J. Franco, M. Cho, J. Watt, K. Chanda, G. Groeseneken, T. Grasser, "Maximizing reliable performance of advanced CMOS circuits—A case study," *Reliability Physics Symposium, 2014 IEEE International* , vol., no., pp.2D.4.1,2D.4.6, 1-5 June 2014

[2] D. Angot, V. Huard, L. Rahhal, A. Cros, X. Federspiel, A. Bajolet, Y. Carminati, M. Saliva, E. Pion, F. Cacho, A. Bravaix, "BTI variability fundamental understandings and impact on digital logic by the use of extensive dataset," *Electron Devices Meeting (IEDM), 2013 IEEE International* , vol., no., pp.15.4.1,15.4.4, 9-11 Dec. 2013.

[3] C. Prasad, M. Agostinelli, J. Hicks, S. Ramey, C. Auth, K. Mistry, S. Natarajan, P. Packan, I. Post, S. Bodapati, M. Giles, S. Gupta, S. Mudanai, K. Kuhn, "Bias temperature instability variation on SiON/Poly, HK/MG and trigate architectures," *Reliability Physics Symposium, 2014 IEEE International* , vol., no., pp.6A.5.1,6A.5.7, 1-5 June 2014.

[4] B. Kaczer, T. Grasser, J. Martin-Martinez, E. Simoen, M. Aoulaiche, P.J. Roussel, G. Groeseneken, "NBTI from the Perspective of Defect States with Widely Distributed Times," *Proc. Int. Rel. Phys. Symp.*, p. 55,2009.

[5] M. Toledano-Luque, B. Kaczer, E. Simoen, R. Degraeve, J. Franco, P.J. Roussel, T. Grasser, G. Groeseneken,"Correlation of single trapping and detrapping effects in drain and gate currents of nanoscaled nFETs and pFETs," *Reliability Physics Symposium (IRPS), 2012 IEEE International* , vol., no., pp.XT.5.1,XT.5.6, 15-19 April 2012

[6] X. Federspiel, M. Rafik, D. Angot, F. Cacho, D. Roy, "Interaction between BTI and HCI degradation in High-K devices," *Reliability Physics Symposium (IRPS), 2013 IEEE International* , vol., no., pp.XT.9.1,XT.9.4, 14-18 April 2013

[7] B. Kaczer, T. Grasser, P.J. Roussel, J. Franco, R. Degraeve, L.A. Ragnarsson, E. Simoen, G. Groeseneken, H. Reisinger, "Origin of NBTI variability in deeply scaled pFETs," *Reliability Physics Symposium (IRPS), 2010 IEEE International* , vol., no., pp.26,32, 2-6 May 2010

[8] P. Weckx, B. Kaczer, C. Chen, J. Franco, E. Bury, K. Chanda, J. Watt, Ph. J. Roussel, F. Catthoor, G. Groeseneken, "Characterization of time-dependent variability using 32k transistor arrays in an advanced HK/MG technology", accepted to Int. Reliab. Phys. Symp. (IRPS 2015).

[9] P. Weckx, B. Kaczer, H. Kukner, P.J. Roussel, P. Raghavan, F. Catthoor, G. Groeseneken, "Non-Monte-Carlo methodology for high-sigma simulations of circuits under workload-dependent BTI degradation—Application to 6T SRAM," *Reliability Physics Symposium, 2014 IEEE International* , vol., no., pp.5D.2.1,5D.2.6, 1-5 June 2014

[10] E. Seevinck, F.J. List, J. Lohstroh, "Static-noise margin analysis of MOS SRAM cells," IEEE Journal of Solid-State Circuits, 1987, pp. 748- 754

[11] P. Weckx, B. Kaczer, M. Toledano-Luque, T. Grasser, P.J. Roussel, H. Kukner, P. Raghavan, F. Catthoor, G. Groeseneken, "Defect-based methodology for workload-dependent circuit lifetime projections - Application to SRAM," *Reliability Physics Symposium (IRPS), 2013 IEEE International* , vol., no., pp.3A.4.1,3A.4.7, 14-18 April 2013

[12] H. Reisinger, T. Grasser, W. Gustin, C. Schlünder, "The statistical analysis of individual defects constituting NBTI and its implications for modeling DC- and AC-stress," *IRPS*, 2010, pp. 7-15

[13] T. Grasser, P. Wagner, H. Reisinger, T. Aichinger, G. Pobegen, M. Nelhiebel, B. Kaczer, "Analytic modeling of the bias temperature instability using capture/emission time maps," *IEDM*, 2011, pp. 1-4

[14] M.J.M. Pelgrom, C.J. Duinmaijer, A.P.G. Welbers, "Matching properties of MOS transistors," *Solid-State Circuits, IEEE Journal of* , vol.24, no.5, pp.1433,1439, Oct 1989

Countering Early Propagation and Routing Imbalance of DPL Designs in a Tree-based FPGA

Emna Amouri, Shivam Bhasin, Yves Mathieu, Tarik Graba, Jean-Luc Danger

Institut TELECOM / Departement COMELEC, 46 rue Barrault, Paris, France

firstname.lastname@telecom-paristech.fr

Abstract—**The Wave Dynamic Differential Logic (WDDL) offers an effective way to resist Side Channel Attacks (SCA). But, it suffers from early propagation and routing imbalance between dual signals. In this paper, we deal first with the EPE problem. We study the security of BCDL logic, which is known to counter early propagation, and we compare it to WDDL logic. We target a custom tree-based FPGA of 2048 cells. Next, we try to solve the routing imbalance problem by performing an adjacent placement and a *timing_balance_driven* routing.**
Side channel analyses are performed on FPGA circuit implementing PRESENT crypto-processor. Experimental results show that both avoiding early propagation and diminishing routing imbalance by controlling placement and routing tools enhance the design security against SCA.

Keywords—*Side Channel Attacks, Dual-rail Precharge Logic (DPL), FPGA, placement, routing.*

I. INTRODUCTION

Since its introduction by Paul Kocher [1], side channel attacks (SCAs) have been presenting a serious threat to cryptographic applications. These attacks rely on the data-dependent power consumption variation in order to extract the secret key. Dual-rail precharge logic (DPL) is a promising countermeasure to protect cryptographic devices against SCA. The principle consists in consuming the same amount of power consumption regardless of data inputs. This is achieved by using a differential logic (Each signal is conveyed by two complementary wires) and precharging the differential signals in every clock cycle. Many implementations of secure dual rail cells have been proposed, specifically for ASICs, such as SABL [3], WDDL [2], STTL [4] and BCDL [5].

Wave dynamic differential logic (WDDL) [2] is a common style of DPL. The components used are limited to positive logic to avoid glitches. Inverters are implemented by cross coupling complementary outputs. WDDL is well suited for FPGAs [2]. However, it suffers from two major vulnerabilities i.e. early propagation effect (EPE) and routing imbalance. EPE is caused by the difference of arrival time between input signals of dual gates. If the transition probability of the gate is unity, the gate will evaluate without waiting for all the signal to acquire the right value. This causes a difference in switching output timing and thus leads to data-dependant power leakage [7]. On the other hand, to have balanced dual nets in FPGAs, efficient control of the placement and the routing of the differential design is required.

The solution to counter EPE is based on synchronization. If a DPL gate waits for all the signal to be VALID before evaluating, EPE will not occur. To counter EPE in FPGA, Bhasin [6] has proposed to modify truth table of basic logic

gates to evaluate only when all the inputs are VALID. Practical results on FPGA show that this technique reduces the leakage by half. Nassar has introduced another DPL countermeasure called BCDL [5] which is capable of resisting EPE by using a global signal to achieve the required synchronization. All the present work have mostly focussed on elimination of EPE with little or no focus on routing imbalance.

To overcome the imbalance routing problem in FPGA, Schaumont [8] suggests to implement a second complementary WDDL module on the FPGA, with identical routing to the direct WDDL circuit. However, this technique causes a twofold area increase compared to WDDL design. Robert P. McEvoy [9] proposes Isolated WDDL. However, the area increase resulting from this method is comparable to Schaumont technique. Baddam and Zwolinski [10] presented a design technique that separates the true part from the false part of the design by implementing an inverter using XOR gate. This technique may be sensitive to attacks based on glitches. All techniques mentioned above suffer from EPE.

Another branch of research is the study of routing algorithms on custom FPGA. Recently, *timing_balance_driven* routing algorithm [13] was shown to be efficient on hierarchical tree-based FPGA architecture, and increased the robustness of cryptographic device against SCA. We take this research a few steps further by investigating the impact of routing on FPGA in absence and presence of EPE.

The FPGA we are targeting is a multilevel hierarchical FPGA which is a novel Tree-based architecture (TFPGA) presented in [11]. It was shown that this architecture is better in terms of area density as compared to common VPR-style MESH architecture [14]. This was an attractive reason to design a prototype of this architecture. On the other hand, with a custom FPGA chip, we know exactly the architecture interconnection and we have a total control on placement and routing tools.

Our Contributions: In this paper, we present two main contributions. Firstly, we design a custom tree-based FPGA of 2048 cells, integrating a specific block in each cell to allow the implementation of BCDL. Original BCDL implementation [5] recommends 5- input LUTs and block memories for proper implementation. We show how BCDL can be adapted to TFPGA with 4-input LUTs and no memories. Next, we deal with balanced dual-rail routing in FPGA, by performing an adjacent placement and a *timing_balance_driven* routing to WDDL and BCDL. Applying these techniques does not cause any area overhead. Our experimental results show that both balancing routings and avoiding EPE significantly reduce the amount of power leakage exploitable by SCA.

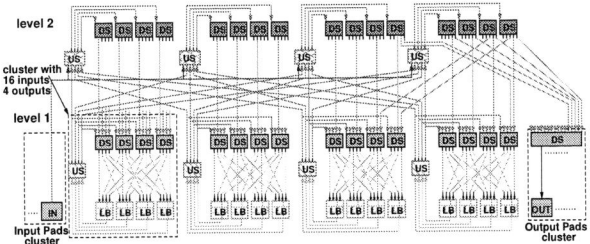

Fig. 1. TFPGA Architecture with 2 Levels of Hierarchy (4x4)

II. HIERARCHICAL TREE-BASED FPGA

The Tree-based FPGA architecture (TFPGA) was designed and evaluated in [11]. Unlike mesh-based architecture where logic and routing resources are arranged in island style, in a tree-based architecture, logic and routing resources are arranged in hierarchical manner.

TFPGA architecture shown in figure 1 is composed of clusters located at different levels. Each cluster contains a switch block to connect local sub-clusters. The cluster of level 1 contains logic blocks (LBs). TFPGA has linear populated switch boxes and unidirectional wires. It unifies two unidirectional networks: downward network and upward network. The downward network is based on the Butterfly Fat Tree (BFT) topology. Each switch box is divided into downward mini switch boxes (DS) which connect signals to local sub-clusters inputs. The upward network comprises upward mini switch boxes (US). These USs allow sub-clusters outputs to be connected to other sub-clusters in the same cluster and to clusters in other levels of hierarchy. As shown in Figure 1, input pads and output pads are connected to US and DS respectively. Thus, input pads can reach all LBs of the architecture, and output pads can be reached by all LBs from different paths.

Figure 1 shows TFPGA architecture with 2 levels of hierarchy and *Arity* 4x4 (the *level2* cluster contains $k = 4$ sub-clusters and each *level1* cluster contains $k = 4$ LBs). We have designed a prototype of this architecture, with 4 levels of hierarchy and arity 8x8x8x4 (2048 LBs), fabricated with ST $65nm$ process and using only standard cells.

III. BCDL PRINCIPLE

The strategy of BCDL [5] is to synchronize the gate inputs before starting the evaluation and precharge phases. To eliminate the early propagation effect and to avoid glitches, the synchronization should follow the two conditions:

- Evaluation phase should start only after all the input signals are VALID.

- The precharge phase should start before the first (fastest) input becomes NULL.

a) Synchronization: The synchronization is done using a specific cell, as shown in fig. 2. $Sync$ is the signal authorizing the evaluation. It raises up to '1' when all signals are in valid state. This signal is defined by the following equation.

$$Sync(x,y) = \begin{cases} 1, if\ x \neq (0,0)\ and\ y \neq (0,0) \\ 0, otherwise. \end{cases}$$

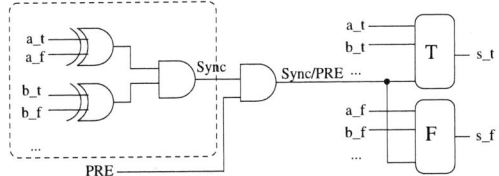

Fig. 2. BCDL n-input cell

b) Global Precharge: A global precharge signal named PRE is used to induce the precharge state globally. It forces all cell outputs to '0'. This signal must always arrive before any input. The synchronization is then ensured by the $Sync/PRE$ signal which is the "AND" result between $Sync$ and PRE signals.

The computation is done as follows:

- When PRE signal falls to '0' and thus the signal $Sync/PRE$ falls also to '0', the precharge phase is forced, independently from the inputs.

- When $Sync/PRE$ raises to '1', this indicates that the precharge phase is completed and that all input signals are VALID, thus inducing the beginning of the evaluation phase.

IV. IMPLEMENTATION OF BCDL ON TREE-BASED FPGA WITH A SPECIFIC LUT STRUCTURE

In [5], authors implemented the S-Boxes in the embedded RAM of the FPGA, and used 2 5-input LUT to implement a 2-input BCDL function, thus integrating the synchronization scheme in the True and False cells. The Tree-based FPGA architecture is composed of LUTs with 4 inputs and does not contain memories. But, each LUT contains specific gates to permit the implementation of BCDL. In this section, we explain how we adapted the implementation of BCDL to our FPGA architecture.

A. Optimized Synchronization

In BCDL design, differential input signals of synchronization cells $((a_t, a_f), (b_t, b_f))$ shown in figure 2) must be perfectly balanced. Otherwise, $Sync$ signal evaluates at different times depending on input values. Subsequently, dual gates outputs timings are different which leads to information leakage. To get rid of this problem, we do not use synchronization cells. We create one synchronization signal $Sync$ which is common to all dual gates. This signal is intentionally delayed with regards to the input signals of each gate, since it has to be the slowest signal among gate inputs.

B. Specific Synchronization block in LUT structure

In TFPGA architecture, LUT structure is a tree of multiplexers as shown in fig. 3. "OR" and "AND" gates are put at the front of the multiplexers tree. The precharge signal PRE is a global signal in TFPGA, having similar routing as the clock signal. It is generated in a way that guarantees a minimal skew with respect to the clock. EN_PRE is a configuration point enabling the precharge signal. It is set to '1' if we want to use the LUT in BCDL mode, and it is put to '0' otherwise. The "AND" gate is integrated in the LUT structure to generate $Sync/PRE$ signal for BCDL design. So, the synchronization signal $Sync$ must be plugged on the first

978-1-4799-7670-6/15 $31.00 © 2015 IEEE

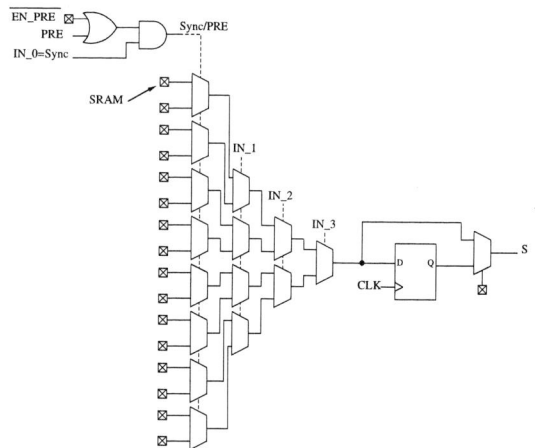

Fig. 3.   4-LUT structure

pin of LUT. This constraint is added to our routing tool and is guaranteed. The AND gate output ($Sync/PRE$) is chosen to be on the first column of the multiplexers tree to guarantee that the synchronization signal is the slowest signal at LUT-level. This avoids local early evaluation.

C. Reduced Complexity

As the $Sync/PRE$ signal is the first to switch before the precharge phase and it is the latest to switch in the evaluation phase (after all input signals are VALID), there is no glitches in LUTs. Therefore, BCDL design is not limited to positive functions as in WDDL design. As TFPGA LUTs have 4 inputs, we can use all possible functions with 3-inputs, which reduces the BCDL design complexity. The fourth input is used for the synchronization signal.

V.   CONSTRAINED PLACEMENT AND ROUTING

A. Adjacent Placement

keeping dual gates in the same cluster of TFPGA can be favorable to obtain routing nets as symmetrical as possible. To achieve this constrained placement, we place first true gates using only the half of the LUTs in each cluster. Then, each false gate is placed adjacent to the original gate.

B. Timing_Balance_Driven Routing Algorithm

To improve dual-rail balance, we propose a routing algorithm which goal is to balance the propagation delays of routing resources used by two dual connections and also to balance the number of routing resources used. In this part, we present enhancements made to the PathFinder routing algorithm [14] to be aware of timing and resources imbalances. PathFinder is an iterative rip-up algorithm based only on the congestion negotiation.
FPGA routing resources delays are inserted in the routing tool. Thus, the router can compute the timing imbalance between two routed dual signals Consider the connection to sink j of net i. We define the cost of using a routing resource n in the new router as:

$$Cost(n) = (1 - Delay\_Crit(i,j)) \cdot Cong\_cost(n)$$
$$+ Delay\_Crit(i,j) \cdot Res\_diff\_delay(i,n,j)$$
$$\cdot Diff\_res\_nb(i,n,j) \qquad (1)$$

Cong_cost(n) is the congestion cost of a routing resource n.

TABLE I.   ROUTING RESULTS OF PRESENT BCDL

P. & R. Strategy	($\Delta delay$) (ps)			Res_imbalanced connections nb.
	Max	Mean	Std Dev	
Adjacent P. & unconstrained R.	1391	97	148	88
Adjacent P. & Timing_Balance_Driven R.	163	29	29	0

It takes into account the number of nets sharing this resource, and the history of congestion on it [14].
Diff_res_nb(i,n,j) is the difference between the estimated total resouces number that will be used to reach the sink j of net i from the net driver and using the node n and the number of used interconnect resources in the routing of the dual signal.
Res_diff_delay(i,n,j) is the difference between the delay of a resource n used to route a connection to sink j of net i, and the delay of the resource n' used by the complementary connection. n and n' must be wires of the same type (downward or upward) and situated at the same hierarchical level in the tree-based architecture.
Delay_Crit(i,j) is the delay imbalance criticality between the connection j of net i and the dual connection. It can be formulated as:

$$Delay\_Crit(i,j) = min(0.9, \frac{Connection\_diff\_delay(i,j)}{Max\_connection\_diff\_delay})$$
(2)

where:
- $Max\_connection\_diff\_delay$ is the maximum $Connection\_diff\_delay(i,j)$ among all routed connections in the design.
- $Connection\_diff\_delay(i,j)$ is the timing imbalance between complementary connections j and j'.
The connection criticality is a fractional number between 0 and 1. High connection criticality means that the real and dual connections have an important delay imbalance.

We can see from equation 1 that the congestion-delay trade-off of each connection is controlled by how critical it is. The router performs many iterations until all routing resources conflicts are resolved. In a routing iteration, once the original net is routed, the router stores informations about the routing. Then, during the routing of the dual net, the router computes the cost of interconnect resources n and tries to find the path with the minimum differential delay and the minimum differential routing resources number, considering the resource congestion.

C. Routing Results

To evaluate the complementary networks balance, we compute, for all dual connections of all dual nets, the absolute difference $\Delta delay = |delay(true) - delay(false)|$. $Delay(true)$ and $delay(false)$ are interconnect latencies of true and false signals respectively. We compute also the number of imbalanced connections in terms of used routing resources number. The table I shows the routing results of WDDL and BCDL designs with unconstrained routing (Congestion Pathfinder algorithm) and the $timing\_balance\_driven$ routing. Both routings are applied after an adjacent placement. The new router succeeded to route all dual connections with the same number of routing resources ($Res\_imbalanced$ $connections$ $nb.$ $=0$). Moreover, it achieves 70% of average timing improvement compared to results obtained with unconstrained routing.

978-1-4799-7670-6/15 $31.00 © 2015 IEEE

TABLE II.   SIDE CHANNEL ATTACK RESULTS ON PRESENT UNPROTECTED MODULE

	Minimum Traces to Disclose (MTD) the Secret Key															
SBox	1	2	3	4	5	6	7	8	9	10	11	12	13	14	15	16
PRESENT Unprotected	1206	1108	694	694	500	1310	1498	694	500	1000	596	213	896	1305	800	1213

TABLE III.   SIDE CHANNEL ATTACK RESULTS ON PRESENT WDDL AND BCDL MODULES

		Security Gain															
	SBox	1	2	3	4	5	6	7	8	9	10	11	12	13	14	15	16
WDDL	Adjacent P. & unconstrained R.	11	13	5	85	–	26	–	–	–	–	16	–	17	–	–	–
	Adjacent P. & Timing_Balance_Driven R.	–	–	–	–	–	16	36	–	–	9	2	–	–	–	–	–
BCDL	Adjacent P. & unconstrained R.	–	–	–	121	–	27	–	–	–	68	59	768	–	–	–	–
	Adjacent P. & Timing_Balance_Driven R.	–	–	–	–	–	229	16	–	600	–	108	–	–	–	–	–

VI. SECURITY EXPERIMENTAL RESULTS

We implemented on TFPGA the unprotected PRESENT (as a reference) and both WDDL and BCDL. These DPL designs are implemented in TFPGA using an unconstrained routing and using a *timing_balance_driven* routing. Both routings are applied with an adjacent placement. We captured 300,000 traces for each implementation and we performed CPA attacks [15].

We compute the security gain (SG), i.e. the ratio between the number of Measurements To Disclose (MTD) the secret key of WDDL/BCDL protected modules to corresponding single-rail unprotected module. Table II summarizes the results of a CPA on PRESENT unprotected module. With the acquired traces, we retrieve all the 16 bytes of the secret key using only 1500 traces. Next we performed attacks on each of WDDL and BCDL implementations. We do not recover all the 16 bytes of the key from any DPL implementation. Table III shows the security gain computed for each SBOX of the PRESENT implementations. For example, the security gain for the SBOX 1 with WDDL design implemented with an adjacent placement and an unconstrained routing is equal to 11. This means that the number of traces required to retrieve the nibble 1 is equal to 13266 (1206 x 11). " − " sign means that the particular byte of the secret key is not disclosed at the end of the attack. It can be seen from results that BCDL implementation is more robust against SCA compared to WDDL. In fact, applying an unconstrained routing, we can retreive 8 bytes of the secret key for WDDL design and only 5 bytes for BCDL design. On the other hand, we can note that the *timing_balance_driven* routing enhances the security of both dual-rail designs. In fact, the number of bytes attacked is reduced from 8 to 4 for WDDL and from 5 to 4 for BCDL. But, the security of BCDL design is increased by a factor of 238 on average compared to unprotected design, whereas the average SG of WDDL design is equal to 15. Based on these results, it can be said that BCDL implementation provides higher robustness than WDDL one. We cannot directly compare our results with the original paper of BCDL and WDDL because the FPGA architecture, measurement setup and environment are not the same which totally changes the SNR of the side-channel acquisition.

VII. CONCLUSION

Side-channel resistance and implementation aspects of DPL countermeasures have been a hot-topic for some time now. In this paper, we studied the security of BCDL countermeasure on a custom hierarchical Tree-based FPGA (TFPGA). BCDL is known to counter EPE, thanks to the synchronization using a global precharge signal. We implemented a BCDL PRESENT cryptographic design in TFPGA and compare its side-channel resistance to unprotected and WDDL design. First, we tested the two DPL variants without any routing constraints. Experimental results show that BCDL logic has increased robustness compared to WDDL. However, routing imbalance still causes information leakage. Next, we developed and tested a *timing_balance_driven* routing to reduce dual-rail timing imbalance. This routing technique can be applied to any dual-rail logic. We learn from the obtained results that the proposed constrained routing algorithm improves the security of cryptographic designs. Again with the constrained routing, BCDL logic proves to be more robust than WDDL againt SCA. It increases robustness by a factor of 15 on average from WDDL.

Although countering EPE and reducing timing imbalance, BCDL design is still attackable. This indicates that there are other sources of leakage. Since we have a complete knowledge of the FPGA circuit, we plan to carry out deep investigations in order to identify these sources of leakage.

REFERENCES

[1] P. Kocher, J. Jaffe, B. Jun., *Differential Power Analysis*, Proc. of CRYPTO 99, ser. LNCS, vol. 1666, pp. 388-397.

[2] K. Tiri and I. Verbauwhede, *A Logic Level Design Methodology for a Secure DPA Resistant ASIC or FPGA Implementation*, Proc. DATE 2004.

[3] K. Tiri, M. Akmal and I. Verbauwhede, *A Dynamic and Differential CMOS Logic with Signal-Independent Power Consumption to Withstand Differential Power Analysis on Smart Cards*, ESSCIRC 2002.

[4] A. Razafindraibe, M. Robert and P. Maurine, *Improvement of dual rail logic as a countermeasure against DPA*, VLSI - SoC 2007, Atlanta, USA.

[5] M. NASSAR et al. *BCDL: A High Speed Balanced DPL for FPGA with Global Precharge and no Early Evaluation*. Proc. DATE 2010.

[6] S. Bhasin et al., *Countering early evaluation: an approach towards robust dual-rail precharge logic*. WESS 2010.

[7] D. Suzuki and M Saeki. *Security Evaluation of DPA Countermeasures Using Dual-Rail Pre-charge Logic Style*. CHES 2006.

[8] P. Yu and P. Schaumont, *Secure FPGA circuits using controlled placement and routing*, Proc. of CODES+ISSS 2007, Salzburg, Austria.

[9] R. P. McEvoy et al., *Isolated WDDL: A Hiding Countermeasure for Differential Power Analysis on FPGAs*, ACM TRETS 2009, vol. 2.

[10] K. Baddam and M. Zwolinski, *Divided backend duplication methodology for balanced dual rail routing*,CHES 2008.

[11] Z. Marrakchi et al., *FPGA Interconnect Topologies Exploration*, International Journal of Reconfigurable Computing, Volume 2009.

[12] A. Bogdanov et al. *PRESENT: An Ultra-Lightweight Block Cipher*. CHES 2007, Vienna, Austria, September 10-13, 2007.

[13] E. Amouri et al., *Balancing WDDL Dual-Rail Logic in a Tree-based FPGA to Enhance Physical Security*, FPL 2014.

[14] V. Betz, A. Marquardt and J. Rose, *Architecture and CAD for deep-submicron fpgas*, Kluer Academic Publishers, January 1999.

[15] B. Éric, C. Clavier and F. Olivier, *Correlation Power Analysis with a Leakage Model*, Proc. of CHES 2004, Cambridge, MA, USA.

978-1-4799-7670-6/15 $31.00 © 2015 IEEE

FinFET Stressor Efficiency on Alternative Wafer and Channel Orientations for the 14 nm Node and Below

G. Eneman, A. De Keersgieter, A. Mocuta, N. Collaert, A. Thean

Imec, Kapeldreef 75, 3001 Heverlee, Belgium

Abstract— **This simulation work studies whether optimal wafer and channel orientations exist that maximize the mobility of 10 nm-node strained-silicon FinFETs. For NFinFETs, strain-relaxed buffers or source/drain stressors yield the highest mobilities on rotated-notch wafers. For PFinFETs, industry-standard directions give the highest mobilities when using $Si_{1-y}C_y$ strain-relaxed buffers as a stress booster. Using {110} substrates leads to strained mobilities that are in between what can be obtained by industry-standard and rotated-notch directions.**

Keywords—strain; stress; FinFETs; Strain-Relaxed Buffers; TCAD

I. INTRODUCTION

Stressor techniques are essential performance booster elements since their introduction in planar CMOS technologies in the 90 nm node [1]. In FinFETs, most stressors remain effective, however with some differences w.r.t. planar FETs: the most notable examples are Strain-Relaxed Buffers (SRBs) or virtual buffers, that become uniaxial stress elements rather than biaxial [2]. Stressed silicon-nitride Contact Etch-Stop-Layers (CESL) display a complex stress profile when deposited on fins [3], rather than the simpler lateral and vertical stress that is reported for planar technologies [4].

Most studies for stressors focus on 'Industry-Standard' transistors, that are fabricated on a {100}-oriented silicon wafer, and have the channel in a <110> crystallographic direction. Whether stressors are promising for alternative wafer and channel directions is not a trivial question: Firstly, the unstrained electron and hole mobilities depend on the crystal orientation of the channel plane that forms the inversion layer, and on the direction of the electrical conduction. Secondly, when moving from planar to fin technologies, the conduction takes place both at the top of the fin and at the fin sidewalls, which may have different crystal orientations. Thirdly, the effect of stress on mobility strongly depends on the crystal orientation. Finally, the stress generated by stress techniques depends on the device orientation due to the elastic anisotropy of silicon. The purpose of this work is to perform a systematic study of stressor effectiveness for different wafer and channel orientations, taking the above considerations into account.

II. SETUP

A. Unstrained Mobility: Orientation Dependence

Inversion-layer mobility depends on the surface and channel orientation. In this paper, unstrained mobility values were taken from [5] at an inversion charge density N_{INV} of 1×10^{13} cm^{-2}.

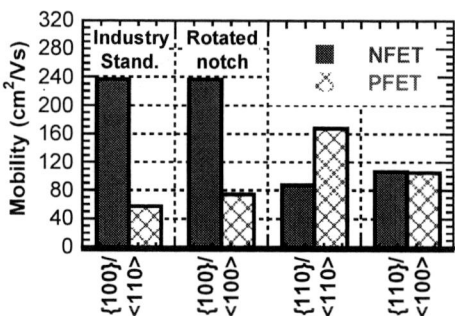

Figure 1. Unstrained silicon electron and hole mobilities for different surface orientations ({100} or {110}) and channel orientations (<110> or <100>), taken from [5]. Inversion charge density N_{INV} is 1×10^{13} cm^{-2}.

Data from [5] is used except for the following directions: {100}/<100> NFET mobility is assumed to be identical to {100}/<110> NFET mobility, in line with experimental results [6]. {100}/<100> PFET mobility is taken 30% higher than the {100}/<110> PFET mobility to account for the anisotropy in band structure leading to a higher performance for the <100>-oriented devices [6].

Fig. 1 shows electron and hole mobility versus substrate and channel orientation, which is at the same time the mobility for planar transistors fabricated in this direction. As this figure indicates, there is a large difference between NFET and PFET mobility for planar transistors fabricated on {100} substrates, irrespective of the channel direction. NFETs have a lower mobility on {110}- than on {100}-substrates. On the other hand, for {110}-substrates there is a smaller difference between NFETs and PFETs, which may have the advantage to lead to more balanced circuits.

B. FinFET Geometry Effect

FinFETs have a channel at the top and sides of the fin, which may have a different surface orientation. The direction of a FinFET is fully determined by the orientation of the starting substrate {mnp}, and the channel direction <ijk>. Four fin directions are studied in this work and are shown in Fig. 2, along with the crystal directions of the top-surface, the sidewall, and the channel.

978-1-4799-7670-6/15 $31.00 © 2015 IEEE

The average unstrained fin mobility $\mu_{FinFET,0}$ is estimated as a weighted average between the mobility of the top- and sidewall surface (μ_{top} and μ_{side} respectively):

$$\mu_{FinFET,0} = \frac{\mu_{top}F_W}{F_W + 2F_H} + \frac{2\mu_{side}F_H}{F_W + 2F_H} \qquad (1)$$

$$= \mu_{top} \cdot \frac{1}{1 + 2 \cdot AR} + \mu_{side} \cdot \frac{2 \cdot AR}{1 + 2 \cdot AR}$$

F_W and F_H are the fin width and height, respectively. AR is the fin's aspect ratio ($AR = F_H/F_W$). Assuming e.g. a F_W of 10 nm and a F_H of 30 nm gives an AR of 3 which leads to the unstrained NFET and PFET mobility $\mu_{FinFET,0}$ in Fig. 3.

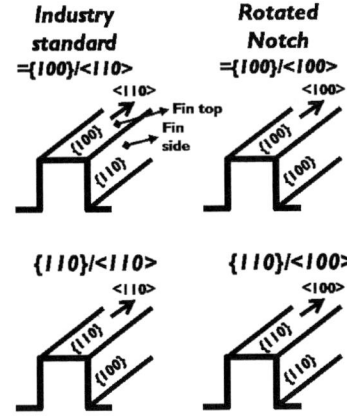

Figure 2. Overview of the FinFET orientations, defined by {mnp}/<ijk>, where {mnp} is the wafer orientation and <ijk> is the channel direction.

Fig. 3 shows that unstrained fins fabricated in a rotated-notch or a {110}/<110> direction have similar mobilities than industry-standard planar FETs (Fig. 1) thanks to their dominant sidewall conduction on a {100} plane. Fins in an industry-standard or {110}/<100> direction have a reduced electron mobility.

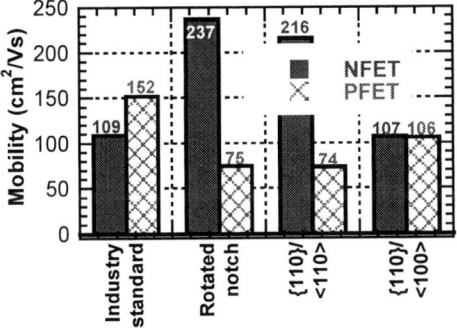

Figure 3. Unstrained FinFET mobility $\mu_{FinFET,0}$ for the different wafer/channel orientations defined in Fig. 2. A F_W of 10 nm and a F_H of 30 nm (AR of 3) is assumed.

C. Piezoresistance

To first order, the sensitivity of mobility to stress depends on the direction of the applied stress, the direction of the current flow and the carrier type (electrons or holes). The stressors studied here, SRB's and source/drain epilayers, generate predominantly longitudinal stress, i.e. running from source to drain. Fig. 4 shows the longitudinal piezoresistances π_{ijk} for substrate orientations {mnp} and channel directions <ijk>, corresponding to:

$$\delta\mu_{ijk} / \mu_{ijk} = \pi_{ijk} \cdot \sigma_{ijk} \qquad (2)$$

μ_{ijk} is the mobility of carriers in the <ijk> direction, while σ_{ijk} is the magnitude of stress in the <ijk> direction. Tensile stresses have positive value, compressive stresses are negative. As indicated in Fig. 4, NFETs in a <100> direction have strong stress sensitivity. For PFETs, <110> channels have the highest stress sensitivity.

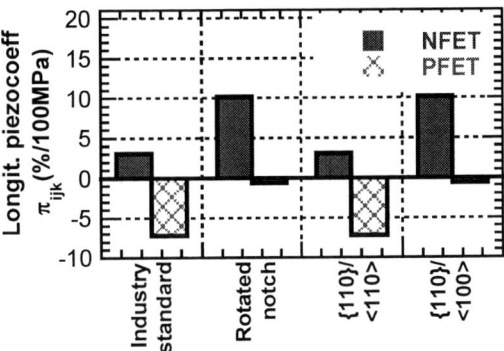

Fig. 4. Longitudinal piezoresistance coefficients π_{ijk} for substrate orientations {mnp} and channel directions <ijk>, based on data from [7]. To first order, π_{ijk} does not depend on {mnp}.

III. RESULTS

A. Strain-Relaxed Buffers

Strain-relaxed or virtual buffers (SRB's) generate channel stress when the buffer has a different relaxed lattice constant than the channel material. For wide, planar architectures the stress generated by this technique is biaxial, however for tall fins, only the longitudinal component remains [2]. NFETs require tensile stress for mobility enhancement (Fig. 4), therefore SRB's with a larger lattice constant than Si, like $Si_{1-x}Ge_x$ SRB's, are required. For Si PFETs, a $Si_{1-y}C_y$ SRB can be used, as the compressive stress generated by this SRB leads to PFET mobility enhancement. $Si_{1-y}C_y$ SRB's are technically challenging as it is difficult to maintain a high substitutional carbon concentration in $Si_{1-y}C_y$ throughout the complete process flow.

$Si_{1-x}Ge_x$ and $Si_{1-y}C_y$ SRB's have the following lattice mismatch $\varepsilon_\parallel$ with a silicon channel:

978-1-4799-7670-6/15 $31.00 © 2015 IEEE

- $Si_{1-x}Ge_x$ SRB: $\varepsilon_{\parallel}$=0.42% per 10% of Ge in SRB

- $Si_{1-y}C_y$ SRB: $\varepsilon_{\parallel}$=-0.45% per 1% of C in SRB

The stress $\sigma_{Channel}$ generated by the SRB in the channel is approximately uniaxial, and depends on the channel direction. For channels in a <110> direction:

$$\sigma_{channel} = \frac{2\varepsilon_{\parallel}}{S_{11}+S_{12}+S_{44}/2} = 169\,GPa\cdot\varepsilon_{\parallel} \qquad (3)$$

S_{11}, S_{12} and S_{44} are the compliance constants of silicon. For channels in a <100> direction:

$$\sigma_{channel} = \frac{\varepsilon_{\parallel}}{S_{11}} = 130\,GPa\cdot\varepsilon_{\parallel} \qquad (4)$$

Combining equations (1-4) leads to an estimate of the orientation-dependent FinFET mobility with SRB stress as:

$$\mu_{FinFET} = \mu_{FinFET,0}\cdot\left(1 + \pi_{ijk}\cdot\sigma_{channel}\right) \qquad (5)$$

Fig. 5 shows the strained mobility of n-type silicon-channel fins on a $Si_{1-x}Ge_x$ SRB as a function of the SRB composition. Rotated-notch n-fins are found to yield the highest mobility under SRB stress, thanks to their high unstrained mobility (Fig. 3) as well as their high stress sensitivity (Fig. 4). Industry-standard n-fins have fairly low unstrained mobility (Fig. 3) and the smallest stress sensitivity of all the studied orientations (Fig. 4), as a consequence the strained mobility is the lowest of all orientations. On {110} substrates, the unstrained n-fin mobility for <110> and <100> channels is similar to rotated-notch resp. industry-standard fins (Fig. 3). When grown on a $Si_{1-x}Ge_x$ SRB, fins on {110} wafers are in between the rotated-notch and industry-standard orientations (Fig. 5) because of the difference in piezoresistance.

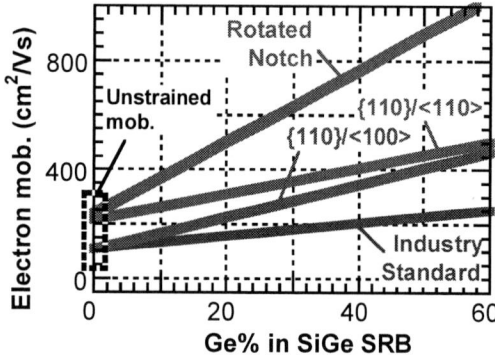

Figure 5. Strained Si NFinFET mobility versus composition of the underlying $Si_{1-x}Ge_x$ SRB. A F_W of 10 nm and a F_H of 30 nm (AR of 3) is assumed.

For PFinFETs, the strained mobility of Si-channel on $Si_{1-y}C_y$ SRB's is shown in Fig. 6. Industry-standard p-fins have a high unstrained mobility (Fig. 3) and high stress sensitivity (Fig. 4), so the highest strained mobility is found for this

orientation. Rotated-notch p-fins have the poorest strained mobility due to their low unstrained mobility and their small stress sensitivity. Similar to n-fins, fabricating strained p-fins on {110} substrates leads to mobilities between rotated-notch and industry-standard, with strained channels in the {110}/<110> direction significantly outperforming the channels in the {110}/<100> direction.

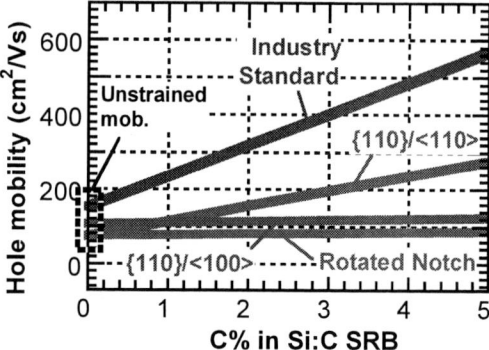

Figure 6. Strained Si PFinFET mobility versus composition of the underlying $Si_{1-y}C_y$ SRB. A F_W of 10 nm and a F_H of 30 nm (AR of 3) is assumed.

B. Source-Drain Stressors

NFinFETs require $Si_{1-y}C_y$ source/drain (S/D) stressors for mobility boost, while PFinFET mobility is improved by $Si_{1-x}Ge_x$ S/D stressors. The effectiveness of source/drain stressors strongly depends on dimensions like gate length, fin height, spacer thickness, source/drain epi thickness, as well as gate and fin pitch. Stress simulations were performed with Sentaurus-Process [8] on 10 nm-node fins for the industry-standard direction [2]. For this technology, source/drain stressors are found to generate a dominant longitudinal channel stress:

- $Si_{1-x}Ge_x$ S/D: $\sigma_{Channel}$=-248 MPa per 10% of Ge in S/D

- $Si_{1-y}C_y$ S/D: $\sigma_{Channel}$=262 MPa per 1% of C in S/D

According to the model from [9], the channel stress generated by a source/drain stressor should be about 23% lower for <100> channels than for <110> channels due to the difference in apparent Young's modulus. Using this correction for the <100> channels, the strained FinFET mobility with S/D stressors can be calculated using (5).

Fig. 7 shows the strained mobility for NFinFETs with $Si_{1-y}C_y$ S/D stressors. Qualitatively, conclusions are similar to the results for SRB stressors: in the absence of stressors, the highest mobility is found for rotated-notch and {110}/<110> n-fins. When using S/D stressors, the highest mobility is obtained for rotated-notch n-fins, while industry-standard are the least interesting to be combined with S/D stressors. NFinFETs on {110} substrates are in between rotated-notch and industry-standard orientations. As reported in [2], SRB's are more efficient than S/D stressors. This is confirmed in this work for all orientations: higher absolute mobilities are found when SRB's are used (Fig. 5) than for S/D stressors (Fig. 7).

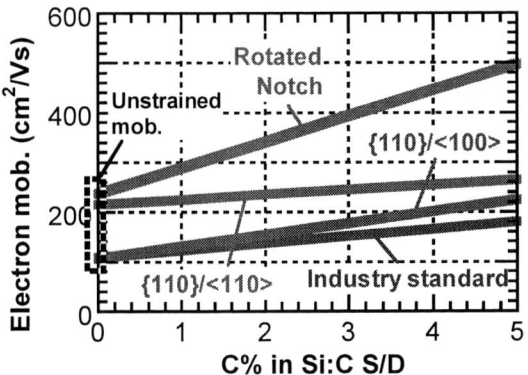

Figure 7. Strained Si NFinFET mobility versus composition of the $Si_{1-y}C_y$ S/D stressor. A F_W of 10 nm and a F_H of 30 nm (AR of 3) is assumed.

For PFinFETs, using $Si_{1-x}Ge_x$ S/D stressors gives almost no mobility benefit for rotated-notch and {110}/<110> directions (Fig. 8) as in these directions the hole mobility is largely independent on stress (Fig. 4). Industry-standard and {110}/<110> p-fins are the most interesting to be combined with S/D stressors since they have the strongest stress sensitivity. Overall the highest strained mobility can be obtained for industry-standard PFinFETs, similar to when using SRB's.

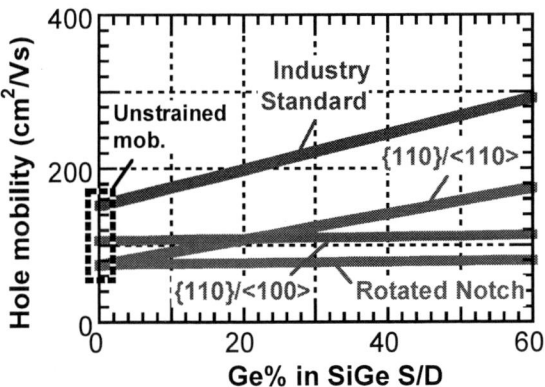

Figure 8. Strained Si PFinFET mobility versus composition of the $Si_{1-x}Ge_x$ S/D stressor. A F_W of 10 nm and a F_H of 30 nm (AR of 3) is assumed.

IV. CONCLUSIONS

This work uses stress simulations of tight-pitch, 10 nm-node FinFETs to study whether optimal wafer and channel orientations exist that maximize the mobility of 10 nm-node

silicon-channel FinFETs with SRB's and source/drain stressors. The top and sidewall of the fins have different unstrained mobilities, moreover stressor effectiveness is orientation-dependent. If SRB's or S/D layers are used to boost mobility, the most promising orientation for NFETs is rotated-notch, while the lowest mobilities are found for industry-standard directions. Unfortunately, for PFETs the opposite trend is found: the highest mobilities are predicted for industry-standard and the lowest for rotated-notch. Therefore there is no optimal device orientation for NFETs and PFETs for combined maximized mobility.

On the other hand, if equal electron and hole mobility is desired for layout efficiency, several possibilities can be identified. One option is using unstrained {110}/<100> n- and pFETs. Another option can be found for the Industry-standard direction: in this case using a $Si_{1-x}Ge_x$ SRB for the nFinFET with about 20% Ge is expected to yield similar mobility as the unstrained PFinFET counterpart.

REFERENCES

[1] T. Ghani, M. Armstrong, C. Auth, M. Bost, P. Charvat, G. Glass, T. Hoffmann, K. Johnson, C. Kenyon, J. Klaus, B. McIntyre, K. Mistry, A. Murthy, J. Sandford, M. Silberstein, S. Sivakumar, P. Smith, K. Zawadzki, S. Thompson, and M. Bohr, "A 90 nm high volume manufacturing logic technology featuring novel 45 nm gate length strained silicon CMOS transistors," in IEDM Tech. Dig., 2003, pp. 978–980.

[2] G. Eneman, D. Brunco, L. Witters, B. Vincent, P. Favia, A. Hikavyy, A. De Keersgieter, J. Mitard, R. Loo, A. Veloso, O. Richard, H. Bender, S. Lee, M. Van Dal, N. Kabir, W. Vandervorst, M. Caymax, N. Horiguchi, N. Collaert, A. Thean, "Stress simulations for optimal mobility group IV p- and nMOS FinFETs for the 14 nm node and beyond", in IEDM Tech. Dig., 2012, pp. 131.

[3] G. Eneman, N. Collaert, A. Veloso, A. De Keersgieter, K. De Meyer, T. Hoffmann, "On the efficiency of stress techniques in gate-last n-type bulk finfets", ESSDERC Proc., 2011, pp.115

[4] G. Eneman, P. Verheyen, A. De Keersgieter, M. Jurczak, K. De Meyer, "Scalability of stress induced by contact etch stop layers: a simulation study", IEEE Trans. Electr. Dev. 54 (6), pp. 1446 (2007).

[5] M. Yang, V. W. C. Chan, K. K. Chan, L. Shi, D. M. Fried, J. H. Stathis, A. I. Chou, E. Gusev, J. A. Ott, L. E. Burns, M. V. Fischetti, M. Ieong, "Hybrid-orientation technology (HOT): opportunities and challenges", IEEE Trans. Electr. Dev. 53 (5), pp. 965 (2006).

[6] H. Sayama, Y. Nishida, H. Oda, T. Oishi, S. Shimizu, T. Kunikiyo, K. Sonoda, Y. Inoue, M. Inuishi, "Effect of <100> channel direction for high performance SCE immune pMOSFET with less than 0.15μm gate length", in IEDM Tech. Dig., 1999, pp. 657.

[7] Smith C., "Piezoresistance effect in germanium and silicon", Phys Rev. 94, pp.42 (1954).

[8] Sentaurus Process Reference Manual, F-2011.09 (2011)

[9] G. Eneman, E. Simoen, P. Verheyen, K. De Meyer, "Gate influence on the layout sensitivity of $Si_{1-x}Ge_x$ S/D and $Si_{1-y}C_y$ S/D transistors, including an analytical model", IEEE Trans. Electr. Dev. 55 (10), pp. 2703 (2008).

INVITED

Trapping induced parasitic effects in GaN-HEMT for power switching applications

Gaudenzio Meneghesso, Matteo Meneghini,
Enrico Zanoni

University of Padova, Department of Information
Engineering
Via Gradenigo 6/B 35131
Padova, Italy

Piet Vanmeerbeek, Peter Moens
ON Semiconductor
Westerring 15, B-9700 Oudenaarde Belgium

Abstract—**This paper summarizes our recent results on the analysis of the trapping-induced parasitic effects in GaN-based high electron mobility transistors for power switching applications. More specifically, we demonstrate the following relevant mechanisms: (i) dynamic Ron shows a significant increase when the devices are operated at high temperature levels; this effect is ascribed to a stronger trapping in the buffer region; (ii) the kinetics of buffer-related trapping processes can be effectively investigated by means of backgating tests, carried out at various temperature levels; (iii) buffer trapping is strongly correlated to drain-buffer vertical leakage. The reduction of vertical leakage is an important step towards the reduction of high temperature dynamic Ron. Finally, we demonstrate that by proper epitaxial design and device optimization it is possible to fabricate devices with very low dynamic Ron (measured at 150 °C, V_{DS}=500 V)**

Keywords—*GaN, gallium nitride, transistor, HEMT, trapping, defect, leakage, breakdown;*

I. INTRODUCTION

Conversion losses are responsible for more than 10 % of global electricity consumption; in terms of efficiency, the transistor is one of the most critical components of a power converter. Resistive losses related to the on-resistance of the device, and switching losses related to the charging/discharging of the parasitic capacitances can be reduced only through the design of optimized devices, based on innovative semiconductor materials. Over the last few years, gallium nitride (GaN) has emerged as an excellent material for the fabrication of efficient power converters: the high electron mobility of the bi-dimensional electron GaN allows to minimize the on-resistance (Ron) of high-electron mobility transistors. In addition, the low (on-resistance)x(gate charge) product (Ron*Qg) permits to significantly reduce the switching losses compared to conventional silicon devices.

Another key advantage of GaN is its high breakdown voltage (in excess of 3 MV/cm); thanks to this important property it is possible to fabricate small devices with breakdown voltages in excess of 1 kV, that are expected to find wide applications in kV range power converters, e.g. for application in the photovoltaics and automotive fields.

Despite the high potential of GaN-based HEMTs, several factors still limit the dynamic performance of these devices, favoring the so-called dynamic-Ron issue. Dynamic Ron is the recoverable increase in the on-resistance, induced by the exposure to high bias in the off-state.

Several mechanisms can be responsible for dynamic Ron in power HEMTs:

1. The trapping of electrons in the gate-drain surface [1, 2], that results in virtual-gate effects, which can severely reduce the conductivity of the channel and increase device on-resistance. This problem can be almost completely solved through the optimization of the surface treatments, and through suitable passivation layers

2. The trapping of electrons in the gate insulator of MIS-HEMTs, which may result in relevant threshold voltage shift, with consequent instabilities of the electrical behavior of the devices [3]

3. The trapping of electrons in the buffer; this process may be induced by the presence of hot electrons [4], or by the high drain-substrate voltage difference [5, 6]

This paper contributes to the understanding of the buffer-related trapping processes in AlGaN/GaN-based Metal-Insulator-Semiconductor (MIS) HEMTs. The results described below indicate that: (i) buffer trapping may constitute a prevalent parasitic mechanism that limits the dynamic behavior of the devices; (ii) backgating measurements constitute a reliable approach to investigate such trapping processes; (iii) buffer-related trapping increases with temperature, and is strongly related to vertical (drain to bulk) leakage; (iv) by optimizing buffer leakage and epitaxial growth it is possible to fabricate devices with negligible dynamic Ron up to 500 V, 150 °C.

The results presented in this paper give a significant contribution to the understanding of the origin of dynamic Ron,

978-1-4799-7670-6/15 $31.00 © 2015 IEEE

and on the possible strategies that can be used to solve this problem.

II. EXPERIMENTAL DETAILS

The study was carried out on 650 V GaN-based transistors with a MIS-HEMT structure; the devices were grown on a p-type silicon substrate. Two different generations of devices were evaluated, with high (1st generation) and low (2nd generation) level of buffer (vertical) leakage and identical structure of the AlGaN/GaN heterointerface.

The channel is formed thanks to an $Al_{0.25}Ga_{0.75}N$ barrier, while the gate insulator is constituted by in-situ grown SiN.

The devices were submitted to an extensive characterization, based on double-pulsed measurements (at room temperature and high temperature), backgating measurements (as a function of temperature) and dc leakage evaluation. The results of this analysis are summarized in the following.

III. RESULTS AND DISCUSSION

Figure 1 reports the results of pulsed measurements carried out on one of the analyzed devices starting from two different quiescent bias points: $(V_G , V_D)=(0 V, 0 V)$, which induces negligible trapping effects, and $(V_G , V_D)=(-0 V, 100 V)$, which can induce measurable trapping processes. The results indicate that the devices do not suffer from relevant trapping when they are tested at room temperature (Figure 1 (a)). On the other hand, exposing the results to higher temperatures may result in a significant increase in dynamic Ron, as shown in Figure 1 (b).

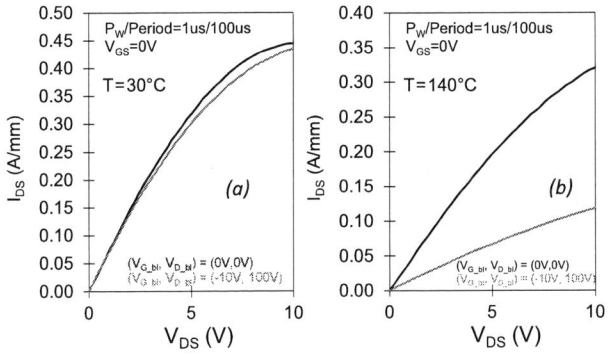

Figure 1: results of pulsed I_D-V_D measurements carried out on one of the analyzed samples at (a) room temperature and (b) high temperature

When the devices are submitted to high off-state bias (e.g. $(V_G , V_D)=(-0 V, 100 V)$), dynamic Ron may originate either from trapping at the surface (in the gate-drain access region), or from trapping in the buffer. A reliable method for distinguishing between these two processes consists in carrying out backgating measurements. During a backgating test, the device is exposed to a high (negative) substrate bias, with no bias applied to the front contact. As a consequence, no significant lateral trapping is induced between gate and drain

(no surface trapping is favored), and all the changes in dynamic Ron can be ascribed to buffer trapping processes. In our investigation, we carried out the backgating tests by using the on-the-fly method [7]. We applied a negative bias for a long time to the devices; every second we quickly pulsed the device to the on-state, for a rapid evaluation of the on-resistance, and then switched back the samples to the backgating condition. These measurements – that were carried out at increasing negative bias levels – provide information on the kinetics of the buffer-related trapping processes.

Typical results of backgating measurements are shown in Figure 2; as can be noticed, exposing the devices to a high backgating bias results in a significant increase in Ron. Increasing the backgating bias speeds up the trapping kinetics, and increases the maximum variation of on-resistance. Since backgating tests favor buffer trapping without inducing any significant surface trapping, these results indicate that the dominant trapping process in our transistors occurs in the buffer.

Further tests [7] demonstrated that the trapping rate strongly depends on the vertical leakage current, flowing from the drain to the substrate of the devices. These results support the hypothesis that the traps located in the buffer can be filled by the vertical leakage, thus favoring an increase in dynamic Ron.

This hypothesis was confirmed by fabricating a second generation of samples, with improved epitaxial structure and reduced vertical leakage. Figure 3 reports the vertical leakage curves measured on representative devices from each wafer. With the second generation of samples the leakage has been reduced by several orders of magnitude.

As a consequence of the reduction of vertical leakage current components, the buffer-related trapping processes are significantly reduced; backgating tests carried out on the devices of 2nd generation (Figure 4) demonstrated a negligible increase in Ron after exposure to high backgating voltages (compare Figure 2 and Figure 4), and the good performance of the developed technology.

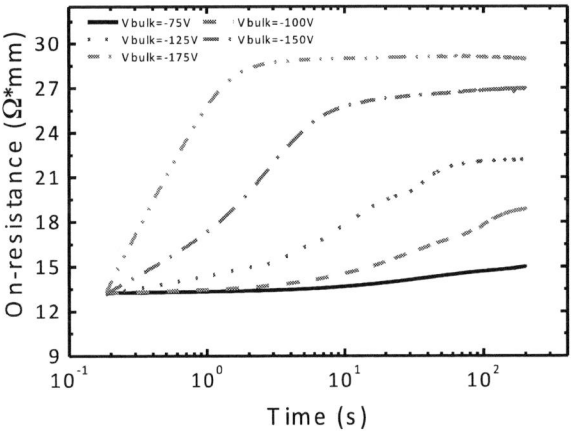

978-1-4799-7670-6/15 $31.00 © 2015 IEEE   111

Figure 2: results of backgating tests carried out at increasing (negative) voltages applied to the device (measurements were carried out on a device of 1st generation)

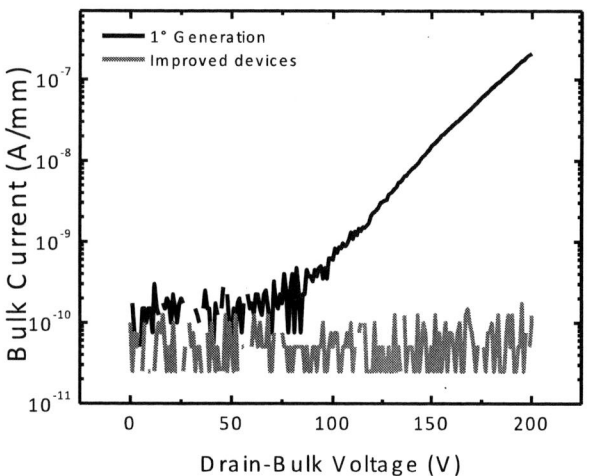

Figure 3: vertical leakage measurements carried out on a device of 1st generation and on a device of 2nd generation

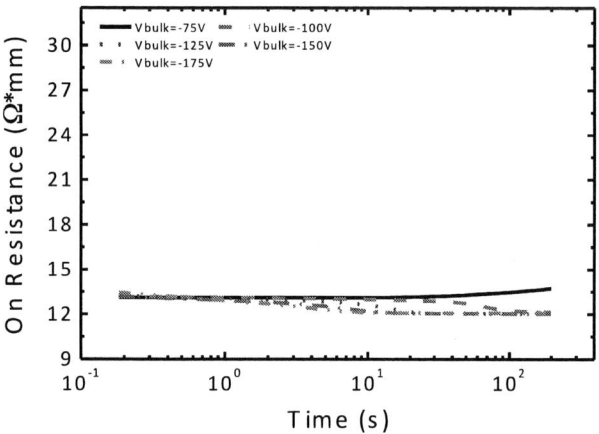

Figure 4: results of backgating tests carried out at increasing (negative) voltages applied to the device (measurements were carried out on a device of 2nd generation)

The optimization of buffer leakage allowed us to fabricate devices with negligible dynamic Ron in a wide temperature and voltage range. Figure 5 and 6 report the results of pulsed characterization carried out with trapping voltages up to 500 V, at room temperature (Figure 5) and at high temperature (150 °C, Figure 6). The pulsed measurements were carried out with an on-state time of 20 μs and a period of 2 ms. The results demonstrate the good stability of this technology, compared to the previous ones.

IV. CONCLUSIONS

With this paper we have described our latest results on the analysis of the trapping processes in AlGaN/GaN-based MIS-HEMTs.

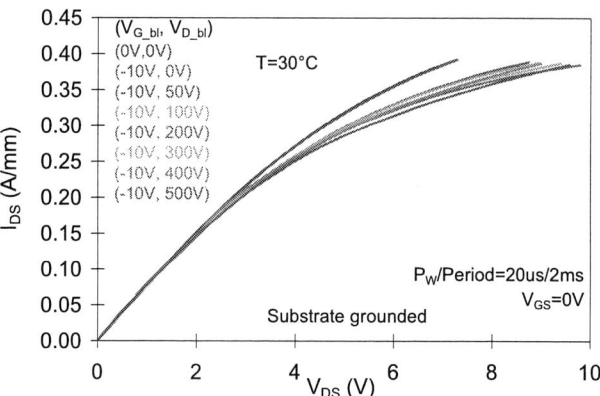

Figure 5: results of pulsed I_D-V_D measurements carried out on one of the improved samples at room temperature, up to a trapping voltage of 500 V

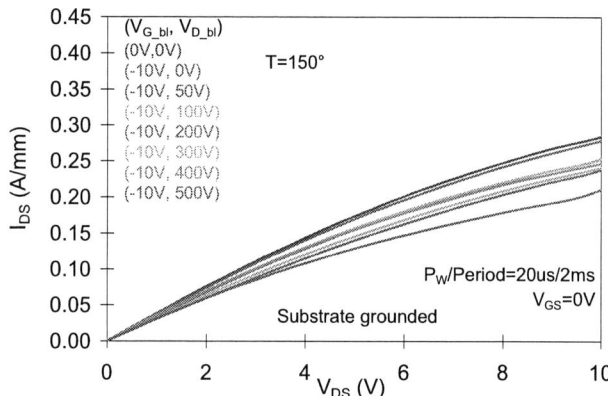

Figure 6: results of pulsed I_D-V_D measurements carried out on one of the improved samples at high temperature (150 °C), up to a trapping voltage of 500 V

The experimental data collected within this work demonstrate that when submitted to high off-state bias, GaN-based transistors can show measurable trapping processes. The increase in dynamic Ron is stronger when the devices are operated at high temperatures, demonstrating the existence of thermally-activated trapping processes.

Backgating tests indicated that buffer-related trapping plays a dominant role in determining the overall dynamic Ron increase; a significant reduction of trapping effects can be obtained through the optimization of the epitaxial structure and the reduction of the vertical leakage current components.

978-1-4799-7670-6/15 $31.00 © 2015 IEEE

Based on these results, we have demonstrated and fabricated a 2nd generation of devices with negligible trapping effects up to 500 V (trapping was evaluated at 150 °C, i.e. in the worst case scenario).

ACKNOWLEDGEMENTS

. This work was supported in part by the European Commission through the ENIAC project "Energy Efficient Converters using GaN Power Devices" (E2COGaN).

REFERENCES

[1] R. Vetury, N. Q. Zhang, S. Keller, and U. K. Mishra, "The impact of surface states on the DC and RF characteristics of AlGaN/GaN HFETs," IEEE Trans. Electron Devices, vol. 48, no. 3, pp. 560–566, Mar. 2001

[2] W. Saito, T. Nitta, Y. Kakiuchi, Y. Saito, K. Tsuda, I. Omura, and M. Yamaguchi, "Suppression of dynamic on-resistance increase and gate charge measurements in high-voltage GaN-HEMTs with optimized fieldplate structure,"IEEE Trans. Electron Devices, vol. 54, no. 8, pp. 1825–1830, Aug. 2007

[3] D. Bisi, M. Meneghini, C. de Santi, A. Chini, M. Damman, P. Brueckner, M. Mikulla, G. Meneghesso, and E. Zanoni, "Deep-Level Characterization in GaN HEMTs-Part I: Advantages and Limitations of Drain Current Transient Measurements", IEEE Transactions on Electron Devices 60 (10), pp. 3166-3175, 2013

[4] M. Meneghini, D. Bisi, D. Marcon, S. Stoffels, M. Van Hove, T.-L. Wu, S. Decoutere, G. Meneghesso, and E. Zanoni, "Trapping in GaN-based metal-insulator-semiconductor transistors: Role of high drain bias and hot electrons", Applied Physics Letters 104, 143505 (2014)

[5] D. Bisi, M. Meneghini F. A. Marino, D. Marcon, S. Stoffels, M. Van Hove, S. Decoutere, G. Meneghesso, and E. Zanoni, "Kinetics of Buffer-Related RON-Increase in GaN-on-Silicon MIS-HEMTs," Electron Device Letters, IEEE , 35,10,1004, (2014)

[6] M. Meneghini, D. Bisi, D. Marcon, S. Stoffels, M. Van Hove, T.-L. Wu, S. Decoutere, G. Meneghesso, and E. Zanoni, "Trapping and Reliability Assessment in D-Mode GaN-Based MIS-HEMTs for Power Applications", Power Electronics, IEEE Transactions on , vol.29, no.5, pp.2199-2207, May 2014

[7] M. Meneghini, P. Vanmeerbeek, R. Silvestri, S. Dalcanale, A. Banerjee, D. Bisi, E. Zanoni, G. Meneghesso, P. Moens, "Temperature-Dependent Dynamic Ron in GaN-Based MIS-HEMTs: Role of Surface Traps and Buffer Leakage," Electron Devices, IEEE Transactions on , vol.62, no.3, pp.782-787, March 2015

Plasma Induced Damage Investigation in the Fully Depleted SOI Technology

M. Akbal, G. Ribes, M. Guillermet and L. Vallier

Abstract—**Plasma induced damage (PID) in the Fully Depleted SOI devices was studied for two etch plasma processes. Different antenna test structures were used in order to show that plasma non-uniformity can occur between the device nodes inducing severe charging damage. Also, the PID behavior disparity between the test structures related to the antennas architecture was observed. These different behaviors can be compared to degradation mechanism as hot carrier injection (HCI) or bias temperature instability (BTI) depending of charging distribution over the device node.**

Index Terms— **Plasma induced damage (PID), Fully Depleted silicon-on-insulator (FDSOI) technology, Plasma non-uniformity.**

I. INTRODUCTION

The plasma induced damage is considered as one of the most serious reliability issues [1] [2] [3] [4] and has been widely studied in the standard bulk CMOS devices [5]. Fully Depleted SOI (FDSOI) technology is admitted as a serious alternative to CMOS scaling [6]. The purpose of this paper is to understand how the plasma induced damage occurs in FDSOI devices. The particular architecture of this device induces a new plasma-wafer interaction behavior. Indeed, in this case the substrate is isolated, thus the field across the oxide during the plasma processing is determined by the differential charging between the gate and the diffusion nodes [7]. Based on this understanding, structures with antenna connected to gate and to the drain/source nodes are designed. This structure should enable us to detect a possible potential difference between the device nodes. In addition, in order to compare FDSOI PID configuration and bulk configuration, a test structure with a protection diode connected to the drain node is designed. In this case only the charges collected by the gate antenna can induce damage.

II. EXPERIMENTAL DETAILS

The n and p MOSFET IO devices used in this work are fabricated by using advanced CMOS FDSOI 14nm technology [8]. A high-k gate stacks (HfO2/SiO2) with an electrical thickness CET= 4.2nm is used as gate dielectric materials. The device size is WxL=1µmx0.15µm.

Fig.1 shows the antenna test structures used in this work to investigate the plasma damage in the FDSOI technology. The antenna attached to the device node consist of a plate of metal with an area of 1950µm² which give an antenna ratio denoted AR=13000. The MOSFET with a dual protection diode (gate and drain nodes protected) [7] is used as a reference since the protection diode can eliminate the PID damage.

The samples are exposed to two different etch plasma processes, process (A): non-uniform process and process (B): uniform process during 420s at 20°c. The capacitively coupled plasma reactor (CCP) is used.

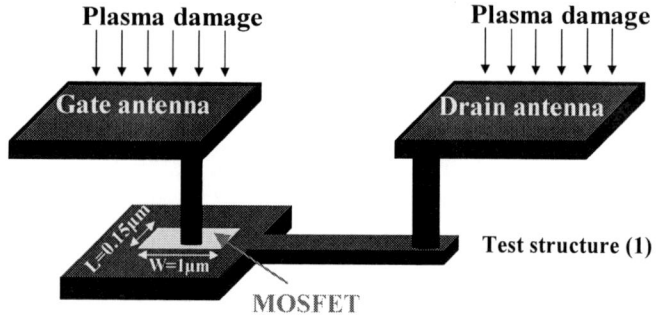

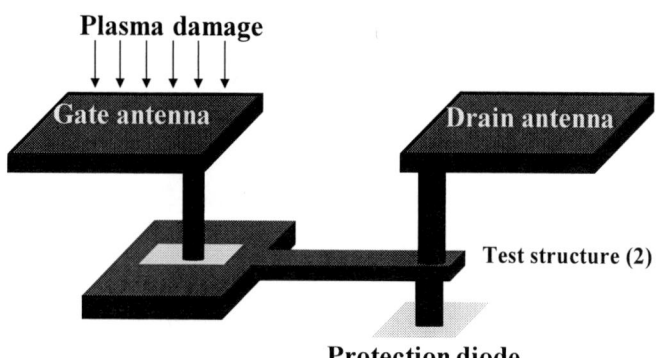

Fig. 1. Schematic representation of the antenna test structures (1) and test structure (2) used in this work to investigate the plasma induced damage in the fully depleted SOI devices.

M. Akbal is with STMicroelectronics, 38926 Crolles, France and also with LTM/CNRS, 38054 Grenoble, France (e-mail: madjid.akbal@st.com).
G. Ribes is with STMicroelectronics and L. Vallier is with LTM/CNRS. M. Guillermet is with CEA/LETI, 38054 Grenoble, France.
This work has been partly funded by the European Community and French government under projects Places2Be and Nano2017.

978-1-4799-7670-6/15 $31.00 © 2015 IEEE

III. EXPERIMENTAL RESULTS & DISCUSSION

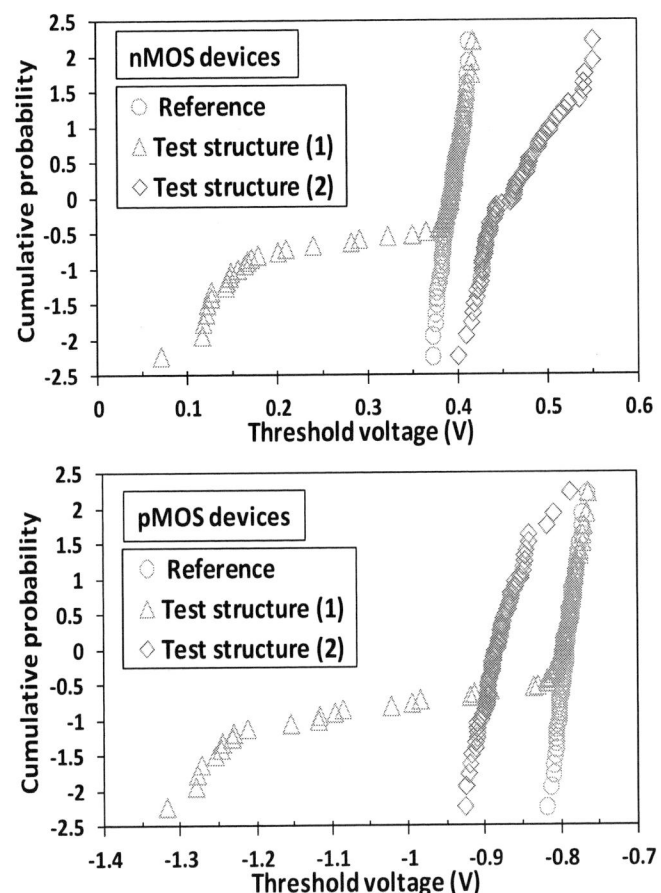

Fig. 2. Threshold voltage distribution of the nMOS and the pMOS test structures exposed to the non-uniform etch plasma process (A).

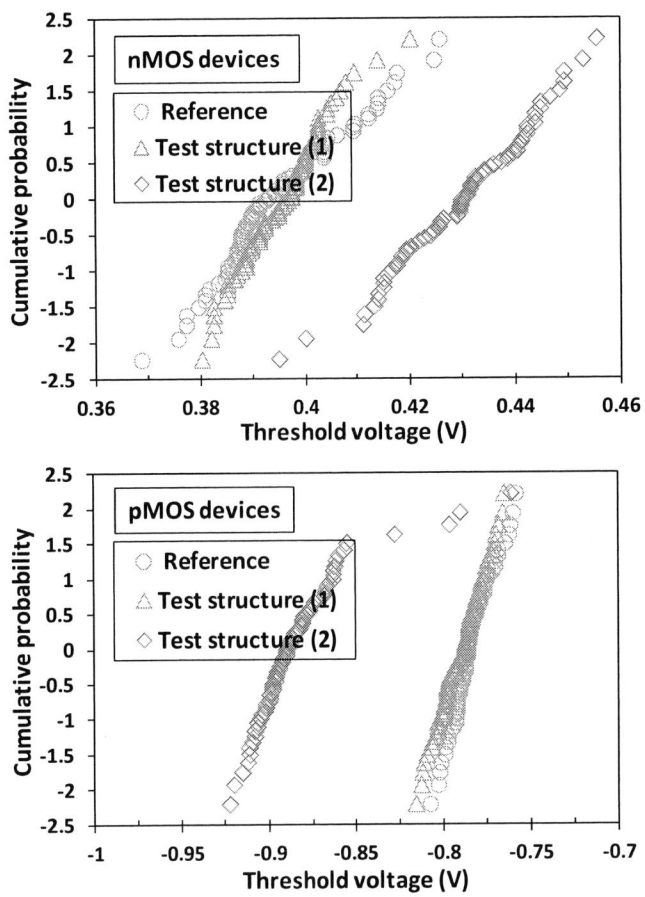

Fig. 3. Threshold voltage distribution of the nMOS and the pMOS test structures exposed to the uniform etch plasma process (B).

Fig.2 and **Fig.3** shows the threshold voltage distribution of the devices after the exposition to plasma processes. First, we focus on the results obtained from the test structure (1). This structure (**Fig.1**) shows two different behaviors. Indeed, the samples exposed to the plasma process (A) shows severe plasma damage unlike to those exposed to the plasma process (B). In fact a significant shift is observed with respect to the reference device. Hence, we can expect that a potential difference between the gate and the drain/source antennas has been probably applied during the plasma process.

Moreover, we can also note that part of the devices exposed to the plasma process (A) is not damaged. Indeed, they show a similar threshold voltage distribution as reference device. **Fig.4** shows the map of the threshold voltage for the nMOS and pMOS test structures (1) exposed to the plasma process (A). The damages are located in wafer center resulting from the plasma non-uniformity.

In addition, we also note that the trapped charge sign resulting of the threshold voltage shift is the same between the nMOS and the pMOS test structures. Indeed, the devices show a negative threshold voltage shift ($\Delta VT<0$) after the exposition to plasma process probably induced by hole trapping in the device gate oxide. This mechanism is due to a negative voltage stress during the plasma process [9][10].

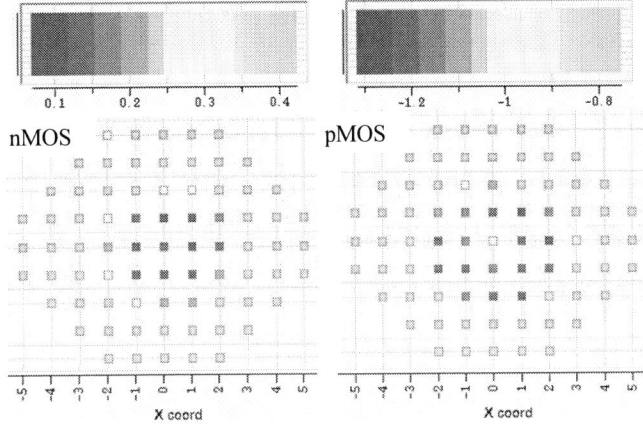

Fig. 4. Map of the threshold voltage for the nMOS and pMOS test structure (1) after the exposition to etch plasma process (A).

In opposition to the test structure (1), the test structure (2) adopts a similar behavior between the process (A) and the process (B). The samples are damaged in these two cases. In addition, no edge-center effect is detected. Thus, we can conclude that this test structure is not sensitive to the plasma local non-uniformity.

978-1-4799-7670-6/15 $31.00 © 2015 IEEE

The samples are exposed to two different plasma etch processes. Non-uniform process (A) and uniform process (B). The experimental results demonstrates that the test structure (1) shows severe plasma damage induced by the etch plasma process (A). These damages are probably induced by differential charging occurred between the device nodes during the processing [7]. Indeed, the gate and the drain/source antennas are identical. Lai et al[11] demonstrate that the plasma damage level is impacted by the antenna spacing on partially depleted devices. In order to verify this effect on FDSOI devices, the test structures with AR fixed to 13000 and different spacing between the gate and the diffusion node antenna varied from 70μm (illustrated in **Fig.1**) to 490 μm are designed and exposed to plasma process (A) such as illustrated in **Fig.5**.

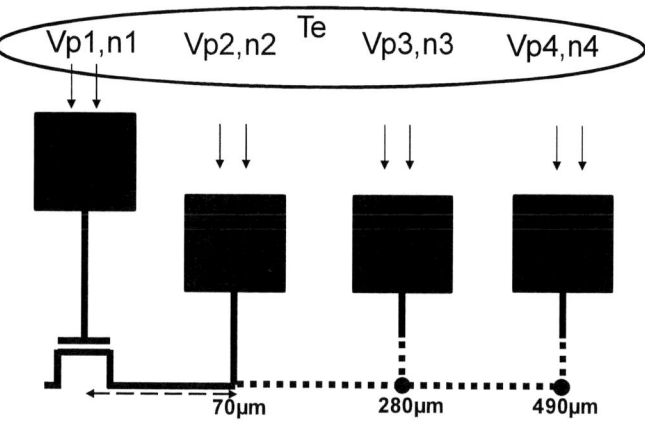

Fig. 5. Schematic representation of the antenna test structures used to investigate the plasma non-uniformity effect between the device nodes by varying the antenna spacing from 70μm to 490μm.

Fig.6 shows the threshold voltage shift of the nMOS and pMOS test structures shown in **Fig.5** after the processing. The plasma damage dependence with antenna spacing is detected. Thus, in addition of the edge-center effect, the plasma non-uniformity can occur between the device nodes inducing important plasma damage. Indeed, the potential difference between the antennas is a function of the plasma density, hence as a function of the plasma process uniformity such as summarized by the equation (1) [12].

$$Vp1-Vp2=\frac{KTe}{e}\ln\left(\frac{n1}{n2}\right)<Vp1-Vp3=\frac{KTe}{e}\ln\left(\frac{n1}{n3}\right)<Vp1-Vp4=\frac{KTe}{e}\ln\left(\frac{n1}{n4}\right)$$

With Vp is the plasma potential, Te and n are the electrons temperature and the plasma density respectively and e the elementary charge.

Fig.7 shows the voltage stress profile through the wafer induced by the plasma processes. First, plasma process (B) which induces a uniform voltage stress. Second, the process (A) induces a uniform voltage stress at wafer edge and a non-uniform negative voltage stress at the wafer center. Indeed, the damaged nMOS and pMOS test structures (1) localized in the wafer center shows a negative threshold voltage shift. The **Fig.7** enables explain the behavior

difference between the test structure (1) and test structure (2). The test structure (2) is protected at the drain node. Consequently, the degradation is driven only by the plasma potential at the gate antenna. Hence the gate node of this structure sees a voltage stress in the case of process (A) and (B) modulated by the potential level accumulated on the gate during the process.

Thus, we can conclude that the test structure (1) is sensitive to plasma local variations which occur between the device nodes while the test structure (2) is sensitive to the plasma variation over the wafer.

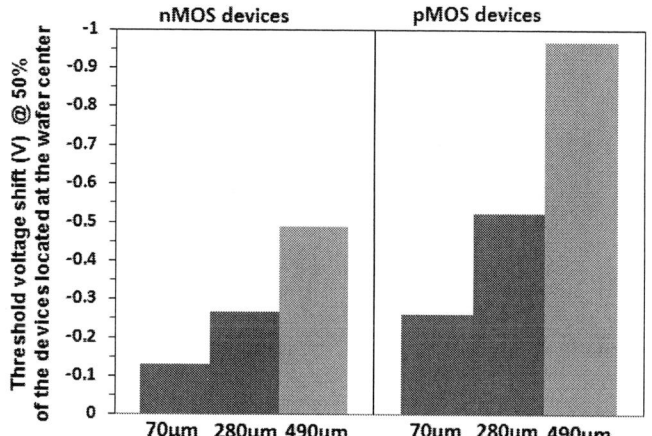

Fig. 6. Median threshold voltage shift evolutions as a function of the antenna spacing for the nMOS and pMOS test structure induced by the plasma process etch (A).

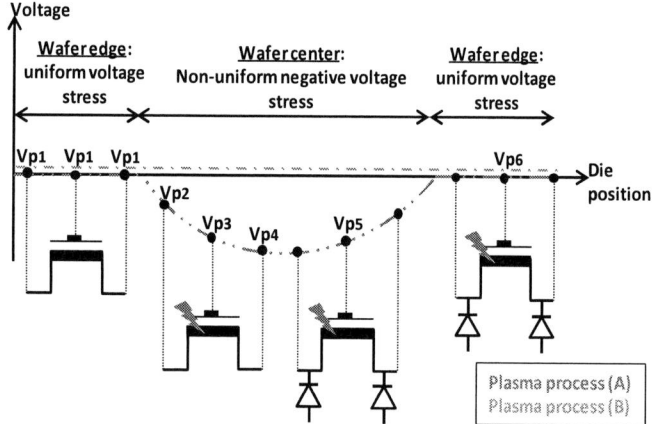

Fig. 7. Schematic representation of the voltage stress profile through the wafer and the plasma stress seen by test structures (1) and (2) with the plasma process (A) and process (B).

It is now interesting to link the plasma damage level to the test structures characteristics. Indeed, the samples have shown different plasma damage level as a function of the plasma processes. The test structure (1) exposed to the plasma process (A) shows severe plasma damage with respect to the test structure (2). In the test structure (1) the damage occurs when the differential charging are collected by the gate and

978-1-4799-7670-6/15 $31.00 © 2015 IEEE

the source and the drain node. Hence an electric field is applied between drain and source in opposition to BTI stress. This mechanism may be compared to a hot carrier injection stress (HCI). Unlike, in the test structure (2) the damage level is driven by the plasma potential at the gate antenna which is equivalent to bias temperature instability stress (BTI) such as illustrated in the **Fig.8**.

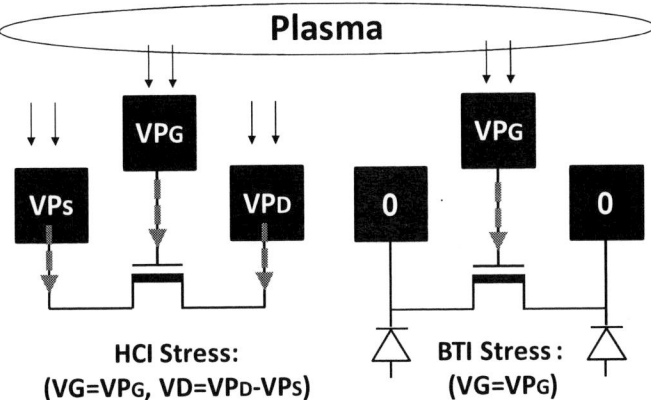

Fig. 8. Schematic representation of the comparison between the reliability effects (HCI and BTI stress) and the PID behavior in the FDSOI devices

The BTI stress is critical at high temperature in opposition to the HCI which can induce important threshold voltage shift at low temperature [13]. Electrical stress was performed by using the reference devices at the same condition than the etch plasma process (20°C during 420s). **Fig.9** shows the threshold voltage shift after the stress. Indeed, the hot carrier injection stress induces an important shift compared to the BTI stress. This result can be explained by the temperature of the stress and is consistent with the fact that the test structure (1) is more degraded than structure (2) in case of non-uniform processes like process (A).

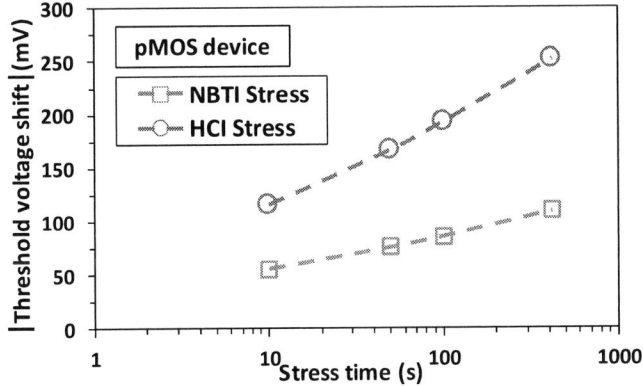

Fig. 9. Time evolution of threshold voltage for the FDSOI pMOS device under HCI stress (Vg=-1.98V, Vd=-2.5V) and BTI stress (Vg=-2.5V) at 20°C

IV. CONCLUSION

In this work the behavior of the Fully Depleted SOI (FDSOI) test structures for two different plasma processes has been investigated. It was demonstrated that the plasma

non-uniformity can induce severe plasma damage. Indeed, in addition of the edge-center effect a local non-uniformity can occur between the device nodes causing severe degradation. The damage level induced by this mechanism depends of the spacing between the gate and the drain/source antennas. Indeed, the damages increase if the antenna spacing increases. This mechanism can be compared to HCI stress.

Antenna test structures with the diffusion node protected has been also studied. This structure is similar to standard bulk test structure in terms of plasma damage and this PID configuration can be compared to BTI stress. Hence FDSOI devices without PID protection may be stressed in HCI configuration in opposition to Bulk devices.

The understanding of these different PID stress configurations is important because they are usual in product design. Based on this understanding robust design rule for the FDSOI technology can be established and reliable product can be designed.

REFERENCES

[1] K.P. Cheung, C.P. Chang, "Plasma-charging damage: a physical model", Journal of applied physics, 1994, pp. 4415-4426.

[2] C.C. Chen, H.C. Lin, C.Y. Chang, M.S. Liang, Chien, Chao-Hsin, S.K Hsien, T.Y. Huang, T.S. Chao, "Plasma-Induced Charging Damage in Ultrathin (3-nm) Gate Oxides", IEEE Transactions on Electron Devices, 2000, pp. 1355-1360.

[3] P.J. Tzeng, J.C. Li, C.C. Yeh, K.S. chang-Liao, "Reduction and Non-uniformity of High Density Plasma Process Induced Electrical Degradation in MOS devices", International Plasma Process Induced Damage, 1999, pp. 100-103.

[4] J-P. Carrere, J-C. Oberlin, M. Haond, "Topographical Dependence of Charging and New Phenomenon During Inductively Coupled Plasma (ICP) CVD Process", IEEE International Symposium on Plasma Process-Induced Damage, 2000, pp. 164-167.

[5] K.J. Kuhn, "Process Technology Variation", IEEE Transactions on Electron Devices, 2011, Vol. 58, N° 8, pp. 2197-2208.

[6] N. Planes, O. Weber, V. Barral, S. Haendler, D. Noblet, "28nm FDSOI Technology Platform for High-Speed Low-Voltage Digital Applications", Symposium on VLSI Technology, 2012, pp. 133-134.

[7] M. Akbal, G. Ribes, L. Vallier, "New insight in plasma charging impact on gate oxide breakdown in FDSOI technology", submitted to International Reliability Physics Symposium 2015.

[8] O. Weber, E. Josse, F. Andrieu, A. Cros, E. Richard, "14nm FDSOI Technology for High Speed and Energy Efficient Applications", Symposium on VLSI Technology, 2014, pp. 1-2.

[9] K. Eriguchi and M. Niwa, "Temperature and stress polarity-dependent dielectric breakdown in ultrathin gate oxides", Applied Physics Letters 73, 1998, pp. 1985-1987.

[10] K. Eriguchi, M. Kamei, K. Okada, H. Ohta and K. Ono, "Threshold Voltage Shift Instability Induced by Plasma Charging Damage in MOSFETs with High-k Dielectric", IEEE International Conference on Integrated Circuit Design and Technology and Tutorial, 2008, pp. 97-100.

[11] W. Lai, D. Harmon, T. Hook, V. Ontalus, J. Gambino, "Ultra-thin Gate Dielectric Plasma Charging Damage in SOI Technology" International Reliability Physics Symposium, 2006, pp. 370-373.

[12] V. Vahedi, N. Benjamin, A. Perry, "Topographic Dependence of Plasma Charging Induced Device Damage", IEEE International Symposium on Plasma Process-Induced Damage, 1997, pp. 41-44.

[13] M. Dai, C. Gao, K. Yap, Y. Shan, Z. Cao, K. Liao, L. Wang, B. Cheng, S. Lui, "A Model With Temperature-Dependent Exponent for Hot-Carrier Injection in High-Voltage nMOSFETs Involving Hot-Hole Injection and Dispersion", IEEE Transactions on Electron Devices 2008, pp. 1255-1258.

978-1-4799-7670-6/15 $31.00 © 2015 IEEE

Plasma-induced photon irradiation damage on low-k dielectrics enhanced by Cu-line layout

Taro Ikeda, Akira Tanihara, Nobuhiko Yamamoto, and Shigeru Kasai

Technology Development Center,
Tokyo Electron Yamanashi Limited,
650 Mitsuzawa, Hosaka-cho, Nirasaki 407-0192, Japan
e-mail: taro.ikeda@tel.com

Koji Eriguchi and Kouichi Ono

Graduate School of Engineering,
Kyoto University,
Kyoto daigaku-Katsura, Nishikyo-ku, Kyoto 615-8540,
Japan

Abstract—We demonstrate experimentally and theoretically the existence of circuit-layout-dependent low-*k* damage by plasma radiation. Circuit-layout-dependent low-*k* damage apparently occurs in nitrogen (N₂) plasma, not in argon (Ar) plasma. Using an electromagnetic simulation and the dispersion analysis, we reveal that *E*-field in the low-*k* film is enhanced for specific Cu-line layouts in the case of N₂ plasma. The results of electromagnetic simulations and dispersion analysis are consistent with the obtained experimental results. We propose a new low-*k* damage model, where "near-field" by the irradiated copper lines plays an important role in the damage creation. The near-field enhances *E*-field in the low-*k* film, accelerating the bond-breakage, i.e., the dielectric constant increase. The present model framework is useful for optimizing an integrated circuit layout, simultaneously minimizing the plasma radiation damage.

Keywords— low-k, plasma damage, Cu, layout, near-field

I. INTRODUCTION

Cu and low-*k* dielectrics [1, 2] have been used for ultra-large scale integrated circuits (ULSIs) in order to reduce *RC*-delay. It is widely known that the dielectric constant of low-*k* is a key to the *RC*-delay, determining LSI performance. However, these low-*k* films are sensitive to plasma-induced damage (PID)— plasma-induced charging damage (PCD) [3-5], plasma-induced physical damage (PPD) [5-7], and plasma-induced radiation damage (PRD) [8-11]. Historically, blanket samples have been used for the PRD evaluations. For verifying PRD accurately, low-*k* samples with practical circuit layout should be used because the increase in the interline *k*-value (~ capacitance *C*) directly corresponds to that in the *RC* delay, resulting in LSI performance degradation. A preliminary study was reported on an experimental evidence of the presence of PRD using low-k structures with Cu lines [12]. In this study, we investigate the PRD using low-*k* samples with various Cu lines to clarify how circuit layout impacts on PRD creation in different plasma sources, focusing on "near-filed" mechanisms in detail. We conduct electrical measurements for samples with various Cu-line layouts, unified with electromagnetic simulations (EMS) and a surface plasmon dispersion analysis. The PRD mechanism is discussed based on the near-field radiation from (electrons in) Cu lines. We present that the PRD (by interline *C*) is enhanced in a specific layout and a plasma source against scaling, which may be induced by the near-field. The present findings suggest that one should implement the obtained circuit-layout-dependent PRD model into future high-performance LSI designs.

II. EXPERIMENTAL

A. Test structures and plasma radiation

As seen in Fig. 1(a), comb-shaped Cu lines embedded in low-*k* films (SiCN / low-*k* / SiCN) were fabricated on a 12-in. p-type Si(100) substrate. Samples were cut into square chips of 25 × 25 mm². The ratio of Cu line to low-*k* space widths was unity for all line-and-space (L/S) patterns. The L/S patterns investigated were 60, 120, and 180 nm. The total length of each Cu line was 3 cm. We prepared Cu probing pads for *I–V* and *C–V* tests of the interline low-*k*. A passivation layer was deposited on the Cu lines and low-*k* films. As shown in Fig. 1(b), a quartz cover of 100 mm in diameter was placed on the samples for preventing particle influxes (ions, electrons, radicals) from plasma.

A surface-wave plasma reactor was used for creating PRD on the samples. The source power was 400 W. N₂ or Ar gas was used. The temperatures of samples were monitored to be lower than 125 °C. Table I shows the detail of process conditions.

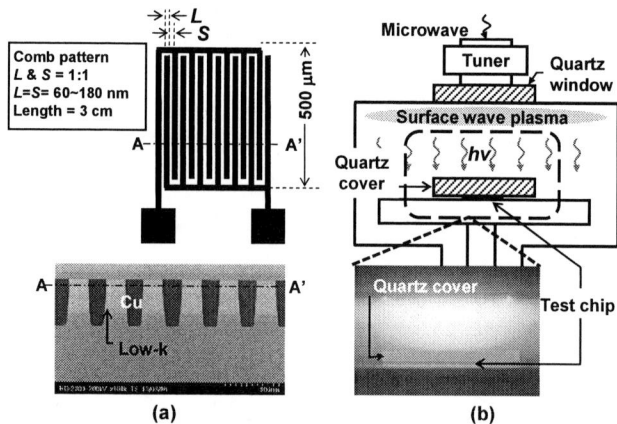

Fig. 1. (a) Cu-line layout and SEM cross sectional view of a sample and (b) surface wave plasma system used in this study.

TABLE I. PLASMA CONDITION USED IN THIS STUDY.

Gas	: N$_2$ = 1000 [sccm] Ar = 1000 [sccm]
Pressure	: 20 [Pa]
Source Power	: 400 [W]
Process Time	: 10, 100 [min]
Substrate temp	: <125 [°C]

Figure 2 shows the optical emission spectra from typical Ar and N$_2$ plasmas generated in the reactor. The arrows in this figure correspond to the wavelengths studied in an electromagnetic simulation discussed later in this article.

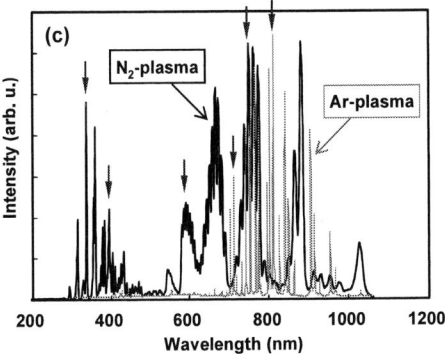

Fig. 2. Optical emission spectra from Ar and N$_2$ plasmas.

B. Measurement methods

Interline capacitance between Cu lines (C) was measured at room temperature. The dielectric constant of low-k films in various layouts was determined from the measured C. In determining the dielectric constant, an electromagnetic simulation (EMS) was carried out to determine the relationship between the C and the dielectric constant (ε) in advance as shown in Fig. 3. We assumed that the C obeyed the obtained C–ε relation for all PRD cases.

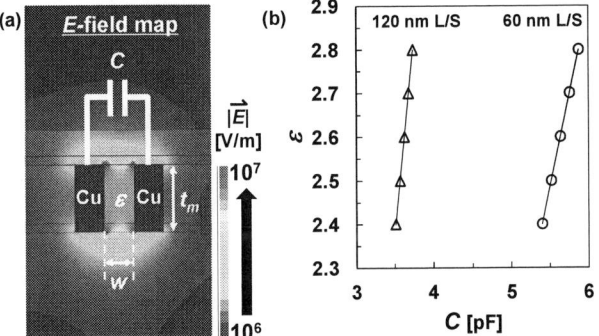

Fig. 3. (a) Contour mapping of E-field in the vicinity of the structure (V = 1 V, t_m = 270 nm, W = 60 ~ 180 nm) and (b) C–ε relation obtained by EMS.

III. RESULTS AND DISCUSSION

A. Leakage current analysis

As shown in Fig. 4(a), the leakage current through low-k films (I_{leak}) decreased by N$_2$ plasma exposure for both 60- and 120-nm-L/S samples. This result indicates the presence of low-k damage. In the analogy with the model [13, 14], when electron trap sites are created by PRD in low-k films, electrons are trapped, thus I_{leak} decreases during I–V measurement as shown in Fig. 4(b). Hence, the I_{leak} decrease observed in Fig. 4(a) corresponds to low-k damage creation.

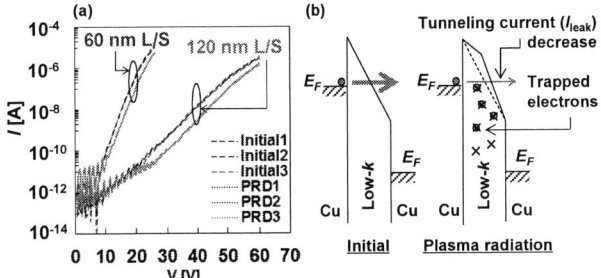

Fig. 4 (a) I–V characteristics for various damaged samples (10-, 60-, and 100-min exposures) and (b) the mechanism for I_{leak} decrease by trapped electrons.

B. Dielectric constant analysis

Figure 5 shows $\varepsilon_{dam}/\varepsilon_i$ as a function of L/S for N$_2$ plasma exposure. The $\varepsilon_{dam}/\varepsilon_i$ is the ratio of dielectric constants before (ε_i) and after (ε_{dam}) plasma radiation. As seen, $\varepsilon_{dam}/\varepsilon_i$ increases with L/S and the radiation time. The time-dependent $\varepsilon_{dam}/\varepsilon_i$ increase is attributed to the creation of PRD in the low-k films. In contrast, $\varepsilon_{dam}/\varepsilon_i$ is approximately constant for both L/S for Ar plasma exposure even for 10 min, as shown in Fig. 6. These L/S- and plasma-dependent PRD are the main results of this study—PRD is enhanced by the layouts and plasma sources against scaling. As discussed below, this enhancement may be induced by near-field [15] created by Cu lines under (N$_2$) plasma irradiation. The near-field enhances E-fields in the interline low-k films, leading to creation of traps (broken-bond states). The creation of traps (energy levels) makes the band-gap narrower, which increases the dielectric constant. Regarding the Ar plasma case, the E-field is not enhanced enough to create traps due to the lower near-field.

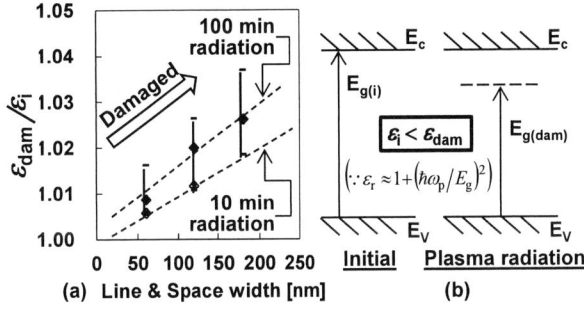

Fig. 5. (a) Dielectric constant ratio ε_i / ε_{dam} for the case of N$_2$ plasma radiation as a function of L/S, and (b) the mechanism of ε-increase by low-k damage. Five devices were tested under each condition.

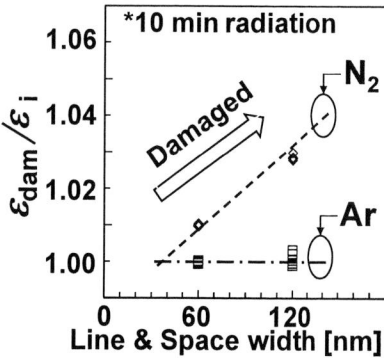

Fig. 6. Dielectric constant ratio $\varepsilon_i/\varepsilon_{dam}$ for N_2 and Ar plasma radiations as a function of L/S. Five devices were tested under each condition.

C. EMS results—near-field effect

To clarify the ε-increase above, we performed an EMS. In the EMS, the dielectric constant of Cu lines was introduced. Figures 7 show the simulated E-fields. To emulate N_2 and Ar plasma radiation, we selected the wavelengths on the basis of the observed major peaks in the spectra from N_2 or Ar plasmas in Fig. 2. The wavelengths of 337, 381, and 590 nm are the peaks from N_2 plasma, while those of 701, 766, and 811 nm, from Ar plasma. As seen in Fig. 7, in the case of N_2 plasma, E-fields in low-k films are stronger in 120- than 60-nm-L/S and blanket. The results reveal that E-fields are enhanced not only by the presence of Cu line but also by the specific L/S. In the case of Ar plasma, on the other, E-fields are not apparently enhanced. The enhanced E-field in the N_2 plasma exposure accelerates the degradation of low-k films (increasing the dielectric constant). The EMS predictions are consistent with the experimental results. The correlations between the E-field enhancement and Cu-line layout are characterized by the dispersion relation of surface plasmons for the Cu/low-k/Cu geometry as discussed in the next subsection.

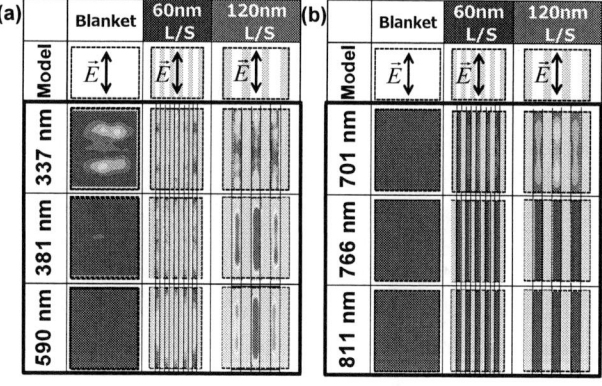

Fig. 7. Contour mapping of E_0^2 for blanket, 60- and 120-nm L/S at (a) $\lambda_0 = 337$, 381 and 590 nm (N_2) and (b) $\lambda_0 = 701$, 766 and 811 nm (Ar). In the EMS, the dielectric constant of Cu lines was introduced. (ex. $\varepsilon_{Cu} = -25.73 + j\,0.354$ at $\lambda_0 = 590$ nm)

D. Dispersion relation for Cu/low-k/Cu L/S geometry

To identify the coupling of incident photons and the near-field formed by Cu lines enhancing the E-field in low-k films, the dispersion relation for the L/S geometry is analyzed. Letting a space width between Cu lines d, low-k dielectric constant ε_1, and the embedded Cu lines ε_2, the dispersion relations take the forms [16] as

$$\text{Antisymmetric}: \varepsilon_1 k_{z2} + \varepsilon_2 k_{z1} \tanh\left(\frac{-ik_{z1}d}{2}\right) = 0\,, \tag{1a}$$

$$\text{Symmetric}: \varepsilon_1 k_{z2} + \varepsilon_2 k_{z1} \coth\left(\frac{-ik_{z1}d}{2}\right) = 0\,, \tag{1b}$$

where k_{z1} and k_{z2} is defined by

$$k_{z1,2}^2 = \varepsilon_{1,2}\left(\frac{\omega}{c}\right)^2 - k_x^2\,, \tag{2a}$$

$$k_x^2 = \left(\frac{\omega}{c}\right)^2 \left(\frac{\varepsilon_1 \varepsilon_2}{\varepsilon_1 + \varepsilon_2}\right)\,. \tag{2b}$$

Herein, the E-field in the low-k ($E_x^{(1)}$) and that in the Cu line ($E_x^{(2)}$) are given respectively by

$$E_x^{(1)} = E_0 e^{i(k_x x - k_{z1}|z| - \omega t)}, \qquad E_x^{(2)} = E_0 e^{i(k_x x - k_{z2}|z| - \omega t)}. \tag{3}$$

Accordingly, each solution for the antisymmetric or symmetric mode gives a distinct coupling mode of plasmons in Cu lines and reflects the E-field symmetry. Figure 8 plots the dispersion relations with the bulk plasma frequency ω_p of 1.64×10^{16} s^{-1}. Figure 9 illustrates spatial distributions of energy density in Cu and low-k systems. The density is defined by

$$u_{low-k} = \frac{1}{2}\vec{E} \cdot \vec{D}^* = \frac{1}{2}\varepsilon_1 \vec{E} \cdot \vec{E}^*. \tag{4}$$

While, within Cu, the density is derived [17] as

$$u_{Cu} = \frac{1}{2}\text{Re}\left[\frac{d(\omega \varepsilon_2)}{d\omega}\right]\vec{E} \cdot \vec{E}^*. \tag{5}$$

Figures 8 and 9 indicate that the E-field is significantly enhanced by standing waves in the low-k film only in the symmetric mode. Hence, the enhancement of incident E-field observed in Fig. 7 is explained by the effects of the presence of Cu lines (layout) with the 60- and/or 120-nm L/S.

E. Damage creation model in low-k under photon irradiation

Finally, Fig. 10 presents the low-k damage mechanism enhanced by near field. As seen, the probability W_{dam} (from bonding Ψ_i to anti-bonding states Ψ_{dam}) can be described by the perturbation term in the Hamiltonian as

$$H' = -(e/m)\vec{A} \cdot \vec{p}\,, \tag{6.1}$$

and

$$W_{dam} \propto \left|\left\langle \Psi_{dam}|H'|\Psi_i \right\rangle\right|^2 \propto \left|\vec{A}\right|^2 \propto \left|E_0\right|^2\,, \tag{6.2}$$

where e is the elementary charge, m is the mass of electron, $\vec{p}$ is the momentum of electrons transiting from bonding to anti-bonding states, and $\vec{A}$ is the vector potential. Thus, W_{dam} is dependent on the generated E-field. When Cu lines are irradiated to photons during plasma processing, the near-field is formed in low-k films depending on the wavelength and the layout. For a specific layout and a plasma condition, the E_0^2 in interline low-k films with Cu lines increases as seen in Fig. 10, compared to such a blanket wafer. This mechanism results in an increase in W_{dam}. We speculate that this near-field effect induces the enhancement of low-k damage as observed here in this study against scaling.

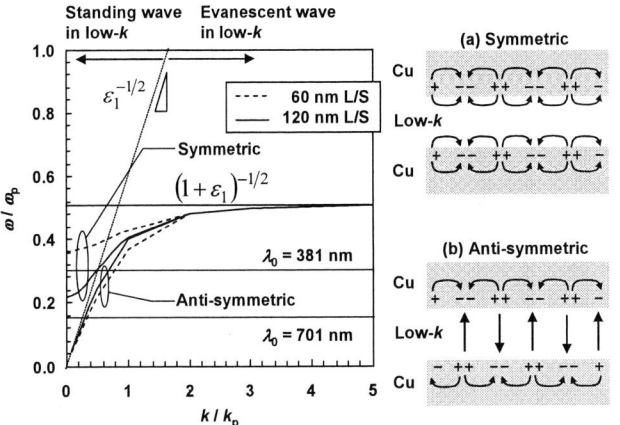

Fig. 8. Dispersions for Cu/low-k/Cu geometries for 60- and 120-nm L/S. Dispersion calculated is plotted in the left panel. Distributions of induced charges and E-field for (a) symmetric mode and (b) antisymmetric mode are shown on the right.

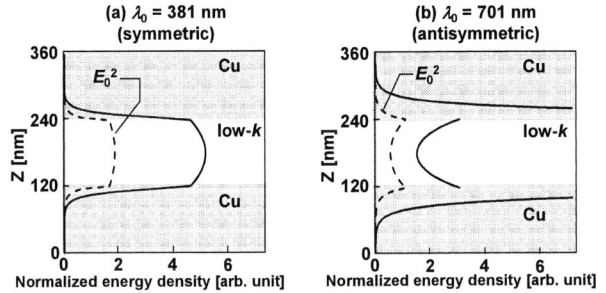

Fig. 9. Spatial distributions of the energy density for 120-nm-L/S in the Cu/low-k/Cu geometry for the metal and the dielectric areas, calculated by Eqs. (4) and (5). (a) symmetric mode and (b) antisymmetric mode. For reference, the spatial distribution of the E_0^2 by the EMS is also included (dashed). The free space wavelength is set to (a) $\lambda_0 = 381$ nm (N$_2$) or (b) $\lambda_0 = 701$ nm (Ar), where E_0^2 is maximal for EMS.

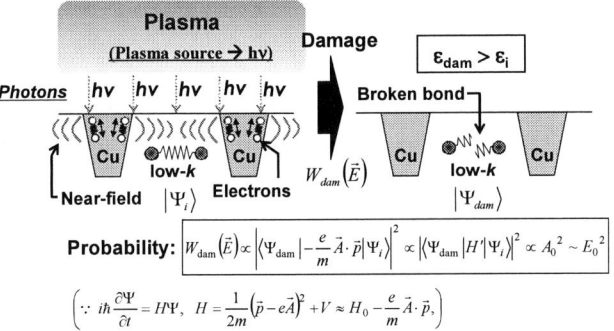

Fig. 10. Mechanism of layout-enhanced low-k damage. Depending on the Cu line layout and the energy of photons, the E-field formed in the low-k is increased.

IV. CONCLUSIONS

By measuring I–V and C–V characteristics of the devices with various Cu-line layouts exposed to N$_2$ or Ar plasma radiation, we clarified the enhancement of interline low-k degradation by

PRD. Using an EMS and dispersion relation analysis, we revealed that the E-fields between low-k films are enhanced by specific Cu-line layouts and plasma process conditions. These EMS predictions were in good agreement with experimental data and the dispersion-relation combined with the experimentally obtained emission spectra. A new PRD creation mechanism was proposed, where the near-field accelerates bond-breakage in low-k films. It is confirmed that one should implement the proposed model into future LSI layout and plasma-process designs to minimize PRD.

REFERENCES

[1] SIA, *The International Technology Roadmap for Semiconductors, 2012 update*, 2012.

[2] K. Maex, M. R. Baklanov, D. Shamiryan, F. lacopi, S. H. Brongersma, and Z. S. Yanovitskaya, "Low dielectric constant materials for microelectronics," *J. Appl. Phys.*, vol. 93, pp. 8793-8841, 2003.

[3] K. P. Cheung, *Plasma Charging Damage*. Heidelberg: Springer, 2001.

[4] A. Martin, "Review on the reliability characterization of plasma-induced damage," *J. Vac. Sci. & Technol. B*, vol. 27, pp. 426-434, 2009.

[5] K. Eriguchi and K. Ono, "Quantitative and comparative characterizations of plasma process-induced damage in advanced metal–oxide–semiconductor devices," *J. Phys. D*, vol. 41, p. 024002, Jan. 2008.

[6] G. S. Oehrlein, "Dry etching damage of silicon: A review," *Materials Sci. Eng. B*, vol. 4, pp. 441-450, 1989.

[7] K. Eriguchi, Y. Takao, and K. Ono, "A new aspect of plasma-induced physical damage in three-dimensional scaled structures — Sidewall damage by stochastic straggling and sputtering," *Proc. Int. Conf. on Integrated Circuit Design & Technol.*, pp. 1-4, 2014.

[8] J. Lee and D. B. Graves, "Synergistic damage effects of vacuum ultraviolet photons and O$_2$ in SiCOH ultra-low- k dielectric films," *J. Phys. D: Appl. Phys.*, vol. 43, p. 425201, 2010.

[9] H. Shi, H. Huang, J. Bao, J. Liu, P. S. Ho, Y. Zhou, J. T. Pender, M. D. Armacost, and D. Kyser, "Role of ions, photons, and radicals in inducing plasma damage to ultra low-k dielectrics," *J. Vac. Sci. & Technol. B*, vol. 30, p. 011206, 2012.

[10] G. S. Upadhyaya, J. L. Shohet, and J. B. Kruger, "Direct measurement of topography-dependent charging of patterned oxide/semiconductor structures," *Appl. Phys. Lett.*, vol. 91, p. 182108, 2007.

[11] C. Cismaru, J. L. Shohet, and J. P. McVittie, "Plasma vacuum ultraviolet emission in a high density etcher " *Proc. Int. Symp. Plasma Process-Induced Damage* pp. 192-5, 1999

[12] T. Ikeda, K. Eriguchi, A. Tanihara, S. Kasai, and K. Ono, "Experimental evidence of layout-dependent low-k damage during plasma processing - Role of "near-field" in damage creation -," *Proc. Symp. Dry Process*, pp. 131-132, 2014.

[13] I.-C. Chen, S. E. Holland, and C. Hu, "Electrical breakdown in thin gate and tunneling oxides," *IEEE Trans. Electron Devices*, vol. ED-32, pp. 413-422, 1985.

[14] K. Eriguchi and Y. Kosaka, "Correlation between two time-dependent dielectric breakdown measurements for the gate oxides damaged by plasma processing," *IEEE Trans. Electron Devices*, vol. 45, pp. 160-164, Jan 1998.

[15] M. Ohtsu and K. Kobayashi, *Optical Near Fields*. Berlin: Springer-Verlag, 2003.

[16] J. A. Dionne, L. A. Sweatlock, H. A. Atwater, and A. Polman, "Planar metal plasmon waveguides: frequency-dependent dispersion, propagation, localization, and loss beyond the free electron model," *Phys. Rev. B*, vol. 72, p. 075405, 2005.

[17] L. D. Landau and E. M. Lifshitz, *Electrodynamics of Continuous Media*, 2nd ed. Oxford: Butterworth-Heinenann, 1984.

[18] R. J. Glauber, and M. Lewenstein, "Quantum optics of dielectric media," *Phys. Rev. A*, vol. 43, pp. 467-491, 1991.

Surface Orientation Dependence of Ion Bombardment Damage during Plasma Processing

Yukimasa Okada, Koji Eriguchi*, and Kouichi Ono
Graduate School of Engineering
Kyoto University
Kyoto 615-8540, Japan
*eriguchi@kuaero.kyoto-u.ac.jp

Abstract—**Geometrical transition to three dimensional (3D) or Si nanowire (SNW) MOSFETs imposes critical issues regarding process technologies. High energy ion bombardment damage in 3D MOSFETs has been considered inevitable because of the fundamental nature of plasma process. In this study, we further investigated plasma-induced physical damage (PPD) on Si substrates with different surface orientations—(100), (111), and (110) to emulate PPD of future 3D and SNW devices. A classical molecular dynamics simulation implies that the channeling of incident ions is expected in a substrate with the (110) plane. However, spectroscopic ellipsometry identified thinner damaged layers in the case of (110) plane for higher ion energies (> 500 eV) and the pseudo-extinction coefficient k was smaller for the (110) plane. A capacitance–voltage measurement confirmed that the damaged layer consisted of SiO_2. Thus, the same Si loss leading to Si recess that degrades device performance is presumable on both of the planes. The present findings provide key guidelines for designing future SNW devices exposed to plasma.**

Keywords—*plasma-induced physical damage, surface orientation, ellipsometry, capacitance–voltage*

I. INTRODUCTION

Recently, three dimensional (3D) devices such as FinFET [1] and Si-nanowire (SNW) devices [2] have attracted much attention, where, in terms of Si crystal plane, three major orientations—(100), (111), and (110)—may be used as the channel plane of devices. Fabrication of such devices imposes various new device-related issues [1]. Historically, the dependence of carrier mobility on the surface orientation has been one of emerging topics in device design fields [2-6]. Regarding reliability of gate dielectric materials, there have been tremendous efforts devoted to clarification of impacts of the orientations on time-dependent dielectric breakdown (TDDB) lifetime [7]. Surface and interface properties governed by the orientations have been also studied intensively. However, besides the above device-related topics, there have been few discussions on process-related issues except a thermal oxidation process [8].

For example, it has been reported that, in 3D structures, plasma-induced physical damage (PPD) [9, 10]—one of plasma process-induced damage—creates defects in a fin-bulk [11] due to bombardment of species sputtered at the reacting surface and stochastic straggling of impinging species [12-14]

This work was financially supported in part by a Grant-in-Aid for Scientific Research 25630293 from the JSPS.

in a Si bulk. These created defects play roles as carrier trapping sites, thus degrade the device performance [15]. Therefore, it is extremely important to clarify the dependence of PPD on the surface orientation, because the difference by the orientation is considered to be no more negligible in the ultimately-scaled regime. In this article, we compare PPD creation mechanisms in Si wafers with different surface orientations using spectroscopic ellipsometry (SE), surface analysis, and electrical characterizations. A molecular dynamics (MD) simulation was also employed. Although the (110) surface is subject to a channeling process, the present-day analyses identify no clear effects regarding the channeling [13, 16]. The SE analysis only assigned a smaller thickness increase on the (110) surface of damaged samples for high ion energy regions. However, due to a stochastic effect, the thickness increase is a weak function of the surface orientation. These are in contrast to the "channeling" picture. These results may be attributed to the detection limit of the employed analysis techniques. The obtained findings suggest that careful attentions should be paid to results by PPD analysis techniques used in manufacturing lines. Possible mechanisms for these issues are discussed by the PPD range theory and MD simulation results.

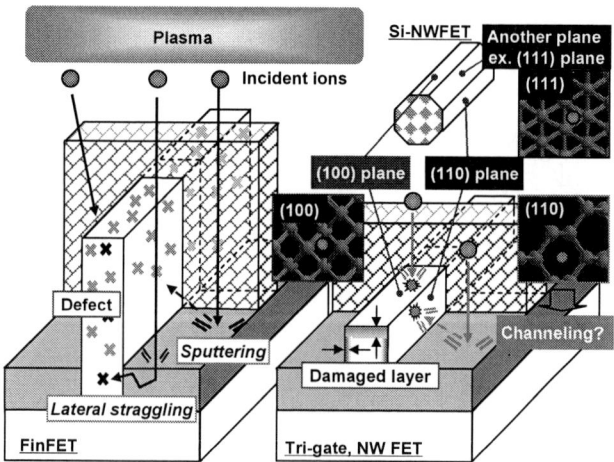

Fig. 1. Illustrations of PPD creation—(a) lateral straggling and sputtering as dominant mechanisms in a FinFET, and (b) a scenario for surface-orientation dependent PPD mechanisms in Si nanowire devices.

978-1-4799-7670-6/15 $31.00 © 2015 IEEE

II. Damage Creation Mechanisms Depending on Surface Orientations

Figure 1 shows schematic illustrations of a scenario for how PPD mechanisms depend on the surface orientation of a device exposed to plasma. Si substrate damage in a planar FET was modeled on the basis of the so-called range theory[10, 12]. The distribution range of injected ions can be determined from the stopping power. The damaged thickness (d_{dam}) is described by the average energy of incident ions (E_{ion}) as,

$$d_{dam} = A \cdot (E_{ion})^{\alpha}, \qquad (1)$$

where A and α are process- and material-dependent parameters. (typically $\alpha = 0.3$ [10]). In FinFET and SNW devices, one should take into account the stochastic effects and sputtering process [14] as illustrated in Fig. 1.

Figure 2 shows MD simulation results for the (100), (111), and (110) surfaces. The potential model presented by Wilson et al. [17] and the Stillinger–Weber function [18] were used for Si–Ar and Si–Si system, respectively. An atom was injected at normal incidence with a monochromatic energy E_{ion} [19, 20]. The MD code used here was originally developed by Ohta and Hamaguchi [19]. Further details of the MD scheme were described elsewhere [20]. One thousand Ar atoms were injected. Note that the sizes of simulation domains are not the same with each other due to the periodic boundary conditions. As seen, the surface amorphous layer was formed by the impacts. Underneath the layer, defects such as interstitials and displaced Si atoms ("dumbbell") were created [14]. Note that the surface areal density differs from the planes—the (110) surface has the highest density (N_{surf}).

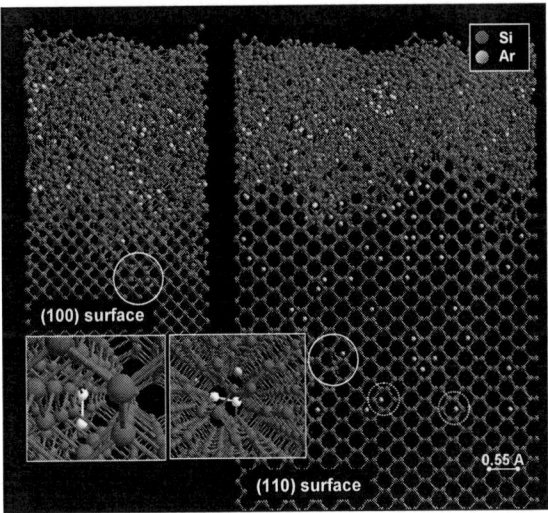

Fig. 2. Examples of MD simulation results for the final structures with (100) (left) or (110) (right) surface orientation. For the (110) case, Ar interstitial structures are observable deeper in the Si substrate due to channeling processes.

Figure 2 indicates the thickness of the layers (assigned as damaged layer thickness by SE or TEM analyses) is almost the same with each other, although a slightly thinner layer and a rough interface between the layer and the substrate are observable in the case of (110) surface for higher E_{ion} case.

Primarily due to the straggling along the incidence of ions, a stochastic effect suppresses the thickness difference. However, the interstitial Ar atoms are distributed widely in the deep Si substrate in the (110) case, compared to the other surfaces. This observation is consistent with previous reports [16, 21]. Thus the MD simulation predicts the difference in PPD mechanisms induced by the surface orientation as follows; for the case of the (110) surface, (A) a slightly thinner surface layer is formed (may be due to the highest N_{surf}), and (B) interstitial defects are located in the deep Si substrate. Figure 3 shows the normalized profiles of the number of Si and those of incident Ar atoms are compared among various planes. In particular, it is confirmed that Ar atoms are widely distributed due to channeling processes in the case of (110).

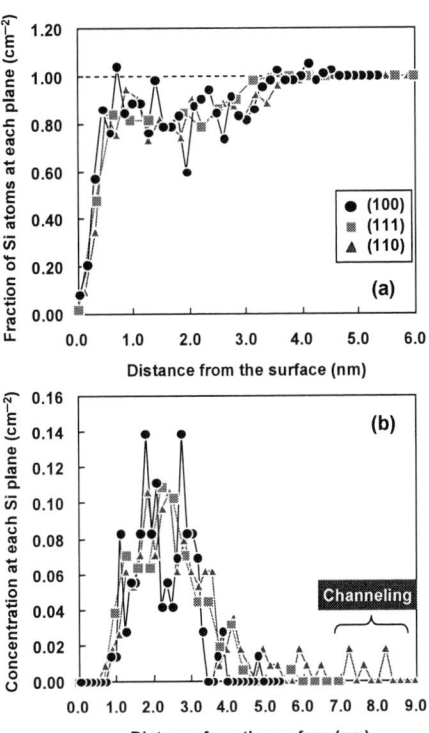

Fig. 3. "Normalized" depth profiles of (a) displaced or interstitial Si atoms and (b) incident Ar atoms for various surface orientations predicted by MD simulations. ($E_{ion} = 200eV$). As seen from (b), the channeling is observed for the (110) case.

III. Experimental Detials

A. Sample structure and plasma treatments

N-type Si chips were cut into a quarter of the size of 4-in. wafer with different orientations of the planes—(100), (111), and (110)—and mounted on a wafer stage of an inductively coupled plasma reactor. The structure of the chips consisted of native oxide layer and the substrate. We varied an input bias power from 10 to 240 W. The self-dc bias voltage (V_{dc}) during the plasma exposure was determined from an oscilloscope, giving the average energy of incident ions approximately from 16 to 1110 eV. The chips with different orientations were exposed to plasmas simultaneously for 2 min to eliminate uncertainty among the plasma treatments. The sample without plasma exposure was the control (Ref).

B. Characterization methods

To identify damaged structures, in particular, the thickness, spectroscopic ellipsometry (SE) was employed. For the SE analysis, an optimized optical model—consisting of four layer, i.e., SiO_2/SiO_2:Si/Si sub.—was used [22]. A surface layer (SL) is assumed to be thin SiO_2. The thickness (d_{SL}) is determined. The interfacial layer (IL) between SL and Si substrate is introduced as a composite of SiO_2 and crystalline silicon with the Bruggeman effective medium approximation (EMA) [22] with the thickness (d_{IL}) and volume fraction being used as the fitting parameters. Note that d_{dam} is equivalent to ($d_{SL} + d_{IL}$). The electrical thickness was determined by a capacitance–voltage (C–V) measurement.

IV. RESULTS AND DISCUSSION

Fig. 4 shows d_{dam} and d_{IL} of the damaged samples with various orientations as a function of E_{ion} (= V_p − V_{dc}). As seen, both d_{dam} and d_{IL} increase with an increase in E_{ion}. Note that one can see no clear difference in the d_{IL} among the orientations, while the thinner d_{dam} is observable for the (110) surfaces in the larger E_{ion} regions. This is in consistent with the MD simulations. However, the difference is small in the low E_{ion} regions [presumably thinner d_{dam} for the (110) surface], which may be due to the stochastic straggling. Fig. 5 shows the pseudo-extinction coefficients extracted from SE results corresponding to the pseudo-absorption coefficient of the damaged structures. Thus, the (110) samples imply less PPD in terms of k. This is in contrast to channeling-process-based pictures. It is speculated that, owing to the thinner d_{dam}, the lower k value may be obtained for the (110) surface. Therefore, SE may identify lower PPD in the case of (110) surfaces.

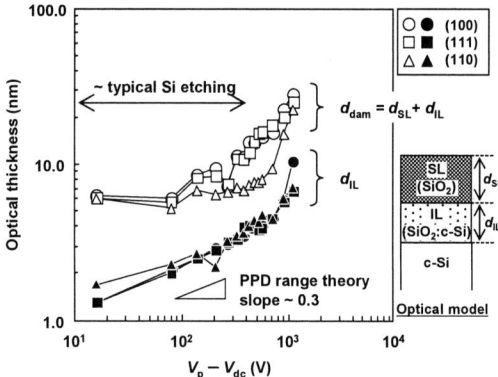

Fig. 4. (a) Damaged layer thicknesses as a function of the average energy of incident ions (E_{ion}) on the (100), (111), and (110) surfaces. As seen, a power-law dependence is seen for $E_{ion} < 500$ eV. For higher energies, the (110) structures show thinner damaged layer d_{dam} (= $d_{SL} + d_{IL}$), which may be attributed to channeling processes.

To confirm the obtained slightly thinner d_{dam} on the (110) surface, the surface roughness (RMS) was compared using an atomic force microscope. Fig. 6 shows the surface structure, indicating larger RMS for the (110) surface. If one includes the

RMS in assigning d_{SL} with the EMA, d_{dam} will become further thinner. Thus, the slight difference in the low E_{ion} regions can be confirmed.

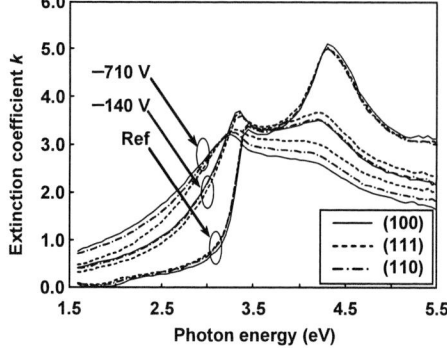

Fig. 5. Pseudo-extinction coefficients as a function of E_{ion} on the (100), (111), and (110) surfaces.

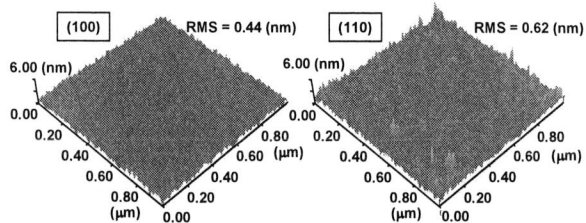

Fig. 6. Surface structures by an atomic force microscope analysis.

To further investigate the damaged structures, we performed a C–V measurement to assign the dielectric constant (ε) of the damage layer. Fig. 7 shows the relationship between d_{SL} and d_{dam} by SE and the electrical thickness (EOT) by the C–V in three different planes for lower E_{ion} cases. As seen, although d_{SL} itself is found to be thinner than EOT, but d_{dam} is approximately equal to EOT. This result clarifies that the surface damaged layer d_{dam} (= $d_{SL} + d_{IL}$) assigned by SE is equivalent to the layer of ε = 3.9 (SiO_2) by the C–V. Thus, one should consider the damaged layer with a thickness of d_{dam} is stripped off by the subsequent wet-etching after a plasma step.

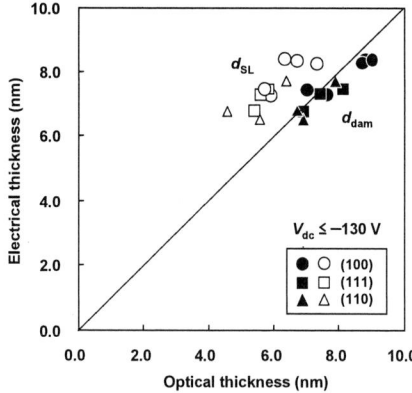

Fig. 7. Relationship between the electrical and optical thicknesses of the damaged layers formed on the (100), (111), and (110) surfaces.

Finally, let us consider a scenario of PPD in SNW devices regarding; (A) an in-line damage identification, (B) a number of latent defects, and (C) Si loss (Si recess [9]) by a wet-etching after the PPD. Fig. 8 illustrates the scenario. Fig. 8 shows the predicted damaged structures with amorphous regions and the profiles of created defects (dumbbells and interstitials) for the (100) and (110) surfaces. Due to the channeling of incident ions, the distribution of defects extends deeper in the case of (110) surface. In the higher E_{ion}, the difference becomes larger as seen. Regarding (A), an in-line monitoring method defines d_{dam} on the basis of the concentration of defects (as indicated as "SE criterion"). The difference in the assigned d_{dam} becomes larger as E_{ion} increases. The difference from the planes is also illustrated in the top right panel. The in-line monitoring method can not identify all the defects due to the detection limit. This is the second inevitable nature of PPD analysis—(B). Regarding (C), i.e., Si recess and the residual defects, a wet-etch depth (d_{wet}) is deterministic in accordance with the defect density, thus, the d_{wet} depends on the profiles (in this case, d_{wet} is larger for the (100) surface, resulting in a larger d_R). Hence, the number of residual defects is almost the same irrespective of the surface orientation. Note that the predicted PPD may degrade the device performance such as threshold voltage and drain current. [14, 15]

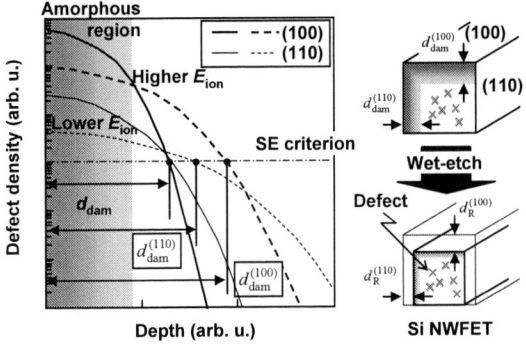

Fig. 8. Profiles of defects created by PPD in (100) and (110) surfaces for lower and higher E_{ion} cases. SE identifies the d_{dam} within the detection limit. The number of residual defects is independent of the surface orientaitions.

V. CONCLUSION

We have studied the effects of surface orientations on PPD creation, which are important for SNW device fabrication. In the typical E_{ion} used in plasma etching, SE assigned a thinner damaged layer thickness on the (110) surface, in particular, for the lower E_{ion}. This observation is in sharp contrast to widely predicted pictures based on the channeling of incident ions and results by the MD simulations. Due to the detection limits and wet-etch criterion after the PPD creation, one must accept larger Si recess on the (100) plane and orientation-independent residual defects. The presented scenario becomes a key guideline for designing fabrication process of SNW devices.

REFERENCES

[1] K. J. Kuhn, M. D. Giles, D. Becher, P. Kolar, A. Kornfeld, R. Kotlyar, S. T. Ma, A. Maheshwari, and S. Mudanai, "Process Technology Variation," *IEEE Trans. Electron Devices*, vol. 58, pp. 2197-2208.

[2] M. Saitoh, Y. Nakabayashi, K. Uchida, and T. Numata, "Short-Channel Performance Improvement by Raised Source/Drain Extensions With Thin Spacers in Trigate Silicon Nanowire MOSFETs," *IEEE Electron Device Lett.*, vol. 32, pp. 273-275, 2011.

[3] M. Saitoh, Y. Nakabayashi, K. Ota, K. Uchida, and T. Numata, "Performance Improvement by Stress Memorization Technique in Trigate Silicon Nanowire MOSFETs," *IEEE Electron Device Lett.*, vol. 33, pp. 8-10, 2012.

[4] K. Uchida, A. Kinoshita, and M. Saitoh, "Carrier Transport in (110) nMOSFETs: Subband Structures, Non-Parabolicity, Mobility Characteristics, and Uniaxial Stress Engineering," *IEDM Tech. Dig.*, pp. 1-3, 11-13 Dec. 2006 2006.

[5] M. Saitoh, N. Yasutake, Y. Nakabayashi, K. Uchida, and T. Numata, "Understanding of strain effects on high-field carrier velocity in (100) and (110) CMOSFETs under quasi-ballistic transport," *IEDM Tech. Dig.*, pp. 1-4, 7-9 Dec. 2009 2009.

[6] K. Uchida, M. Saitoh, and S. Kobayashi, "Carrier transport and stress engineering in advanced nanoscale transistors from (100) and (110) transistors to carbon nanotube FETs and beyond," *IEDM Tech. Dig.*, pp. 1-4, 15-17 Dec. 2008 2008.

[7] Y. Mitani and A. Toriumi, "Re-consideration of Influence of Silicon Wafer Surface Orientation on Gate Oxide Reliability from TDDB Statistics Point of View," *Proc. Int. Rel. Phys. Symp.*, pp. 299-305, 2010.

[8] T. Ohmi, K. Matsumoto, K. Nakamura, K. Makihara, J. Takano, and K. Yamamoto, "Influence of silicon wafer surface orientation on very thin oxide quality," *J. Appl. Phys.*, vol. 77, pp. 1159-1164, 1995.

[9] T. Ohchi, S. Kobayashi, M. Fukasawa, K. Kugimiya, T. Kinoshita, T. Takizawa, S. Hamaguchi, Y. Kamide, and T. Tatsumi, "Reducing Damage to Si Substrates during Gate Etching Processes," *Jpn. J. Appl. Phys.*, vol. 47, pp. 5324-5326, Jul. 2008.

[10] K. Eriguchi, Y. Nakakubo, A. Matsuda, Y. Takao, and K. Ono, "Model for Bias Frequency Effects on Plasma-Damaged Layer Formation in Si Substrates," *Jpn. J. Appl. Phys.*, vol. 49, p. 056203, 2010.

[11] A. Matsuda, Y. Nakakubo, Y. Takao, K. Eriguchi, and K. Ono, "Atomistic simulations of plasma process-induced Si substrate damage - Effects of substrate bias-power frequency," *Proc. Int. Conf. on Integrated Circuit Design & Technol.*, pp. 191-194, 29-31 May 2013 2013.

[12] J. Lindhard, M. Scharff, and H. E. Schiott, "Range Concepts and Heavy Ion Ranges," *Mat. Fys. Medd. K. Dan. Vidensk. Selsk.*, vol. 33, pp. 1-41, 1963.

[13] S. M. Sze, *Semiconductor Devices, Physics and Technology*, 2nd ed. Hoboken, NJ: John Wiley & Sons, Inc., 2002.

[14] K. Eriguchi, Y. Takao, and K. Ono, "A new aspect of plasma-induced physical damage in three-dimensional scaled structures — Sidewall damage by stochastic straggling and sputtering," *Proc. Int. Conf. on Integrated Circuit Design & Technol.*, pp. 1-4, 2014.

[15] K. Eriguchi, Y. Nakakubo, A. Matsuda, Y. Takao, and K. Ono, "Plasma-Induced Defect-Site Generation in Si Substrate and Its Impact on Performance Degradation in Scaled MOSFETs," *IEEE Electron Device Lett.*, vol. 30, pp. 1275-1277, 2009.

[16] A. H. Al-Bayati, K. G. Orrman-Rossiter, R. Badheka, and D. G. Armour, "Radiation damage in silicon (001) due to low energy (60-510 eV) argon ion bombardment," *Surf. Sci.*, vol. 237, pp. 213-231, 1990.

[17] W. D. Wilson, L. G. Haggmark, and J. P. Biersack, "Calculations of nuclear stopping, ranges, and straggling in the low-energy region," *Phys. Rev. B*, vol. 15, pp. 2458-2468, 1977.

[18] F. H. Stillinger and T. A. Weber, "Computer simulation of local order in condensed phases of silicon," *Phys. Rev. B*, vol. 31, pp. 5262-5271, 1985.

[19] H. Ohta and S. Hamaguchi, "Molecular dynamics simulation of silicon and silicon dioxide etching by energetic halogen beams," *J. Vac. Sci. & Technol.*, vol. A19, pp. 2373-2381, 2001.

[20] H. Ohta, A. Iwakawa, K. Eriguchi, and K. Ono, "An interatomic potential model for molecular dynamics simulation of silicon etching by Br+-containing plasmas," *J. Appl. Phys.*, vol. 104, p. 073302, 2008.

[21] H. Hensel and H. M. Urbassek, "Implantation and damage under low-energy Si self-bombardment," *Phys. Rev. B*, vol. 57, pp. 4756-4763, 1998.

[22] A. Matsuda, Y. Nakakubo, Y. Takao, K. Eriguchi, and K. Ono, "Modeling of ion-bombardment damage on Si surfaces for in-line analysis," *Thin Solid Films*, vol. 518, pp. 3481-3486, 2010.

High-Speed Analog-to-Digital Converters in downscaled CMOS

(Invited Paper)

Annachiara Spagnolo*, Bob Verbruggen[†], Stefano D'Amico[‡] and Piet Wambacq*[§]

* imec, Leuven, Belgium
[†] Xilinx, Dublin, Ireland
[‡] Università del Salento, Lecce, Italy
[§] Vrije Universiteit Brussel - Brussels, Belgium

Abstract—High data-rate communications need high speed analog-to-digital converters. Recent flash and time interleaved SAR converters implemented in downscaled CMOS technologies have achieved GS/s conversion rates with very low power consumption. Flash ADCs can reach high speed with a single channel but the resolution is limited by exponential complexity and power consumption. SAR ADCs are well suited for higher resolution but, due to the sequential operation, require either massive interleaving or very fast technologies to achieve high speed. Hybrid architectures combine the advantages of different architectures to achieve the optimum compromise for a given resolution. In this paper the trade-offs between power, area and complexity for high speed designs are discussed and the potential of hybrid architectures is investigated.

I. INTRODUCTION

The development of modern high data-rate communications, such as 60GHz or serial links, have boosted the interest in high speed analog-to-digital converters. Thanks to the availability of fast devices, analog-to-digital converters implemented in downscaled CMOS have achieved conversion rates in the GS/s range [1]. Technology scaling also plays an important role in the efficiency of analog-to-digital converters. Indeed, whereas very high resolution designs are limited by noise and have to cope with the reduced voltage headroom that comes with downscaling, in low-to-moderate resolution converters most of the power budget is commonly consumed by circuits which are not limited by noise. As a result, the efficiency of these converters has significantly improved over years thanks to the reduces parasitics and the lower supply voltage of downscaled CMOS processes [2]. Fig. 1 shows the Walden Figure of Merit of analog-to-digital converters featuring more than 1GS/s conversion rate and less than 50dB SNDR versus technology feature size. The trend clearly shows that the power consumption of these ADCs scales very well with technology. Nevertheless, the link between efficiency and technology is not straightforward. Indeed, several architecture and circuit level techniques have been proposed in the recent years to improve the conversion efficiency. In addition, many architectures are digitally assisted to relax requirements that severely impact the power consumption, as matching and linearity [2]. The cost of this digital assistance in terms of area and power consumptioin obviously also goes down in downscaled CMOS.

In the following sections, a selection of recently published

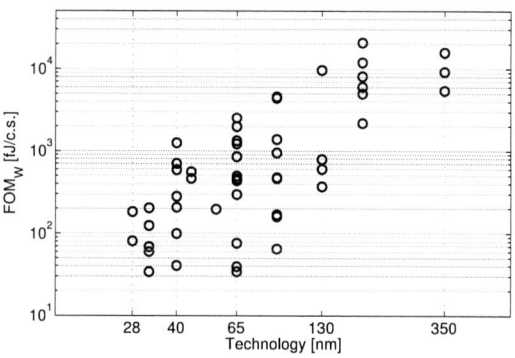

Fig. 1. Figure of Merit versus technology feature size of high speed and low-to-medium resolution ADCs presented at ISSCC and VLSI from 1997 to 2015 [1].

high speed ADCs in downscaled CMOS is discussed, focusing on the trade-offs between power, area and complexity associated with each architecture.

II. FLASH CONVERTERS

Flash ADCs are well suited for high speed applications. Indeed, due to the parallel operation, they can achieve GS/s conversion rates with no need of time interleaving. However, the main disadvantage is the power consumption, which increases exponentially with the resolution.

Many architecture or circuit level techniques that reduce the complexity and the power consumption of flash converters have been reported. Folding and interpolation are weel known techniques and have been recently used in high speed ADCs. For instance, in [3], a 5b folding flash ADC is reported where a 1b folding front-end is used to halve the number of comparisons done in the flash sub-converter. Another example is the 6b flash converter of [4] which leverages interpolation to halve the number of comparators. Here, the core of the ADC only consists of 30 comparators with built-in offset and 29 offset-averaging SR latches which implement the intermediated threshold between adjacent comparators.

In addition, most state-of-the-art flash ADCs use digital threshold calibration to relax the matching requirements of the comparator. This can significantly reduce the power consumption [3], [4], [5] at the cost of the area allocated for the digital controllers and memory which can be very small in downscaled CMOS technologies.

An efficient offset calibration technique that exploits the randomness of process mismatch is reported in [5]. In this high speed 6b ADC, each comparator consists of a programmable preamplifier and a regenerative latch. The preamplifier is constructed of 12 selectable differential input pairs with minimum size devices. The calibration algorithm chooses one combination of the input pairs among 2^{12} available subsets to compensate the mismatch. Thanks to the calibration technique, the ADC achieves a conversion rate of 5GS/s with less than 60fJ per conversion step in a 32nm SOI CMOS process.

As shown by the above examples, several techniques can be used to improve the efficiency of flash ADCs and make this architecture a valid option for low resolution applications. However, the exponential complexity and power consumption remain the main limitation of flash ADCs, since they become prohibitive if resolutions higher that 6b are targeted.

III. HIGH SPEED SAR CONVERTERS

SAR converters are commonly preferred in medium or high resolution applications for their power efficiency and low complexity. The disadvantage of SAR converters is the conversion speed. Indeed, due to the sequential operation, the maximum speed scales linearly with the resolution, resulting in a large number of interleaved channels if GS/s rates are targeted. Nevertheless, recent state-of-the-art SAR ADCs have achieved very high conversion rate thanks to speed optimization techniques [6], [7], [8], [9].

An effective way to increase the conversion rate of a SAR ADC is the use of multi-bit per cycle architectures. For instance, in [7] two bits are resolved in each SAR step by doing three comparisons at the same time. Compared to a single-bit per cycle converter, the maximum speed is doubled, but this has a non negligible cost in terms of power consumption due to the larger number of comparisons required in each conversion and the increased complexity of the DAC. This additional cost is efficiently spent in the 3b/cycle SAR ADC reported in [8]. This 4× interleaved 6b ADC uses interpolation to significantly reduce the complexity of the DAC and the number of comparators, yielding very low power consumption at 5GS/s.

A very fast single-bit per cycle SAR is reported in [9]. With a single channel, this 8b ADC achieves a conversion rate of 1.2GS/s, which results in 8.8GS/s and 90GS/s for the 8× and 64× interleaved implementations of [10] and [11], respectively. Implemented in 32nm CMOS SOI technology, this design benefits of course from fast devices and reduced parasitics which allow for higher speed, but it also relies on a very effective speed optimization. In asynchronous SAR ADCs, the conversion time is only limited by the comparator regeneration time and the fixed time slot that is provided for DAC settling and comparator reset. In [9] timing requirements for the comparator reset are avoided by using the alternate comparators technique. Two comparators run in an alternate way, such that while one of them is active, the other one

is in reset mode. This removes the reset time from the conversion timeline and ensures, for each comparator, enough reset time to avoid memory issues due to the floating body effect of which SOI devices suffer. In addition, redundancy is used to relax the requirements for the DAC settling, which also results in time saving. Due to the redundancy and the alternate comparators, the time budget is dominated by only the comparator regeneration, which is also optimized for speed [9]. As a result, the conversion speed takes advantage of an extremely fast comparator regeneration that, anyway, would be difficult to achieve in a standard CMOS process.

The above examples have shown that efficient SAR ADCs can be optimized for high speed, although either fast technologies or multi-bit per cycle schemes in combination with time interleaving are needed to achieve GS/s rates.

IV. HYBRID CONVERTERS

ADC designers have to cope with the trade-offs between power, area and complexity. Several architectures are available and suitable for high speed, but each of them has its pros and cons. In the previous Sections, a few examples of flash and SAR converters have been considered and it has been shown that flash ADCs are well suited for high speed as long as the required resolution is below 6b [3], [4], [5]. Conversely, SAR are a valuable option if higher resolutions are targeted, but fast technologies or aggressive speed optimization are needed to avoid massive interleaving [6], [7], [10]. Hybrid converters combine the advantages of different architectures to find the optimum trade-off for a given resolution by

- minimizing the number of active comparators per conversion to limit the power consumption,
- minimizing the calibration effort to limit the area and the complexity,
- maximizing the channel speed to limit the number of interleaved channels.

In this section, two hybrid ADCs are discussed. In both ADCs threshold calibration is used to limit the power consumption of the comparator. Further, a linear front-end reduces the calibration effort.

A. Folding/PLBS converter

In [12], [13] a 6b time interleaved Pipelined Binary Search (PLBS) converter with folding front-end is reported. The channel architecture is shown in Fig. 2.

The front-end is a 1b folding stage and the core of the ADC consists of a 3b PLBS sub-converter, terminated with an array of eight 2b flash converters. Each input-referred threshold of the PLBS tree is implemented by a dedicated comparator with individually calibrated built-in offset, yielding relaxed matching constraints and low power consumption. Further, as in a comparator-based asynchronous binary search (CABS) architecture [14], only one path in the tree is clocked at each conversion.

In addition, each comparator of the PLBS sub-converter, is merged with an amplifier for a 1b/stage pipelining. This pipelined binary search architecture overcomes the

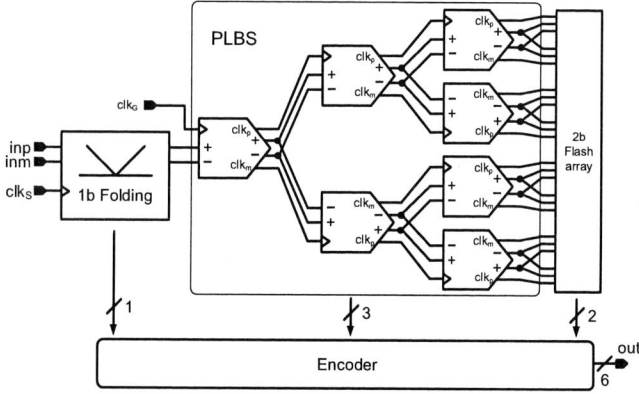

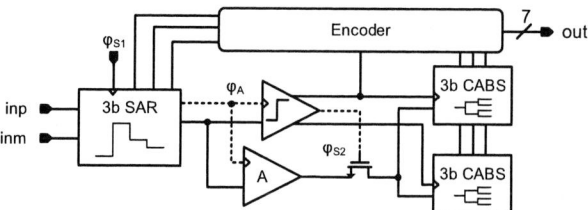

Fig. 4. Architecture of the pipelined SAR/CABS ADC [15].

Fig. 2. Block diagram of the 6b Folding/PLBS ADC [13].

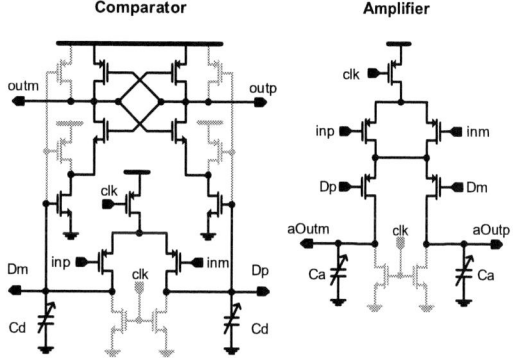

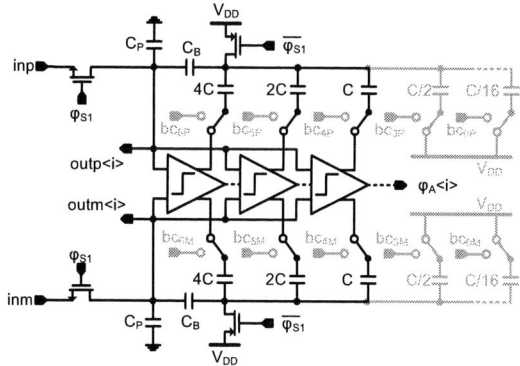

Fig. 5. Simplified schematic of the SAR front-end [15].

Fig. 3. Schematic of the comparator and dynamic amplifier with grayed out reset transistors [13].

energy per conversion step is as low as 48fJ.

In this 6b architecture, a total of 7 comparators is clocked in each conversion (an extra one due to the 2b flash termination). Thus, the power consumption is very low and depends linearly on the resolution, as in a SAR converter. Further, due to the pipelined operation the single channel achieves very high speed and only 4 interleaved channels are required. The calibration effort is exponential with the resolution, but for a 6b implementation it is fairly low and consists of 128 comparators and 28 amplifiers for the full ADC.

typical limitation of pipelined converters, i.e. the linearity requirements of the amplifier which often result in high power consumption. The underlying assumption is that linearity requirements can be avoided since each input-referred threshold is individually calibrated. This allows for a very low power dynamic amplifier.

The schematics of the comparator and the amplifier are shown in Fig. 3. The built-in offset is implemented with an unbalanced input pair and it is calibrated with the digitally controlled capacitors C_d. The dynamic amplifier has at its input the internal comparator nodes D_p and D_m and its gain can be also calibrated by acting on the slew rate of the output nodes, which can be adjusted by the capacitors C_a.

As any fully calibrated ADC, the disadvantage of the PLBS conversion is the exponential calibration effort. To cope with this, a 1b linear front-end is used to halves the number of threshold to be calibrated compared to a fully PLBS implementation. The fron-end consists of a folding stage with passive charge sharing [12].

An optimized timing scheme is reported in [13] which increases the speed of [12] by 60%. In [13] the timing of the folding stage is based on a clock with 25% duty cycle. This relaxes the timing requirements of the folding stage that were the speed bottleneck of [12]. With four interleaved channels the prototype implemented in 40nm CMOS LP achieves a conversion rate of 3.5GS/s, while consuming 4.1mW. The

B. SAR/CABS converter

The calibration effort of a PLBS converter would be prohibitive for resolutions higher than 6b. For a 7b implementation, a different time interleaved hybrid converter has been proposed in [15]. The basic structure is similar to [13]. Indeed, also in this case, a linear front-end is used before a fully calibrated sub-converter to reduce the number of the thresholds to be calibrated in each channel. The channel architecture is shown in Fig. 4.

The first stage of each channel consists of a SAR converter that determines the 3 MSBs. An amplifier amplifies the SAR residue while a comparator detects is sign. The amplified residue is held by a simple NMOS switch at the input of the next stage to pipeline its operation and increase the channel speed. The last stage consists of two 3b CABS sub-converters [14], which are clocked based on the sign of the SAR residue. Since each input referred threshold of the CABS stage is individually calibrated, linearity requirements of the amplifier are avoided. Thus, the low power dynamic amplifier with embedded comparator of Fig. 3 is also used in this ADC.

In this architecture, the calibration effort in each channel consists of 16 comparators, which results in a total of 128

978-1-4799-7670-6/15 $31.00 © 2015 IEEE   128

TABLE I
STATE OF THE ART COMPARISON

Ref.	[5]	[10]	[13]	[15]
Architecture	Flash	TI SAR	TI PLBS	TI SAR/CABS
Process	32nm SOI	32nm SOI	40nm LP	40nm GP
Supply (V)	0.85	1	1.1	0.9
Resolution (bit)	6	8	6	7
Speed (GS/s)	5.0	8.8	3.5	3.5
Power (mW)	8.5	35	4.1	6.2
SNDR$_{min}$ (dB)	30.9	37.0	29.5	34.7
E$_{Q,max}$ (fJ/c.s.)	59	69	48	40
E$_{Q,min}$ (fJ/c.s.)	45	58	41	27

comparators for the $8\times$ interleaved ADC. Thus, this ADC performs a 7b conversion with the calibration effort of a single channel flash ADC and a power consumption that is comparable to a SAR ADC, since only 7 comparisons are done at each conversion. Further, due to the pipelined operation, the speed of the individual channel is only limited by the number of sequential comparisons that are done in the first stage.

Each comparator of the CABS stage implements a different threshold with unbalance input pair, as in [14], and has relaxed matching constraints. The calibration is done in foreground by applying at the comparator input the desired threshold and driving the average output of the comparator toward 0.5.

Whereas in other calibrated ADCs [4], [5], [13], [14] the references required for the calibration are generated by a dedicated DAC, this is done here by reusing the DAC of the SAR stage and extending its resolution.

The SAR front-end is shown in Fig 5. The core of the SAR consists of a 3b capacitive charge redistribution DAC with embedded passive T&H. The grayed out part is inactive during normal operation. Three different comparators are used to remove the reset time from the critical path and save some logic. In calibration mode, the bottom plates of the DAC array are switched to the control lines bc_{xP} and bc_{xM}, and the full 7b DAC is used to generate the desired reference voltage at the amplifier input. Since the same DAC is used, a good matching between calibration and normal operation is achieved at a negligible cost of power consumption.

In this way, the pipelined SAR/CABS architecture makes a compromise between speed and complexity. The power consumption is kept low due to the binary search nature and a the calibration that is efficiently performed by using the built-in calibration references. The prototype implemented in 40nm CMOS GP achieves a conversion rate of 3.5GS/s with eight interleaved channels and less than 40fJ per conversion step.

V. COMPARISON AND CONCLUSION

The performance of the main ADCs discussed in this paper [5], [10], [13], [15] is summarized in Table I. The resolution ranges from 6 to 8 bit and the conversion rate from 3.5 to 8.8GS/s. The power consumption of these ADCs is very low and results in values of the energy per conversion step below

70fJ. This proves that very heterogeneous architectures can be used for low power high speed conversion. Indeed, despite the exponential power consumption, flash architectures are a valuable option as long as threshold calibration allows for minimum size comparators. Conversely, highly efficient SAR ADCs can be optimized to achieve very high speed with a low number of interleaved channels and an affordable cost of circuit overhead and power consumption. Hybrid converters make a compromise between speed and complexity for the lowest power consumption. Indeed, both the folding/PLBS and the SAR/CABS achieve in a standard CMOS process competitive speed at improved efficiency compared to ADCs implemented in the most advanced CMOS SOI technologies.

REFERENCES

[1] B. Murmann, " ADC Performance Survey 1997-2015. [Online]. Available: http://web.stanford.edu/ murmann/adcsurvey.html."

[2] ——, "A/D converter trends: Power dissipation, scaling and digitally assisted architectures," in *Custom Integrated Circuits Conference, 2008. CICC 2008. IEEE*, Sept 2008, pp. 105–112.

[3] B. Verbruggen, J. Craninckx, M. Kuijk, P. Wambacq, and G. Van der Plas, "A 2.2mW 5b 1.75GS/s Folding Flash ADC in 90nm Digital CMOS," in *Solid-State Circuits Conference, 2008. ISSCC 2008. Digest of Technical Papers. IEEE International*, Feb 2008, pp. 252–611.

[4] Y.-S. Shu, "A 6b 3GS/s 11mW fully dynamic flash ADC in 40nm CMOS with reduced number of comparators," in *VLSI Circuits (VLSIC), 2012 Symposium on*, June 2012, pp. 26–27.

[5] V. H.-C. Chen and L. Pileggi, "An 8.5mW 5GS/s 6b Flash ADC with Dynamic Offset Calibration in 32nm CMOS SOI," in *VLSI Circuits (VLSI)*, 2013.

[6] V. Tripathi and B. Murmann, "An 8-bit 450-MS/s single-bit/cycle SAR ADC in 65-nm CMOS," in *ESSCIRC (ESSCIRC), 2013 Proceedings of the*, Sept 2013, pp. 117–120.

[7] H. Wei, C.-H. Chan, U.-F. Chio, S.-W. Sin, U. Seng-Pan, R. Martins, and F. Maloberti, "A 0.024mm^2 8b 400MS/s SAR ADC with 2b/cycle and resistive DAC in 65nm CMOS," in *Solid-State Circuits Conference Digest of Technical Papers (ISSCC), 2011 IEEE International*, Feb 2011, pp. 188–190.

[8] C.-H. Chan, Y. Zhu, S.-W. Sin, U. Seng-Pan, and R. Martins, "26.5 A 5.5mW 6b 5GS/s 4x-Interleaved 3b/cycle SAR ADC in 65nm CMOS," in *Solid- State Circuits Conference - (ISSCC), 2015 IEEE International*, Feb 2015, pp. 1–3.

[9] L. Kull *et al.*, "A 3.1mW 8b 1.2GS/s single-channel asynchronous SAR ADC with alternate comparators for enhanced speed in 32nm digital SOI CMOS," in *Solid-State Circuits Conference Digest of Technical Papers (ISSCC), 2013 IEEE International*, Feb 2013, pp. 468–469.

[10] ——, "A 35mW 8b 8.8GS/s SAR ADC with Low-Power Capacitive Reference Buffers in 32nm Digital SOI CMOS," in *VLSI Circuits (VLSIC)*, 2013.

[11] ——, "A 90GS/s 8b 667mW 64x interleaved SAR ADC in 32nm digital SOI CMOS," in *Solid-State Circuits Conference Digest of Technical Papers (ISSCC), 2014 IEEE International*, Feb 2014, pp. 378–379.

[12] B. Verbruggen, J. Craninckx, M. Kuijk, P. Wambacq, and G. Van der Plas, "A 2.6mW 6b 2.2GS/s 4-times interleaved fully dynamic pipelined ADC in 40nm digital CMOS," in *IEEE International Solid-State Circuits Conference, (ISSCC)*, 2010.

[13] A. Spagnolo, B. Verbruggen, P. Wambacq, and S. D'Amico, "A 4.1-mW 3.5-GS/s 6-Bit Time-Interleaved ADC in 40-nm CMOS," *IEEE Transactions on Circuits and Systems II: Express Briefs*, vol. 61, no. 7, pp. 466–470, July 2014.

[14] G. Van der Plas and B. Verbruggen, "A 150MS/s 133 uW 7b ADC in 90nm digital CMOS Using a Comparator-Based Asynchronous Binary-Search sub-ADC," in *Solid-State Circuits Conference, 2008. ISSCC 2008. Digest of Technical Papers. IEEE International*, Feb 2008, pp. 242–610.

[15] A. Spagnolo, B. Verbruggen, S. D'Amico, and P. Wambacq, "A 6.2mW 7b 3.5GS/s time interleaved 2-stage pipelined ADC in 40nm CMOS," in *European Solid State Circuits Conference (ESSCIRC), ESSCIRC 2014 - 40th*, Sept 2014, pp. 75–78.

Overview of methods to increase linearity of high-performance ADC

Hua Fan*, Kehong Liu*, Airong Liu+, Lishan Lv*, Zhiliang Qiao* and Qiang Li*

* University of Electronic Science and Technology of China, Chengdu, China
Email: *fanhua7531@163.com
+Institute of Semiconductors, Chinese Academy of Sciences, Beijing, China.
Email: +arliu@semi.ac.cn

Abstract—**This paper makes a review of main methods to improve the linearity of ADCs. Methods are collected in view of mainstream techniques. A model of redundant noise-shaping Pipelined Flash-SAR ADC is proposed, which is the combination of Flash, Pipeline, SAR and $\Sigma\Delta$ ADC, the effectiveness of the model is demonstrated by simulation. Also, application of dynamic element matching (DEM) linearization techniques to the $\Sigma\Delta$ ADC is verified in a third-order four-bit continuous-time $\Sigma\Delta$ modulator with a standard deviation of 1%. Finally, auxiliary DAC calibration technique to improve the linearity of ADC is verified by a concrete 12-bit SAR ADC chip.**

Index Terms—Analog-to-Digital Converter, successive approximation, Pipeline, Flash, $\Sigma\Delta$ ADC.

I. INTRODUCTION

An ADC is a common building block of modern electronics equipment. A high linearity and high absolute resolution analog-to-digital converter (ADC) serves as an important role in instrumentation and measurement(I&M), system-on-a-chip (SoC) for 3G receivers, digital X-ray imaging systems and radar applications, etc. These applications always require the ADC to achieve over 10-bit resolution at high sampling rate. In particular, Bio-medical and smart sensors applications often require Analog-to-digital converters with high absolute resolution (beyond 16 bits) and linearity as well as low power to improve battery-life and the system portability, because the frequent operations of ADC results in large power consumption. As shown in Fig.1, the biomedical electronics detecting system usually includes electrodes, an amplifier, a low-pass filter, a sample and hold circuit and an ADC. In this system, at first, the sensor converts signals to digital output, ADC is one of the key analog building blocks to convert the biomedical signals to digital signals used in the back-end microprocessor to be analyzed and processed [1]. Because the amplitude of these signals is quite small, ADC with high resolution is demanded. If the resolution of the ADC is too low, the difference between these signals can not be distinguished, which might affect the diagnosis of doctors [1].

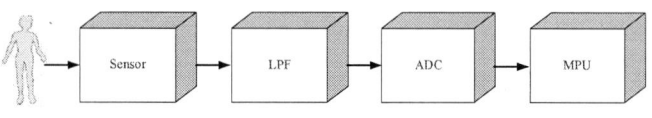

Fig. 1. wireless health monitoring system.

Different architectures are in use today to meet a wide range of requirements, e.g. sampling rate, linearity, noise and power consumption. Popular architectures include Flash ADC, Successive approximation register (SAR) ADC, $\Sigma\Delta$ ADC and pipeline ADC, all architectures have their own advantages and disadvantages, making them suitable for different ranges of resolutions and sampling rate. Recent ADC publications show a trend towards implementation in digital CMOS process with smaller feature sizes and better energy efficiencies in order to better exploit the advanced sub-100-nm CMOS technology, meanwhile, the ADC is also desired to be designed with the same technology and supply voltage as those used for the digital system, as a result, the demands of low analog complexity and low power consumption make the SAR ADC more well suited to nanometer CMOS process than other types of ADCs. It is noteworthy that the ADC with the lowest FOM, published so far, is a SAR ADC [2]. But SAR ADC suffers from serial disadvantage that has prevented it from being used in high-speed, high-resolution applications. To overcome the low-speed disadvantage of the SAR architecture, time-interleaved SAR ADCs have been proposed, which can operate in GHz range, but their resolutions have been limited within 10 bits. On the other hand, hybrid architecture and other calibration techniques are always utilized to achieve higher resolutions and speed.

II. METHODS TO IMPROVE PERFORMANCE OF ADC

A. Hybrid architecture

The design of wide-band high-gain OPAMPs in low-voltage nanometer CMOS processes limits the Opamp-based ADCs such as pipeline ADC, where high-gain OPAMPs are required in the initial stages to reduce errors due to finite OPAMP gain. In order to offset the single ADC design challenges imposed by rapidly emerging deep submicron CMOS technologies, a potential candidate to replace conventional single ADC is being investigated, referred to as hybrid ADC architecture. For example, two or more SAR ADCs can be pipelined for better energy-efficiency [3]. The power consumption of 12-bit 210MS/s pipelined SAR ADC proposed in [4] is 5.3 mW, while the 10-bit 200MS/s pipeline ADC in [5] consumes 45.4mW. Besides, the 11-bit 410MS/s pipelined SAR ADC in [6] consumes less power than the 11-bit 300MS/s pipeline

978-1-4799-7670-6/15 $31.00 © 2015 IEEE   130

ADC in [7]. Table I summarizes the recently published hybrid ADCs.

TABLE I
RECENTLY PUBLISHED HYBRID ADCS

Reference	Tech. (nm)	ENOB	fs (MS/s)	FOM (fJ/step)	Architecture
JSSC14 [8]	65	10.67	50	57	P_SAR
ESSCIRC14 [9]	40	11	20	4	P_SAR
VLSIC13 [10]	22	10.67	240	259.6	$\Sigma\Delta$_SAR
TCASI13 [11]	90	8.69	150	24.7	F_SAR
JSSC12 [12]	65	10	90	35.8	$\Sigma\Delta$_SAR
ISSCC12 [13]	40	9.5	250	10	P_SAR
JSSC11 [14]	65	10.7	50	52	P_SAR
JSSC11 [15]	65	8.86	40	65	P_SAR
JSSC10 [16]	180	12.23	80	750	$\Sigma\Delta$_Pipe

*P_SAR represents Pipelined_SAR ADC
*F_SAR represents Flash_SAR ADC
*$\Sigma\Delta$_SAR represents Noise-Shaping SAR ADC
*$\Sigma\Delta$_Pipe represents $\Sigma\Delta$_Pipeline ADC

Generally speaking, Pipelined SAR ADC is well suited for implementing ADCs over wide ranges of speed and resolution. Here, a 14-bit pipelined SAR ADC is modeled in SIMULINK, which is divided into 7-bit(first stage) and 8-bit(second stage) SAR ADC, a large resolution in the first stage is useful to decrease the total power consumption of the ADC. As shown in Fig.2, the 14-bit ADC achieves a signal-to-noise ratio (SNR) of 86.28 dB, signal-to-noise-and-distortion ratio (SNDR) of 85.99 dB (13.99 ENOB) and a spurious free dynamic range (SFDR) of 105.42 dB. It is obvious that the resolution of Nyquist ADC is not possible to surpass its design value.

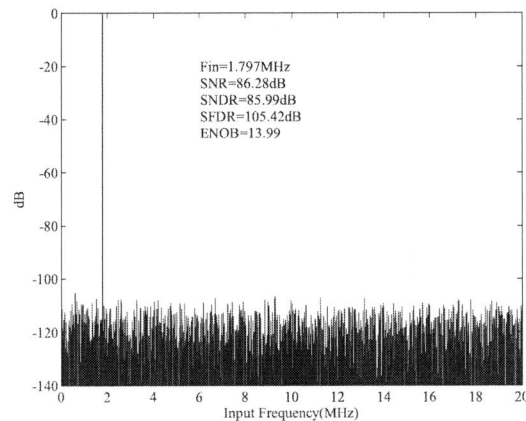

Fig. 2.   FFT of 14-bit Pipelined Flash-SAR ADC.

To further improve the speed of Pipelined SAR ADC, two Flash-SAR ADCs can be pipelined. In [17], we have already proposed a novel Flash-SAR ADC, then the details can be referred to [17]. However, higher resolution (beyond 14 bits) is still difficult to achieve. To further enhance the accuracy of pipelined Flash-SAR ADC, this work presents a novel Noise-Shaping Pipelined Flash-SAR ADC model by introducing noise-shaping technique proposed in [18] and [19]. In [18], all the stages need delay unit, while in this work, delay is only needed between stages. As shown in Fig.3, it consists of one N_1-bit front-end Flash-SAR stage and one N_2-bit back-end Flash-SAR stage. At first, the MSB bits are resolved by the front-end Flash-SAR stage, then the residue is extracted, the back-end Flash-SAR stage resolves the remaining LSBs. In this Noise-Shaping Pipelined Flash-SAR ADC model, the extracted back-end Flash-SAR quantization noise Q_2 is firstly scaled by $\frac{1}{G}$, then, feedback to the input of the ADC and appears at the final digital output. The feedback path provides the first order noise shaping for scaled $Q_2(\frac{Q_2}{G})$.

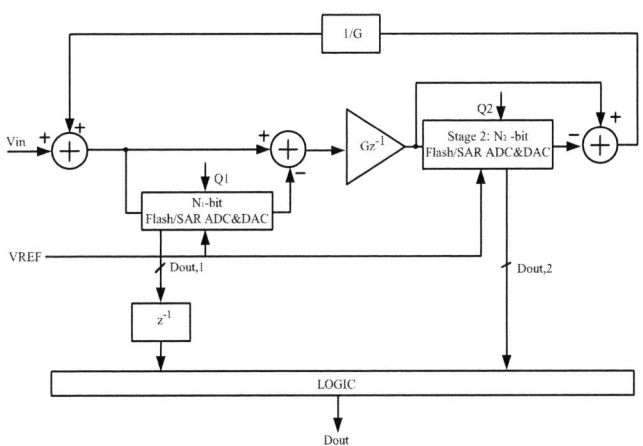

Fig. 3.   Noise-Shaping Pipelined Flash-SAR ADC architecture.

The output transfer function of the model is derived as follows:

$$D_{out,1} = V_{in} + Q_1 - \frac{Q_2}{G}, \tag{1}$$

$$D_{out,2} = Q_2 - Q_1 G z^{-1}, \tag{2}$$

$$D_{out} = D_{out,1} z^{-1} + \frac{D_{out,2}}{G}, \tag{3}$$

(1) and (2) are substituted in (3), D_{out} will be

$$D_{out} = V_{in} z^{-1} + \frac{Q_2}{G}(1 - z^{-1}), \tag{4}$$

It is apparent from (4) that the quantization noise of the first stage Flash-SAR ADC is ideally canceled out and the final digital output codes only include $\frac{Q_2}{G}$, which is the scaled quantization noise of the second stage Flash-SAR ADC.

Noise-Shaping Pipelined Flash-SAR ADC in Fig.3 is modeled in SIMULINK, and Fig.4 shows the output PSD. The bandwidth is 20MHz, oversampling ratio is 64. Compared to Fig.2, the Noise-Shaping technique improves the SNR from 86.28 dB to 99.31 dB, SNDR from 85.99 dB to 98.98 dB and SFDR from 105.42 dB to 115.68 dB.

This architecture may be an attractive choice for the high-speed high-accuracy applications since the complexity of the first order noise shaping is minimal, and no extra amplifier is required for realizing first-order noise shaping.

978-1-4799-7670-6/15 $31.00 © 2015 IEEE

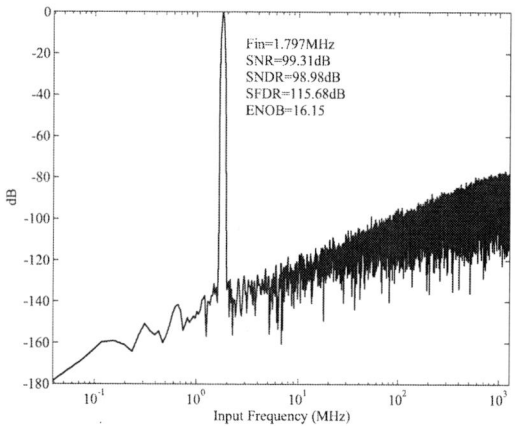

Fig. 4. FFT of Noise-Shaping Pipelined Flash-SAR ADC architecture.

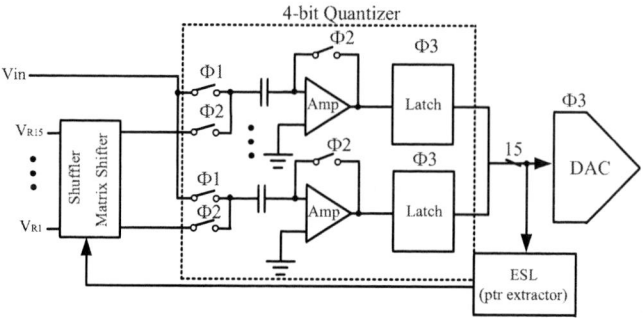

Fig. 5. DWA techniques [21].

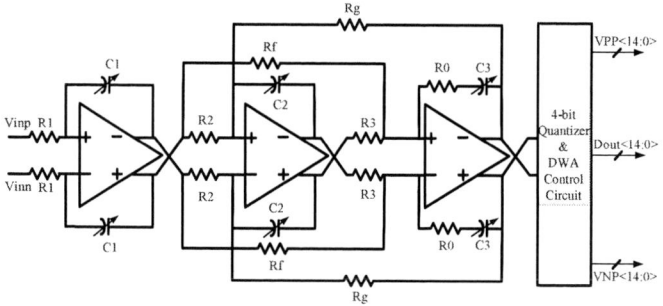

Fig. 6. continuous-time $\Sigma\Delta$ modulator.

B. Dynamic Element Matching

One of the main factor limiting the accuracy and linearity of the ADC is the element mismatch. Dynamic element matching (DEM) is an useful approach to improve the linearity of the DAC in $\Sigma\Delta$ ADC. Different from calibration techniques that require an exact measurement of each unit element to calibrate the mismatch errors, DEM does not need the actual mismatch information of the unit element. Therefore, DEM is less sensitive to small matching variations due to age and temperature influence [20]. The common algorithms include Randomization, Clocked Averaging(CLA), Individual level averaging(ILA), data weighted averaging (DWA), etc. Randomization converts the distortions into white noise, which can effectively depress harmonics, but may upraise the noise floor, while the CLA can keep the noise floor, but the distortions caused by DAC may drop in the effective band, which may increase the demands on anti-aliasing filter. ILA may upraise the noise floor and convergence of error is relatively slow, which may decrease the efficiency of the algorithm. In DWA algorithm, each element in the feedback DAC will be used the same times, which leads to the errors of DAC quickly average to zero. Moreover, DWA in Fig.5 proposed in [21] is simple to realize and can provide the first-order shaping for the error, therefore, DWA is more advantageous than the other algorithms.

In this work, we have realized a 4-bit DWA circuit unit in Cadence, and then is verified in a third-order four-bit continuous-time $\Sigma\Delta$ modulator with a 1% standard deviation added to the current sources, as shown in Fig.6. At first, the 15-bit thermometer codes of the quantizer will be rotated by a barrel shifter, and then be converted to feedback DAC switch control signal. As shown in Fig.7, DEM improves the SFDR from 65.14 dB to 75.08 dB.

C. DAC calibration

Least mean square(LMS) is able to correct errors caused by capacitor mismatch and can improve conversion accuracy and speed. In general, with the LMS algorithm, the analog signal path is not disturbed during calibration and thus can maintain maximum sampling rate. But the LMS algorithm is limited to convergency problem, and the actual LMS calibration was usually done via off-chip software approach [22]. To ease system complexity, and obviate the need of complicated post-processing, auxiliary DAC calibration is an energy-efficient method to calibrate capacitor mismatch and easy to realize on-chip [23]. With auxiliary DAC calibration technique, the mismatch of each capacitor is measured and calibrated by an auxiliary calibration ADC. In [24], we have realized a 12-bit time-domain SAR ADC with on-chip auxiliary DAC calibration designed in a 0.18 μm, 6M1P CMOS process. Measurement results show that, the ADC, at 12-bit, 100kS/s, achieves a Nyquist SFDR of 90.36 dB. To the best of our knowledge, 90.36 dB SFDR is the highest SFDR among all the time-domain SAR ADCs with 12-bit resolution. As a result, auxiliary DAC calibration is an very effective and simple method to improve linearity of ADC.

III. CONCLUSION

This work presented a novel model of redundant noise-shaping Pipelined Flash-SAR ADC, because of the first order noise-shaping technique, SFDR is improved about 10 dB compared to which without noise-shaping. Moreover, a 4-bit DWA circuit unit is realized in Cadence, and then is verified in a third-order four-bit continuous-time $\Sigma\Delta$ modulator with a standard deviation of 1%. Simulation results demonstrate that DEM improves the SFDR from 65.14 dB to 75.08 dB. Finally, auxiliary DAC calibration is verified in a 12-bit 100kS/s SAR ADC designed in a 0.18 μm, 6M1P CMOS process.

978-1-4799-7670-6/15 $31.00 © 2015 IEEE

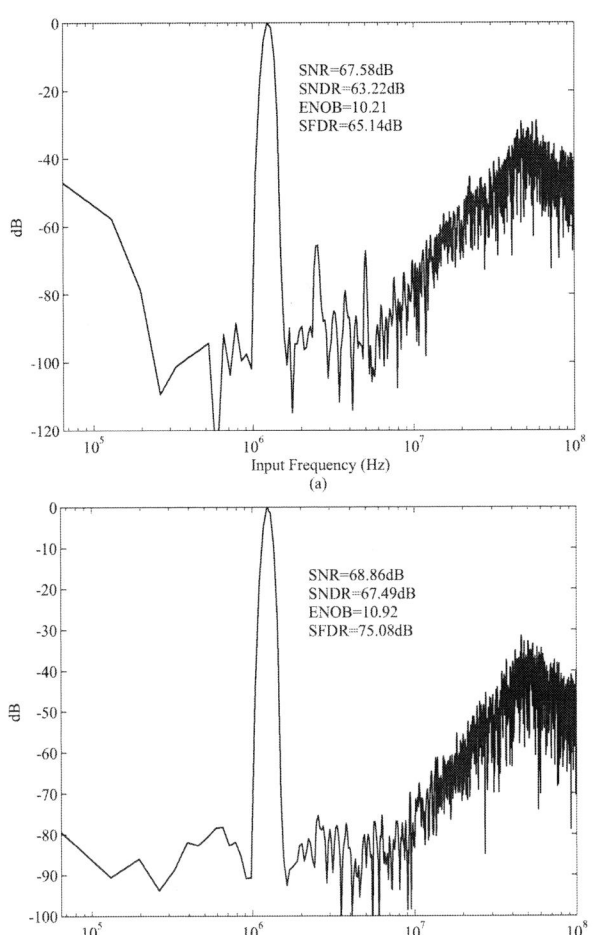

Fig. 7. FFT of the third-order $\Sigma\Delta$ modulator before (a) and after (b) DWA DEM

IV. ACKNOWLEDGMENTS

This work was supported by the National Natural Science Foundation of China (NSFC) under Grant 61401066 and the Fundamental Research Funds for the Central Universities under Grant ZYGX2014J029 as well as supported by a scholarship from the China Scholarship Council(CSC).

REFERENCES

[1] H.-C. Chow and W.-T. Lin, "High resolution successive approximation adc for low power biomedical applications," *Procedia Engineering*, vol. 50, pp. 275–283, 2012.

[2] Tai, Hung-Yen and Hu, Yao-Sheng and Chen, Hung-Wei and Chen, Hsin-Shu, "A 0.85 fJ/conversion-step 10b 200kS/s subranging SAR ADC in 40nm CMOS," in *Digest of Technical Papers of IEEE International Solid-State Circuits Conference(ISSCC)*. IEEE, 2014, pp. 196–197.

[3] F. van der Goes, C. Ward, S. Astgimath, H. Yan, J. Riley, J. Mulder, S. Wang, and K. Bult, "A 1.5 mW 68dB SNDR 80MS/s 2× interleaved SAR-assisted pipelined ADC in 28nm CMOS," in *Digest of Technical Papers of IEEE International Solid-State Circuits Conference(ISSCC)*. IEEE, 2014, pp. 200–201.

[4] C.-Y. Lin and T.-C. Lee, "A 12-bit 210-MS/s 5.3-mW pipelined-SAR ADC with a passive residue transfer technique," in *2014 Symposium on VLSI Circuits (VLSIC)*. IEEE, 2014, pp. 1–2.

[5] C.-J. Tseng, Y.-C. Hsieh, C.-H. Yang, and H.-S. Chen, "A 10-Bit 200 MS/s Capacitor-Sharing Pipeline ADC," *IEEE Transactions on Circuits and Systems I: Regular Papers*, vol. 60, no. 11, pp. 2902–2910, 2013.

[6] B. Verbruggen, M. Iriguchi, M. de la Guia Solaz, G. Glorieux, K. Deguchi, B. Malki, and J. Craninckx, "A 2.1 mW 11b 410 Ms/s dynamic pipelined SAR ADC with background calibration in 28nm digital CMOS," in *2013 Symposium on VLSI Circuits (VLSIC)*. IEEE, 2013, pp. C268–C269.

[7] Miki, T. and Morie, T. and Ozeki, T. and Dosho, S., "An 11-b 300-MS/s Double-Sampling Pipelined ADC With On-Chip Digital Calibration for Memory Effects," in *2011 Symposium on VLSI Circuits (VLSIC)*. IEEE, 2011, pp. 122–123.

[8] Kuppambatti, Jayanth and Kinget, Peter R, "Current Reference Pre-Charging Techniques for Low-Power Zero-Crossing Pipeline-SAR AD-Cs," *IEEE Journal of Solid-State Circuits*, vol. 49, no. 3, pp. 683–694, 2014.

[9] B. Malki, B. Verbruggen, P. Wambacq, K. Deguchi, M. Iriguchi, and J. Craninckx, "A complementary dynamic residue amplifier for a 67 dB SNDR 1.36 mW 170 MS/s pipelined SAR ADC," in *European Solid-State Circuits Conference(ESSCIRC)*. IEEE, 2014, pp. 215–218.

[10] C. C. Lee, E. Alpman, S. Weaver, C.-Y. Lu, and J. Rizk, "A 66dB SNDR 15MHz BW SAR assisted $\Delta\Sigma$ ADC in 22nm tri-gate CMOS," in *2013 Symposium on VLSI Circuits (VLSIC)*. IEEE, 2013, pp. C64–C65.

[11] Y.-Z. Lin, C.-C. Liu, G.-Y. Huang, Y.-T. Shyu, Y.-T. Liu, and S.-J. Chang, "A 9-Bit 150-MS/s subrange ADC based on SAR architecture in 90-nm CMOS," *IEEE Transactions on Circuits and Systems I: Regular Papers*, vol. 60, no. 3, pp. 570–581, 2013.

[12] J. A. Fredenburg and M. P. Flynn, "A 90-MS/s 11-MHz-Bandwidth 62-dB SNDR Noise-Shaping SAR ADC," *IEEE Journal of Solid-State Circuits*, vol. 47, no. 12, pp. 2898–2904, 2012.

[13] Verbruggen, B. and Iriguchi, M. and Craninckx, J., "A 1.7 mW 11b 250MS/s 2× interleaved fully dynamic pipelined SAR ADC in 40nm digital CMOS," in *Digest of Technical Papers of IEEE International Solid-State Circuits Conference(ISSCC)*. IEEE, 2012, pp. 466–468.

[14] Lee, C.C. and Flynn, M.P., "A SAR-Assisted Two-Stage Pipeline ADC," *IEEE Journal of Solid-State Circuits*, vol. 46, no. 4, pp. 859–869, 2011.

[15] Furuta, M. and Nozawa, M. and Itakura, T., "A 10-bit, 40-MS/s, 1.21 mW Pipelined SAR ADC Using Single-Ended 1.5-bit/cycle Conversion," *IEEE journal of solid-state circuits*, vol. 46, no. 6, pp. 1360–1370, 2011.

[16] O. Rajaee, T. Musah, N. Maghari, S. Takeuchi, M. Aniya, K. Hamashita, and U.-K. Moon, "Design of a 79 dB 80 MHz 8X-OSR Hybrid Delta-Sigma/Pipelined ADC," *IEEE Journal of Solid-State Circuits*, vol. 45, no. 4, pp. 719–730, 2010.

[17] H. Fan, Q. Wei, F. Qiao, and H. Yang, "A novel redundant pipelined successive approximation register adc," *IEICE Electronics Express*, vol. 10, no. 5, pp. 20 130 047–20 130 047, 2013.

[18] O. Rajaee, S. Takeuchi, M. Aniya, K. Hamashita, and U.-K. Moon, "Low-OSR over-ranging hybrid ADC incorporating noise-shaped two-step quantizer," *IEEE Journal of Solid-State Circuits*, vol. 46, no. 11, pp. 2458–2468, 2011.

[19] Z. Chen, P. Zhang, H. Wei, S.-W. Sin, U. Seng-Pan, R. Martins, and Z. Wang, "Noise shaping implementation in two-step/SAR ADC architectures based on delayed quantization error," in *2011 IEEE 54th International Midwest Symposium on Circuits and Systems (MWSCAS)*. IEEE, 2011, pp. 1–4.

[20] R. T. Baird and T. S. Fiez, "Linearity enhancement of multibit $\Delta\Sigma$ A/D and D/A converters using data weighted averaging," *IEEE Transactions on Circuits and Systems II: Analog and Digital Signal Processing*, vol. 42, no. 12, pp. 753–762, 1995.

[21] J.-G. Jo, J. Noh, and C. Yoo, "A 20-MHz bandwidth continuous-time sigma-delta modulator with jitter immunity improved full clock period SCR (FSCR) DAC and high-speed DWA," *IEEE Journal of Solid-State Circuits*, vol. 46, no. 11, pp. 2469–2477, 2011.

[22] Liu, W. and Huang, P. and Chiu, Y., "A 12-bit, 45-MS/s, 3-mW Redundant Successive-Approximation-Register Analog-to-Digital Converter With Digital Calibration," *IEEE Journal of Solid-State Circuits*, no. 99, pp. 1–1, 2011.

[23] J. G. Kauffman, P. Witte, M. Lehmann, J. Becker, Y. Manoli, and M. Ortmanns, "A 72 dB DR, CT $\Delta\Sigma$ Modulator Using Digitally Estimated, Auxiliary DAC Linearization Achieving 88 fJ/conv-step in a 25 MHz BW," *IEEE Journal of Solid-State Circuits*, vol. 49, no. 2, pp. 392–404, 2014.

[24] H. Fan, X. Han, Q. Wei, and H. Yang, "A 12-bit self-calibrating SAR ADC achieving a Nyquist 90.4-dB SFDR," *Analog Integrated Circuits and Signal Processing*, vol. 74, no. 1, pp. 239–254, 2013.

Optimal Design to Maximize Efficiency of Single-Inductor Multiple-Output Buck Converters in Discontinuous Conduction Mode for IoT Applications

Yoshitaka Yamauchi, Yuki Yanagihara, Hiroshi Fuketa, Takayasu Sakurai, and Makoto Takamiya

University of Tokyo
Tokyo, Japan

Abstract—To clarify the design guide of single-inductor multiple-output (SIMO) buck converters in discontinuous conduction mode (DCM) targeted for small-size and low-power IoT applications, equations of optimal design parameters (transistor size, inductance, and switching frequency) to maximize the power conversion efficiency are derived for the first time. The analytical optimal designs are verified with SPICE simulations. The maximum efficiency of SIMO DCM buck converters is lower than that of conventional single-inductor single-output (SISO) DCM buck converters, because the power loss due to the energy distribution switches is added. The efficiency degradation is analytically explained for the first time.

Keywords—*buck converter; discontinuous conduction mode; single-inductor multiple-output*

I. INTRODUCTION

In the applications of Internet of Things (IoT), tiny and energy efficient IoT nodes are required for physical data collection. The requirements for power-management ICs are the multiple output voltages, small output current (μA – mA), and small size. Figs. 1 (a) and 2 show a conventional single-inductor single-output (SISO) buck converter and a single-inductor multiple-output (SIMO) buck converter, respectively. Instead of the SISO buck converter in continuous conduction mode (CCM), the SIMO buck converter in discontinuous conduction mode (DCM) is an excellent candidate to meet the requirements for IoT, because the SIMO buck converter can provide multiple output voltages using only one inductor and DCM is better suited for the small output current. In DCM, the inductor current (I_L) is intermittently zero within each switching cycle as shown in Fig. 1 (b). The design guide to maximize the power conversion efficiency of SISO and SIMO DCM buck converters, however, is not reported, though that of the SISO CCM buck converter is reported in [1]. In addition, the efficiency degradation of the SIMO DCM buck converter over the SISO DCM buck converter is not theoretically clarified.

Therefore, in this paper, the optimal designs of transistor size, inductance, and switching frequency to maximize the power conversion efficiency of both SISO and SIMO DCM buck converters are proposed, and the analytical optimal designs are verified with SPICE simulations. Then, the maximum efficiency of SISO and SIMO DCM buck converters are compared and analyzed.

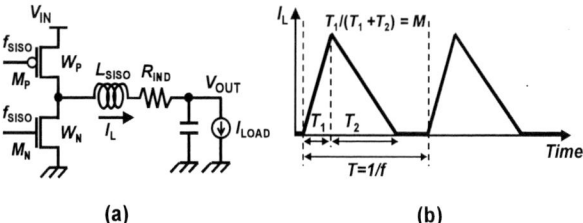

Fig. 1. Single-inductor single-output (SISO) buck converter. (a) Circuit. (b) Inductor current waveform in DCM.

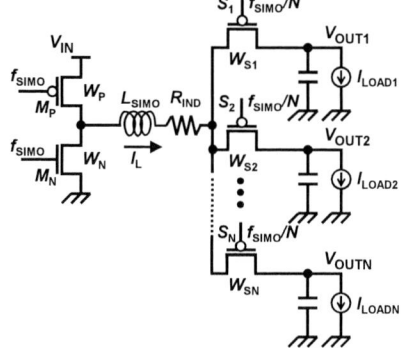

Fig. 2. Single-inductor multiple-output (SIMO) buck converter.

II. OPTIMAL DESIGN OF SISO DCM BUCK CONVERTER

To obtain the optimal design of the SISO DCM buck converter in Fig. 1 (a), the optimal design of the SISO CCM buck converter in [1] is expanded. In the section, equations (12) - (18) are newly proposed in this paper, while equations (1) - (11) are similar to [1]. The effective ON resistance (R_{EFF}) and effective switching capacitance (C_{EFF}) of the power transistors (M_P and M_N in Fig. 1 (a)) are as follows.

$$R_{EFF} = \frac{R_N}{W_N}\left(1-M\right) + \frac{R_P}{W_P}M \qquad (1)$$

$$C_{EFF} = W_N C_N + W_P C_P \qquad (2)$$

where W_N, W_P, R_N, R_P, and C_N, C_P are the gate widths, ON resistance per unit gate width, and the switched capacitance per unit gate width of nMOS (M_N) and pMOS (M_P), respectively. $M (= V_{OUT}/V_{IN})$ is the conversion ratio, where V_{IN}

978-1-4799-7670-6/15 $31.00 © 2015 IEEE

and V_{OUT} are input and output voltages, respectively. By minimizing C_{EFF} at constant R_{EFF}, the optimal gate width ratio ($\alpha_{OPT} = W_P / W_N$) is determined as follows.

$$\alpha_{OPT} = \frac{W_P}{W_N} = \sqrt{\frac{M R_P C_N}{(1-M) R_N C_P}} \tag{3}$$

Here, the total gate width ($W_{TOTAL} = W_N + W_P$) is defined. By substituting (3) into (1) and (2), each of R_{EFF} and C_{EFF} is a function of W_{TOTAL} as follows.

$$R_{EFF} = \frac{R_{AVE}}{W_{TOTAL}} \tag{4}$$

$$R_{AVE} = (1 + \alpha_{OPT}) \left[(1-M) R_N + \frac{M R_P}{\alpha_{OPT}} \right] \tag{5}$$

$$C_{EFF} = W_{TOTAL} C_{AVE} \tag{6}$$

$$C_{AVE} = \frac{C_N + \alpha_{OPT} C_P}{1 + \alpha_{OPT}} \tag{7}$$

The power loss ($P_{LOSS,SISO}$) in a buck converter consists of three kinds of loss: the switching loss ($P_{CAP,SISO}$), the resistive loss ($P_{RES,SISO}$) of M_N and M_P, and the resistive loss ($P_{IND,SISO}$) of the inductor [1].

$$P_{LOSS,SISO} = P_{CAP,SISO} + P_{RES,SISO} + P_{IND,SISO} \tag{8}$$

where

$$P_{CAP,SISO} = f_{SISO} C_{EFF} V_{IN}^2 = f_{SISO} W_{TOTAL} C_{AVE} V_{IN}^2 \tag{9}$$

$$P_{RES,SISO} = R_{EFF} I_{RMS,SISO}^2 = \frac{R_{AVE}}{W_{TOTAL}} I_{RMS,SISO}^2 \tag{10}$$

$$P_{IND,SISO} = R_{IND} I_{RMS,SISO}^2 = \frac{L_{SISO}}{\tau_L} I_{RMS,SISO}^2 \tag{11}$$

$$I_{RMS,SISO}^2 = \frac{2}{3} I_{LOAD} \sqrt{\frac{2 V_{IN} M (1-M) I_{LOAD}}{f_{SISO} L_{SISO}}} \tag{12}$$

As shown in Fig. 1 (a), f_{SISO}, L_{SISO}, R_{IND}, I_{LOAD} are the switching frequency, the inductance, the parasitic resistance of the inductor, and the load current, respectively. τ_L is the figure of merit of an inductor technology [1-2] and $I_{RMS,SISO}$ is the effective value of I_L. The difference of this work from [1] is the discontinuous I_L shown in Fig. 1 (b) and (12).

In the design of the SISO DCM buck converter, three design parameters (W_{TOTAL}, L_{SISO}, f_{SISO}) are available. To maximize the power conversion efficiency, $P_{LOSS,SISO}$ should be minimized by setting the derivatives with respect to each parameter to zero [1]. Then the optimal parameters are derived as follows.

$$W_{TOTAL,OPT} = W_{SISO} \tag{13}$$

$$L_{SISO,OPT} = \frac{R_{AVE} \tau_L}{W_{SISO}} \tag{14}$$

$$f_{SISO,OPT} = \frac{2 I_{LOAD}}{W_{SISO} V_{IN}} \sqrt[3]{\frac{M (1-M) R_{AVE}}{9 \tau_L C_{AVE}^2}} \tag{15}$$

W_{SISO} is an arbitrary parameter (total gate width). In the optimal design, $P_{CAP,SISO} = P_{RES,SISO} = P_{IND,SISO}$, and the minimum $P_{LOSS,SISO}$ ($P_{LOSS,SISO,MIN}$), a newly defined loss ratio ($LR_{SISO,MIN}$), and the maximum efficiency ($\eta_{SISO,MAX}$) are shown as follows.

$$P_{LOSS,SISO,MIN} = V_{IN} I_{LOAD} \sqrt[3]{\frac{24 R_{AVE} C_{AVE} M (1-M)}{\tau_L}} \tag{16}$$

$$LR_{SISO,MIN} = \frac{P_{LOSS,SISO,MIN}}{P_{OUT}} = \sqrt[3]{\frac{24 R_{AVE} C_{AVE} (1-M)}{\tau_L M^2}} \tag{17}$$

$$\eta_{SISO,MAX} = \frac{1}{1 + \sqrt[3]{\dfrac{24 R_{AVE} C_{AVE} (1-M)}{\tau_L M^2}}} \tag{18}$$

III. OPTIMAL DESIGN OF SIMO DCM BUCK CONVERTER

In this section, the optimal design of the SIMO DCM buck converter with N outputs in Fig. 2 is derived. Then, the analytical optimal designs are verified with SPICE simulations. In this paper, the time-multiplexing control [3-4] shown in Fig. 3 is used for the SIMO DCM buck converter, because the cross regulation between outputs is very small.

A. Analysis of Optimal Design

To simplify the analysis and to understand the essence of the SIMO DCM buck converter, N outputs in Fig. 2 are assumed to be equal. Specifically, $V_{OUT1} = V_{OUT2} = \dots = V_{OUTN} = V_{OUT}$, $I_{LOAD1} = I_{LOAD2} = \dots = I_{LOADN} = I_{LOAD}$, and $W_{S1} = W_{S2} = \dots = W_{SN} = W_S$. In this assumption, the circuit in Fig. 2 can be transformed to the circuit in Fig. 4. Compared to the circuit of the SISO DCM buck converter in Fig. 1 (a), an energy distribution switch (S) is added and the load current increases N times. The power loss ($P_{LOSS,SIMO}$) is represented as follows.

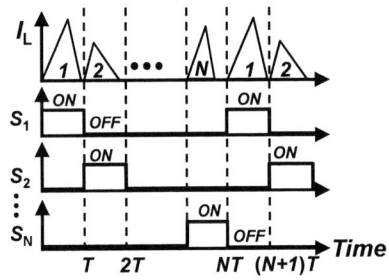

Fig. 3. Waveforms of SIMO DCM buck converter.

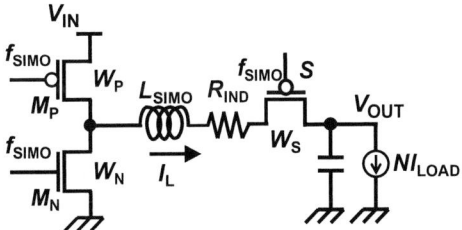

Fig. 4. Equivalent circuit of SIMO DCM buck converter with equal outputs.

978-1-4799-7670-6/15 $31.00 © 2015 IEEE

$$P_{\text{LOSS,SIMO}} = P_{\text{CAP,SIMO}} + P_{\text{RES,SIMO}} + P_{\text{IND,SIMO}} \qquad (19)$$

where

$$P_{\text{CAP,SIMO}} = f_{\text{SIMO}} W_{\text{TOTAL}} C_{\text{AVE}} V_{\text{IN}}^2 + f_{\text{SIMO}} W_{\text{S}} C_{\text{P}} V_{\text{IN}}^2 \qquad (20)$$

$$P_{\text{RES,SIMO}} = \frac{R_{\text{AVE}}}{W_{\text{TOTAL}}} I_{\text{RMS,SIMO}}^2 + \frac{R_{\text{S}}}{W_{\text{S}}} I_{\text{RMS,SIMO}}^2 \qquad (21)$$

$$P_{\text{IND,SIMO}} = R_{\text{IND}} I_{\text{RMS,SIMO}}^2 = \frac{L_{\text{SIMO}}}{\tau_{\text{L}}} I_{\text{RMS,SIMO}}^2 \qquad (22)$$

$$I_{\text{RMS,SIMO}}^2 = \frac{2}{3}(NI_{\text{LOAD}})\sqrt{\frac{2V_{\text{IN}}M(1-M)NI_{\text{LOAD}}}{f_{\text{SIMO}}L_{\text{SIMO}}}} \qquad (23)$$

$$R_{\text{S}} = R_{\text{P}}\frac{V_{\text{IN}}-V_{\text{TH}}}{MV_{\text{IN}}-V_{\text{TH}}} \qquad (24)$$

W_{S} is the gate width of S, R_{S} is ON resistance of S per unit gate width, and V_{TH} is the threshold voltage of S. R_{S} instead of R_{P} is added, because the gate-source voltages of M_{P} and S are different in Fig. 4. In (20) and (21), the switching loss and resistive loss due to S are added. In the design of the SIMO DCM buck converter, four design parameters (W_{TOTAL}, W_{S}, L_{SIMO}, f_{SIMO}) are available. Compared with the previous section, W_{S} is newly added. The optimal design parameters are derived as follows.

$$W_{\text{TOTAL,OPT}} = W_{\text{SIMO}} \qquad (25)$$

$$W_{\text{S,OPT}} = \sqrt{\frac{R_{\text{S}}C_{\text{AVE}}}{R_{\text{AVE}}C_{\text{P}}}}W_{\text{SIMO}} \qquad (26)$$

$$L_{\text{SIMO,OPT}} = \frac{R_{\text{AVE}}\tau_{\text{L}}}{W_{\text{SIMO}}}\left(1+\sqrt{\frac{R_{\text{S}}C_{\text{P}}}{R_{\text{AVE}}C_{\text{AVE}}}}\right) \qquad (27)$$

$$f_{\text{SIMO,OPT}} = \frac{2NI_{\text{LOAD}}}{V_{\text{IN}}W_{\text{SIMO}}}\sqrt[3]{\frac{M(1-M)}{9\tau_{\text{L}}C_{\text{AVE}}^2}\Big/\left(1+\sqrt{\frac{R_{\text{S}}C_{\text{P}}}{R_{\text{AVE}}C_{\text{AVE}}}}\right)} \qquad (28)$$

W_{SIMO} is an arbitrary parameter (total gate width). In the optimal design, $P_{\text{CAP,SIMO}} = P_{\text{RES,SIMO}} = P_{\text{IND,SIMO}}$, and the minimum $P_{\text{LOSS,SIMO}}$ ($P_{\text{LOSS,SIMO,MIN}}$), the loss ratio

($LR_{\text{SIMO,MIN}}$), and the maximum efficiency ($\eta_{\text{SIMO,MAX}}$) are shown as follows

$$P_{\text{LOSS,SIMO,MIN}} = V_{\text{IN}}\left(NI_{\text{LOAD}}\right)$$
$$\times \sqrt[3]{\frac{24R_{\text{AVE}}C_{\text{AVE}}M(1-M)}{\tau_{\text{L}}}\left(1+\sqrt{\frac{R_{\text{S}}C_{\text{P}}}{R_{\text{AVE}}C_{\text{AVE}}}}\right)^2} \qquad (29)$$

$$LR_{\text{SIMO,MIN}} = \frac{P_{\text{LOSS,SIMO,MIN}}}{P_{\text{OUT}}}$$
$$= \sqrt[3]{\frac{24R_{\text{AVE}}C_{\text{AVE}}(1-M)}{\tau_{\text{L}}M^2}\left(1+\sqrt{\frac{R_{\text{S}}C_{\text{P}}}{R_{\text{AVE}}C_{\text{AVE}}}}\right)^2} \qquad (30)$$

$$\eta_{\text{SIMO,MAX}} = \frac{1}{1+\sqrt[3]{\frac{24R_{\text{AVE}}C_{\text{AVE}}(1-M)}{\tau_{\text{L}}M^2}\left(1+\sqrt{\frac{R_{\text{S}}C_{\text{P}}}{R_{\text{AVE}}C_{\text{AVE}}}}\right)^2}} \qquad (31)$$

Table I summarizes the derived equations. Similar to the SISO DCM buck converter, the optimal design for SISO DCM buck converter has one degree of freedom (W_{SIMO}). When $R_{\text{S}} = 0$, (27) – (31) are equal to (14) – (18), respectively, which is reasonable.

B. Verification with SPICE Simulations

To check the validity of the derived equations in this paper, the optimal design of a SIMO DCM buck converter with two outputs in Fig. 2 is performed using 1.8V, 180nm CMOS process and the analytically derived $P_{\text{LOSS,SIMO}}$ is compared with SPICE simulations. Fig. 5 shows the calculated and SPICE simulated $P_{\text{LOSS,SIMO}}$. In Figs. 5 (a) –(d), one of the four design parameters (W_{TOTAL}, W_{S}, L_{SIMO}, f_{SIMO}) are varied, respectively, and all the other parameters are optimum values shown in Table II. The calculated results are consistent with the SPICE simulated results, which shows the validity of the equations in this paper. The minimum $P_{\text{LOSS,SIMO}}$ in each Figs. 5 (a) –(d) are identical, which is the evidence of the minimum loss (= maximum efficiency) design.

TABLE I SUMMARY OF EQUATIONS FOR BUCK CONVERTERS.

	[1]	This Work	
Operation mode	CCM	DCM	
Topology	SISO		SIMO
$W_{\text{TOTAL,OPT}}$	$\frac{R_{\text{AVE}}I_{\text{LOAD}}}{V_{\text{IN}}}\sqrt[3]{\frac{8\tau_{\text{L}}}{3R_{\text{AVE}}C_{\text{AVE}}M(1-M)}}$	W_{SISO} (13)	W_{SIMO} (25)
L_{OPT}	$\frac{V_{\text{IN}}}{2I_{\text{LOAD}}}\sqrt[3]{3R_{\text{AVE}}C_{\text{AVE}}\tau_{\text{L}}^2 M(1-M)}$	$\frac{R_{\text{AVE}}\tau_{\text{L}}}{W_{\text{SISO}}}$ (14)	$\frac{R_{\text{AVE}}\tau_{\text{L}}}{W_{\text{SIMO}}}\left(1+\sqrt{\frac{R_{\text{S}}C_{\text{P}}}{R_{\text{AVE}}C_{\text{AVE}}}}\right)$ (27)
f_{OPT}	$\sqrt[3]{\frac{M^2(1-M)^2}{3R_{\text{AVE}}C_{\text{AVE}}\tau_{\text{L}}^2}}$	$\frac{2I_{\text{LOAD}}}{W_{\text{SISO}}V_{\text{IN}}}\sqrt[3]{\frac{M(1-M)R_{\text{AVE}}}{9\tau_{\text{L}}C_{\text{AVE}}^2}}$ (15)	$\frac{2NI_{\text{LOAD}}}{W_{\text{SIMO}}V_{\text{IN}}}\sqrt[3]{\frac{M(1-M)R_{\text{AVE}}}{9\tau_{\text{L}}C_{\text{AVE}}^2}\Big/\left(1+\sqrt{\frac{R_{\text{S}}C_{\text{P}}}{R_{\text{AVE}}C_{\text{AVE}}}}\right)}$ (28)
$W_{\text{S, OPT}}$			$\sqrt{\frac{R_{\text{S}}C_{\text{AVE}}}{R_{\text{AVE}}C_{\text{P}}}}W_{\text{SIMO}}$ (26)
$P_{\text{LOSS, MIN}}$	$V_{\text{IN}}I_{\text{LOAD}}\sqrt[3]{\frac{24R_{\text{AVE}}C_{\text{AVE}}M(1-M)}{\tau_{\text{L}}}}$	$V_{\text{IN}}I_{\text{LOAD}}\sqrt[3]{\frac{24R_{\text{AVE}}C_{\text{AVE}}M(1-M)}{\tau_{\text{L}}}}$ (16)	$V_{\text{IN}}\left(NI_{\text{LOAD}}\right)\sqrt[3]{\frac{24R_{\text{AVE}}C_{\text{AVE}}M(1-M)}{\tau_{\text{L}}}\left(1+\sqrt{\frac{R_{\text{S}}C_{\text{P}}}{R_{\text{AVE}}C_{\text{AVE}}}}\right)^2}$ (29)
$\frac{P_{\text{LOSS,MIN}}}{P_{\text{OUT}}}$	$\sqrt[3]{\frac{24R_{\text{AVE}}C_{\text{AVE}}(1-M)}{\tau_{\text{L}}M^2}}$	$\sqrt[3]{\frac{24R_{\text{AVE}}C_{\text{AVE}}(1-M)}{\tau_{\text{L}}M^2}}$ (17)	$\sqrt[3]{\frac{24R_{\text{AVE}}C_{\text{AVE}}(1-M)}{\tau_{\text{L}}M^2}\left(1+\sqrt{\frac{R_{\text{S}}C_{\text{P}}}{R_{\text{AVE}}C_{\text{AVE}}}}\right)^2}$ (30)
$\eta_{\text{MAX}} = \frac{1}{1+\frac{P_{\text{LOSS,MIN}}}{P_{\text{OUT}}}}$	$\frac{1}{1+\sqrt[3]{\frac{24R_{\text{AVE}}C_{\text{AVE}}(1-M)}{\tau_{\text{L}}M^2}}}$	$\frac{1}{1+\sqrt[3]{\frac{24R_{\text{AVE}}C_{\text{AVE}}(1-M)}{\tau_{\text{L}}M^2}}}$ (18)	$\frac{1}{1+\sqrt[3]{\frac{24R_{\text{AVE}}C_{\text{AVE}}(1-M)}{\tau_{\text{L}}M^2}\left(1+\sqrt{\frac{R_{\text{S}}C_{\text{P}}}{R_{\text{AVE}}C_{\text{AVE}}}}\right)^2}}$ (31)

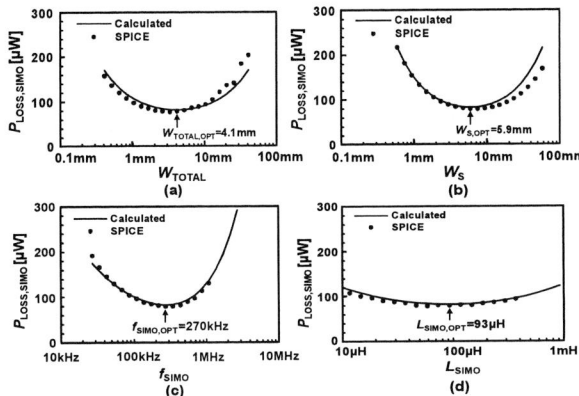

Fig. 5. Calculated and SPICE simulated $P_{LOSS,SIMO}$. In (a) –(d), one of four design parameters (W_{TOTAL}, W_S, L_{SIMO}, f_{SIMO}) are varied, respectively.

TABLE II PARAMETERS USED IN FIG. 5

M	0.5	τ_L	38 [μH/Ω]
V_{IN}	1.8 [V]	I_{LOAD}	1.0 [mA]
V_{OUT}	0.9 [V]	α_{OPT}	1.9
R_N	900 [Ω·μm]	$W_{TOTAL,OPT}$	4.1 [mm]
R_P	3600 [Ω·μm]	$W_{S,OPT}$	5.9 [mm]
C_N	2.8 [fF/μm]	$f_{SIMO,OPT}$	270 [kHz]
C_P	3.2 [fF/um]	$L_{SIMO,OPT}$	93 [μH]
R_S	8600 [Ω·μm]	$\sqrt{\dfrac{R_S C_P}{R_{AVE} C_{AVE}}}$	1.5
R_{AVE}	4000 [Ω·μm]	$\left(1+\sqrt{\dfrac{R_S C_P}{R_{AVE} C_{AVE}}}\right)^{\frac{2}{3}}$	1.8
C_{AVE}	3.0 [fF/μm]		

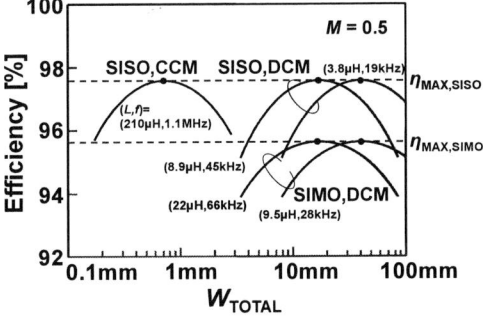

Fig. 6. Calculated W_{TOTAL} dependence of efficiency of SISO CCM, SISO DCM, and SIMO DCM buck converters.

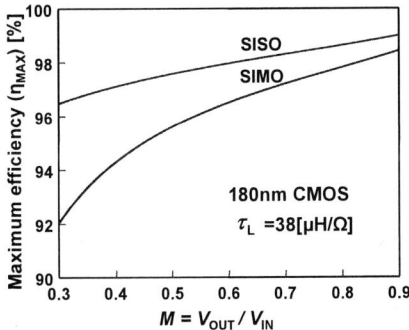

Fig. 7. Calculated M dependence of η_{MAX} of SISO and SIMO DCM buck converters.

IV. COMPARISON OF MAXIMUM EFFICIENCY OF SISO AND SIMO DCM BUCK CONVERTERS

Fig. 6 shows the calculated W_{TOTAL} dependence of the efficiency of SISO CCM, SISO DCM, and SIMO DCM buck converters. Parameters except W_{TOTAL}, L, and f are common as shown in Table II among the converters. In SISO CCM, the optimal W_{TOTAL} is uniquely determined, while the optimal W_{TOTAL} of SISO DCM and SIMO DCM has one degree of freedom. η_{MAX} of SISO CCM, however, is equal to η_{MAX} of SISO DCM as shown in Table I. η_{MAX} of SIMO DCM is lower than η_{MAX} of SISO DCM as shown in (18) and (31). Fig. 7 shows the calculated M dependence of η_{MAX} of SISO and SIMO DCM buck converters using (18) and (31). Parameters except M are similar to Table II. When M is increased from 0.3 to 0.9, the efficiency difference between SISO and SIMO decreases from 4.5% to 0.6%. Direct comparison of (18) and (31) results in a complicated equation, while comparison of (17) and (30) results in a simple equation as follows.

$$RLR = \frac{LR_{SIMO,MIN}}{LR_{SISO,MIN}} = \left(1+\sqrt{\frac{R_S C_P}{R_{AVE} C_{AVE}}}\right)^{\frac{2}{3}} \quad (32)$$

The ratio of the loss ratio (RLR) is a function of M and transistor-related parameters and does not depend on inductor-related parameters. The efficiency degradation between SISO and SIMO is universally explained by RLR. RLR is 2.4, 1.8, and 1.6 at M = 0.3, 0.5, and 0.9, respectively in 1.8V, 180-nm CMOS process. RLR decreases as M is increased, because S in Fig. 4 is pMOS and R_S decreases.

V. CONCLUSIONS

The design guides to maximize the power conversion efficiency of both SISO and SIMO DCM buck converters are proposed as shown in Table I. In the conventional SISO CCM buck converter, the optimal design parameters are uniquely determined, while the optimal design parameters in the SISO and SIMO DCM buck converters are not unique and have some flexibility. The maximum efficiency of the SIMO DCM buck converter is lower than that of the SISO DCM buck converter. The efficiency degradation mechanism is universally explained by the proposed RLR in (32).

REFERENCES

[1] G. Schrom, P. Hazucha, F. Paillet, D. S. Gardner, S. T. Moon, and T. Karnik, "Optimal Design of Monolithic Integrated DC-DC Converters," IEEE International Conference on Integrated Circuit Design and Technology, pp. 65-67, 2006.

[2] M. Takamiya, K. Onizuka, and T. Sakurai, "3D-Structured On-Chip Buck Converter for Distributed Power Supply System in SiPs," IEEE International Conference on IC Design and Technology, pp. 33-36, 2008.

[3] D. Ma, W. Ki, C. Tsui, and P. Mok, "Single-Inductor Multiple-Output Switching Converters with Time-Multiplexing Control in Discontinuous Conduction Mode," IEEE Journal of Solid-State Circuits, Vol.38, No.1, pp. 89 - 100, Jan. 2003.

[4] Y. Zhang and D. Ma, "A Fast-Response Hybrid SIMO Power Converter with Adaptive Current Compensation and Minimized Cross-Regulation," IEEE Journal of Solid-State Circuits, Vol.49, No.5, pp. 1242 - 1255, May. 2014

Thermal experimental and modeling analysis of high power 3D packages

H. Oprins, V. Cherman, G. Van der Plas, F. Maggioni[1], J. De Vos, E. Beyne

Imec, Leuven, Belgium

[1]Also with Dept. Mechanical Engineering, K.U. Leuven, Leuven, Belgium

E-mail: herman.oprins@imec.be

Abstract— **In this paper, we present the experimental and modeling characterization of 3D packages for high power applications using a dedicated stackable test chip. An advanced CMOS test chip with programmable power distribution and a 32x32 array of temperature sensors has been designed, fabricated, stacked and packaged in bare die 3D packages. The package thermal measurements have been used to characterize the thermal behavior and to successfully validate the thermal finite element modeling results.**

Keywords—3DIC; thermal analysis; thermal measurements; high power

I. INTRODUCTION

3D-TSV technology refers to the vertical stacking of integrated circuits, which are interconnected using through-Si vias (TSVs). Using the 3rd dimension offers designers increased system integration at lower cost, with reduced footprint and with shorter interconnect lengths [1]. However, thermal management issues are much more pronounced in 3D systems compared to conventional 2D system. Moreover, the thermal coupling between the chips in the 3D systems is typically very strong. Therefore, a detailed thermal analysis of the 3D system is required to avoid excessive temperatures in the different chips in the 3D package. Different types of stackable thermal test chips with integrated heaters and temperature sensors have been reported in literature for the experimental study of the thermal behavior of 3D packages and their cooling needs, the effect of the package and integrated lid [2], or to study a specific application such as the combination of logic and memory chip in a 3D package [3, 4]. These test chips typically have a fixed location of the heaters, which are either organized as hot spots to mimic the local heating of certain blocks on the chip layout or as a large heater that covers most of the chip area to emulate uniform heating. The thermal analysis using uniform heating is relevant to obtain information on the package thermal resistance and on the interfaces in the package, whereas hot spot measurements provide information on the lateral spreading and the detrimental impact of reduced die thickness. For a complete thermal analysis of the 3D package, both heater configurations should be considered.

The purpose of the work presented in this paper is to develop a versatile test vehicle and a modeling framework to analyze the thermal behavior of 3D packages for high power applications. The dedicated stackable test chip is able to dissipate up to 80W and features a programmable power dissipation map, which enables the generation of different user defined power dissipation patterns and a 32x32 sensor array to obtain an on-chip temperature map measurement with high spatial resolution. These capabilities allow mimicking applications of chip power dissipation patterns, such as multi-core CPU, GPU, memory and to characterize the thermal properties of the 3D package. The thermal measurement results in combination with the modeling results can be used to study the global and local thermal effect of 3D technology specific structures and of cooling solutions.

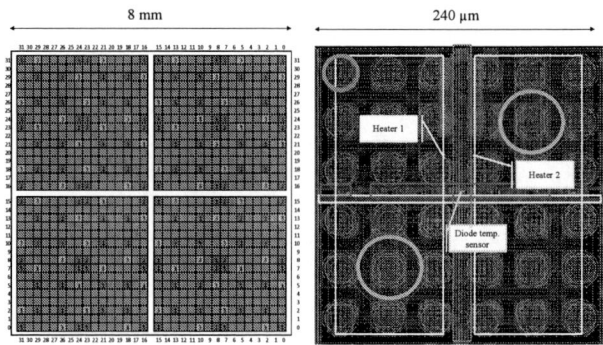

Fig. 1. Left: Location of thermal cells (blue squares) in test chip layout. Right: Thermal cell detail with temperature sensor and heaters.

II. TEST MATERIAL

A. Test chip descrition

An advanced stackable CMOS test chip has been designed and fabricated for the experimental analysis of thermal and mechanical aspects of 3D system integration. This 8x8 mm2 test chip is schematically shown in Fig. 1. The test chip layout consists of an array of 32x32 basic cells which are 240x240 μm^2 in size. All these cells contain a diode as temperature sensor in the center of the cell. The voltage-temperature sensitivity of this diode is -1.55 mV/°C for a current of 5µA. 832 cells in the 32x32 array contain two 200x100 μm^2 metal meander heaters in the BEOL. A transition gate in each of those cells controls the activation of the heater cells individually, allowing the application of a custom power dissipation pattern in the test chip, ranging from localized power dissipation to quasi-uniform power dissipation, covering

75% of the chip. Due to the use of the transition gate, the voltage applied to the test chip is limited to 1-1.5V. At 1V, the power dissipation of one heater cell 100mW. The target power dissipation of the test chip is 80W. This means that a current of 80A is required to power the test chip.

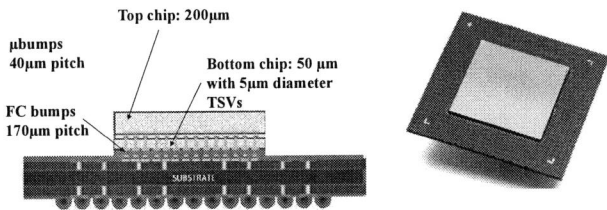

Fig. 2. Bare die 3D package schematic and photograph.

The 3D package consists of two stacked test chip. The first test chip (referred to as Tier 1) is 200μm thick. The second test chip (Tier 2) is thinned down to 50μm and contains 5μm diameter TSVs. This second tier is stacked to the first one in a back-to-face configuration using thermo-compression bonding [5]. The μbumps consist of a 5 μm thick Cu pad and 3.5μm thick Sn bump both with a diameter of 15μm at the side of Tier 1, and a Cu pad with a thickness of 7μm and a diameter of 25μm at the side of Tier 2. The pitch of the μbumps is 40μm, this means that a 240x240μm cell is connected with a 6x6 array of μbumps. The Tier 2 chip contains 5μm thick 50μm in diameter copper pillars plated on the front side passivation layer to be connected to the package substrate. In the area of a cell, two of these Cu pillars are present with a pitch of 170μm. The two-die stack is packaged face down in a 14x14 mm^2 flip chip BGA package with a 4 layer packaging laminate substrate with a thickness of 330μm. The package is an exposed bare die package, which allow direct access to the backside of the top chip in the package for cooling. Fig. 2 shows a schematic of the package cross section and a photograph of the fabricated package.

B. Measurement environments

To be able to perform high power thermal measurements with the fabricated 3D packages, several challenges for the measurements set-up needed to be addressed:

- High power levels up to 100W at 1V and 100A need to be supplied to the 3D package.
- This power needs to be adjustable and controllable.
- The actual power in the test chip needs to be measured during the operation of the test chip.
- The use of very high currents up to 100A will cause significant Joule heating in the connections and traces in PCB and package substrate. This heat generation needs to be limited.
- The measurement data of the 1024 temperature sensors per die of the 3D package need to be read out. The temperature measurement in the 32x32 array of sensors results in a temperature distribution map of the whole chip area, shown in Fig. 3.
- The packages need to assembled is such a way that advanced air or liquid cooling solutions can be connected to

the exposed backside of the package in order to evaluate the chip level impact of these cooling solutions.

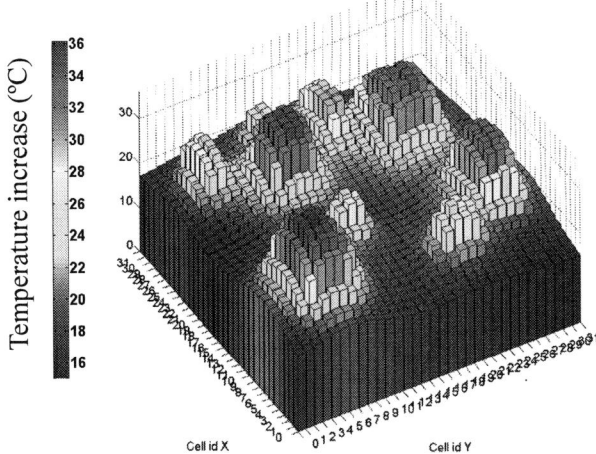

Fig. 3. Full chip temperature measurement using the 32x32 array of temperature sensors.

Two test configurations have been developed to evaluate the thermal behavior of 3D packages for high power applications. The first configuration is a measurement socket with is directly cooled by a high performance cooler. In this socket, there is a direct connection between the backside of the top chip in the 3D exposed package and the heat sink. On top of the heat sink an 80 x 80 mm^2 fan is mounted to provide the forced convection cooling of the socket, resulting in a low overall thermal resistance for this socket configuration. This socket is referred to as high power socket. Using this socket, the main part of the heat generated in the package will be removed through the top side of the package. A photograph of this set-up is shown in Fig. 4. In the second configuration, the bare die package is directly solder to the PCB. The drawback is that this connection is permanent and therefore one test board is required for each package to be tested. The advantage is that this situation represents a more realistic environment to test the packages. The soldered bare die packages, shown in Fig. 5 allow evaluating advanced active cooling solutions, such as forced convection air cooling, heat pipes or single phase or two phase liquid cooling since they provide direct access to the back side of the top chip.

The read-out of all the sensors is performed by a logic controller on the test chip, which controls the scanning of the test structures array. The connections of this controller to the package is made using 12 FC bumps. All the other FC bumps are used to provide the power and ground connections to the test chip to deliver the high power and current. In the package substrate, all power and ground connections have been connected in parallel and the PCB has been designed with wide power and ground traces to create a low resistance. The PCB has been bolted directly to the power supply to limit losses in the connection wires. In the die stack, loop back Kelvin structures are implemented to test the die-die connectivity, and thus the μbump yield for each cell. These structures can be used to estimate the resistance of individual heater cells and detect possible non-functioning heater cells. Fig. 6 shows the

978-1-4799-7670-6/15 $31.00 © 2015 IEEE
139

distribution of the resistance of these structures for the two dies in the 3D package. Four point measurements are performed for the current and voltage supplied during the thermal measurements to assess the actual dissipated power in the 3D package. Using both the socket configuration and soldered configuration, a chosen power dissipation profiles can be applied to the top and bottom chip such as uniform power dissipation, localized hot spots or defined power dissipation maps. At the same time, the actual power dissipation and the whole chip temperature distribution is measured in both chips. This means that the presented test chip can be used for the experimental characterization of the lateral and vertical heat transfer in the 3D package, of the thermal die-to-die coupling, characterization of the inter die thermal resistance and of the local and global impact of cooling solutions applied to the package. Other applications of this test chip include the analysis of the mechanical impact of stacking [6] and packaging [7], measurement of temperature induced stress [8] and thermal [9] and mechanical [7] model validation.

Fig. 4. First test configuration: measurment socket with attached heat sink and fan.

 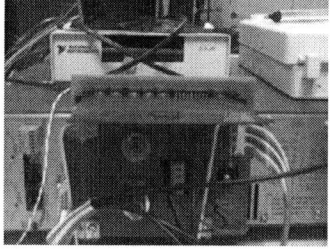

Fig. 5. Second test configuration: the bare die 3D package is solderd to the PCB. The PCB is bolted directly to the power supply to eliminate Joule heating in the connections.

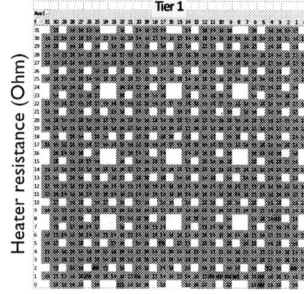 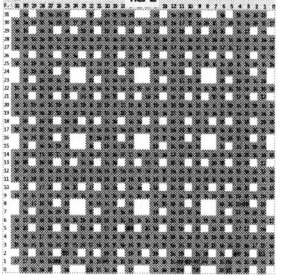

Fig. 6. Resistance measurement using the loop-back structures.

III. THERMAL MODEL

For the analysis of the thermal behavior of the 3D packages, the temperature measurements are combined with a simulation study. Detailed parameterized 3D finite element models (FEM) representing the socket and soldered package configurations have been constructed by using a general purposed finite element software package. By means of steady state thermal simulations, the temperature distribution inside the package configuration is modeled for the different power dissipation patterns. The detailed models include the PCB, the solder balls, the package laminate, Si interposer, both chips in the die stack and the interconnections between the chips and the package. To reduce the complexity of the thermal models, several structures have been replaced by elements with equivalent thermal properties. The µbumps and Cu pillars arrays embedded in underfill material are replaced by a material with equivalent in plane and out of plane thermal properties. Using a unit cell modeling approach for a µbump with a certain geometry and pitch the equivalent in plane and out of plane thermal conductivity can be found taking into account the local thermal spreading and constriction resistances. Experimental values for the equivalent resistance of such mixed layers of connections and underfill material can be extracted from the temperature measurements with uniform power dissipation.

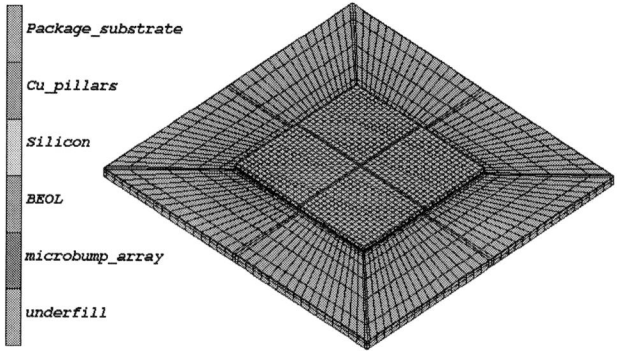

Fig. 7. Finite element model of the bare die 3D package.

IV. MEASUREMENT AND MODELING RESULTS

Using the test vehicle, power patterns can be dissipated in either chip of the 3D package while the temperature map is measured in both chips. This allows making different combinations of power application and temperature measurement in the different tiers, depending on the required analysis. In this section we report the temperature distribution in both top and bottom chip, for power dissipation in the bottom chip of the 3D package mounted in the test socket with heat sink. In this configuration, the heat generated in the test chip is remove through the top side of the package. For power dissipation in the bottom chip, the heat flow path is quasi one dimensional from the bottom chip, through the die-die interface and the top chip towards the heat sink. This test case allows characterization of the thermal properties of the die-die interface and of the thermal coupling between top and bottom chip in the 3D package.

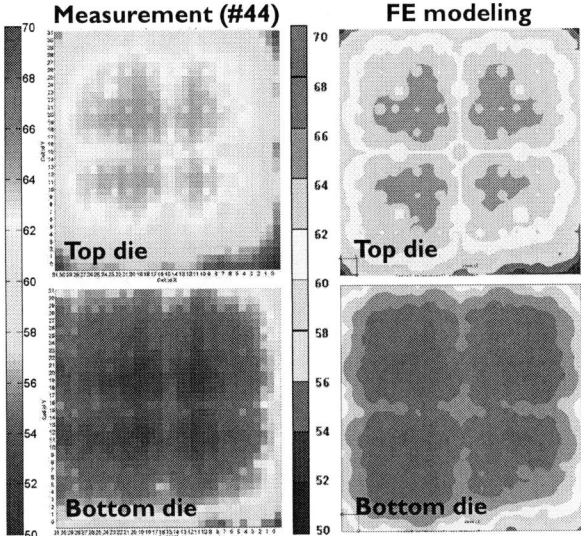

Fig. 8. Measurement and modeling results for the temperature distribution in the top and bottom chip of the 3D package for a quasi-uniform power dissipation (51W) in the bottom chip.

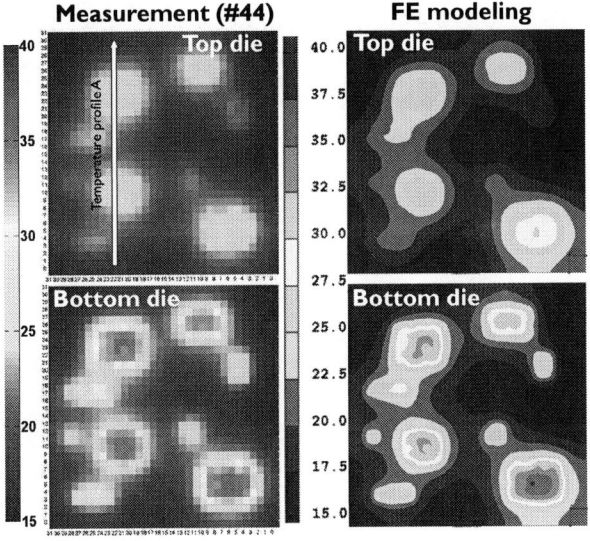

Fig. 9. Measurement and modeling results for the temperature distribution in the top and bottom chip of the 3D package for a power map with localized heat sources in the bottom chip.

For the analysis in Fig. 8, power is dissipated in all 832 thermal cells of the bottom chip. For this quasi-uniform power dissipation, the measured voltage and current on the package are 1.2V and 42.5A respectively, resulting in an actual power dissipation of 51W. Fig. 8 shows the measurement and modeling results for the temperature distribution. In the second test case, the power in the bottom chip is concentrated in localized hot spots. The measured power for this power dissipation map is 15.3W (1.5V, 10.2A). In Fig. 9, the measurement results are shown for the temperature distribution in the top and bottom chip. These test cases demonstrate that the test vehicle is capable of dissipating a high power in a programmable on-chip power map and measure the actual dissipated power. With the presented measurement set-up it is possible to measure the temperature distribution across the chip surface for both chip in the 3D package and to visualize the temperature distribution map as 32x32 thermal pixels. This temperature maps has a good spatial resolution allowing to identify the locations of the individual heater cells. A very good agreement between the modeling results and temperature measurements has been achieved for both the case of uniform power dissipation as well as for localized heat sources.

V. CONCLUSIONS

In this paper, we present an advanced stackable CMOS test chip that can used for the thermal of 3D stacked IC packages for high power applications. The test chip features 832 individually controllable heater cells to create a programmable power dissipation distribution pattern and uses a 32x32 array of temperature sensors to measure the chip temperature distribution map with a power up to 80W. 3D stacks consisting of two test chips (200µm thick top die and 50µm bottom die with TSVs) have been packaged in a bare die BGA package, allowing to accurately assess the thermal performance for different power dissipation patterns, extract the thermal self-heating and thermal coupling resistances and validate thermal modeling methods.

ACKNOWLEDGMENTS

This work was performed within the frame of imec's Industrial Affiliation Program (IIAP) on 3D System Integration. The authors would like to thank the program partners and the 3D team for their support.

REFERENCES

[1] Garrou, Ph., "Handbook of 3D-Integration: Technology and Applications of 3D Integrated Circuits", Wiley-VCH, 2008.

[2] Sikka, K.; Wakil, J.; Toy, H.; Hsichang Liu; , "An efficient lid design for cooling stacked flip-chip 3D packages," ITherm 2012, pp.606-611, May 30 2012-June 1 2012.

[3] Santos, C., et al., "System-level thermal modeling for 3D circuits: Characterization with a 65nm memory-on-logic circuit," 3D Systems Integration Conference (3DIC), 2013 IEEE International , pp.1,6, 2-4 Oct. 2013.

[4] Oprins, H., et al.; "Numerical and experimental characterization of the thermal behavior of a packaged DRAM-on-logic stack," IEEE 62nd Electronic Components and Technology Conference (ECTC), pp.1081-1088, May 29 2012-June 1 2012.

[5] De Vos, J., et al. "High density 20µm pitch CuSn microbump process for high-end 3D applications," Electronic Components and Technology Conference (ECTC), 2011 IEEE 61st. IEEE, 2011.

[6] Cherman, V. et al. "3D stacking induced mechanical stress effects". Proceedings 64th Electronic Components and Technology Conference – ECTC, May 2014, Lake Buena Vista, FL, USA. pp. 309-315.

[7] Cherman, V. et al. Effects of packaging on mechanical stress in 3D-ICs", 2015 IEEE 65nd Electronic Components and Technology Conference (ECTC), 2015, May 2015

[8] Salahouelhadj, A.; Gonzalez, M., Oprins, H.; Die thickness impact on thermo-mechanical stress in 3D packages, Eurosime 2015, 19-22 April.

[9] Oprins, H., et al., "Experimental thermal characterization and thermal model validation of 3D packages using a programmable thermal test chip" ECTC 2015 IEEE 65nd Electronic Components and Technology Conference (ECTC), 2015.

Metallization scheme optimization of plastic-encapsulated electronic power devices

Jan Ackaert, Tony Colpaert
Corporate R&D ON Semiconductors
Oudenaarde, Belgium
jan.ackaert@onsemi.com

Aditi Malik
Corporate assembly R&D
ON Semiconductors
Phoenix AZ. USA

Mario Gonzalez
Imec
Leuven, Belgium

Abstract— **Deformations of metal interconnects, cracks in interlayer dielectrics and passivation layers in combination with plastic packaging are still a major reliability concern for integrated circuit power semiconductors. In order to describe and understand the failure mechanism and its root cause, already a lot of work has been done in the past. However for the first time it is demonstrated that stress induced in the inter layer dielectric (ILD) can be the main cause of failure for a power switching device. The impact of metallization scheme on the amount of electrical failures of a power switching device was investigated in detail. It was found that with replacing a single layer metallization by a double layer metallization, the number of electrical failures reduced drastically. This improvement was achieved by two mechanisms. First, the stress induced by the molding compound (MC) through the metallization on the ILD under the metallization reduced significantly. Secondly the stress in the passivation located at the foot of the bottom metallization was relocated to the foot of the top passivation. At that location it is far less likely the passivation cracks would propagate into the ILD. These observations were confirmed by 3-D FEM simulations. The simulations enabled to locate and quantify the critical stress levels leading to electrical failures. As a result, the improved metallization scheme could lead to a distinct reduction of the principal stress at the most critical positions and, consequently, to an improvement of the reliability of the devices.**

Keywords—powerdevice; interlayerdielectric; interconnects; stress

I. INTRODUCTION

Deformations of metal interconnect, cracks in interlayer dielectrics and passivation layers in combination with plastic packaging are still a major reliability concern for integrated circuit power semiconductors In order to describe and understand the failure mechanism and its root cause, already a lot of work has been done in the past [1]–[7].

The main root cause of the issues is the different nature of the silicon chip, the leadframe and the molding compound (MC) embedding this silicon. In general the MC-plastic is softer (low young's modulus) and has a high coefficient of thermal expansion (CTE). In contradiction, the silicon is hard and has a much lower CTE. The leadframe is hard and has a high CTE. This mismatch of the material properties leads to a thermo-mechanically induced stress in operational conditions with varying thermal conditions. Worst case conditions are observed at the outer edges and in the corner regions of the silicon devices; there, the thermo-mechanical stress reaches its maximum value. Interconnects of power devices are often made of aluminum. To protect the aluminum during mechanical handling, from corrosion and to limit moisture penetration in the underlaying interlayer dielectrics (ILD), integrated circuits are coated with plasma silicon nitride (SiN). Again there is a significant mismatch between these two

materials. The aluminum is soft and has a high CTE, while the SiN is hard and brittle and has a low CTE.

Together with the wide and thick metal layers as used in general on power devices, this combination leads to cracks in the SiN. After temperature cycling stressing, cracks mainly occur at the corners and edges of the device where the thermo-mechanical stress conditions are the most extreme. (Fig. 1)

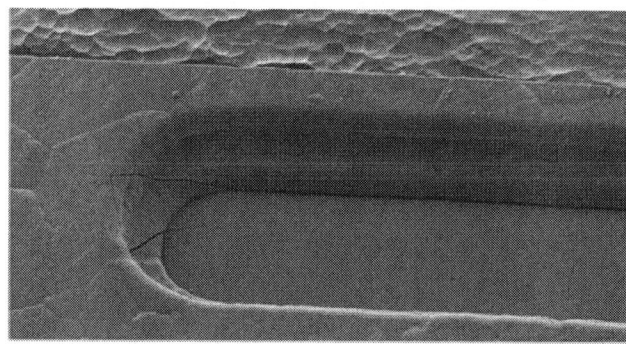

Fig. 1 Typical passivation cracks on large Al areas of a typical power IC metallization layer coated with SiN after 1000 TCs between −65°C and +175°C.

In a number of cases, cracks propagate into the ILD layers and even into the Si underneath thereby causing electrical shorts. On top of that, moisture can penetrate the IC from outside causing corrosion or other moisture related defects. Since the failure mechanism needs time to develop, these failures must be considered as a severe reliability hazard. To improve the quality and minimize the risk of reliability issues, all components, MC, passivation and interconnect including all possible interactions must be taken in account. In the 1970's a protective coating of the Al metal lines was introduced to prevent corrosion [1] and mechanical handling damage.

Thicker layers were shown to be more stable concerning cracking [5]. Also the impact of layout has been investigated [8]–[12]. This resulted in the introduction of slotted broad metal lines and of 45° lines in corner regions. However, in the case of thick metal stacks still a considerable number of cracks in the passivation and in the ILD were observed after temperature cycling (TC). So far, less is known on the impact of the stress induced by the MC on ILD under the metallization. It is part of this paper to bring more insights on this subject.

Within a power device, the stress distribution is governed by two construction features: while the top side of the chip is conventionally covered by the MC, the back side is attached to a solid Cu heat slug. Fig. 2 shows an illustration of a typical

978-1-4799-7670-6/15 $31.00 © 2015 IEEE

power device with the source connection in the center and the isolation rings on the edges.

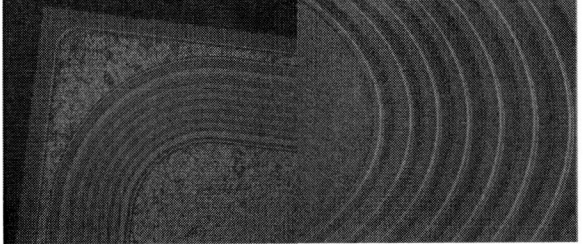

Fig. 2 Typical electronic power device with the in the center the source contact region and on the edges the isolation rings.

The main features of power ICs are thick and broad metal plates for the wiring of power transistors, which cover a major part of the chip area. The thickness of the top metallization, which is also called power metallization, lies in the range between 3.5 and 5 μm. Finite element simulation is a cost and time effective approach and is often performed to address these reliability issues. Tremendous growth in computer power has made it possible to perform large scale simulations, and hence, plenty of work has been reported in this field [13], [14].
Earlier, the impact of the metal stack, passivation layer, layout and MC has been studied in detail with the help of a finite element model (FEM). The conclusion was that the whole counterforce against the shear stress during TC, is built up by the edges of the passivation layer only [13]. In a further study, it was concluded that the shape of the metal edge profile could increase the resistance to crack formation significantly [15]. Still this was insufficient to eliminate failures completely. In that study, the impact of a trench profile on the Cu leadframe at the perimeter of the Si die is investigated in detail. The study is based on both experimental data and FEM simulations.

II. THE FINITE ELEMENT MODELING

Earlier two modeling approaches have been used successfully for resolving passivation issues by our team: Ansys and Msc.Marc. [15] [17]
ANSYS was used to create a full TO220 package model and all the assembly steps. One reliability temperature cycle was simulated to validate the experimentally found cracking locations in the passivation and the device. Stress analysis was done in the complete package during the various assembly steps and reliability temp cycle. Approximations were done to reduce the solver time though the exact metal planar layout was considered. Fig. 3 shows a half-symmetry model of the package showing the lead, die, die attach and the MC. Since all the details of the metal layers couldn't be included in the full package, submodeling technique was used where smaller models with details of metal layers, ILD and passivation layers were built. The boundary conditions on these submodels were interpolated from the full package model. Several such sub models were created at various locations through the entire die, for example, around the rounded metal corners. The results for one such model is shown in Fig. 3 and Fig. 10. Since no creep properties of the solder or any other material has been considered only one temperature cycle has been simulated.

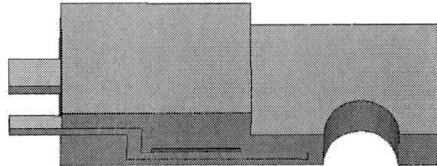

Fig. 3 Simplified cross section of the layout as taken in account by FEM simulating the thermo-mechanical stresses in the passivation layer.

Msc.Marc was used to locate stress concentration on a micro scale. The device itself was manufactured with a source-, gate- and isolation interconnects as a 4μm thick Al layer. The metallization is protected with a 1.3μm thick SiN layer. The edges of the metallization have been manufactured with a specific partially rounded profile to minimize passivation stress concentrations [15]. Simplifications in the model were carried out to reduce the solver time. Therefore, mold compound, lead frame and die attach were included in the model, however, only silicon and power metal were taken into account for the chip. Advanced global-local meshing features were needed to cope with the large package dimension and the small dimensions of the passivation structure. A view of the meshed model is shown in Fig. 4 (the passivation is partially removed to show the cross-sectional view of the model).

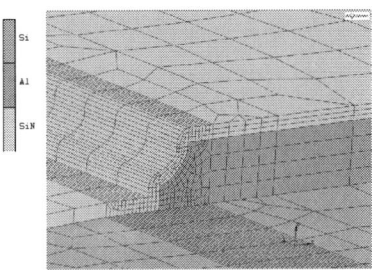

Fig. 4 Finite element model simulating the thermo-mechanical stress in the passivation layer.

In this FEM, consisting of about 100000 elements, the complete temperature cycling load (operational + environmental temperature changes) is applied and the mechanically induced stresses are calculated over the whole temperature profile. These stresses are induced by mismatch in CTE between the different materials (see Table 1). In addition to the linear properties, the aluminum is simulated as elasto-plastic material with a yield stress of 250 MPa.
All the material properties have been considered as a function of temperature. Table 1 reports the typical values at room temperature. Plasticity has been considered for the metal and the solder layers.

TABLE I.

Material	CTE (ppm/°C)	E-modulus (GPa)
Al	24	70
SiN (passiv.)	2.8	143
MC below Tg	10	24
Leadframe	16	120
Silicon	2.6	169

978-1-4799-7670-6/15 $31.00 © 2015 IEEE   143

III. THE MODEL VERIFICATION

Suo [5] and Huang et al. [6], [7] have shown that below the glass transition temperature Tg, the MC gains a high Young's modulus. Due to shrinking relative to the silicon, the MC causes a shear stress in the passivation film covering the metal plates. This shear stress τ caused by the CTE mismatch between the MC and Si is maximal near the chip edges and decreases in the direction of the chip center. It was shown that, after typically 250 temperature cycles (TCs), the yielded metal layer does not transfer the shear stress to the silicon chip any longer. The only balancing force is now established by the thin brittle passivation layer anchored at the edges of the metallization plate[13].

In order to verify the simulations, the locations of the maximal simulated stress are compared with the locations of the actual failing sites on samples that have been exposed to TC. Fig. 5 shows the location and magnitude of the stress as simulated by Ansys. Surprisingly, the simulations revealed that the stress concentrations are already present at room temperature without the elevated or reduced temperature conditions as with TC.

Fig. 5 The simulation reveals that stress the stress levels (red) are present in the oxide under the metal are significantly higher than the stress in the passivation of the metallization itself.

Fig. 6 shows picture of failure analysis on a sample after TC

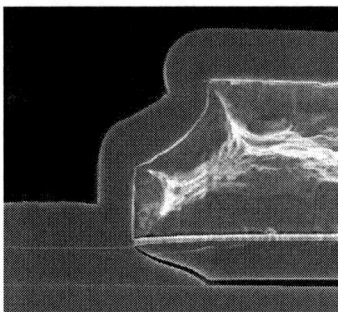

Fig. 6 Failure analysis after 1000 TC's reveals a failure in the oxide under the metal. The passivation itself remains intact.

The simulation predicts the maximal stress levels in the Inter Layer Dielectric (ILD) under the metal tracks. Comparing the location of the predicted maximal stress on Fig. 5 with the location of the cracks in the ILD on Fig. 6, it is obvious that the locations with maximal stress coincide very well with the locations of the actual ILD cracks. Since the simulation predicts so well the zones where passivation damage can be expected after thermal cycling, we can conclude that the simulated model is a reliable tool to use for enhancement of the package in order to reduce the impact on the passivation.

IV. THE EXPERIMENT

Simulations and failure analysis showed that the stress induced by the mold compound during the thermal cycling is transferred to the oxide layer under the metal. Earlier experiment on reducing the impact of stress induced by the

mold compound demonstrated the beneficial impact of introducing a sloped topography [15]. A sloped topography is distributing the stress of a wider area. This reduces the risk of stress concentrations to exceed locally the strength of the layer. Fig. 6 shows that the metal layer is already processed with a sloped edge; therefore extra measures need to be introduced to redistribute the stress induced on the edges of the metallization by the mold compound. A solution was found in introducing a second metal layer. Adding an extra metal layer provides the extra degree of freedom to limit the area with thick metal only where it is required for supporting a high electric current. In other areas where metallization is required for electric field distributions, metal thickness can be limited to a minimum. Fig.7 shows a SEM picture of a cross section of the modified double layer metallization scheme.

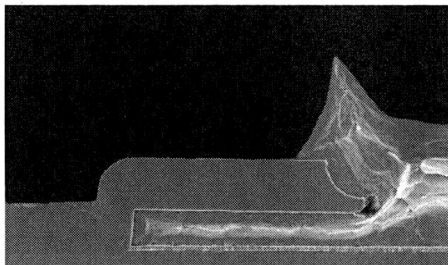

Fig. 7 In the modified metallization scheme, the 4μm thick metal layer has been split up in a 1μm thick bottom layer and a 3μm top layer with an intermediate passivation layer.

On the two types of metallization as described above, samples were verified on electrical breakdown after 24 and 168 hours. Fig. 8 shows a significant degradation of the breakdown with the single layer metal samples. Fig. 9 shows a stable breakdown distribution over the duration of the test with the single layer metal samples.

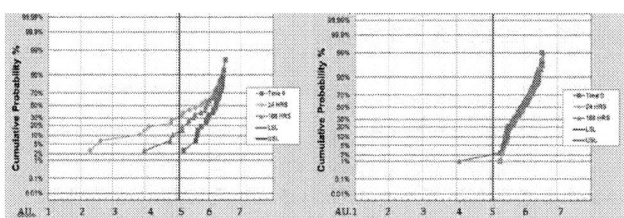

Fig. 8 The number of electrical breakdown failures as a function of the stress time for single layer metallization samples

Fig 9 The number of electrical breakdown failures as a function of the stress time for double layer metallization samples

As one can observe in Fig. 8 and 9, with the introduction of the second metal layer the number of electrical breakdown failures has been drastically reduced.

V. RESULTS AND DISCUSSION

Earlier in this paper it was discussed that during TC, the shear stress induced by the MC has to be absorbed by the edges of the passivation. The shear stress is reaching a maximum with the cold cycles of the TC [6]. In earlier work it was shown that the edge profile of the metal interconnect has a significantly impact on the sensitivity for passivation crack formation local stress concentrations [15]. With introducing the second metal layer we observe a further significant reduction electrical failures indicating that the force induced by the mold compound on the different layers of the device is reduced resulting in a drastic reduction of the failures after reliability stressing. On the isolation rings, topography was reduced from 4 to 1μm. As shown in fig. 6, in the source region, the edge of the topography was spread out over a wide band avoiding local stress concentrations.

To verify these hypotheses, FEM simulations were used to compare the maximal stress in the passivation at room temperature as well as hot and cold conditions of TC. Fig. 10 compares the max principal stresses in the device for the two metallization schemes considered. The stress has been compared at room temperature as well as for the extreme cold and hot temperature conditions of TC.

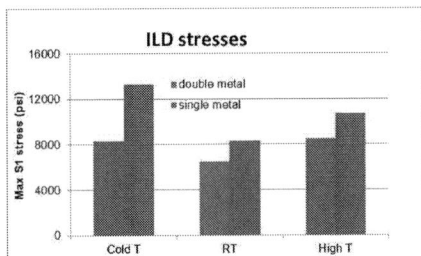

Fig. 10 The maximal stress in the device with single and double metallization at room temperature and for the extreme cold and hot temperature conditions of TC.

As can be observed in Fig. 10, for all three temperature conditions, maximal stress is reduced significantly. Earlier studies show that MC induced stress is maximal at cold TC conditions [15][16]. This is also demonstrated in this study: for the cold TC condition, an extreme stress condition for the single metal layout can be observed. With the double metal layout, this extreme stress condition is no longer present. With the double metal layout stress level at all temperature conditions have been reduced to acceptable moderate levels. This is in line with the observed improvement of the electrical results.

Next to the Ansys simulations used above, Msc.Marc, was used to locate stress concentration on a micro scale in the passivation layer on top of the bottom metal layer. In this simulation the two metallization schemes have been compared. Fig. 11 and Fig.12 show the results of this comparison for a room temperature condition.

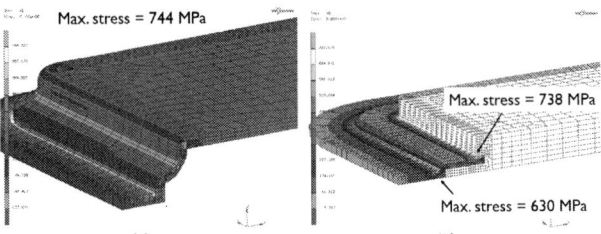

Fig. 11 With the single metal layout stress is focused in the passivation at the foot of the metallization close to the ILD.

Fig. 12 With the double metal layout, stress is distributed over foot of both metal levels avoiding a stress concentration in a single location close to the ILD.

With Msc.Marc FEM it was possible to locate and quantify the stress that is induced by the MC. For the single metal approach, stress in the passivation is concentrated at the foot of the metallization. From there cracks can easily propagate in the ILD causing severe damage to the device. With the double metal approach, stress at the foot of the bottom metal layer is significantly reduced. The main part of the stress is located at the foot of the top metal layer. At the location it is far less likely that a crack propagates into the ILD. Also this simulation is in line with the observed improvement of the electrical results.

VI. CONCLUSIONS

The impact of metallization scheme on the amount of electrical failures of a power switching device was investigated in detail. It was found that with replacing a single layer metallization by a double layer metallization, the number of electrical failures reduced drastically. This improvement was achieved by two mechanisms. First, the stress induced by the MC through the metallization on the ILD under the metallization reduced significantly. Secondly the stress in the passivation located at the foot of the bottom metallization was relocated to the foot of the top passivation. At that location it is far less likely the passivation cracks propagate into the ILD. These observations were confirmed by 3-D FEM simulations. The simulations enabled to locate and quantify the critical stress levels leading to electrical failures. As a result, the improved metallization scheme could led to a distinct reduction of the principal stress at the most critical positions and, consequently, to an improvement of the reliability of the devices.

VII. REFERENCES

[1] H. Inayoshi, K. Nishi, S. Okikawa, and N. Nakashima, "Moisture induced aluminum corrosion and stress on the chip in plastic encapsulated LSIs," in Proc. IEEE 17th IRPS, 1979, pp. 113–117.
[2] M. Isagawa, Y. Iwasaki, and T. Sutoh, "Deformation of Al metallization in plastic encapsulated semiconductor devices caused by thermal shock," in Proc. IEEE 18th IRPS, 1980, pp. 171–177.
[3] R. J. Usell, Jr. and S. A. Smiley, "Experimental and mathematical determination of mechanical strains within plastic IC packages and their effect on devices during environmental tests," in Proc. IEEE 19th IRPS, 1981, pp. 65–73.
[4] S. Okikawa, M. Sakimoto, M. Tanaka, T. Sato, T. Toya, and Y. Hara, "Stress analysis of passivation film crack for plastic molded LSI caused by thermal stress," in Proc. IEEE ISTFA, 1983, pp. 275–280.
[5] Z. Suo, "Reliability of interconnect structures," in Comprehensive Structural Integrity, vol. 8,W. Gerberich,W. Yang, I. Milne, R. O. Ritchie, and B. Karihaloo, Eds. Amsterdam, The Netherlands: Elsevier, 2002, pp. 1–61.
[6] M. Huang, Z. Suo, and Q. Ma, "Plastic ratcheting induced cracks in thin film structures," J. Mech. Phys. Solids, vol. 50, no. 5, May 2002, pp. 1079–1098,.
[7] M. Huang, Z. Suo, and Q. Ma, "Metal film crawling in interconnect structures caused by cyclic temperatures," Acta Mater., vol. 49, no. 15, Sep. 2001, pp. 3039–3049.
[8] C. G. Shirley and R. C. Blish, II, "Thin film cracking and wire ball shear in plastic dips due to temperature cycle and thermal shock," in Proc. IEEE 25th IRPS, 1987, pp. 238–249.
[9] R. C. Blish, II and P. R. Vaney, "Failure rate model for thin film cracking in plastic ICs," in Proc. IEEE 29th IRPS, 1991, pp. 22–29.
[10] S. Okikawa, T. Toida, M. Inatsu, and M. Tanimoto, "Stress analysis for passivation and interlevel-insulation film cracks in multilayer aluminum structures for plastic packaged LSI," in Proc. IEEE ISTFA, 1988, pp. 75–81.
[11] C. F. Dunn and J. W. McPherson, "Temperature cycling acceleration factors for aluminum metallization failure in VLSI applications," in Proc. IEEE 28th IRPS, 1990, pp. 252–258.
[12] Zhang GQ, van Driel WD, Fan XJ.Mechanics in microelectronics. Springer; 2006.
[13] Alpern P, Nelle P, Barti E, Gunther H, Kessler A, Tilgner R, et al. On the way to zero defect of plastic-encapsulated electronic power devices—part I to part III, vol. 9, 2009, pp. 269–295.
[14] O. Selig, P. Alpern, K. Müller, and R. Tilgner, "Thermomechanical assessment of molding compounds," IEEE Trans. Compon., Hybrids, Manuf. Technol., vol. 15, no. 4, Aug. 1992, pp. 519–523.
[15] J. Ackaert. On the impact of the edge profile of interconnects on the occurrence of passivation cracks of plastic-encapsulated electronic power devices. ICICDT April 2011.
[16] J. Ackaert. Impact of the leadframe profile on the occurrence of passivation cracks of plastic-encapsulated electronic power devices. ICICDT April 2013.

I/O thick oxide device integration using Diffusion and Gate Replacement (D&GR) gate stack integration

R. Ritzenthaler, T. Schram, M. J. Cho, A. Mocuta, N. Horiguchi, A. V.-Y. Thean
imec, Kapeldreef 75,3001 Leuven, Belgium
contact: romain.ritzenthaler@imec.be

A. Spessot, C. Caillat, M. Aoulaiche, P. Fazan
Micron Technology Belgium, imec Campus,
Kapeldreef 75,3001 Leuven, Belgium

K.B. Noh, Y. Son
Assignee at imec from SK-Hynix

Abstract— **In this work, the potential of the recently demonstrated D&GR (Diffusion & Gate Replacement, [5]) for thick oxide I/O devices integration is investigated. A D&GR integration flow compliant with EOT requirements for I/O devices is demonstrated, with no penalty with regard to HKMG Non D&GR flow in terms of short channel effects and intrinsic transistor performance. Threshold voltage tuning options from 150 up to 300 mV are demonstrated, and one preferred integration route (keeping the same work function shifters for both thin and thick oxide devices) is highlighted. Finally, it is also shown that HKMG I/O devices (D&GR and non D&GR) do not suffer from reverse narrow gate width effects.**

Keywords— *Al_2O_3 capping layers, Diffusion and Gate Replacement (D&GR), DRAM periphery transistors, High-k, I/O devices, Metal gate, Mg capping layers.*

I. INTRODUCTION

The expansion of mobile electronics use in our daily lives and the onset of the ''Internet of Things'' (IoT) era highlight the need for low power devices. For tablets and smartphones, mobile DRAM is used with the objective of reducing the overall power consumption (mostly through a diminution of the supply voltage) while keeping acceptable DRAM performance [1]. The requirements of low leakage/low power performance in mobile DRAM are forcing the adoption in the DRAM periphery transistors of characteristics already used by logic devices: High-k dielectrics combined with Metal Gate (HKMG) [2]. DRAM periphery transistors must also be compatible with DRAM process specifications, which implies a process cost as low as possible and resistance to an extra-long "DRAM Anneal" (DA). Such anneal mimics the high thermal budget steps required by the fabrication of the DRAM memory elements (storage capacitor, access transistor, Back End of Line (BEOL)), and should be seen as additional constraint for periphery transistor design.

In order to meet the threshold voltage (V_{TH}) targets, work function shifter (eWF) species are commonly used in the metal gate [3]. A solution for DRAM peripheral transistors based on HfO2 coupled with Al_2O_3 capping for PMOS devices and with a TiN/Mg/TiN stack together with As Ion implantation for NMOS has been demonstrated [4]. Recently, another approach called D&GR ("Diffusion and Gate Replacement") has also

been proposed [5]. Such an innovative approach allows to obtain similar gate stack heights and has been demonstrated for thin oxide transistors. It consists in the deposition of Mg- and Al-based layers into the gate stacks of N-and P-MOS side (**Fig.1.1** and **Fig. 1.2**), followed by an anneal to drive V_{TH} shifters into the High-k (**Fig. 1.3**). Layers down to the High-k are then removed (**Fig. 1.4**), which is followed by the redeposition of a fresh metal gate (**Fig. 1.5**).

In this work, the device options for D&GR thick oxides (I/O) integration are analyzed.

II. DEVICE INTEGRATION

After STI formation and well doping, an interfacial layer is grown (using In Situ Steam Generation). The physical thickness of this layer is about 1nm for thin oxide devices, and 5nm for I/O devices. A 2nm HfO_2 high-k layer is subsequently deposited, followed by the D&GR module (a complete description of process window and process conditions can be found in [5]). Gate stacks for NMOS and PMOS are completed by a Poly-Si electrode.

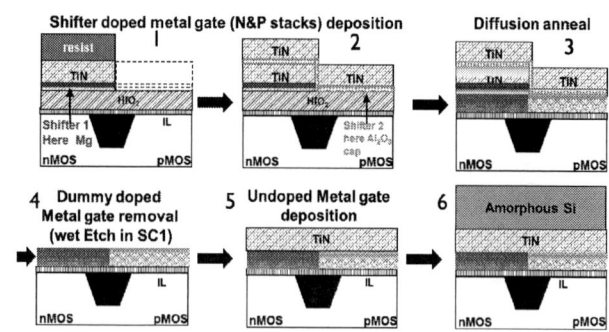

Fig. 1: Illustration of D&GR gate stack module. A Mg-based TiN sandwich, followed by an Al_2O_3 cap layer are used for N and PMOS V_{TH} shifting, respectively.

After gate patterning, junctions are formed with lightly doped drain (LDD), Halo, Spacer deposition, heavily doped drain (HDD) implantations and activation. After the Source/Drain silicidation, the DRAM anneal is added (thermal

978-1-4799-7670-6/15 $31.00 © 2015 IEEE

budget in the 600°C to 750°C, ≥1h range). The process flow is completed with conventional BEOL.

Regarding I/O devices, different integration strategies were conceived and evaluated: keeping the same eWF shifters as for the thin oxide transistors (**Fig. 2.a**), using a P-type eWF shifter for the NMOS I/O transistor (**Fig. 2.b**), or even removing doped layers before the diffusion anneal for thick oxide devices (**Fig. 2.c**; this flow necessitates one additional mask).

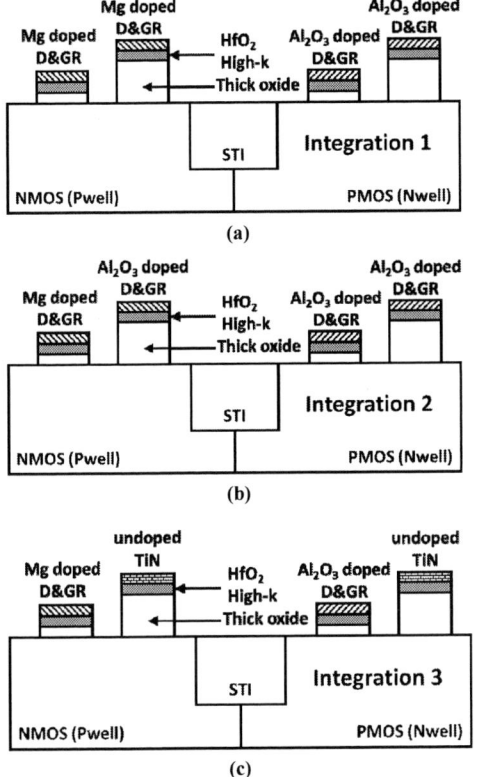

Fig. 2: Schematics of the 3 flows evaluated in this work: identical dopants for I/O transistors (a), cross-use of dopant (b), and removal of V_{TH} shifters (c).

III. DEVICE PERFORMANCE

Capacitances measured in accumulation are shown in **Fig.3** (the 3 integration options highlighted in **Fig.2**, with undoped High-k Metal Gate and Non D&GR HKMG as references). For both devices in N-well and P-wells, EOT values are around 5nm with little differences from one integration scheme to the other. As expected, the use of thin work function shifter layers (Al_2O_3 capping on top of high-k for PMOS, MgO sandwiched in TiN layers for NMOS) is not changing the EOT of I/O devices. Similarly, the D&GR integration induces the deposition of a fresh metal gate at the end of the gate stack module, and do not yield a large variation of EOT if the drive-in anneal thermal is kept under an upper boundary [5] (likely due to a regrowth of the interfacial layer for high temperature diffusion anneals). Therefore, the D&GR integration and its various flavours are not showing dramatic EOT changes.

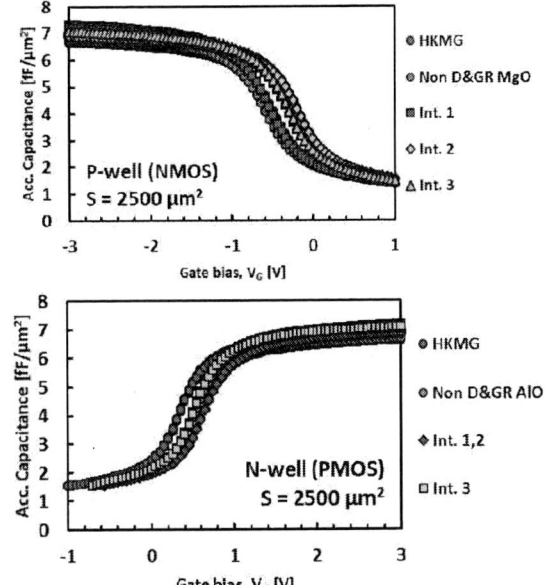

Fig. 3: Accumulation capacitance for undoped High-k Metal Gate, Non D&GR HKMG, and the 3 integration options highlighted in **Fig.2**.

Already apparent in the flat band voltages values (**Fig. 3**), there are however large variations of long channel threshold voltage from one integration scheme to the other (**Figs. 4-5** for NMOS and PMOS devices, respectively).

For NMOS devices, a threshold voltage of 1V is obtained with HKMG without dopants. Using a MgO work function shifter sandwiched between TiN layers (without gate removal, i.e. a Non D&GR integration), a 200 mV V_{TH} shift is demonstrated despite the use of a thick interfacial layer. Regarding D&GR integration, a wide range of threshold voltage values are obtained depending on the integration scheme (**Fig. 4**): adding a layer of PMOS shifters (Al_2O_3 capping layers, "integration 2") is dramatically raising V_{TH}, and removing doped layers before the drive-in anneal ("integration 3") is yielding a small V_{TH} penalty compared to the undoped HKMG reference. On the other hand, the MgO based D&GR integration scheme ("integration 1") is yielding a threshold voltage practically identical to the Non D&GR integration (i.e. leaving the capping layer in place).

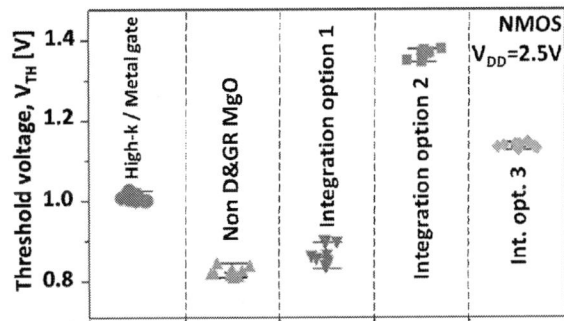

Fig. 4: NMOS Long channel threshold voltage for undoped High-k Metal Gate, Non D&GR (with MgO), and the 3 integration options highlighted in **Fig.2**. V_{DD}=2.5V.

Regarding PMOS devices, a threshold voltage of 1V is obtained with HKMG without capping layers (**Fig. 5**). Similarly to the NMOS case, using an Al_2O_3 capping layer (without gate removal, i.e. a Non D&GR integration), a 150mV V_{TH} shift is demonstrated despite the use of a thick interfacial layer. A similar threshold voltage is obtained by removing the layers ("integration 3"; the physical model involved is still being investigated). The best threshold voltage is obtained by using Al_2O_3 capping D&GR (("integration 1/2"), with 300 mV V_{TH} lowering compared to HKMG without capping layers.

Overall, the integration order (PMOS vs. NMOS first) is based on the risk of residual dopant potentially left on the HfO2 after the initial removal step (**Fig.1**, step 2), which will cause a V_{TH} shift opposite to the desired polarity and increased variability fluctuation. As capping layers leave residues [5], a NMOS first using MgO sandwiched between two TiN layers is preferred in the flow. In light of the obtained threshold voltages, the best integration route regarding associated I/O devices is therefore to use the same shifters for thick and thin oxides. It is also noteworthy that threshold voltage modulations are obtained with the D&GR flow even for thick oxides.

Fig. 5: PMOS Long channel threshold voltage for undoped High-k Metal Gate, Non D&GR (with Al_2O_3), and the 3 integration options highlighted in Fig.2. V_{DD}=2.5V.

Due to identical EOT obtained with the various gate stack integrations, Intrinsic Transistor Performance (ITP) can be directly compared. For both NMOS and PMOS devices (**Fig.6**), no penalty is seen going from undoped HKMG gate stack toward HKMG with eWF shifters (respectively TiN/MgO/TiN gate stack for NMOS and Al_2O_3 capping on top of High-k for PMOS) and the various flavours of D&GR integration (**Fig. 2**). It has been reported in previous work [5] that for thin oxide device work function shift was obtained at the cost of degraded high field mobility. Identical ITP performance suggests that in the case of I/O devices, short channel mobility is not degraded by D&GR integration (using identical junction schemes, it is unlikely for external resistance R_{EXT} to be modified by the gate stack module).

Regarding short channel margin, DIBL values are also comparable regardless of integration scheme (**Fig. 7**). Drive-in anneal is done during the gate stack module, i.e. before the junction implantation. There is therefore no penalty in terms of short channel margin using the D&GR integration scheme.

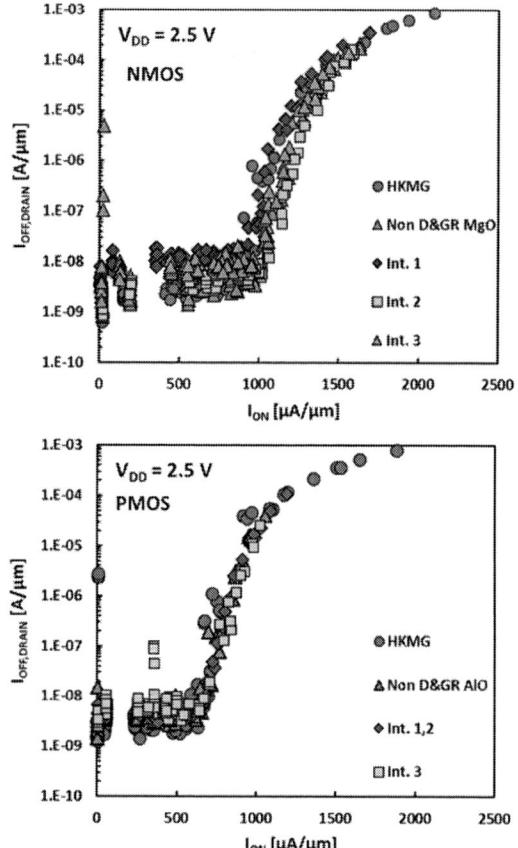

Fig. 6: Intrinsic transistor performance (NMOS and PMOS) for undoped Higk-k Metal Gate, Non D&GR, and the 3 integration options highlighted in **Fig.2**.

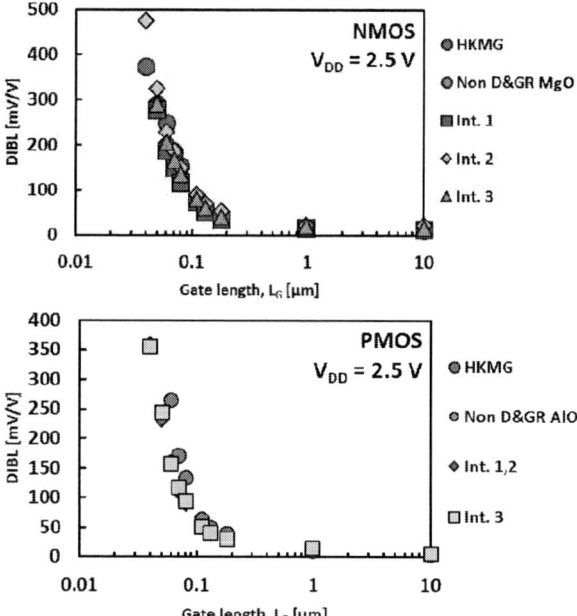

Fig. 7: Short channel effects (DIBL) for undoped High-k Metal Gate, Non D&GR, and the 3 integration options highlighted in **Fig.2**.

Previously published work has shown that Mg-incorporated HKMG gate stacks were prone to reverse narrow width effect [6]. However, no such trends were seen for I/O devices considered in this work (**Fig.8**). For NMOS devices, there is little difference between HKMG without shifters and with MgO (both D&GR and non D&GR). There is also no difference with gate stacks featuring Al_2O_3 capping layer on top of the MgO, highlighting the fact that there is no diffusion of Al_2O_3 through TiN/MgO/TiN layer on the NMOS side. This gate stack configuration corresponds to the situation before the dummy doped metal gate removal, i.e. **Fig. 1.2**. Therefore, there is no indication of a negative effect of the Al_2O_3 capping layer on top of the gate stack for NMOS device. For PMOS devices, a similar situation is observed: there is no reverse narrow width effect occurring and no difference between HKMG without shifters and Al_2O_3 capping layers (both D&GR and non D&GR). Furthermore, the position of the capping layer (below or above high-k) is not yielding any difference in terms of threshold voltage and reverse narrow width effect.

The model presented in [6] explains enhanced narrow width effect with Mg based capping layers by the draining effect of the STI, which decreases the effectiveness of eWF shifters for narrow devices. A possible model to its absence in I/O devices (regardless of D&GR integration) is that the STI draining is considerably hampered by the thick SiO_2 interfacial layer used in this work.

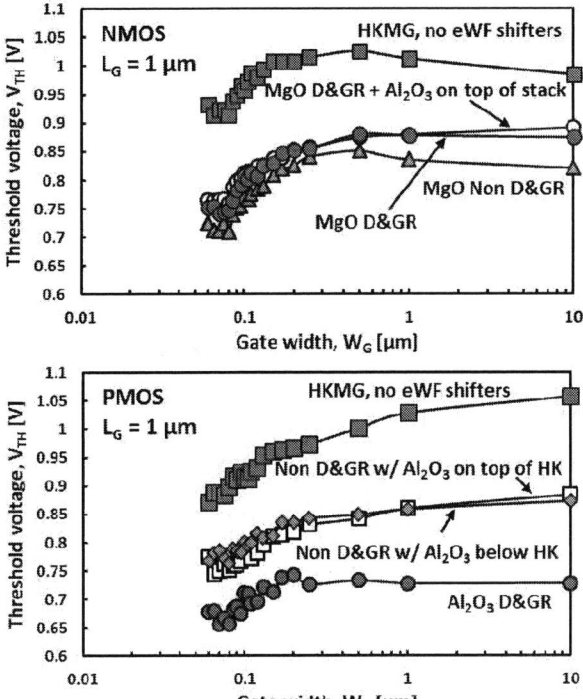

Fig. 8: Threshold voltage V_{TH} vs. gate width W_G for NMOS/PMOS and various I/O integration (D&GR, Non D&GR with work function shifter, High-k without work function shifters). It is shown that the use of capping layers (both D&GR and non D&GR) do not yield to reverse narrow gate width effects with I/O devices.

IV. CONCLUSIONS

In this work, the potential of recently demonstrated D&GR (Diffusion and Gate Replacement, [6]) for thick oxide I/O devices is investigated. A D&GR integration flow compliant with EOT requirements for I/O devices is demonstrated, with no penalty with regard to HKMG Non D&GR flow in terms of short channel effects and intrinsic transistor performance. Threshold voltage tuning options from 150 up to 300 mV are demonstrated with thick oxide D&GR module. One integration route, keeping the same work function shifters for both thin and thick oxide devices, is leading to the best results in terms of threshold voltage lowering. It is also shown that HKMG I/O devices (both D&GR and non D&GR) do not suffer from reverse narrow width effects.

REFERENCES

[1] M. A. Siddiqi "Dynamic RAM: Technology Advancements", CRC press, SBN 978-1439893739, 2012.

[2] S. -Y. Cha, "DRAM Technology - History & Challenges," IEDM 2011 Short Course, 2011.

[3] M. M. Frank, "High-k / Metal Gate Innovations Enabling Continued CMOS Scaling," in Proceedings of ESSDERC 2011, pp. 25-33, 2011.

[4] R. Ritzenthaler, T. Schram, A. Spessot, C. Caillat, H. J. Na, T. Kauerauf, B. Douhard, A. Nazir , S. A. Chew, A.P. Milenin, E. Altamirano-Sanchez, G. Schoofs, J.Albert, F. Sebai, E.Vecchio, V. Paraschiv, M. Aoulaiche, M. J. Cho, W. Vandervorst, S.-G. Lee, P. Fazan, N. Collaert, N. Horiguchi, and A. V.-Y. Thean, "A Simple, Ultra low leakage High-Thermal-Budget-Tolerant HKMG integration for 2x node Low Power DRAM compatible CMOS platform", IEEE Transactions on Electron Devices, vol. 61 no. 8, pp. 2935 – 2943, 2014.

[5] R. Ritzenthaler, T. Schram, A. Spessot, C. Caillat, M. Cho, E. Simoen, M. Aoulaiche, J. Albert, S. A. Chew, K. B. Noh, Y. Son, P. Fazan, N. Horiguchi, and A. Thean, "A new high-k/metal gate CMOS integration scheme (Diffusion and Gate Replacement) suppressing gate height asymmetry and compatible with high-thermal budget memory technologies," International Electron Devices Meeting (IEDM), 2014.

[6] T. Morooka, M. Sato, T. Matsuki, T. Suzuki, K. Shiraishi, A. Uedono, S. Miyazaki, K. Ohmori, K. Yamada, T. Nabatame, T. Chikyow, J. Yugami, K. Ikeda, Y. Ohji, "Suppression of anomalous threshold voltage increase with area scaling for Mg- or La-incorporated high-k/Metal gate nMISFETs in deeply scaled region", VLSI Technology Symposium proceedings, 2010.

Characterization of Onset Tunneling Voltage (V_{onset}) Walkout in High-Voltage Deep Trench Isolation on SOI

Thuy Dao[1], *Senior Member, IEEE*, Mu-Ling Ger[2], and Jiangkai Zuo[2]

[1]*Freescale Semiconductor, Inc., Austin, Texas, USA*
[2]*Freescale Semiconductor, Inc., Tempe, Arizona, USA*

EMAIL: THUY.DAO@FREESCALE.COM

I. INTRODUCTION

Deep trench isolation (DTI) with "walkout" onset tunneling voltage (V_{onset}) can cause serious confusion for performance enhancement and process optimization in technology development. The ordinary breakdown voltage (BV) "walkout" phenomenon occurs when a premature avalanche breakdown injects high energy carriers into an oxide so that subsequent stress may causes an increase in oxide breakdown voltage [1]. However, the increase in the V_{onset} with multiple I-V sweeps for our DTI structures are of very different origins. This is the first report on the characterization of V_{onset} "walkout" in HV deep trench isolation (DTI) on SOI.

II. Deep Trench Test Methodology

HV devices build on SOI substrate can be surrounded by a deep trench ring providing complete isolation on all sides of the device (Fig. 1). Lining the deep trench is a thick oxide film and a poly plug that connected to substrate to serve as a ground plane. During operation the device drain can exceed 100V with respect to the grounded poly plug such that all of the oxide surrounding the device must be able to withstand these high voltages.

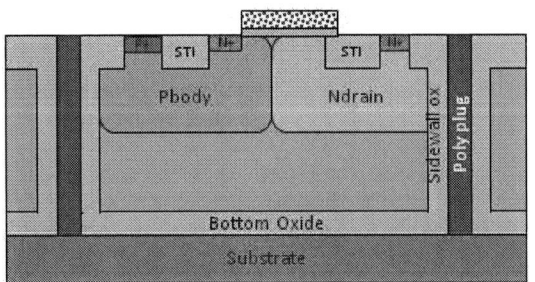

Figure 1: HV Device Cross Section

A test structure consisting of a large array of deep trench rings connected in parallel was created (Fig.2). This structure is tested by ramping the internal body positive or negative with respect to the poly plug [2]. After the initial test, the devices may pass current at a voltage lower than the designed

V_{onset}. But with subsequent stress and passage of current, the V_{onset} then "walkout" to its designed value.

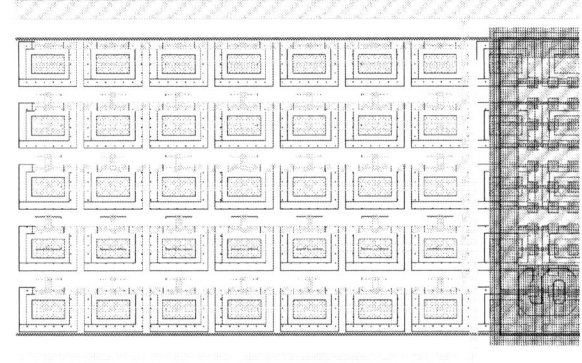

Figure 2: Large Array of Deep Trench Rings

III. Deep Trench Current-Voltage (I-V) Characteristics

The I-V characteristics of the DTI were measured with an Agilent 4156C Semiconductor Parameter Analyzer. A typical example of the repeated I-V sweeps (from 0V to 150V) is shown in Figure 3. There are several phenomena readily observable in this plot. First of all, the 1st sweep has a distinctly different I-V curve than the rest of the sweeps. Secondly, all the curves tend to converge at the final current level from the 1st sweep at the largest bias voltage (i.e. 150V in this case). Thirdly, the onset voltage (V_{onset}) of tunneling (where the tunneling current rises sharply above the noise floor) increases significantly and tends to stabilize after the 1st sweep.

All these observations lead to the conclusion that the I-V curve from the 1st sweep would not provide the true V_{onset} for these DTI structures. This could be challenging if one is to compare V_{onset} values (without looking at IV curves) to optimize the DTI process modules for improving BV. What make things even more interesting is that the seemingly stabilized V_{onset} after the 1st sweep could be enhanced by the final (largest) bias voltage from the sweeps. Figures 4 and 5 illustrate the increased V_{onset} with higher final sweep voltage (about 130V for 0→170V sweeps and about 140V for 0→180V sweeps).

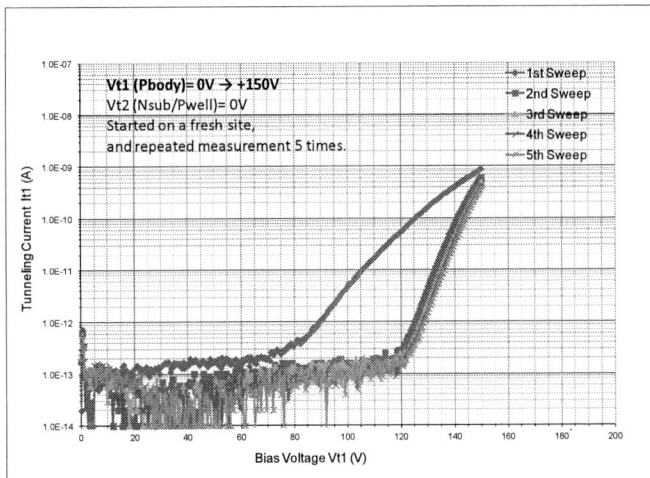

Figure 3: I-V Characteristics of Repeated Sweeps (from 0V to 150V) of the DTI Rings

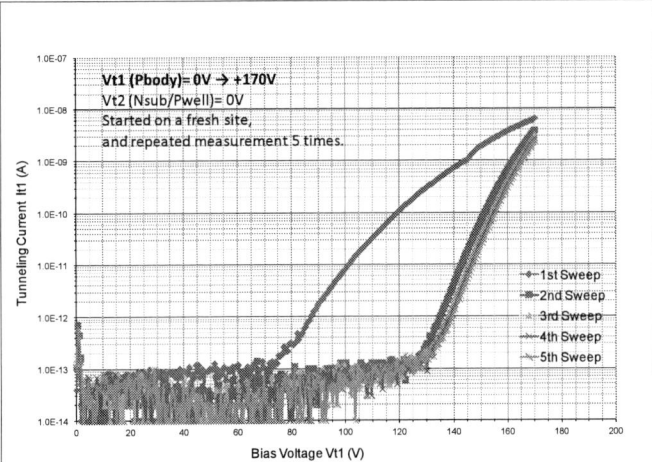

Figure 4: I-V Characteristics of Repeated Sweeps (from 0V to 170V) of the DTI Rings

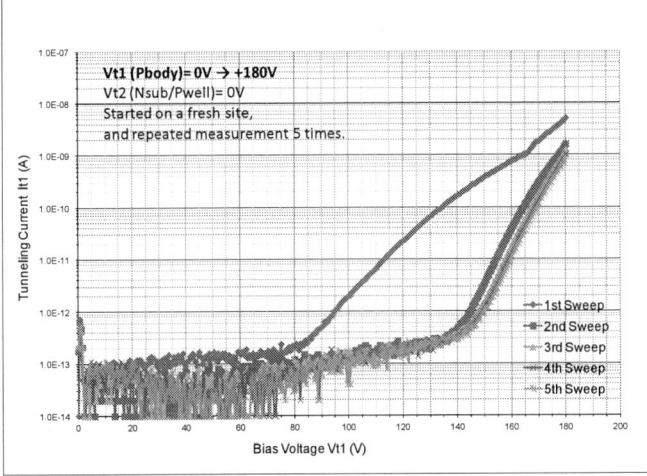

Figure 5: I-V Characteristics of Repeated Sweeps (from 0V to 180V) of the DTI Rings

As was observed and illustrated on the breakdown characteristics of thick silicon oxide [3], the exponential portion (straight line in the semi-log plots Figs.3-5) could be explained well by the theoretical F-N (Fowler-Nordheim) tunneling current. In addition, the 2nd I-V sweep on thick oxide showed a smaller and smaller V_{onset} after the 1st I-V sweep reached 8MV/cm or higher. Thus the authors concluded that the thick SiO_2 film (75~100nm) begins to be damaged at about 8MV/cm stress.

However, The I-V sweeps from these DTI structures showed very different characteristics. From Figs. 3-5, the V_{onset} were actually increasing with repeated sweeping. An indication that the DTI oxide was not damaged. As to the lower V_{onset} and higher tunneling current from the 1st sweep, it may be caused by the electron charge traps in the thick oxide. This might be the reason that subsequent tunneling current was suppressed after the 1st I-V sweep (filling the charge traps). This might also explain for the higher voltage (energy) required for filling the more distant traps.

However, with higher I-V sweep voltages, the DTI thick oxide would eventually started to be damaged. This can be seen in Figure 6. After the 1st sweep to 200V, the subsequent sweeps exhibited three exponential regions. The 1st region (from 80V to 160V) showed a steady increase in the tunneling current. The 2nd region (from 160V to 190V) was the F-N tunneling region. And the 3rd region exhibited a near vertical increased in tunneling current (indicating a breakdown close by). If we compared the region between 80V and 120V (after the 1st sweep) in Figs. 3-6, it was clear that the Fig. 6 showed a relatively high level of leakage current (after the 1st sweep to 200V). In addition, there was a subtle cross-over between the 2nd sweep and the 5th sweep at 180V. Both of these were early symptoms of degradation of the DTI oxide. This may seem to limit the DTI capability to be under 200V (with initial DTI process module). However, this still provided a great margin for the 90V technology requirement.

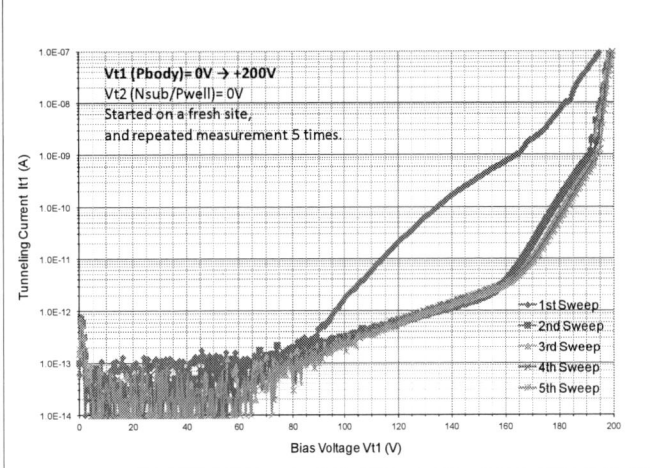

Figure 6: I-V Characteristics of Repeated Sweeps (from 0V to 200V) of the DTI Rings

To avoid any damage to the DTI oxide while still getting the "true" V_{onset}, we have implemented a test procedure to sweep the current first (for filling traps) then followed by the 2nd sweep to record the V_{onset}. This new test methodology provided us with a way to compare the V_{onset} and to optimize the DTI process modules. The test procedures and test results will be discussed in details later in this paper.

IV. Deep Trench Module Processing

The deep trench (DT) module is processed prior to gate transistor formation, providing 'portability' to a next technology generation.

The DT is dry etched into the silicon of SOI substrate stopped on BOX layer. DT etch profile is slightly sloped to enable ISD poly to fill DT without void. This DT is then cleaned and a thermal liner oxide is grown followed by a conformal deposited oxide to cover the sidewalls as well as the bottom of the trench. In this report, the liner oxide step was optimized to improve V_{onset} walkout. Wafers were split at thermal oxide liner step between the baseline dry thermal oxide at 1000^0C and the 950^0C and 1000^0C wet oxide processes.

V. Test Results and Summary

After the process optimization splits, wafers from each split group were tested. First they went through a current sweep ($0nA \rightarrow 2nA$) for filling the charge traps then followed by a moderate I-V ($0V \rightarrow 150V$) sweep across the wafer to compare the DTI performance from each split. The results are shown in Figs.7-9.

From the plots, it might not be obvious how much improvements were observed with the new Wet Ox processes since the tunneling currents just barely risen above the noise floor at 150V with the new processes. But one thing for certain, the tunneling currents at 150V were much smaller with the new Wet Ox processes than with the baseline process.

To obtain the "true" V_{onset} for the DTI with new Wet Ox processes, an additional I-V sweep ($0V \rightarrow 200V$) on the same wafers with the new processes was carried out. (It was later verified that the initial ($0nA \rightarrow 2nA$) sweep was not enough for charge trap filling to obtained the "true" V_{onset} for DTI with new Wet Ox processes.)

With the additional ($0V \rightarrow 200V$) I-V wafer mapping in Figs.10-11, it was observed that not only the V_{onset} Voltages increased significantly, the F-N tunneling currents across the wafer also had a tighter distributions. Comparing Fig.10 and Fig.11 closely, even the minor improvement was discernible by increasing the Wet Ox temperature ($950^\circ C$ to $1000^\circ C$).

In conclusion, a test methodology was developed to measure the "true" V_{onset} of the DTI structures. And the new DTI process module was optimized and demonstrated to have significant improvement on Vonset (>140V) which is more than sufficient for the 90V technology.

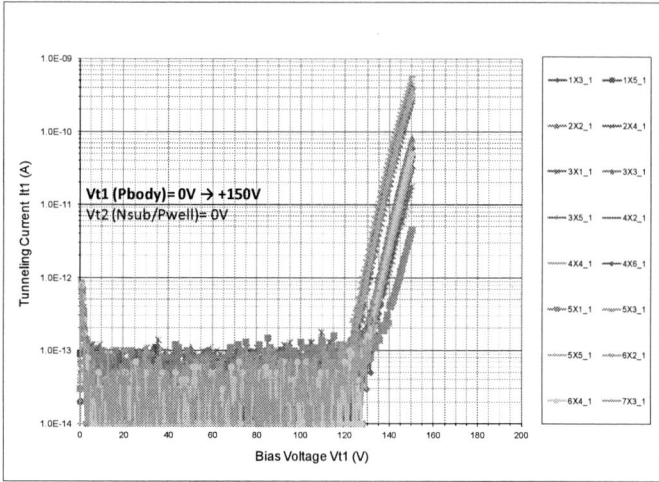

Figure 7: I-V (from 0V to 150V) Wafer Mapping of the DTI Rings (with baseline process)

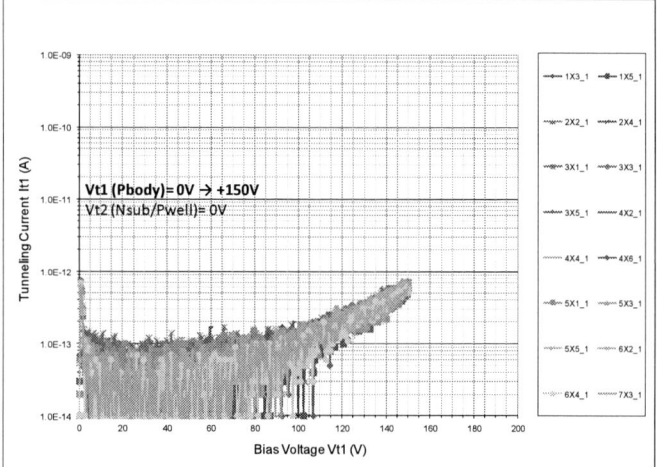

Figure 8: I-V (from 0V to 150V) Wafer Mapping of the DTI Rings (with 950°C Wet Ox process)

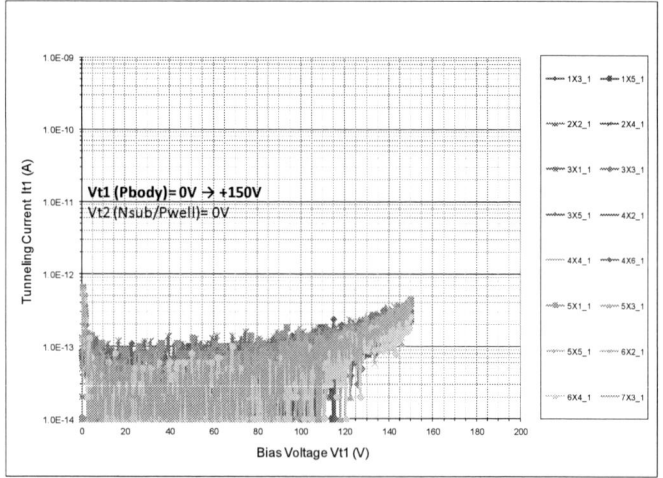

Figure 9: I-V (from 0V to 150V) Wafer Mapping of the DTI Rings (with 1000°C Wet Ox process)

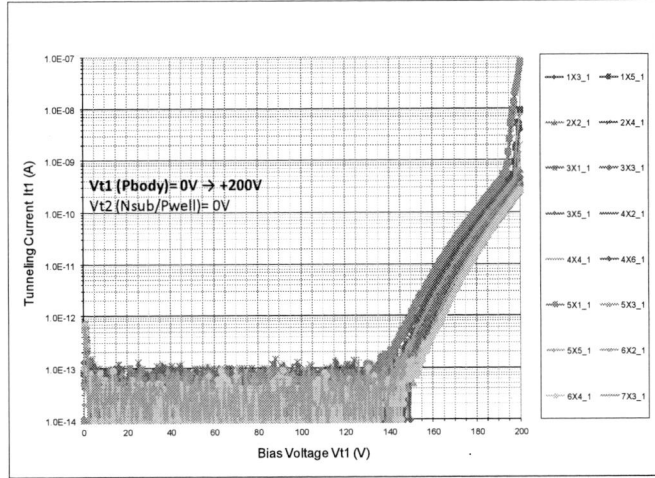

Figure 10: I-V (from 0V to 200V) Wafer Mapping of the DTI Rings (with 950°C Wet Ox process)

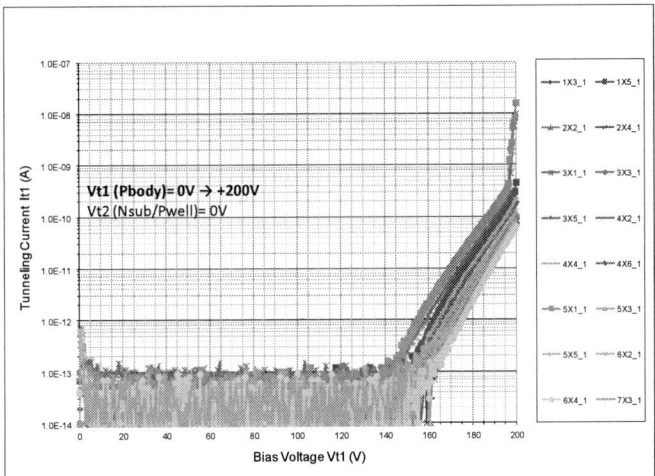

Figure 11: I-V (from 0V to 200V) Wafer Mapping of the DTI Rings (with 1000°C Wet Ox process)

Acknowledgment

The authors thank the process engineers at ATMC for their process development support, the SEM lab engineers for analysis supports, and Mel Miller for his management support.

References

[1] Yuk L. Tsang and John M. Aitken, "Junction Breakdown Instabilities in Deep Trench Isolation Structures". IEEE Transactions on Electron Devices, Vol. 38, NO. 9, September 1991, pp. 2134 - 2138.

[2] Thuy Dao, Todd Roggenbauer, Gordon Boyd, "Improved Deep Trench Isolation Breakdown Voltage for SmartMOS", International Conference on IC Design & Technology 2013 - ICICDT2013, pp. 101 – 103.

[3] K. Nakamura et al., "An Observation of Breakdown Characteristics on Thick Silicon Oxide". Proceedings of 1995 International Symposium on Power Semiconductor Devices & ICs, pp. 374 - 379.

Integrated Front-End/Back-End Simulation of Electromagnetic Fields, Lorentz Force Effects and Fast Current Surges in Microelectronic Protection Devices

Wim Schoenmaker** and Philippe Galy [†]

I. INTRODUCTION

The singled-out purpose of electrostatic discharge (ESD) devices is that these devices should protect electronic circuitry against fast-transient voltage/current spikes. Although the overall signal variation occurs within a nanosecond, the corresponding currents can ramp up to a multi-Ampere level. Fast varying current patterns give rise to equally fast varying induced magnetic fields being proportional to the time rate of change of the current and as a consequence a substantial part of the electric fields can be attributed to the variation of the induced magnetic fields and the full electromagnetic picture is required for understanding the effects of fast-transient input signals. A similar reasoning can be done for high-power switch devices. Continuing the reasoning along these lines: when the fields are varying sufficiently fast, both the induced magnetic fields and electric fields will ultimately have such large components such that the current flow is controlled by both fields. In particular, in semiconductors, the Hall coefficient is directly related to the carrier mobility and therefore *self-induced* Lorentz force modifications could become appreciable. So far, no simulation tools were available to address these concerns and strictly speaking, without actually computing these effects, we have no clue if it is justified to ignore these subtleties all together or that these effects are really a concern. Recently we have set up the full calculation scheme to address these concerns. [1], [2]. Using the corresponding software implementation extended with appropriate field viewing facilities we can now study in great detail the electromagnetic dynamics of fast transient phenomena. We present an implementation that allows the computation of these self-induced electromagnetic field effects for fast transient phenomena. In general, the key ingredients are (1) the semiconductor device equations (2) modifications thereof to account for the Lorentz force (3) the Maxwell equations to compute the electromagnetic fields. All this is done in the time domain since in the frequency domain the "small-signal" analysis up front excludes high current/voltage signals at the ports. The complementary harmonic approach was also constructed The latter is based on a small-signal

*Corresponding author.
* The author is with Magwel NV, Leuven, Belgium. E-mail: wim.schoenmaker@magwel.com
[†]The author is with EPS/TR&D, ST Crolles, France. E-mail: philippe.galy@st.com

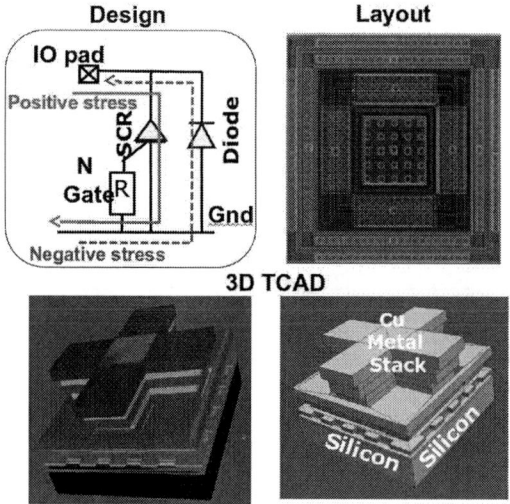

Fig. 1: Design, layout and 3D TCAD view of the ESD protection device.

analysis and consider all degrees of freedom as 'phasors' and an application is given here. We emphasize that the set up is done is such a way that metals, insulators and semiconductors all may be present together in a single problem set up. This requires special attention for constructing the interface conditions between the different materials. However, the benefit of this set up is that now it is possible to address the front-end and the back-end part of the device in an holistic or integrated approach. If back-end current crowding effects are triggered by magnetic fields effects they will impact the current densities inside the front-end part of the design. Such effects can now be studied in detail. We apply this setup to compute the response of an ESD protection device, (see Fig. 1) submitted to a sub-nanosecond transient signal rising up the several amperes. The system of equations that has to be solved corresponds to ~ 250.000 degrees of freedom.

II. TIME-DOMAIN FORMULATION OF THE INTEGRATED FRONT-END/BACK-END SIMULATION

The starting point for capturing electromagnetic phenomena is the system of equations given by Maxwell :

$$\nabla \cdot \mathbf{D} = \rho, \quad \nabla \cdot \mathbf{B} = 0 \tag{1a}$$

$$\nabla \times \mathbf{E} = -\partial_t \mathbf{B}, \quad \nabla \times \mathbf{H} = \mathbf{J} + \partial_t \mathbf{D} \tag{1b}$$

978-1-4799-7670-6/15 $31.00 © 2015 IEEE

$$\mathbf{D} = \epsilon \mathbf{E}, \quad \mathbf{B} = \mu \mathbf{H} \tag{1c}$$

where $\mathbf{D}, \mathbf{E}, \mathbf{B}, \mathbf{H}, \mathbf{J}$ and ρ are the displacement field, electric field, magnetic induction, magnetic field, free current density and charge density, respectively. As a first step we convert these equations using the scalar or Poisson potential V and the vector potential $\mathbf{A}$ in order to ease the connection with compact modeling. Next we apply the current continuity equations for the holes and electrons in semiconducting domains. A new variable, the pseudo-canonical momentum

$$\mathbf{\Pi} = \partial_t \mathbf{A} \text{ with } \mathbf{B} = \nabla \times \mathbf{A} \ , \quad \mathbf{E} = -\nabla V - \mathbf{\Pi} \tag{2}$$

is used to avoid the second-order time derivative. The complete system of equations is then laid out in (3) utilizing a generalized "de Mari" scaling scheme.

$$
\begin{cases}
\frac{1}{\nu} \nabla \cdot [\varepsilon_r (-\nabla V - \mathbf{\Pi})] - \rho = 0, \quad \rho = p - n + N_D \\
\frac{1}{\nu} \nabla \cdot [\varepsilon_r (-\nabla \partial_t V - \partial_t \mathbf{\Pi})] + \begin{Bmatrix} \nabla \cdot [\sigma (-\nabla V - \mathbf{\Pi})] & \text{metal} \\ \nabla \cdot \mathbf{J}_{\text{sc}} & \text{semi} \end{Bmatrix} \\
\qquad\qquad\qquad\qquad\qquad = 0
\end{cases}
\tag{3a}
$$

$$\nabla \cdot \mathbf{J}_n - \partial_t n - R(n, p) = 0 \tag{3b}$$

$$\nabla \cdot \mathbf{J}_p + \partial_t p + R(n, p) = 0 \tag{3c}$$

$$\partial_t \mathbf{A} - \mathbf{\Pi} = 0 \tag{3d}$$

$$
-K\varepsilon_r (-\partial_t \mathbf{\Pi} - \nabla \partial_t V) - K\nabla (\varepsilon_r \partial_t V)
$$
$$
+ [\nabla \times (\nabla \times \mathbf{A}) - \nabla (\nabla \cdot \mathbf{A})]
$$
$$
- K\nu \begin{Bmatrix} \mathbf{J}_c & \text{metal} \\ \mathbf{J}_{\text{sc}} & \text{semi} \end{Bmatrix} = 0 \tag{3e}
$$

In here, $\mathbf{J}_c$ is the conduction current in metal and $\mathbf{J}_{\text{sc}} = \mathbf{J}_n + \mathbf{J}_p$ is the total semiconductor current. K and ν are two dimensionless constants of the scaling method. In the scaling scenario where λ_L and λ_T are the natural units of length and time we find that $K = \mu_0 \epsilon_0 \lambda_L^2 / \lambda_T^2$. For example: for a small-size microelectronic device with a typical length of about one micron and a switching frequency around five GHz, we may set $\lambda_L = 10^{-6}$ meter and $\lambda_T = 10^{-10}$ second. Getting the equations written down in a scaled version allows us to evaluate the relative weight of the various term and this may already give an indication whether or not one may drop out some variables or terms. For above choices we find that $K = 1.112649 \times 10^{-09}$. In order to determine ν we must select a third characteristic variable, e.g. conductance, diffusion length, or carrier density. For example, if we choose the density as 10^{16} m^{-3} as a third natural parameter then $\nu = 6.999487 \times 10^{-3}$.

III. Current densities with Inclusion of the Lorentz Force

The equations (2) and (3) can only be solved if the constitutive relations are given: we must know how the charge densities and current densities depend on the field intensities. In [1] it was shown that the total current with inclusion of the deflection triggered by the Lorentz force is ;

$$
\begin{aligned}
\mathbf{J}(t) &= \mathbf{J}^{\text{EM}}(t) + \mu_{\mathbf{\Pi}} \mathbf{J}^{\text{EM}} \times \mathbf{B}(t) \\
&= \mathbf{J}^{\text{EM}}(t) + \mathbf{J}^{\text{LF}}(t)
\end{aligned}
\tag{4}
$$

variable	mesh entity
V	node
ϕ	node
A	link
Π	link
B	tile
p	dual volume
n	dual volume
ρ	dual volume
D	dual surface
H	dual link

TABLE I: Mapping of continuous variables on the discretization grid

and with $c = p$ or n and s=+1 for holes and -1 for electrons

$$\mathbf{J}^{\text{EM}}(t) = \sigma \mathbf{E}(t) \quad \text{metal}$$
$$\mathbf{J}_c^{\text{EM}}(t) = \mu_c c \mathbf{E} - s \mu_c k T \nabla c \quad \text{semi} , \tag{5}$$

and

$$\mathbf{J}^{\text{LF}}(t) = \mu_H \mathbf{J}^{\text{EM}}(t) \times \mathbf{B}(t) \tag{6}$$

IV. Discretization Steps

The set of equations (3) are submitted to discretization in order to find their solutions. The discretization procedure consists of the following ingredients: First a mesh is constructed with sufficient detail to capture the device geometry as well as possible field variations (skin effects). Next the equations are re-formulated on the mesh exploiting the differential-geometric meaning of the various variables. The corresponding recipe is listed in table I. The equations (4), (5) and (6) show that the total current density consists of two contributions. The usual finite-integration method will lead to the nodal current-balance equations

$$
\sum_j d_{ij} J_{ij}^{\text{EM}}(t) + \sum_j d_{ij} J_{ij}^{\text{LF}}(t)
$$
$$
+ s R_i(t) \Delta v_i + s \partial_t c_i(t) \Delta v_i = 0 \tag{7}
$$

where d_{ij} is the dual area and Δv_i is nodal volume of node i and J_{ij}^{EM} represents the current density expression as obtained without inclusion of the Lorentz force, e.g. for semiconductors it reads[1] with the Bernoulli function $B(x) = x/(\exp(x) - 1)$, the link temperature $T_{ij} = (T_i + T_j)/2$ and $\Pi_{ij} = \mathbf{e} \cdot \mathbf{\Pi}$

$$J_{c,ij}^{\text{EM}} = K_{c,ij}^{\text{DD}} = s \frac{\mu_{c,ij} T_{ij}}{h_{ij}} (c_i B(s X_{ij}) - c_j B(-s X_{ij})) \tag{8}$$

$$X_{ij} = \frac{q}{k T_{ij}} (V_i - V_j + \text{sgn}_{ij} \Pi_{ij}) \tag{9}$$

In here, $\mathbf{e}$ is an *intrinsic* unit vector along the direction of the grid link and $\text{sgn}_{ij} = +1$ if $\mathbf{e}$ points from node i to node neighbor j. If $\mathbf{e}$ points from node j to node neighbor i, we have $\text{sgn}_{ij} = -1$. The variable h_{ij} is the length of the link. Furthermore, J_{ij}^{LF} represent the correction due to the Lorentz force, e.g. for semiconductors it is

$$J_{c,ij}^{\text{LF}}(t) = s \mu_c \left(\mathbf{K}_c^{\text{DD}}(t) \times \mathbf{B}(t) \right) \cdot \mathbf{n}_{ij} \tag{10}$$

[1]We drop the explicit time dependence in the notation but it is tacitly assumed.

978-1-4799-7670-6/15 $31.00 © 2015 IEEE

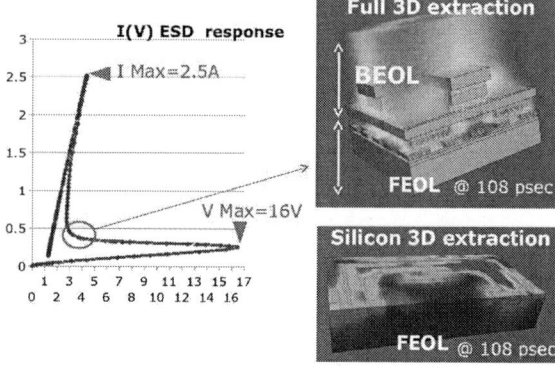

Fig. 2: IV ESD response + 3D extraction of the electric field intensity at 10^{-10} sec. Some reduction in the center is observed due to the skin effect and the value is $\sim 10^3$ V/m.

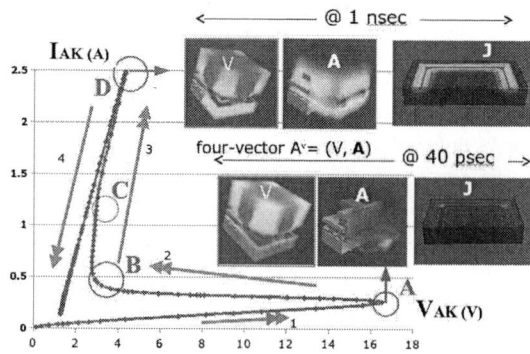

Fig. 3: IV ESD response + 3D Extraction of the vector potential at the point of maximum voltage (A) and maximum current (D) at 0.30×10^{-11} sec.

The vector $\mathbf{n}_{ij}$ is the unit vector along the link $\langle ij \rangle$ pointing from node i to node j and is parallel or anti-parallel to $\mathbf{e}$. The discretization of (10) is elaborated in [1]. The current balance in each node is achieved by summing all contributions from each mesh element and its associated set of links that are attached to the node under consideration. In particular, a contributing link in some mesh element is a boundary segment of two adjacent faces in the element. This observation allows us to determine the Lorentz force contribution.

For the time differentiation we rely on implicit schemes based on backward-difference formulas (BDF). These schemes have a built-in guarantee to be stable.

$$\frac{\partial f}{\partial t} = \frac{1}{h_0} \sum_{i=0} \alpha_{-i} f(t_{-i}) \qquad (11)$$

The parameters α depend on the time step sizes and t_0 corresponds to the latest time instance. Our implementation is limited to the second order BDF. In this case we have

$$\alpha_{-1} = -\frac{t_0 - t_{-2}}{t_{-1} - t_{-2}}$$
$$\alpha_{-2} = -\frac{t_0 - t_{-1}}{t_0 - t_2} \times \frac{t_0 - t_{-1}}{t_{-2} - t - 1}$$
$$\alpha_{-0} = -\alpha_{-1} - \alpha_{-2} \qquad (12)$$

For equal time step sizes the values are $\alpha_{-1} = -2$ respectively $\alpha_{-2} = \frac{1}{2}$ and $\alpha_0 = \frac{3}{2}$.

V. FAST TRANSIENT SIMULATION RESULTS FOR THE ESD PROTECTION DEVICE

In Fig. 1 we have illustrated the ESD protection structure consisting of a silicon-controlled rectifier and diode. The layout is also illustrated and the a 3D view of structure using the integrated front-end back-end simulation TCAD tool is shown. We apply a current drive at the IO pad. The IV response is shown in Fig. 2. The figure also illustrates the electric field at 0.1 nanosec. In Fig. 3 we show the same IV curve but also illustrate the various field intensities at the maximum voltage time instance (A) and maximum current instance (D). The first occurs at 40 picosecond, whereas the second occurs at 1 nanosecond. The magnetic field is shown in Fig. 4. The simulation tool allows us to obtain a very detailed view of

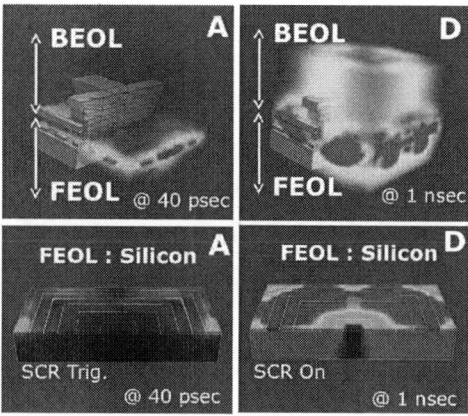

Fig. 4: Magnetic field intensity during ESD in FEOL (bottom pictures) and BEOL (top pictures) at the maximum voltage point (A) and maximum current time point (D) 0.30×10^{-9} sec.

the field intensities and *current densities inside* the device. Such information is relevant to assess the device reliability because the *local* value of the current density determines the electromigration robustness. This current density differs in general from the average current density that is computed as the current per area. Such details are neither visible from the circuit characteristics. In Fig. 5 we show the anode voltage as a function of time. The plot shows the experimental data (marked by "ST"), the calculation without electromagnetic effects, e.g. without inclusion of the self-induced magnetic field and without the inclusion of the Lorentz force deflection (marked by "elec"), as well as the results with the self-induced magnetic fields included but still without the inclusion of the Lorentz force deflection (marked by "EM-xyz) and finally with both the self-induced magnetic field and the Lorentz force deflection included (marked by "EM-xyz+LF"). The difference between the various computed results are in the sub-one percentage range. Similar results are shown for the SRC gate current in Fig. 6, the SRC gate voltage in Fig. 7 and the ground contact current in Fig. 8. The parasitic RC is included in the transient measurements. It reduces the over voltage and the simulations represent the worst case.

978-1-4799-7670-6/15 $31.00 © 2015 IEEE

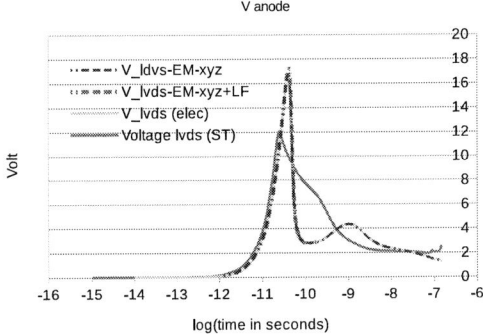

Fig. 5: Voltage of the anode vs. time

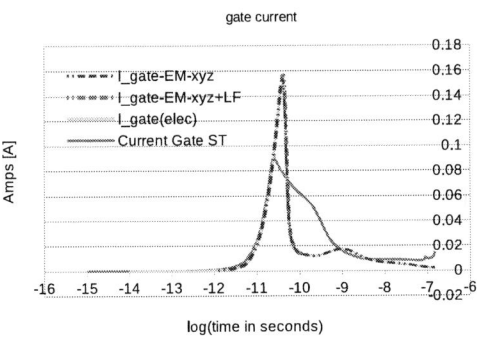

Fig. 6: Current at the SRC gate contact vs time.

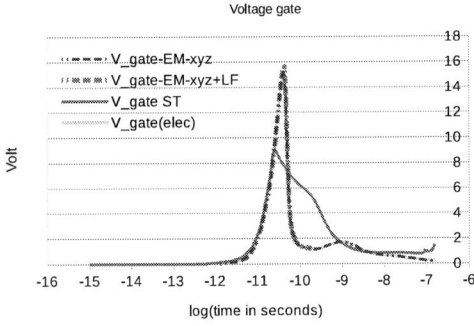

Fig. 7: Voltage of the SRC gate contact vs. time.

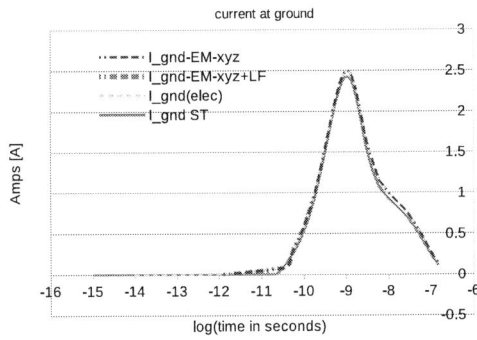

Fig. 8: Current at the ground contact vs. time.

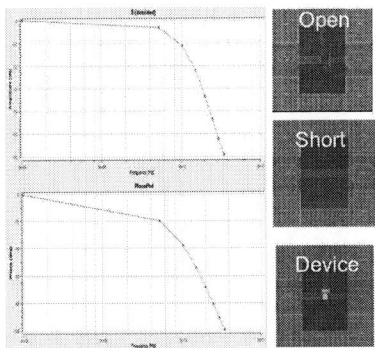

Fig. 9: S parameter result of the FEOL+BEOL design.

VI. SMALL SIGNAL ANALYSIS

With the simulation tool it also is possible to perform an AC analysis in order to compute the S-parameter response of the protection devices. The latter may be used to extract the FEOL+BEOL capacitance [3]. In Fig. 9 we present to result for this S11-parameter up to 25 GHz. Extraction of the capacitance results into a values of 80 fF. The capacitance has also been measured using a three-fold setup of open, closed and device-under-test configuration as illustrated in Fig. 9. The measured value is 85 fF.

VII. CONCLUSION

We presented a full simulation picture of fast-transient current surges including self-induced electromagnetic effects as well as a method to compute the impact of the Lorentz force. We applied the proposed method to an ESD protection structure consisting of an SRC and diode. We observed that the magnetic effects, with and without inclusion of the Lorentz force deflection of the currents have a negligible effect on the circuit-element characteristics. However, the local current densities inside the device are effected. This will impact the device reliability for electromigration robustness. Finally, we emphasize that the simulation techniques of this paper have a much wider applicability than presented here. For example, the simulation tools may be used to analyze magnetic field sensors, integrated passives and large-area substrate effects.

REFERENCES

[1] W. Schoenmaker, Q. Chen, and P. Galy, "Computation of self-induced magnetic field effects including the lorentz force for fast-transient phenomena," *IEEE Transactions on computer aided design of integrated circuits and systems*, pp. 23–34, June 2014.

[2] P. Galy and W. Schoenmaker, "In-depth electromagnetic analysis of esd protection for advanced cmos technology during fast transient and high-current surge," *EEE Transactions on Electron Devices*, pp. 1900–1906, June 2014.

[3] P. Galy, J. Jimenez, W. Schoenmaker, P. Meuris, and O. Dupuis, "Esd rf protections in advanced cmos technologies and its parasitic capacitance evaluation," in *ICICDT, IEEE International Conference on IC Design and Technology, Kaohsiung, Taiwan*, May 2011.

VIII. ACKNOWLEDGMENT

We would like to thank the ESD/RF qualification/ characterization teams and the ESD design team of STM EPS/TR&D as well as the coding and support teams at MAGWEL. Part of this work was financially supported by the EU funded FP7 projects ICESTARS GA214911 and nanoCOPS GA619166.

Impact of local interconnects on ESD design

Mirko Scholz, Shih-Hung Chen,

Geert Hellings, Dimitri Linten

imec

Logic Technology, Device Reliability group

3001 Heverlee, Belgium

Roman Boschke

imec and KU Leuven

Faculty for Engineering, Dept. ESAT

3001 Heverlee, Belgium

Abstract—Local interconnect (LI) as a contact scheme impacts significant the behavior of protection devices under Electro Static Discharge (ESD) stress. The narrow LI reduces the ESD robustness. At the same time, the on-resistance increases. This makes ESD protection design in future technology nodes more challenging, as the ESD design windows continuously shrinks.

Keywords: VLSI, CMOS, interconnects, reliability, component-level ESD, ESD design

I. INTRODUCTION

Since the 32 nm node, local interconnects (LI) are used to connect front-end of line (FEOL) structures [1,2] to the back-end of line (BEOL). They enable a reduced contact resistance and smaller device pitches. At the same time the continuous scaling of the interconnects increases the resistivity and inductance. This results in an increase of the delay and power consumption [3]. The small feature size of the interconnects also impacts their reliability. Previous work [4] investigated the reliability of the dielectric between the LI and a neighboring gate. The study included the investigation of the breakdown voltage and long-term reliability issues.

In this work, we investigate for the first time the impact of LI on the robustness and design of Electro Static Discharge (ESD) protection diodes. The impact of LI-based contact schemes on the robustness of STI and gated ESD diodes is studied. Using LI as contact scheme increases the on-resistance of ESD protection diodes. The results are applied to a design case. It demonstrates how the LI-based contact scheme impacts the ESD protection design in current and future technology nodes.

II. TEST STRUCTURES

The test structures in this study are manufactured in imec's 28 nm CMOS technology platform. Two local interconnect layers, LI1 and LI2 (Fig. 1) exist in this technology. An LI1 strip is 36 nm wide and contacts source/drain or anode/cathode areas. LI2 strips are 38 nm wide and contact the gates and LI1. The pitch between two interconnect strips is 110 nm.

The LI in this technology are tungsten filled trenches which are surrounded by a titanium nitride barrier. NiPtSi silicidation is done locally at the bottom of the LI trenches before the tungsten filling. The distance between an interconnect and a transistor gate is 22 nm. A stack of silicon-nitride (SiN) and silicon oxide is used as dielectric to isolate the gate from the LI1 interconnect. Finally, tungsten plugs (Via0) are used to connect LI2 to the copper backend.

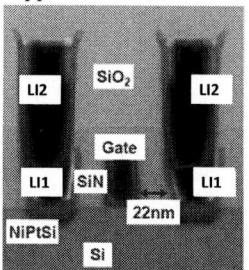

Fig. 1: TEM image of a FEOL device with local interconnects in imec's 28 nm planar CMOS technology: LIx - local interconnects.

The ESD test structures in this work consist of 40 μm wide n-well diodes with a floating gate or a STI region between anode and cathode. The diode length is 140 nm and kept the same for all device variations. The diodes are connected to probe pads through a stack of LI1 and LI2 interconnects, VIA0 and a wide METAL1 line (Fig.2).

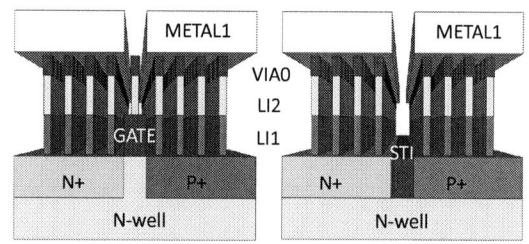

Fig. 2: Cross-section of gated diode (left) and STI diode (right) with LI1, LI2, VIA0 and METAL1.

III. IMPACT ON ESD ROBUSTNESS

In this section the impact of the LI-based contact scheme on the robustness of STI and gated diodes is analyzed with 100 ns Transmission Line Pulse (TLP) characterization [5] and mixed-mode simulations. A one order of magnitude increase in DC leakage current was considered as device failure.

A. Gated diode vs STI diode

Fig. 3 shows the failure current during TLP stress for floating gate (gated) and STI-defined diodes, depending on the number of LI rows. Two observations are made. As expected the intrinsic robustness increases with the number of contact rows. But also STI-defined diodes are about 15 to 20 % more robust than the gated diodes. This suggests that the conduction is more uniform in STI diodes than in gated diodes. Technology calibrated mixed-mode simulations with DECIMM [6] are used to further support this conclusion. The STI-defined and gated diodes are implemented as FEM models in the simulator. The LI are represented with electrodes (ohmic contacts) in the TCAD cross-section.

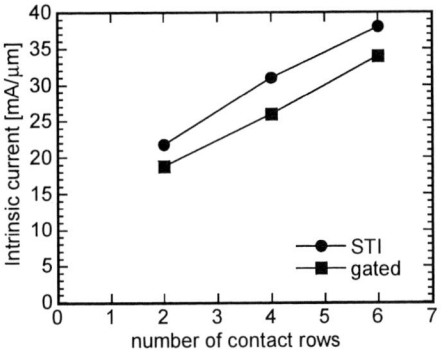

Fig. 3: Width normalized failure current I_{T2} during 100 ns TLP stress for gated and STI-defined diodes depending on the number of LI rows, contact scheme: narrow LI.

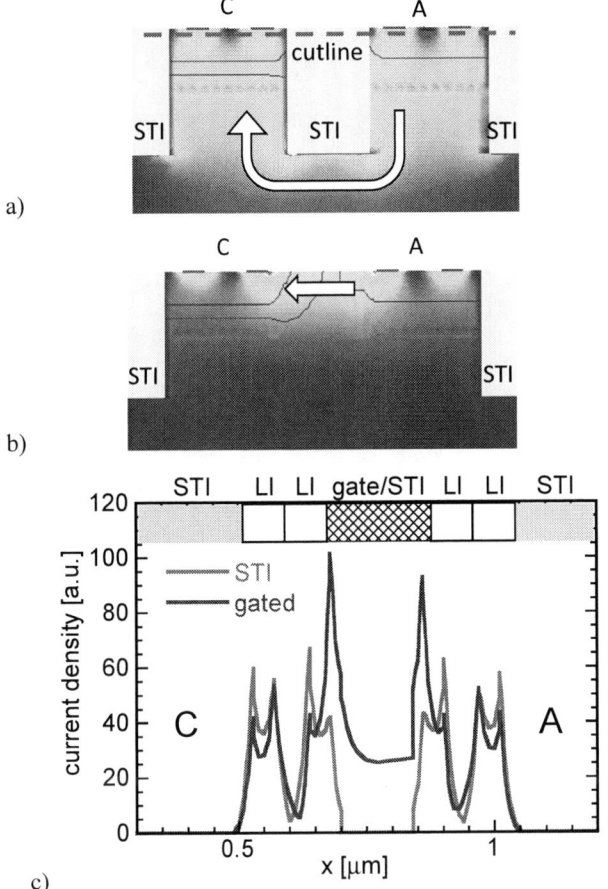

Fig. 4: Simulated spread of current density in STI diode (a), gated diode (b) and current density along cutline (c) for diodes with two contact rows; C: cathode, A: anode; TLP stress level/pre-charge voltage: 20 V.

Figs. 4a and b show the the current density in STI and gated diodes with two LI rows during the same TLP stress. In the STI-defined diodes a similar current density occurs below the various LI contacts. This is because the STI between anode and cathode and the spacing between LI and STI force a vertical current flow into the device. In the gated diode, the current flows mostly horizontal between anode and cathode. This causes a locally higher current density in the in the only a few nanometer thick silicidation layer of the LI closest to the floating gate (Fig. 4c). The silicidation is the weakest layer in the contact scheme and the source for the device failure.

B. Standard LI vs wide LI in gated diodes

Fig. 5a compares the layout of gated diodes with wide and narrow interconnects. The wide interconnect lines occupy the same layout area like the standard "narrow" LI. The connection to METAL1 is done with the same VIA0 tungsten plugs. Thus the *ACTIVE* area of the diodes is the same for the same number of LI rows. The silicidation is always done at the bottom of the interconnect trench. Thus at the bottom of the wide interconnects is one wide silicidation layer. Fig. 5b shows the normalized failure current I_{T2} depending on the number of LI rows and type of interconnect. There is a clear improvement of the robustness when using wide LI. The I_{T2} improves up to 30 % for the wide interconnects. The wide LI allow a more uniform current conduction and better spread of the thermal stress on the silicididation layer during ESD stress.

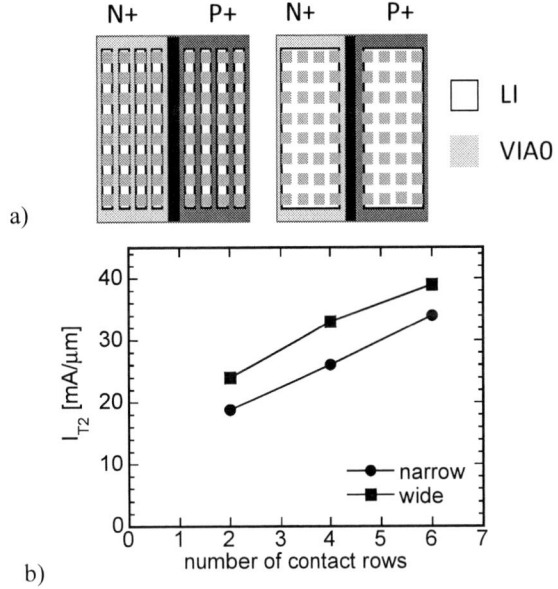

Fig. 5: Layout view for narrow (left) and wide (right) LI designs of gated diodes (a); width normalized failure current I_{T2} during 100 ns TLP stress depending on the number of LI rows for gated diodes, contact scheme: narrow and wide LI (b).

IV. IMPACT ON ON-RESISTANCE

This section investigates the impact of the contact scheme on the on-resistance of the ESD protection diodes. TLP characterization data and TCAD simulations are combined in the analysis. Fig. 6 summarizes the on-resistance for all three studied device variations depending on the number of LI rows. Two observations are made. As expected, the STI diodes have a higher on-resistance. This is due to a longer current conduction path around the STI between anode and cathode during ESD stress. The much shorter current conduction path in the gated diodes reduces the on-resistance. The lowest on-resistance is obtained in the designs where the contact schemes are designed with wide interconnect lines.

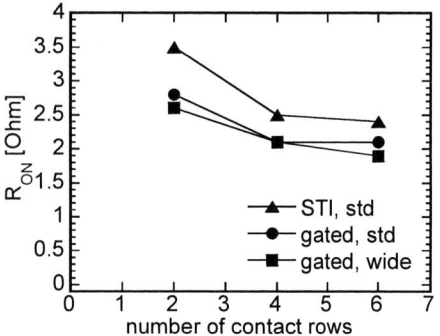

Fig. 6: On-resistance depending on number of LI rows.

The measured TLP I-V curve is compared with the TLP I-V curve extracted from mixed-mode simulations to extract the impact of the LI-based contact scheme on the on-resistance. The LI are represented as ohmic contacts in the simulation setup. The slope of the TLP I-V curve obtained with the simulations is equal to the intrinsic on-resistance of the protection diode independently of the contact scheme used. To obtain the resistance of the contacting scheme, the difference between the simulated and measured on-resistance is calculated.

Fig. 7 compares the measured on-resistance with the extracted resistance of the contact scheme for the three STI diode variations. It shows that the diode on-resistance is dominated by the choice of contact scheme. Only a small part of the on-resistance is attributed to the actual p-n diode.

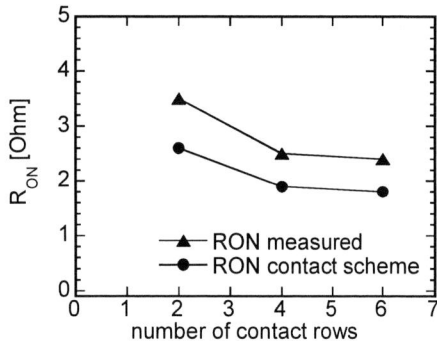

Fig. 7: Estimated on-resistance for STI diode depending on the number of LI rows; diode width: 40 μm

V. DESIGN CASE

This section investigates the impact of the LI on the ESD design with a design case based on TLP measurement data. Although equally important vfTLP measurement data is left out in this work.

First the TLP I-V curves are analyzed by normalizing them to the layout area (Fig. 8). The highest ESD robustness per area is obtained in designs with two LI rows. Because of the same layout area there is no large difference in robustness per layout area between the wide and the standard narrow LI designs for designs with 4 and 6 contact rows.

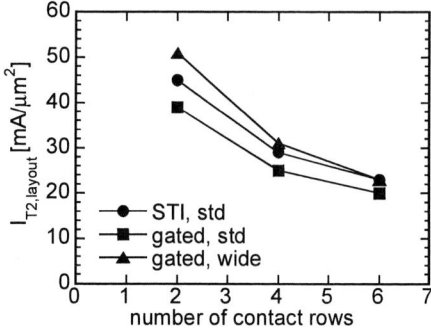

Fig. 8: Layout normalized current density depending on the number of LI rows.

Fig. 9 shows the two protection scenarios which are investigated in the design case. The protection design with one diode is usually used as the standard I/O ESD protection solution. Two diodes in series are used in local clamp ESD protection designs for low operating voltages [7].

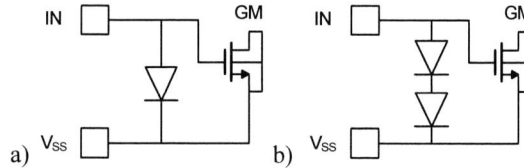

Fig. 9: Protection schemes used in the design case; I/O protection (a) and local clamp protection (b), IN – input, V_{SS}: ground, GM: NMOS gate monitor

In parallel to the ESD protection is the gate oxide of an input transistor or inverter. In this scenario it is represented by a NMOS gate monitor structure. To define the design window the gate monitor structure is characterized standalone with 100 ns TLP testing. The gate stack in the given technology has an effective oxide thickness (EOT) of 1 nm. This results in a gate oxide breakdown voltage during 100 ns TLP stress of 2.8 V in inversion (Fig. 10).

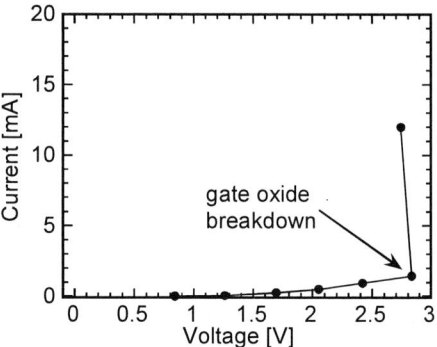

Fig. 10: Measured 100 ns TLP I-V curve during stress of the standalone NMOS gate monitor.

SPICE simulations are used to extract the clamping voltage at a target protection level of 1 kV Human Body Model (HBM) [8]. For the ESD protection design this target level is translated into a TLP failure current of 0.6 A. The SPICE simulation setup (Fig. 11) consists of a 50 Ohm pulsed voltage source to represent the TLP tester and the ESD protection diodes.

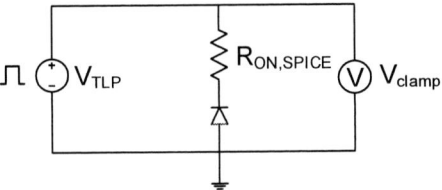

Fig. 11: Spice simulation setup for the extraction of the clamping voltage at a given HBM protection level.

The on-resistance of each diode is modeled by a resistance in series. Its value is calculated based on the device size which is required to pass 0.6 A of TLP stress:

$$I_{T2,\text{int }r} = \frac{I_{T2}}{w} \tag{1}$$

$$R_{ON,\text{int }r} = \frac{R_{ON}}{w} \tag{2}$$

$$R_{ON,SPICE} = R_{ON,\text{int }r} \frac{I_{HBM}}{I_{T2,\text{int }r}} \tag{3}$$

where $I_{T2,int}$ is the intrinsic current density, I_{T2} is the measured failure current, $R_{ON,intr}$ is the intrinsic on-resistance, R_{ON} the measured on-resistance, w the width of the ESD diode, I_{HBM} the failure current at the target protection level and $R_{ON,SPICE}$ the on-resistance of the protection diode when sized for the target protection level.

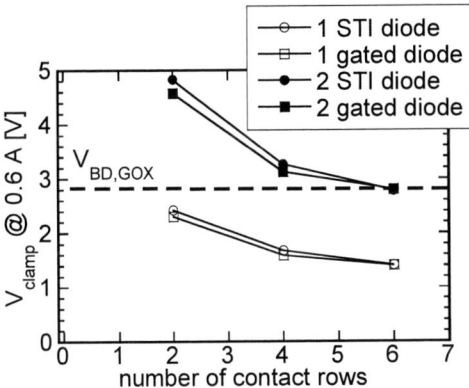

Fig. 12: Simulated clamping voltage at 0.6 A TLP stress (equivalent to 1 kV HBM) depending on the number of LI rows (standard "narrow" LI are only simulated).

In case of one protection diode the design window is met for both type of diodes independent of the number of LI (Fig. 12). When two diodes are used the design window is difficult to meet the ESD protection target level of 1 kV HBM. In this case the width of the protection diodes needs to be increased to lower the on-resistance. Preferably gated or STI diodes with

four or more contact rows are used to obtain an area efficient protection design in the given design window.

VI. CONCLUSIONS

Local Interconnect-based contacting schemes are dominating in sub-32 nm technology nodes. In this work it is shown that the use of LI as contact scheme impacts the robustness of ESD protection structures. STI-defined diodes are more robust than gated diodes. This is due to a more uniform vertical flow of the stress current into the anode and cathode of the STI diodes. The less uniform flow of current in the gated diodes causes an increased current density in the LI closest to the floating gate. A locally higher current density during ESD stress decreases the device robustness. The use of LI as a contacting scheme has also a strong impact on the ESD protection design. It increases the on-resistance of the ESD protection diodes. The available design windows shrinks dramatically. The increase of on-resistance (and clamping voltage) goes along with the decrease of the gate oxide breakdown voltages in current and future technology nodes. The presented design case shows that at least four contact rows are required to obtain an area efficient protection design.

The impact of local interconnects on the ESD design becomes even more important in sub-20 nm FinFET technologies with their even more reduced contact area between the device fins and the local interconnect. The local interconnect properties become the dominating factor in the performance of the ESD protection devices.

ACKNOWLEDGMENT

The authors acknowledge the support from imec's logic insite partners. The authors want to thank Angstrom Design Automation for the collaboration on the development of the mixed-mode simulator DECIMM.

REFERENCES

[1] P. Packan, S. Akbar, M. Armstrong, D. Bergstrom, M. Brazier, H. Deshpande, ..., "High performance 32nm logic technology featuring 2nd generation high-k + metal gate transistors", IEDM Technical Digest, 2009.

[2] C. Auth, C. Allen, A. Blattner, D. Bergstrom, M. Brazier, M. Bost, ..., "A 22 nm high performance and low-power CMOS technology featuring fully-depleted tri-gate transistors, self-aligned contacts and high density MIM capacitors", Proceedings VLSI Technology, 2012.

[3] International Technology Roadmap for Semiconductors (ITRS), 2013.

[4] T. Kauerauf, A. Branka, G. Sorrentino, P. Roussel, S. Demuynck, K. Croes, ..., "Reliability of MOL local interconnects", Proceedings of the IRPS, 2013.

[5] T. Maloney and N. Khurana, "Transmission line pulsing techniques for circuit modeling of ESD Phenomena", Proceedings EOS/ESD Symposium, 1985.

[6] DECIMM, Angstrom Design Automation, http://www.angstromda.com/.

[7] S. Thijs, A. Griffoni, D. Linten, S. Chen, T. Hofmann and G. Groeseneken, "On Gated Diodes for ESD protection in a bulk FinFET CMOS technology", Proceedings. ESD Symposium, 2011.

[8] Industry Council on ESD target levels, "White Paper 1", 2007.

ESD protection diodes in Optical interposer technology

Roman Boschke, Guido Groeseneken

imec and KU Leuven
Faculty for Engineering, Dept. ESAT
3001Heverlee, Belgium
boschke@imec.be

Mirko Scholz, Shih-Hung Chen, Geert Hellings, Peter Verheyen, Dimitri Linten

imec
3001 Heverlee, Belgium

Abstract - **The ESD robustness of planar Si and Ge diodes on Silicon-on-Insulator (SOI) optical interposer is studied by using TLP and vfTLP system. Although Ge diodes show a lower failure current, a superior clamping capability with a resistance lowering behavior, which is attributed to the intrinsic material properties of Ge, makes Ge diodes possess a promising potential for ESD protections.**

I. INTRODUCTION

Integrated optical interconnects [1] use Silicon (Si) ring modulator and co-integrated Germanium (Ge) photodetector. This provides the opportunity to build an ESD solution either in Si or Ge.

The impact on the ESD robustness by introducing a SiGe quantum well in the channel of a planar device has been studied in [2]. Gated diodes show a small decrease in performance due to the smaller band gap of the SiGe, compared to silicon.

Section II describes technology details and the used characterisation methods. In Section III, a direct comparison of ESD diodes in Si and Ge is discussed. Further layout dependencies on Si and Ge ESD diode behaviour are reported in subsection B. The impact of process changes on Ge ESD diode behaviour is discussed in subsection C. Before drawing conclusions, a power modelling of wide range ESD TLP pulse widths on Ge ESD diodes is described in Section IV

II. Device fabrication and characterization

The studied technology is a Silicon-on-Insulator (SOI) technology for integrated optical interconnects [1]. Ge is locally grown epitaxially on top of the SOI film. The thickness of the undoped Ge film is 300 nm. The studied Si ESD diodes are manufactured in the SOI film. Ge ESD diodes are produced in the Ge layer grown on top of the SOI film. For Si devices, a Nickel silicide (NiSi) is formed that lowers the contact resistance. No Nickel Germanide (NiGe) is present for Ge devices, instead Tungsten direct contact is used. Fig. 1 illustrates a cross-section through the anode area of Ge diode. Both Si and Ge ESD diodes in p+/ nwell and n+/ pwell configuration with equivalent doping profiles exist. Diodes with the same geometry for Si and Ge have been characterized. The anode and cathode contact area of all diodes consists of 5 contact rows.

For TLP and vfTLP characterizations, the HANWA W5000 TLP tester is used with 200 ps rise time and pulse length of

5ns, 100 ns or 500 ns. The failure level is judged on a minimum of 10 % leakage increase at 1 V reverse bias.

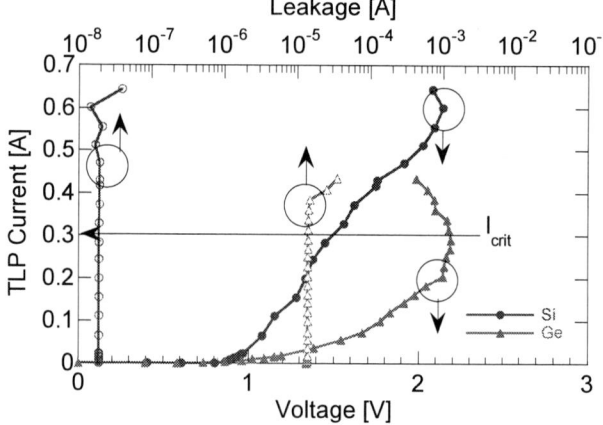

Fig. 1: SEM of SOI technology with locally epitaxial grown Ge islands. Image was taken across the Ge diode's anode.

III. Measurement results

A. Comparison Si vs. Ge in forward and reverse TLP

Fig. 2 and Fig. 3 show the overlay of the forward and reverse biased TLP IV curves for a one fixed diode dimension (L and W), using different materials.

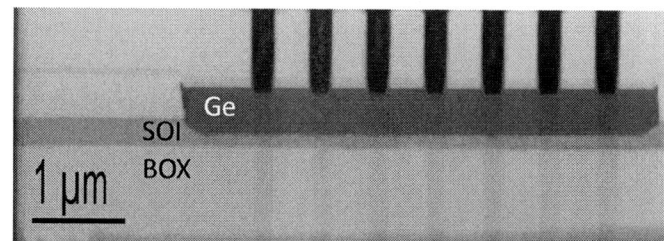

Fig. 2: Forward biased TLP (100 ns) measurement of Si vs. Ge diode (W=50 µm, L=200 nm).

The observed failure current in forward TLP of Ge diodes are around 30% lower compared to the Si diode counterparts.

978-1-4799-7670-6/15 $31.00 © 2015 IEEE

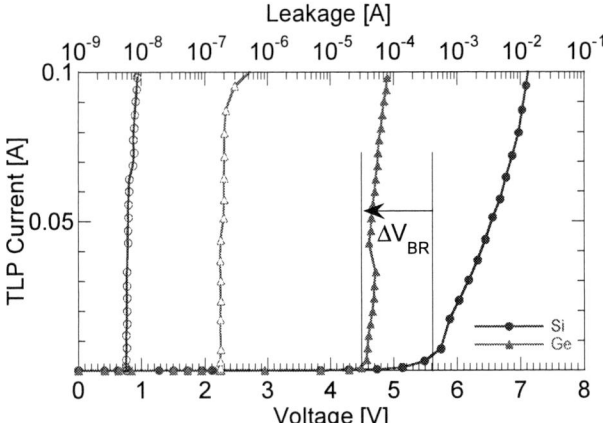

Fig. 3: Reverse biased TLP (100 ns) measurement of Si vs. Ge diode (W=50 μm, L=200 nm).

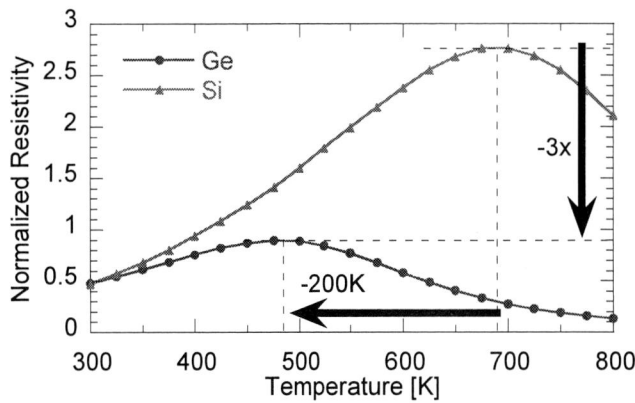

Fig. 4: Estimation of resistance of bare Si and Ge as function of temperature using model of carrier concentration, mobility and bandgap lowering.

Three key observations are made. First, Ge diodes have a significantly higher leakage current compared to its Si counterpart with the same layout. This can be explained by the smaller band gap of Ge (0.66 eV) compared to Si (1.12 eV) [3]. The reverse bias TLP IV curves Fig. 3 show only a breakdown voltage (V_{BR}) shift. The observed shift can be described by the breakdown voltage approximation (1) [4].

$$V_{BR} \cong 60 \cdot \left(\frac{E_g}{1.1}\right)^{\frac{3}{2}} \cdot \left(\frac{N_B}{10^{16}}\right)^{-\frac{3}{4}}, \qquad (1)$$

where E_g is band gap and N_B is background doping in cm^{-3}. The band gap offset between Ge and Si plus an additional difference in background doping can explain the observed ΔV_{BR} (Fig. 3). A second observation is the resistance at low TLP current (<0.1 A) is higher in Ge compared to Si. This can be explained by the presence of NiSi in the Si diode. This reduces the contact resistance to the N+/P+ doping while the absence of NiGe in Ge diode leads to high series resistance. The third observation is that the differential resistance in the Ge diode gets lower with increasing TLP current contrary to Si. Negative differential resistance is observed for Ge diodes. The current level when the device has a zero differential resistance is defined as I_{crit} in Fig. 2. A model was proposed in [5], that describes the resistance lowering by a temperature induced positive feedback mechanisms. The 1000x higher intrinsic carrier concentration at room temperature and more pronounced bandgap lowering with increasing temperature in Ge explain a thermally induced carrier generation. The normalized resistivity as function of temperature for Ge, illustrated in Fig. 4, shows this effect. The resistivity is normalized, that Si and Ge has the same values at 300 K, besides the higher mobility of Ge.

Both Si and Ge shows above 300 K a resistivity increase. However Ge has a 3x lower peak resistivity and 200 K lower peak temperature. Further, the resistivity in Ge drops already at 600 K below the value at room temperature.

An comparative overview of the measured ESD relevant parameters are put together in **Error! Not a valid bookmark self-reference.**.

Table 1: Overview of key parameters of Si and Ge diodes (L=200 nm, W=50 μm).

Material	Si	Ge
I_{t2} [A]	0.6	0.4
Leakage [A]	8E-7	9E-4
V_d [V]	0.3	0.7
V_{BR} [V]	4.5	5.7

B. Layout study for Si and Ge diodes

Fig. 5 and Fig. 6 show TLP IV curves as a function of anode to cathode spacing (L), with a fixed junction width (W) for Si and Ge, respectively. The anode to cathode spacing (L) is varied from 100 nm to 1 μm.

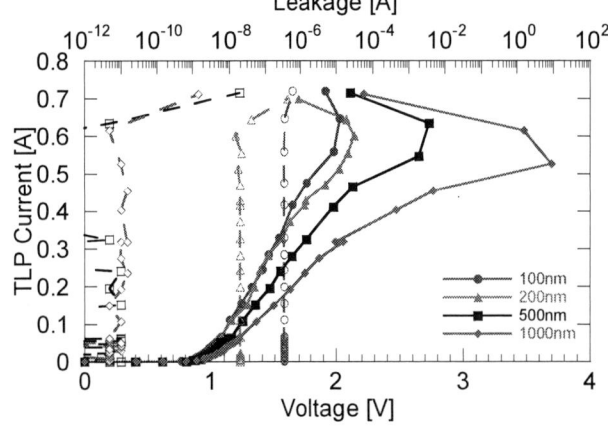

Fig. 5: TLP (100 ns) measurement of L variation of Si diodes (W = 50 μm).

978-1-4799-7670-6/15 $31.00 © 2015 IEEE   163

For Si, the on resistance increases with L and above 500 nm a self-heating induced resistance increase is visible in Fig. 5. For Ge, the high current differential resistance lowering is present for all L variations. I_{crit} is independent of L and the differential resistance gets zero at current density of 6 mA/μm. (see Fig. 6).

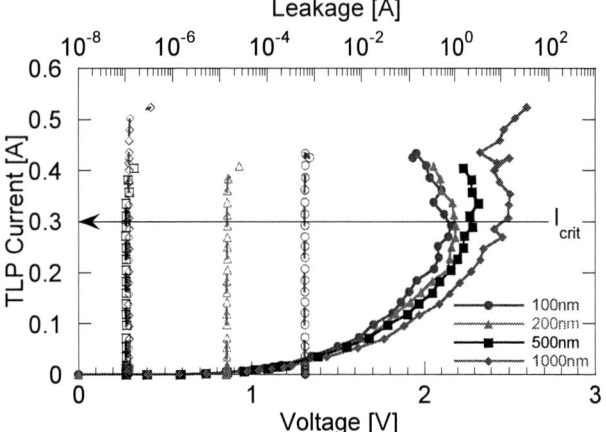

Fig. 6: TLP (100 ns) measurement of L variation of Ge diodes (W=50 μm).

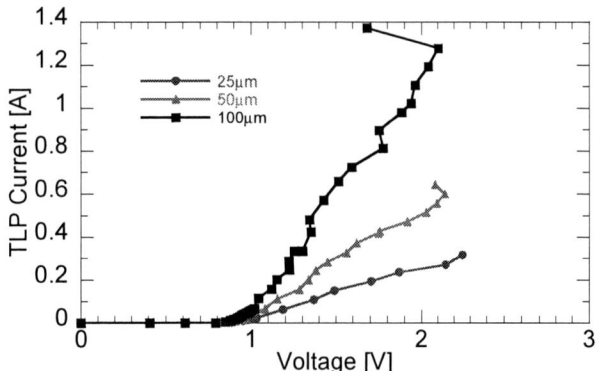

Fig. 7: TLP (100 ns) measurement of W scaling of Si diodes (L=200 nm).

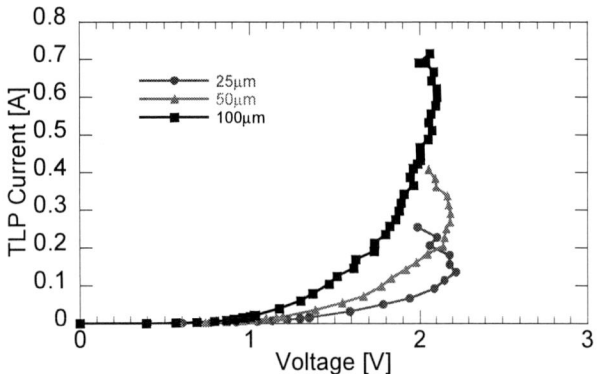

Fig. 8: TLP (100 ns) measurement of W scaling of Ge diodes (L=200 nm).

Fig. 7 and Fig. 8 show the variation of junction width (W) for Si and Ge. The range of junction width (W) variation is 25 μm

to 100 μm. For Si, shown in Fig. 7, the on resistance reduces with W and at W=25 μm the self-heating induced differential resistance increase is visible. For Ge diodes, see Fig. 8, the high current differential resistance lowering is present for all W variations. The low current on-resistance (<0.1 A) reduces with W.

C. Process influence on Germanium behaviour

Two process changes are studied addressing: SOI film doping and junction implant energy.
In Fig. 9, TLP IV curves of Ge diodes with Ge grown on n- or p-doped SOI films are overlaid. No change in TLP IV curve is observed by changing the doping of the SOI film from n- to p-type. Hence the mechanism that leads to lowering of the high current differential resistance occurs only in the Germanium layer.

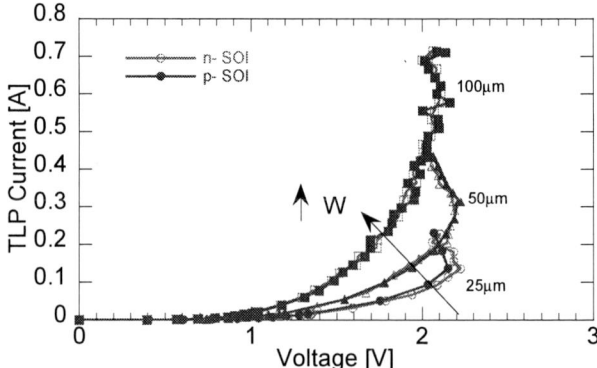

Fig. 9: TLP (100 ns) measurement of W scaling of Ge diodes with n- and p- SOI substrate (L=200 nm).

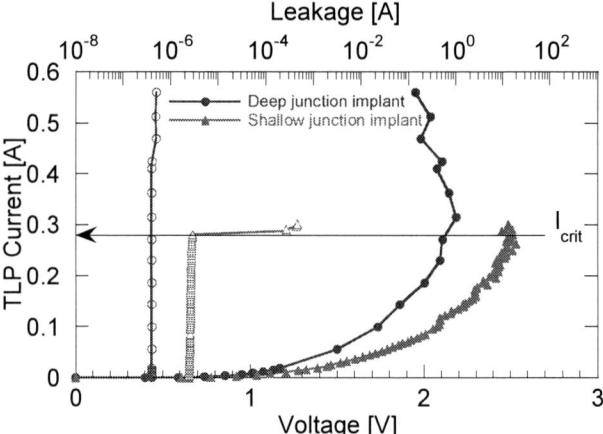

Fig. 10: TLP (100 ns) measurements of Ge diode with different junction implant conditions (W=50 μm, L=200 nm).

Fig. 10 illustrates the modulation of Ge TLP behavior for reduced junction depth of anode by lowering of the junction implant energy. This results in a smaller effective junction area and a higher sheet resistance A less pronounced high current differential resistance lowering was observed for shallower junction implants. However I_{crit} is not changed. Lower I_{t2} and higher V_{t2} for the shallower implant was observed. The shallower implant reduces the volume where

978-1-4799-7670-6/15 $31.00 © 2015 IEEE   164

the current flows. This increases intrinsic current density and leads to earlier failure.

IV. *TLP power density study*

The key conclusion from previous ESD studies of Ge [5] was, that the effect of differential resistance lowering can be expedited with higher dissipated power induced by longer TLP pulses. However it was unknown if the effect is present while vfTLP stress. Therefore Ge diodes have been stressed with pulses from 2 ns up to 500 ns. In Fig. 11 the vfTLP for 5 ns and TLP IV curves for 100 ns and 500 ns are overlayed. It demonstrates that the differential resistance lowering occurs also while vfTLP characterisation. However I_{crit} increases.

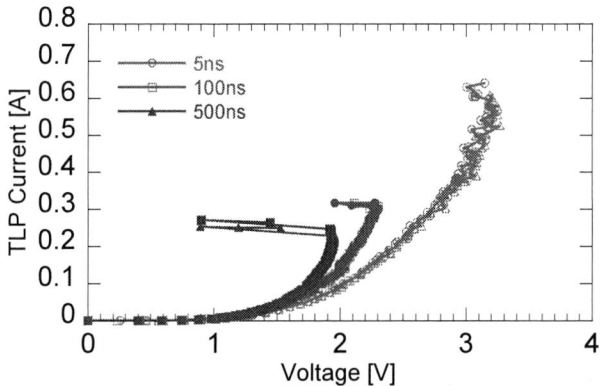

Fig. 11: TLP measurement of Ge diodes (W=50 µm, L=200 nm) using 5 ns, 100 ns and 500 ns TLP pulse length.

The power density (P_{ZDR}) at which the differential resistance become zero can be calculated by Eq. (2)

$$P_{ZDR} = \frac{I_{Crit}}{W} * V_{Crit}, \qquad (2)$$

where I_{crit} and V_{crit} are the current and the voltage when the differential resistance become zero and W is the device width.

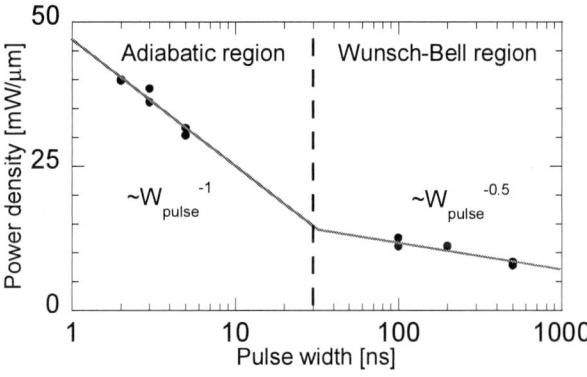

Fig. 12: Power density of zero differential resistance as function of vfTLP and TLP pulse width (W_{pulse}) of a Ge diode (W=50 µm, L=200 nm).

Fig. 12 shows P_{ZDR} as function of TLP pulse width (W_{pulse}). The dependency shows two regions. Below 30 ns. it proportional to W_{pulse}^{-1}. Above 30 ns it shows a $\sim W_{pulse}^{-0.5}$

dependency. The observed dependency for the power at zero resistance in Ge is similar to the power to fail dependency reported by [6] for filament formation in Si. But the Ge diodes can sustain this power density without failing. This could be explained by the lower temperature at which the resistance of Ge lowers, illustrated in Fig. 4 This leads to the conclusion, that the zero differential resistance observed in Ge behaves simular to a filament formed in Si. However it can survive the zero resistance because it occurs at lower temperature.

V. Conclusion

A thermally induced carrier generation that leads to zero-differential resistance at high TLP current is observed for Germanium diodes. This behavior is not known for Si diodes and enables superior clamping capability for Ge diodes. However the Ge diodes show 30% reduced ESD robustness.

The onset of the differential resistance lowering can be suppressed by increasing the series resistance. This finding was confirmed with junction implant energy reduction or increase of L.

Ge diodes show the differential resistance in a wide range of TLP pulse length. This behaviour show similarities to thermal failure observed by [6].

REFERENCES

[1] M. Pantouvaki, et al., "20Gb/s silicon ring modulator co-integrated with a Ge monitor photodetector" European Conference and Exhibition on Optical Communication, pp. 1-3, 2013.

[2] G. Hellings, et al., "ESD Characterization of High Mobility SiGe Quantum Well and Ge Devices for Future CMOS Scaling" EOS/ESD Symposium, pp. 1-6, 2012.

[3] G. Eneman, et al., "Impact of Donor Concentration, Electric Field, and Temperature Effects on the Leakage Current in Germanium p+/n Junctions" Transactions on Electron Devices, Volume: 55 , Issue: 9, pp. 2287-2296, 2008.

[4] S. M. Sze, "Physics of Semiconductor Devices" 2nd ed. New York: John Wiley & Sons, 1999

[5] R.Boschke, et al., "ESD Characterization of Germanium diodes" EOS/ESD Symposium, pp. 1-9, 2014

[6] V.M. Dwyers, et al., "Thermal Failure in Semiconductor Devices" Solid-State Electronics, Vol. 33, No. 5, pp 553-560, 1990.

Preliminary 3D TCAD Electro-thermal Simulations of BIMOS transistor in thin silicon film for ESD protection in FDSOI UTBB CMOS technology

S. Athanasiou[1,2], S. Cristoloveanu[2], Ph. Galy[1]

[1]STMicroelectronics, 850 rue Jean Monnet, 38920 Crolles, France

[2]IMEP, 3 Parvis Louis Néel, CS 50257, 38016 Grenoble Cedex 1, France

Abstract— **The purpose of this paper is to analyze the ESD device electro-thermal behavior of BIMOS transistors integrated in ultrathin silicon film for 28 nm FDSOI UTBB high-k metal gate technology. This evaluation is based on 3D TCAD simulations with classical physical models using Average Current Slope (ACS) method and quasi-static DC stress (Average Voltage Slope (AVS) method). We show how the series resistance and the thermal resistance impact the average and peak temperatures in these devices.**

Index Terms - **BIMOS transistor, ESD protection, FDSOI, CMOS**

I. INTRODUCTION

IT is well known that the ElectroStatic Discharge (ESD) protection is a major challenge in Fully Depleted SOI (FDSOI) Ultra Thin Body & BOX (UTBB) advanced CMOS technologies [1,2]. In this framework, ESD protection made for hybrid bulk (open BOX area) [3,4] shows that the BIMOS transistor solution is more efficient than the standard GGNMOS one [5]. Moreover, BIMOS in thin SOI film (no BOX opening) with body contact and back gate loop further improves the protection [6].

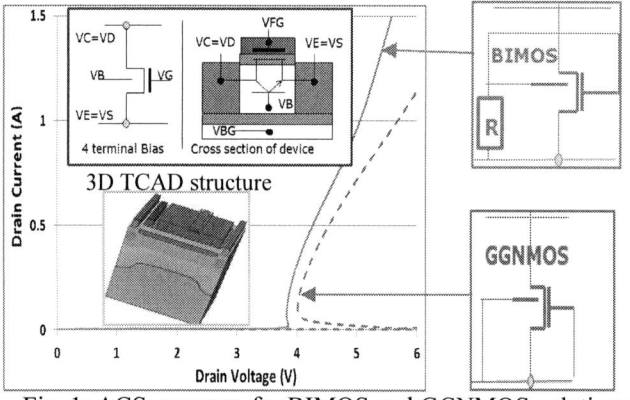

Fig. 1: ACS response for BIMOS and GGNMOS solutions.

As a reminder, BIMOS (Bipolar MOS) is a body-contacted MOS transistor (Fig. 1) which activates both Bipolar and MOS phenomena in the same device simultaneously by biasing the four terminals (source=emitter; drain=collector; body=base and gate). In this design solution, the resistance R is a parameter to tune the trigger voltage (Vt1). The main purpose here is to evaluate the ESD electro-thermal behavior of a BIMOS in thin silicon film and to evaluate its efficiency and intrinsic robustness. We use 3D TCAD electro-thermal simulation model to design and optimize BIMOS transistors integrated directly in the ultrathin film.

II. ESD ISOTHERMAL SIMULATIONS

The design combines the effects of the body contact and the back gate control to explore the isothermal and electro-thermal response under ACS and AVS stresses. The BIMOS design approach is presented on Fig. 2. This preliminary design is the best that can be done with this technology in terms of performance and topology [6]. For the study, the surge is an ACS stress with 1Amp max current, and 100ns rise time while AVS is a 5V stress with a 1ms rise time. This type of ACS simulation is equivalent to the transmission line pulse (TLP) test for human body model (HBM) [7]. Moreover, the simulations are performed with Poison, drift/diffusion and continuity equations. The device is isotherm @ 300K (room temperature) without self heating. Active physical models include doping dependence and high field saturation for mobility degradation, SRH and Auger recombination, as well as avalanche generation model for impact ionization.

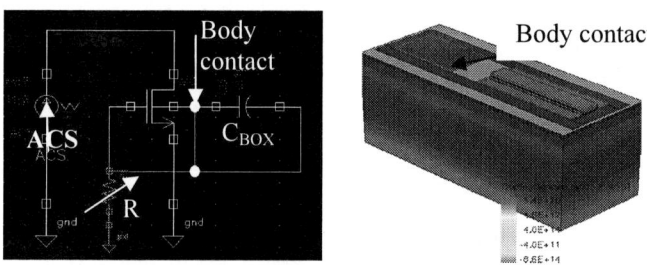

Fig. 2: BIMOS body contact + Back gate loop & topology.

A) Transient response : Isotherm ACS stress

Figure 3 reports the I-V curve of the ESD response (ACS) and clearly shows the V_{t1} and V_{hold}. By extraction : V_{t1}=0.79 V and V_{hold}=0.75V which are very low values. At the end of stress, the current and voltage are the same in all cases. Conduction in the full Si film is involved at this current level.

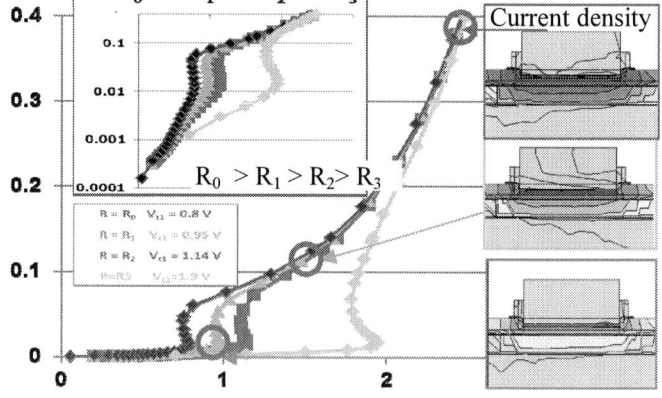

Fig. 3: ACS responses for solution with R parameter.

978-1-4799-7670-6/15 $31.00 © 2015 IEEE

For an accurate view, the current density extractions in 3D topology show that the current flow is spread in the whole structure (see figure 4).

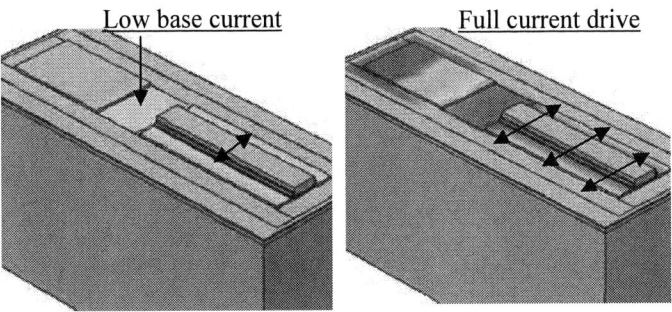

Fig. 4: Current density in thin silicon film during ESD stress.

B) Quasi Static response : Isotherm AVS stress

This section is focused on "DC stress-like" response. It means that the response is for a non-standard ESD stress with a very slow rise time and with low energy (less than 100mA). Figure 5 gives the three I-V curves in quasi static stress (AVS). They look like "Zener" drop voltage. Thus, only R=R_1 and R_2 cases are useful without exhibiting extra leakage current under 1V power voltage.

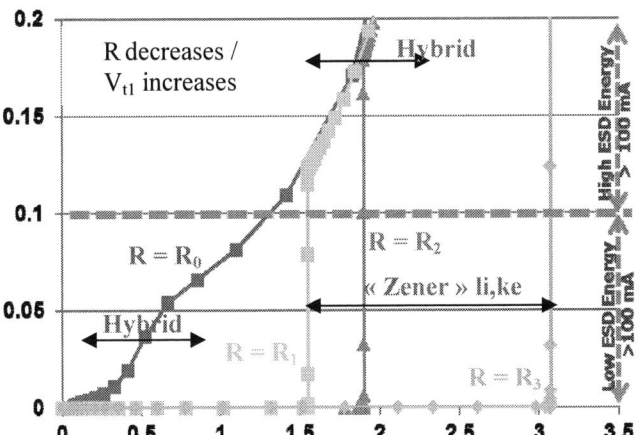

Fig. 5: Comparison of quasi static responses (AVS stress).

Table 1 summarizes the trigger voltage V_{t1} for different values of the resistance R used to tune the sensitivity of triggering in quasi-static (AVS) and transient (ACS) stresses.

R parameter I-V extractions	R_0	R_1	R_2	R_3
V_{t1} AVS condition	N.A.	1.5 V	1.9 V	3.08V
V_{t1} ACS condition	0.8 V	0.95V	1.4 V	2 V

Table 1: Trigger voltage for quasi static and transient stresses

The next section is devoted to electro-thermal simulation of this design and topology in 28nm FDSOI UTBB.

III. ESD ELECTRO THERMAL SIMULATIONS

In this section we describe the device behavior under the same stresses used in isotherm simulations, ACS and AVS. The simulations are done with Poison, drift/diffusion continuity equations as well as lattice temperature; same physical models as in isothermal simulations were used. The device initial thermal condition was 300K. For all simulations R = R_0 to R_3 with $R_0 \gg R_3$ (to evaluate the impact on the design response and robustness). The same device with same resistor connectivity was tested as for isotherm simulations. The effect of contact thermal resistivity was explored to evaluate the impact of the thermal environment on device heating.

A) Transisent response : Electro Thermal ACS stress

We present the results of ACS stress. Figure 6 shows the I-V curves for isothermal and electro-thermal numerical simulation, along with average temperature and maximum temperature within the device. The point of the second (thermal) breakdown is extracted at around 772K for current of 400mA and 2.6V. Beyond this point we observe a large difference between average and maximum temperature due to self-heating: both slopes drastically increase, approaching the silicon melting point at around 1600K. The significant difference between average and maximum temperature indicates the existence of a hot spot inside the device which is verified by the extraction presented later. This is due to the time constant of the ESD stress being less than the thermal time constant.

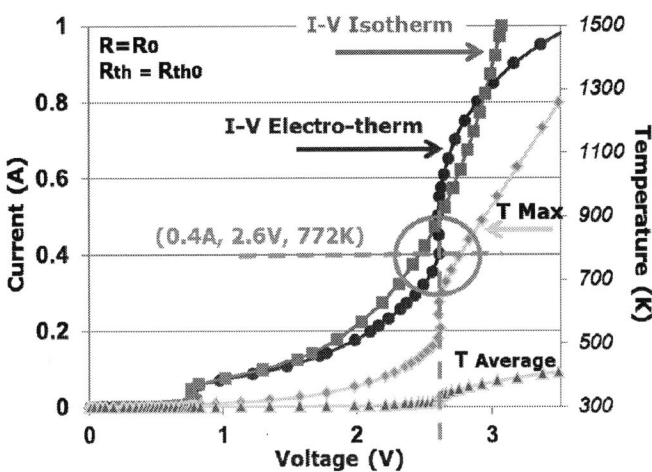

Fig. 6: ACS stress response under thermodynamic simulation, compared for R=R_0 device. Thermal extractions are shown.

2D extraction is performed on different points. Initial hot spot is at 1ns. The point when hot spot reaches a temperature around 350K is 14ns after beginning of ACS stress. Close to the end of the stress at around 100ns the temperature reaches 1000K.

978-1-4799-7670-6/15 $31.00 © 2015 IEEE   167

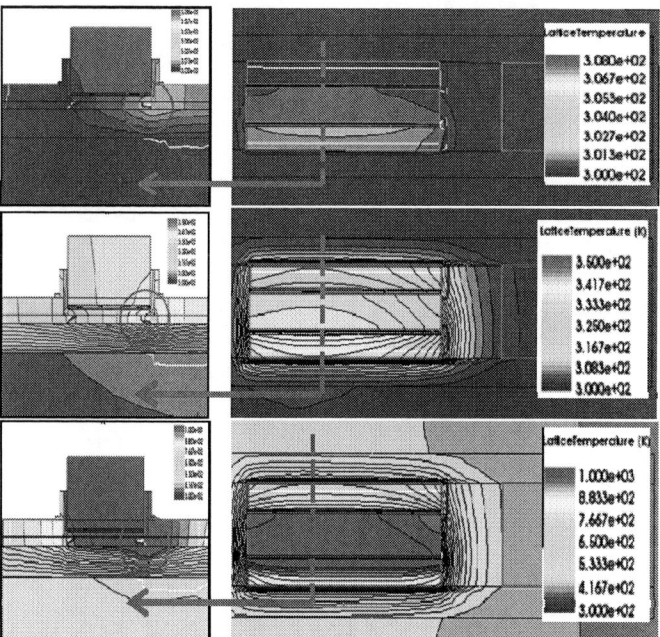

Fig. 7: 2D temperature extraction at 3 points during ACS stress in thermodynamic simulation after 1ns, 14ns and 100ns.

The analysis continues with different values for contact thermal resistivity Rth, during ACS stress. No major impact is observed (see Fig. 8).

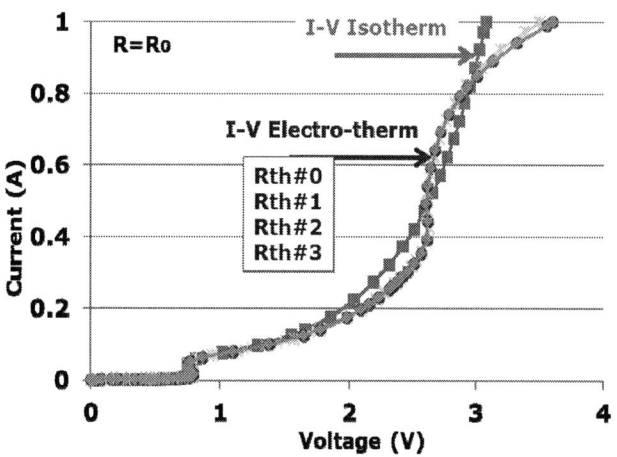

Fig. 8: ACS stress during electrothermal simulation for different contact thermal resistivity values.

Next table gives the Tmax and Taverage evolution for different values of Rth and for $R=R_0$. The same trend is observed for R_1 to R_3.

$(R=R_0)$ Rth para. Temp. extractions	Rth_0	Rth_1	Rth_2	Rth_3
T_{Max} end of ACS	1257K	1354K	1358K	1359K
T_{Ave} end of ACS	408K	445K	446K	447K

Table 2: Maximum and average temperatures for different Rth at the end of stress.

B) Quasi Static response : Electro Thermal AVS stress

AVS stress is equivalent to quasi static event and can reveal useful information about leakage properties of the devices. Figure 9 reports I-V response in isothermal and electro-thermal conditions for various thermal resistances.

For Rth_0 no major impact is observed but for all other Rth values, taking into account package/environmental effects the behavior, is "Zener" like. The drastic change in threshold voltage (1.5V to 0.8V) is induced by the self-heating, by thermal condition and by the value of the design resistance.

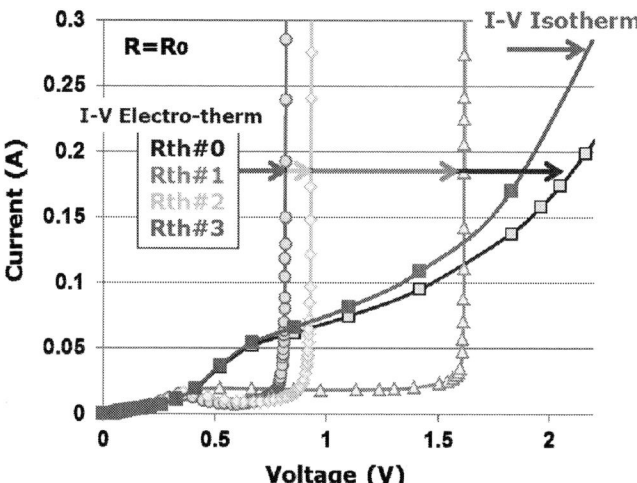

Fig. 9: I-V response to AVS stress for $R=R_0$ and for different Rth values compared to isotherm simulation.

In Figure 10 we analyze the thermal behavior of the characteristics for 2 different Rth values, Rth_0 and Rth_2. Rth_0 exhibits a larger thermal gradient than Rth_2. This result suggests the existence of a hot spot in the first case while a more homogenous self-heating during stress occurs in the second case. This is related to the easier heat dissipation with lower thermal resistance on the thermode contact.

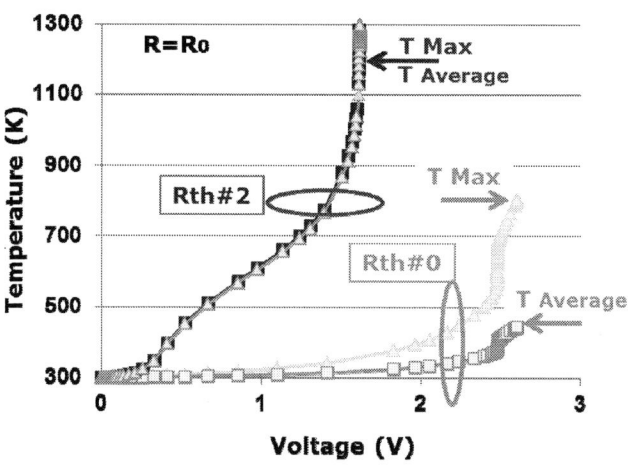

Fig. 10: Average and maximum temperature evolution with R_0 for different Rth values.

Figure 11 shows the 2D extraction of temperature distribution for these 2 cases it is shown that even in the case of the more homogenous heating a smaller hot spot exists and is present in a different area of the device. While hot spot in the Rth_0 case is directly under the area between drain and gate for Rth2 the maximum temperature is moved towards the body contact.

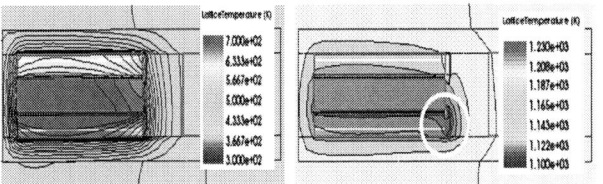

Fig. 11: 2D temperature profiles during AVS for Rth_0 (left) and Rth_2 (right) with R_0 connection.

Focusing on the corner hot spot (1200K), it is noted in Fig. 12 that the current density is higher in this area. An increased current density can be observed also close to the source area of the device. This effect can be possibly attributed to the thickness difference between the epitaxial silicon of the source /drain areas and the thinner undoped intrinsic Si between body contact and the device leading to corner like effects. During this slow event lateral parasitic PIN structures seem to be activating.

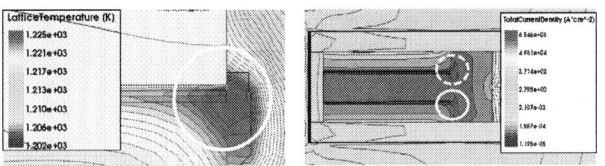

Fig. 12: 2D temperature distribution around corner hot spot and current density AVS for Rth_2& R_0

By changing the resistance R_0 by R_3, the design robustness is proved up to 300 mA as indicated by figure 12. For this value of resistance and modifying the thermal boundary condition the effects on the I-V curves are drastically mitigated.

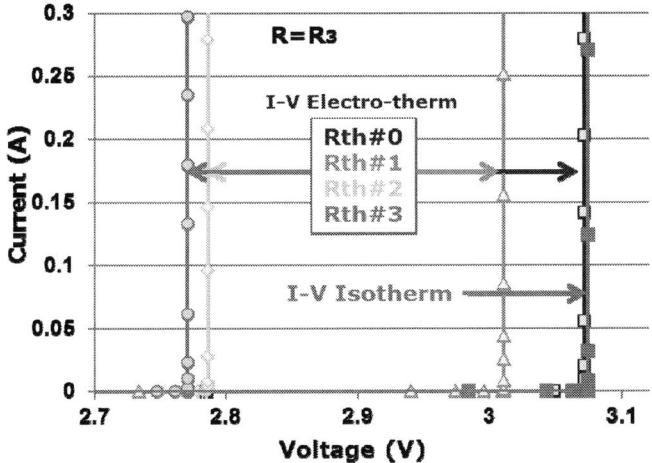

Fig. 13: I-V response to AVS stress for $R=R_3$ and #Rth

Due to the fact that the ESD stress is with low energy and due to the thermal time constant the Tmax and TAverage are the same. Figure 14 reports Tmax and Taverage for all Rth values

and for the same value for the external resistor. It is clearly shown that major differences are observed. Lower Rth allows faster heat dissipation while for higher Rth heat is drastically increased.

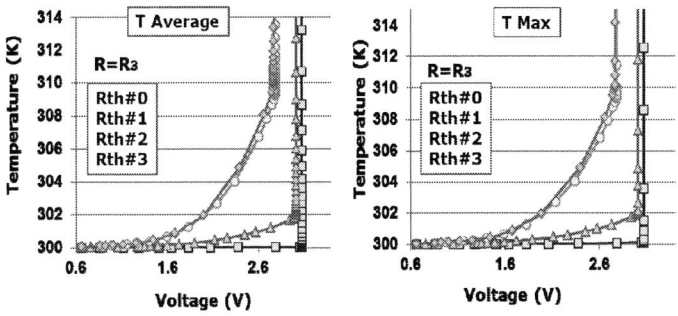

Fig. 14: Average and Maximum temperature comparison for different thermal resistances for $R=R_3$.

CONCLUSION

This paper presents preliminary results on isotherm and electro-thermal numerical simulations for BIMOS transistor high k metal gate designed in thin silicon film 28nm UTBB FDSOI advanced CMOS technology. This design solution is aimed for ESD protection at high and low energy dissipation. Stresses used were ACS and AVS events. Moreover, four R design resistances and four thermal resistances Rth are used as parameters to evaluate the I-V response and the robustness. It is shown that even if the R_0 configuration provides the lowest triggering voltage during a fast ESD event, it exhibits in some cases higher leakage due to the activation of the lateral parasitic structure. This is leading us into using the other design resistances in order to be more robust for the design window under consideration defined as the voltage range between power supply $V_{dd} +10\%$ and breakdown voltage $V_{BR} - 10\%$.

REFERENCES

[1] A. Amerasekera "ESD in Silicon Integrated Circuits", 2nd ed . Hoboken, NJ: Wiley 2002

[2] G. Meneghesso et al. «Turn-on speed of grounded gate NMOS ESD protection transistors ». *Microelectron. Reliab.*, Vol. 36, No. 11/12, pp. 1735-1738, 1996

[3] A. Dray et al. "ESD design challenge in 28nm hybrid FDSOI/bulk advanced CMOS technology". EOS/ESD 2012.

[4] T. Benoist et al "Improved ESD Protection in Advanced FDSOI by Using hybrid SOI/Bulk Co-integration". EOS/ESD 2010, USA.

[5] Ph. Galy et al. " ESD protection with BIMOS transistor for 28nm FDSOI CMOS technology and beyond". EUROSOI 2013.

[6] Ph. Galy et al. "BIMOS transistor in thin silicon film and news solutions for ESD protection in FDSOI UTBB CMOS technology". EUROSOI-ULIS 2015, Italy.

[7] Ph. Galy et al. "Numerical evaluation between Transmission Line Pulse (TLP) and Average Current Slope (ACS) of a submicron gg-nMOS transistor under Electrostatic Discharge (ESD)", Workshop EOS/ESD/EMI, LAAS-CNRS Toulouse, 2002

A High-Speed 2×VDD Output Buffer With PVTL Detection Using 40-nm CMOS Technology

Chua-Chin Wang[†], *Senior Member, IEEE,*
Tsung-Yi Tsai, and Wei Lin

Department of Electrical Engineering
National Sun Yat-Sen University
Kaohsiung, Taiwan 80424
Email: ccwang@ee.nsysu.edu.tw

Abstract—Not only PVT detection techniques but also a leakage compensation design are proposed to carry out 650/500 MHz 2×VDD output buffer in this paper. The proposed 2×VDD output buffer contains a novel PVTL (Process, Voltage, Temperature, Leakage) compensation circuit to resolve the problems in output buffers of nano-scale CMOS technologies. Particularly, the leakage compensation design is realized by an asynchronous control method to control current paths such that the switching loss and the slew rate in the output buffer can be reduced and increased, respectively. The proposed design has been realized and implemented by using a typical 40 nm CMOS process. The data rate is 650/500 MHz given 0.9/1.8 V supply voltage with a 20 pF load. The maximum slew rate is 3.5 (V/ns), and the core area is 0.052 × 0.213 mm².

Index Terms—I/O buffer, PVTL variation, mixed-voltage tolerant, slew rate compensation, gate-oxide reliability.

I. INTRODUCTION

Over the last decade, a great amount of transistors can be integrated in a chip through advanced nano-scale CMOS technology. The benefits of the nanometer technology have greatly changed how engineers design a chip, e.g., lower supply voltage, higher operation speed, and smaller power consumption. However, it's nearly impossible to make every chip fabricated using the same advanced process on a PCB-based system. Therefore, mixed voltage output buffers are needed very often to accommodate different voltage levels in such a scenario [4]-[7].

Several challenges such as offset caused by PVT variations, leakage current, and IC package parasitics, would be encountered while designing the I/O interfaces using nano-scale processes [8], [9]. These challenges would degrade the signal integrity, decrease the slew rate, and cause noise. Most prior designs utilized the logic delay method only to detect PVT variations to reduce the impact on slew rate, which in turn have been proved degrading the speed as well severely [1], [3]. Besides, the logic delay method can only recognize three corners, namely, TT, FF, SS, where the detection of SF and FS corners is left unsolved. Furthermore, the all-PVT corner detection circuit with various input signals is complicated in some other works [2].

Therefore, a high-speed 2×VDD output buffer with PVT detectors, leakage compensation circuit using 40 nm CMOS process is proposed in this study. By using the asynchronous control detection mechanism, the circuit compensates the slew rate and increases the data rate up to 650/500 MHz.

II. 2 × VDD OUTPUT BUFFER CIRCUIT DESIGN

Fig. 1 shows the block diagram of the proposed 2×VDD output buffer comprising 2 major blocks, i.e., PVTL compensation circuit and Output buffer. The details of these two function blocks will be disclosed in the following text.

A. N-PVT Variation Detector

Fig. 2 (a) shows the schematic of the N-PVT variation detector in Fig. 1, which are composed of high skew inverters, inverters, voltage comparators, capacitors and D flip-flops. Notably, the aspect ratio of PMOS and NMOS in the high skew inverter is different from those in a typical inverter, which makes the falling time at K3 larger than its rising time. Therefore, after the signal goes through a voltage comparator and a typical inverter, the duty cycle of the signals at K3b and K2b, is smaller than 50%. When the duty cycle becomes smaller, No[3] and No[2] will both turn into logic 1 to open the corresponding current paths to proceed the compensation.

B. P-PVT Variation Detector

Fig. 2 (b) is the schematic of the P-PVT variation detector in Fig. 1. Unlike the N-PVT variation detector, it is composed of low skew inverters instead of high skew ones, which force the rising time at Q3 to be larger than its falling time. Furthermore, the duty cycles at Q3b and Q2b are larger than 50%, which will turn node Po[3] and Po[2] into logic 1 to shut off the corresponding current paths.

C. Output Buffer

The Output buffer is shown in Fig. 3, where a pre-driver, a Vg1 generator, a VDDIO detector and an Output stage are included. Output buffer is commanded by the control signals from DOUT, Pcode[3:1] and Ncode[3:1], and then generate 6 signals, i.e., PDOUTa, PDOUTb, PDOUTc, Vg4a, Vg4b and Vg4c, which are the gate drives of the output MOSFETs in

†: Prof. C.-C. Wang is the contact author.

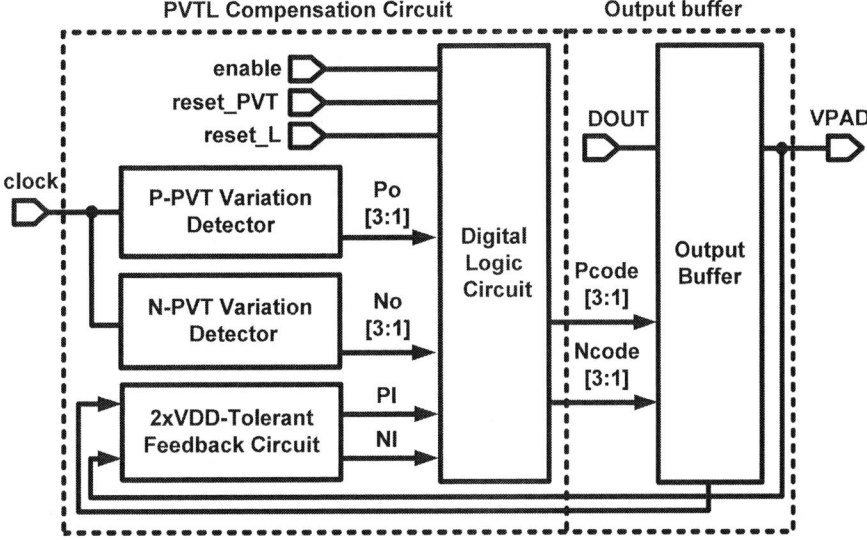

Fig. 1. The block diagram of the proposed output buffer.

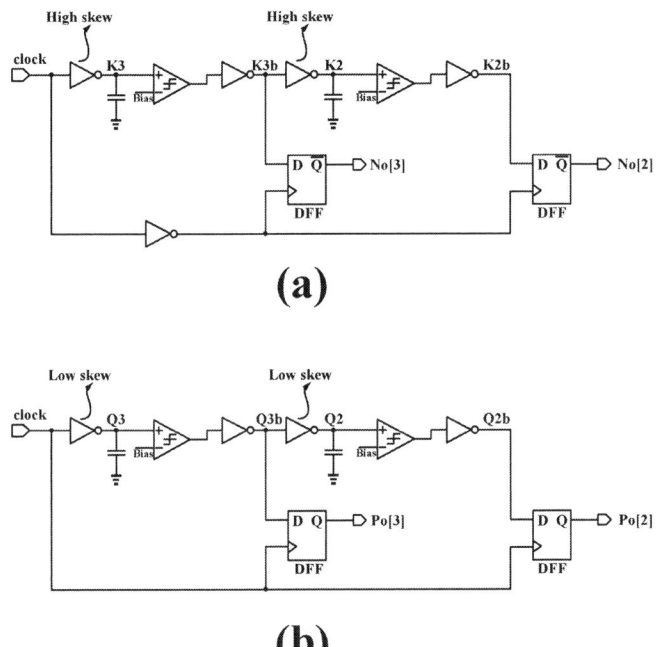

Fig. 2. Schematics of the (a) N-PVT variation detector and (b) P-PVT variation detector.

Output stage. Moreover, VD18 generated by VDDIO detector is designed to be 0.9 V and 0 V, when VDDIO is 1.8 V and 0.9 V, respectively. The description of Output buffer submodules is given as follows.

- Pre-driver : This logic circuit encodes input signals and boosts the driving current.
- Output stage : 3 PMOSs and 3 NMOSs are arranged in parallel to be indirectly driven by Pcode[3:1] and Ncode[3:1] so as to switch on/off current paths at dif-

ferent corners.
- VDDIO detector : According to the different voltages at VDDIO, VD18 generates a corresponding voltage to prevent the oxide layer of P2 from over-stress.
- Vg1 generator : It generates different gate drives at Vg1a, Vg1b, Vg1c to prevent the oxide layer of P1a, P1b and P1c from over-stress.

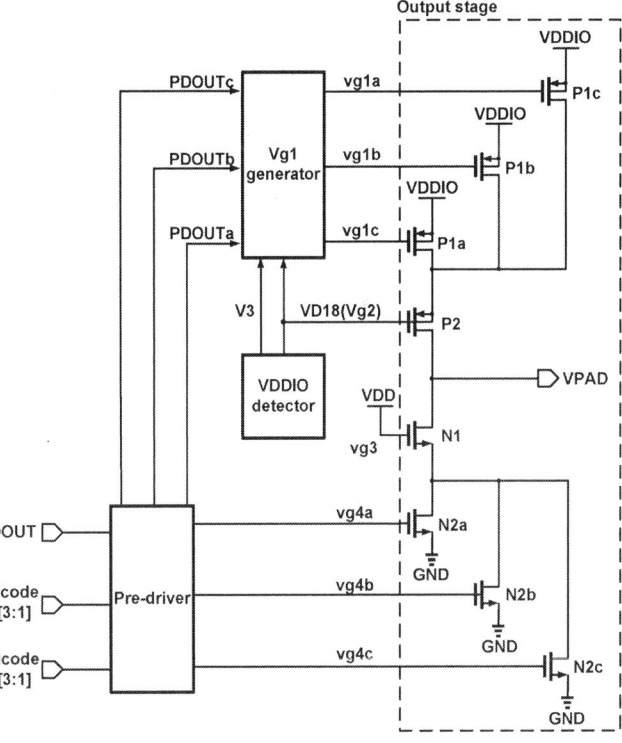

Fig. 3. Schematic of Output Buffer in Fig. 1.

978-1-4799-7670-6/15 $31.00 © 2015 IEEE   171

D. 2×VDD-Tolerant Feedback Circuit

Fig. 4 shows the schematic of the 2×VDD-Tolerant Feedback Circuit. The main function of this circuit is to feedback the output signal which maximum voltage is twice of VDD. Notably, the schmitt trigger delays Pl and Nl and convert them into logic signals, which are coupled to the following Digital Logic Circuit to carry out the leakage compensation.

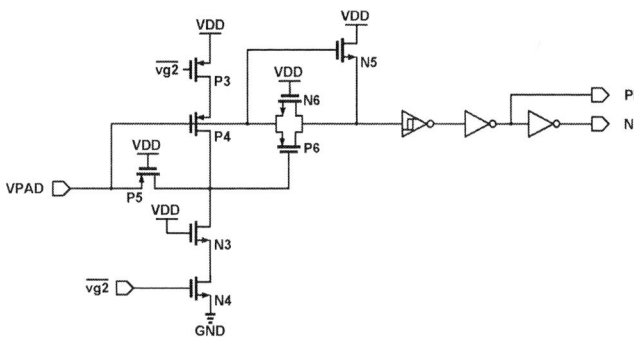

Fig. 4. Schematic of 2×VDD-Tolerant Feedback Circuit.

E. Digital Logic Circuit

Digital Logic Circuit in Fig. 1 is in charge of transmitting the control signals to drive the following Output Buffer. When reset_PVT is logic 0, No[3:1] and Po[3:1] are asserted. By contrast, when reset_PVT is high, Ncode[3:1] and Pcode[3:1] are set to [0,1,1]. When reset_L is logic 0, the leakage compensation is activated to turn Ncode[3:2] and Pcode[3:2] all into logic 0. When reset_L is high, the leakage compensation is closed.

III. SIMULATION AND IMPLEMENTATION

This work is realized and implemented by 40 nm 1P10M CMOS technology without using any thick-oxide device. Fig. 5 shows the layout of this work, where the overall chip size is only 0.597×0.545 mm^2 and a single buffer circuit is only 0.213×0.052 mm^2.

The buffer without compensation and the buffer with PVTL compensation are simulated in Tx mode with VDDIO = 0.9/1.8 V, as shown in Fig. 6 and Fig. 7, respectively. With PVTL compensation, the slew rate is enhanced to 2.095 (V/ns) from 1.813 (V/ns) and 2.998 (V/ns) from 2.488 (V/ns) under the given VDDIO = 0.9/1.8 V, respectively. Besides, the leakage current is reduced to 9.99 (mA) and 14.13 (mA) instead of 16.86 (mA) and 22.62 (mA) given VDDIO = 0.9/1.8 V, respectively. Moreover, the data rate is simulated to be 100 MHz when VDDIO = 0.9/1.8 V, respectively.

Fig. 8 shows output simulation of node VPAD in Fig. 3. The waveforms of all process, temperature and voltage corners are almost aligned at the edges, which justifies the compensation correctness of PVT variations.

The performance of the proposed design is summarized in the Table I. Table II shows the comparison with several prior

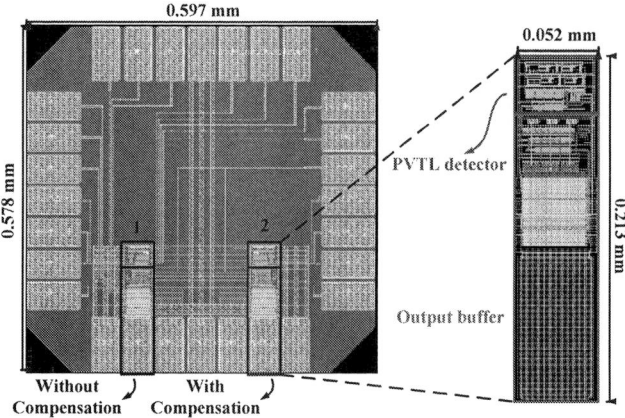

Fig. 5. Layout of the proposed design.

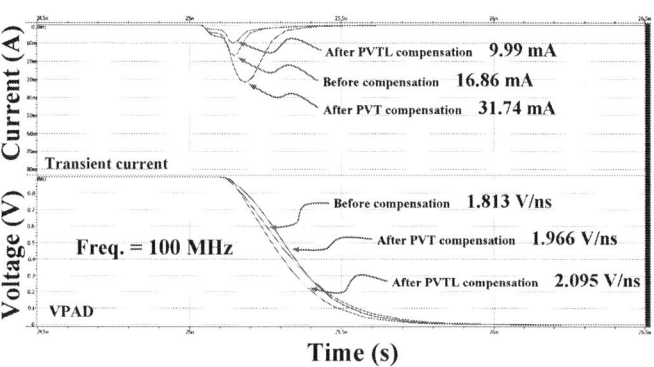

Fig. 6. The simulation of VPAD with VDDIO = 0.9 V.

works. The proposed design is the only one to provide all-PVTL detection and compensation to achieve a better slew rate.

TABLE I
COMPENSATION RESULTS AT DIFFERENT VDDIOS

VDDIO	0.9 V	1.8 V
Maximum data rate (MHz)	650	500
Slew rate improvement (%) (rise/fall)	23% / 28%	30% / 21%
Leakage current without compensation (mA)	16.86	22.62
Leakage current with compensation (mA)	9.99	14.13
Leakage reduction (%)	40.7%	37.5%

IV. CONCLUSION

In this work, the high-speed 2×VDD output buffer with PVTL detection is implemented and realized using a typical 40 nm CMOS process. The data rate is 650/500 MHz when VDDIO = 0.9/1.8 V, respectively. The maximum slew rate is 3.5 (V/ns) when VDDIO = 1.8 V. Notably, the PVT Variation Detector, 2×VDD-Tolerant Feedback Circuit, and Digital

978-1-4799-7670-6/15 $31.00 © 2015 IEEE

TABLE II
PERFORMANCE COMPARISON OF OUTPUT BUFFER

	[4] *ISCAS*	[5] *VLSI-DAT*	[3] *ISSCC*	[6] *ISSCC*	This work
Year	2014	2011	2007	2004	2014
Process (μm)	0.04	0.09	0.18	0.35	0.04
VDD (V)	0.9	1.2	1.8	3.3	0.9
VDDIO (V)	1.8/0.9	2.5/1.8/1.2/0.9	1.8	3.3	1.8/0.9
Maximum date rate (MHz)	500/460	345	500	50	650/500
Slew rate (V/ns)	$0.523 \sim 0.53$	$1.1 \sim 2.5$	$2.1 \sim 3.58$	N/A	$2.4 \sim 3.5$
Power (mW)	3.84	N/A	5.6	13.7	2.2
Core area (mm^2)	0.052×0.254	0.056×0.162	0.009	0.094×0.487	0.052×0.213
PVTL detection	NO	NO	NO	NO	YES

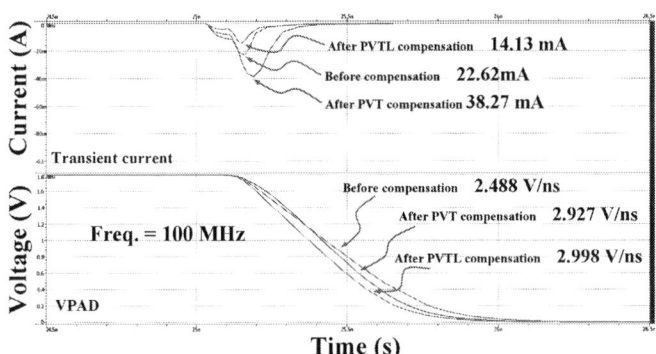

Fig. 7. The simulation VPAD with VDDIO = 1.8 V.

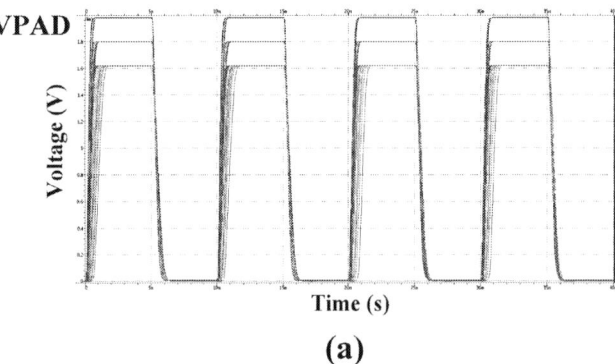

(a)

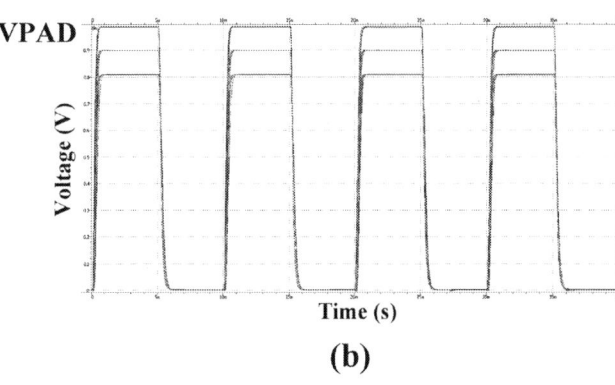

(b)

Fig. 8. The all-corner simulation VPAD when (a) VDDIO=1.8V (b) VDDIO=0.9V.

Logic Circuit are implemented on a single chip, where the core area is only 0.213 × 0.052 mm^2. The area overhead is only 32% for a single output buffer. Therefore, this work is proved on silicon to detect all PVTL corners with compensation.

ACKNOWLEDGMENT

The investigation was partially supported by National Science Council, Taiwan, under grant NSC-102-2221-E-110-083-MY3. The authors would like to express our deepest gratefulness to CIC (Chip Implementation Center) in NARL (Nation Applied Research Laboratories), Taiwan, for the assistance of thoughtful chip fabrication.

REFERENCES

[1] Y.-H. Kwak, I.-H. Jung, and C.-W. Kim, "A Gb/s+ slew-Rate/impedance-controlled output driver with single-cycle compensation time," *IEEE Trans. Circuits Syst. II Exp. Briefs,* vol. 57, no. 2, pp. 120-125, Feb. 2010.

[2] C.-L. Chen, H.-Y. Tseng, R.-C. Kuo, and C.-C. Wang, "On-chip MOS PVT variation monitor for slew rate self-adjusting 2×VDD output buffers," in *Proc. IEEE Int. Conf. on IC Design Technology (ICICDT),* pp. 1-4, May 2012.

[3] Y.-H. Kwak, I. Jung, H.-D. Lee, Y.-J. Choi, Y. Kumar, and C. Kim, "A one cycle lock time slew-rate-controlled output driver," in *Proc. IEEE Int. Solid-State Circuits Conf. (ISSCC),* pp. 408-611, Feb. 2007.

[4] W.-J. Lu, H.-Y. Tseng, and C.-C. Wang, "A High-Speed 2×VDD Output Buffer With PVT Detection Using 40-nm CMOS Technology," in *Proc. IEEE Int. Symposium on Circuits and Systems (ISCAS),* pp. 2079-2082, May 2013.

[5] C.-L. Chen, H.-Y. Tseng, R.-C. Kuo, and C.-C. Wang, "A slew rate self-adjusting 2×VDD output buffer with PVT compensation," in *Proc. IEEE Int. Symposium on VLSI Design, Automation, and Test (VLSI-DAT),* pp. 1-4, Apr. 2012.

[6] M. Bazes, "Output buffer impedance control and noise reduction using a speed-locked loop," in *Proc. IEEE Int. Solid-State Circuits Conf. (ISSCC),* pp. 486-541, Feb. 2004.

[7] C. H. Lim and W. R. Daasch, "Output buffer with self-adjusting slew rate and on-chip compensation," in *Proc. IEEE Symposium on IC/Package Design Integration,* pp. 51-55, Feb. 1998.

[8] S. W. Choi and H. J. Park, "A PVT-insensitive CMOS output driver with constant slew rate," in *Proc. IEEE Asia-Pacific Conf. on Advanced System Integrated Circuits,* pp. 116-119, Aug. 2004.

[9] C.-T. Yeh and M.-D. Ker, "New design of 2 VDD-Tolerant power-rail ESD clamp circuit for mixed-voltage I/O buffers in 65-nm CMOS technology," *IEEE Trans. Circuits Syst. II, Exp. Briefs,* vol. 59, no. 3, pp. 178-182, Jul. 2003.

3D Monolithic Integration: stacking technology and applications

Ionut RADU, Bich-Yen NGUYEN, Gweltaz GAUDIN and Carlos MAZURE

Soitec, Chemin des Franques, Parc Technologique des Fontaines, 38190 Bernin, France

ionut.radu@soitec.com

Abstract- Wafer level stacking of single crystal films enables 3D monolithic integration of electronic devices. The monolithic stacking technology based on Smart Cut[TM] enables front end integration of large variety of devices with nanometer alignment capability; therefore it provides more degree of freedom for the designers and integration for high density and better performance. Several applications can fully take the advantage of using the monolithic 3D stacking technology.

Keywords—wafer stacking, 3D monolithic, Front End integration.

I. INTRODUCTION

3D integration technology is an emerging technology that complements conventional 2D scaling to enable designers to achieve higher levels of integration by allowing multiple die to be stacked vertically. The benefits of 3D IC stacking, such as increasing inter-die communication bandwidth, reducing form factor and lower power consumption are proven and the technology barriers to die stacking are being steadily removed. While several design and technology challenges still need to be solved, currently the biggest challenge for 3D-IC is the manufacturing cost factor which depends on the industry's learning curve.

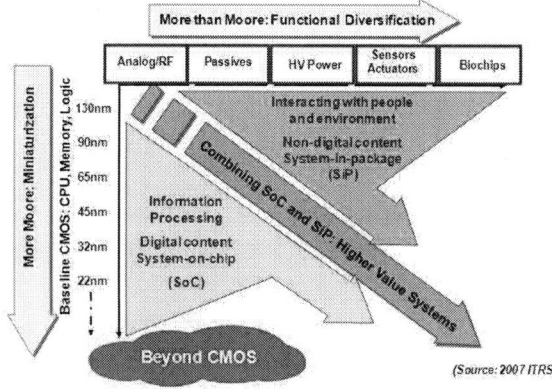

Fig. 1.   3D Integration: Transistor scaling & functional diversification

3D integration aims at providing highly integrated systems by vertically stacking and connecting various materials, technologies, and functional components together. In this emerging field, new technologies and integration schemes will be necessary to meet the associated manufacturing challenges.

The current 3D-TSV technology for integrated circuits is struggling with challenges such as reliability, yield and process cost for high volume manufacturing. 3D integration

approach at the device level has been previously suggested in the early 2000 by many authors [1-3]. The 3D sequential integration, also called monolithic integration, would involve obtaining a thin crystalline template atop an already processed semiconductor substrate (CMOS, DRAM, etc). Hence, a new silicon surface on which new devices can be processed is available for front end of line device integration but at a temperature constrained by the underlying layer composition. The main challenge for the 3D sequential integration scheme resides on the development of low-thermal budget processes for obtaining high quality devices.

Successful demonstration of various 3D sequential integration schemes within CMOS process has already been reported, the new crystalline template being obtained either by amorphous silicon deposition and subsequent crystallization [1] or by low temperature wafer bonding and layer transfer from an initial silicon on insulator (SOI) substrate [3]. In order to make such sequential integration a viable alternative for 3D IC fabrication, the development of low temperature 3D stacking processes compatible with high volume manufacturing is required.

II. LAYER STACKING BY SMART CUT

Smart Cut[TM] technology [4] provides a path to monolithic 3D integration and enables the transfer of a blanket layer of single-crystal Si film onto a processed wafer (Figure 2). In combination with low temperature processing, layer transfer of very thin films (typical thickness is less than 1μm) onto CMOS processed handle wafers is achieved at the wafer level making the technology attractive for very high density vertical integration. Moreover, this integration scheme can be repeated in an iterative mode by using recycling techniques of donor wafer.

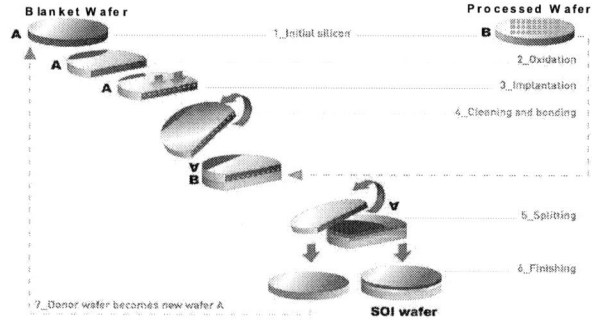

Fig. 2.   Schematic of layer stacking by Smart Cut[TM].

On this new single-crystal Si surface, a second level of devices can be processed and this integration can be repeated in an iterative mode. This technology benefits from existing high-volume manufacturing SOI infrastructure in addition to optimum donor wafer recycling techniques for lower cost-of-ownership. Compared to standard back thinning techniques, the Smart Cut enables ultrathin films (down to few tens of nm) with excellent uniformity, thus simplifying and lowering the cost-of-ownership of the TSV process, or simply replacing TSV with cost effective regular backend vias instead, thus making this technology attractive for applications that require higher interconnect densities. The throughput of this process is ~20-25 wafers/hour because no critical alignment is required during the bonding process.

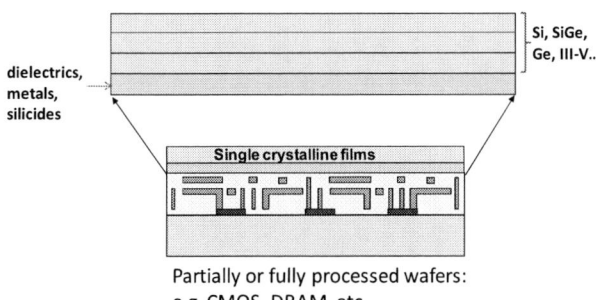

Fig. 3.   Schematic of 3D monolithic stacking technology

There are a number of 3D stacking integration options enabled by Smart Cut - using low temperature oxide bonding [5] and metal-metal non-thermo compression bonding, enabling formation of electrical interconnects during the bonding process [6]. This is based on our ability to develop low temperature bonding processes that are critical for successful Smart Cut.

Typical processing prior wafer stacking includes surface planarization of handle substrate (e.g. chemical-mechanical polishing of CMOS processed wafer) maintaining wafer edge quality and wafer micro-roughness <0.5nm RMS. With the appropriate choice of pre-bonding surface conditioning, high bonding energies can be achieved even after low temperature treatments (below 400°C).

Recently, an original integration scheme was suggested for the low temperature layer transfer technology using the Smart Cut™ technology in combination with low temperature solid-phase epitaxial re-growth process [7,8] offering new possibilities of 3D layer stacking compatible with high density device integration constraints.

In the case of layer transferred PN junctions, the implantation defects are eliminated from the final diode structure by using an appropriate process sequence with temperature not higher than 500°C. Wafer bonding and layer splitting processes are performed before the re-crystallization annealing process. Therefore, the residual EOR defects are removed after the layer splitting by conventional CMP process, resulting in defect-free PN diode. The measurements of the I-V characteristics are shown in Fig. 4.

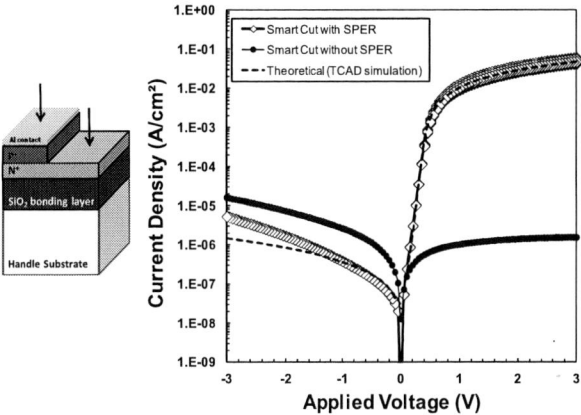

Fig. 4.   I(V) diode characteristics after the low temperature layer transfer process [8].

III.   APPLICATIONS

While performance, power, area or density (PPA) FOM is a driver, cost is a limiting factor and timing depends on overcoming numerous challenges which are application specific. 3D vertical stacking can improve system performance of a microprocessor by reducing interconnect length (RC delay), thus improving latency, power consumption, and bandwidth.

The 3D vertical integration can be categorized into several levels in terms of partitioning granularity with increasing number of vertical connections:

- core level integration (e.g. core + memory stack) providing high access bandwidth
- block-level integration, where functional blocks are partitioned into different tiers. The number of vertical connections is usually more than core + memory stacking.
- gate-level integration, where tiers are partitioned based on each single gate.
- Transistor-level integration, which partitions the transistors into different tiers.

While large Through Silicon Via (TSV) is used to connect at the core and block level and limited by reliability, yield, density and cost, The 3D monolithic integration technology enables high density of vertical connections than TSV for stacking at gate- and transistor-level, therefore it provides more degree of freedom for the designers and integration for high density and better performance. Still more research and development is needed for successfully using the 3rd dimension integration for future SoC product.

Benefits of 3D monolithic integration could be foreseen in case of vertical devices but also using conventional planar devices:

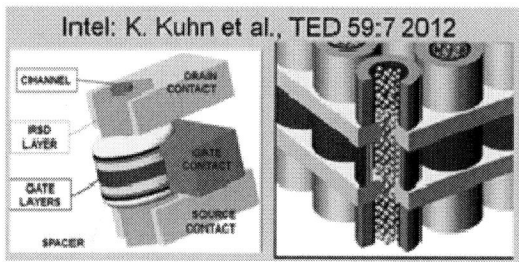

Fig. 5.   Vertical transistor device [8].

NAND Flash scaling problems in 2-D have given rise to several monolithic 3-D process and device architectures that involve vertical cylindrical channels (Figure. 5). The transition to 3D NAND cell enables a path for 'Effective Sub10nm' scaling.

Overview of 3D NAND Designs				
3D Cell (Supplier)	p-BiCS (A)	TCAT (B)	Smart (C)	(D)
Structure				
	Tanaka. H. VLSI 2007	J. Jang, VLSI 2009	Choi, IEDM 2012	G. Hawk, FMS 2011
Charge Storage	SONOS	SONOS	SONOS	-
Key Issues	High Etching A/R Memory Hole RIE	Very High Etching A/R Memory Hole RIE WL Separation	High Etching A/R Memory Hole RIE WL Separation	-

Fig. 6.   Schematic of vertical NAND memory designs [9]

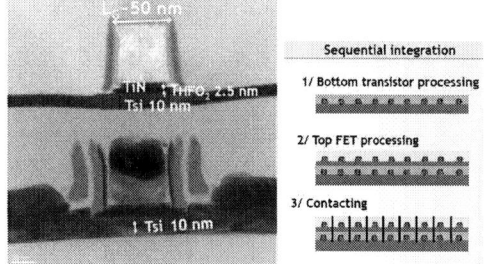

Fig. 7.   Monolithic integration of planar transistors [10]

IV. SUMMARY

Wafer level stacking of single crystal films enables 3D monolithic integration of electronic devices. This stacking technology provides an alternative path to conventional scaling and enables front end integration of large variety of devices with nanometer alignment capability.

Low temperature 3D wafer stacking technology for very dense device integration is presented in this paper. It is shown that good electrical properties are obtained after the layer transfer process without exceeding 500°C annealing.

An optimized process integration scheme is introduced based on the Smart Cut™ technology in combination with low temperature solid-phase epitaxial re-growth process leading to minimum diode leakage after the low temperature layer transfer. This novel 3D stacking technology can be repeated in an iterative mode at the wafer level, providing new possibilities for 3D sequential integrations of advanced CMOS devices.

Potential applications of 3D monolithic integration using planar and vertical devices are discussed.

ACKNOWLEDGMENT

The authors would like to thank colleagues from Soitec and CEA-LETI for helpful discussions.

REFERENCES

[1]   S-M. Jung et al., Three Dimensionally Stacked NAND Flash Memory Technology Using Stacking Single Crystal Si Layers on ILD and TANOS Structure for Beyond 30nm Node, Technical Digest of International Electron Devices Meeting (IEDM) 2006, pp1.

[2]   H. Liu, M. Kumar, J-K-O Sin, A novel 3-D BiCMOS technology using selective epitaxy growth (SEG) and lateral solid phase epitaxial (LSPE), IEEE Electron Device Letter, 2002, Volume 23, Issue 3, pp 151 – 153.

[3]   P. Batude et al., Advances in 3D CMOS Sequential Integration, IEEE international electron devices meeting (IEDM) 2009, p. 1–4.

[4]   M. Bruel, "Silicon-on-Insulator Material Technology," Electron. Lett. 31, 1201 (1995).

[5]   M. Sadaka and L. Di Cioccio, Building Blocks for Wafer Level 3D Integration, Solid State Technology, 2009, 52, p.20.

[6]   L. Di Cioccio et al., Wafer Level 3D Stacking using Smart CutTM and Metal-Metal Direct Bonding Technology, presented at Semiconductor Wafer Bonding 12: Science, Technology, and Applications, Electrochem. Soc. Meeting, 2012, Honolulu, Oct. 7-12.

[7]   G. Gaudin et al., Physical and Electrical Properties of Thin Doped Silicon films Obtained by Low Temperature Smart Cut and Solid Phase Epitaxy, ECS Journal of Solid State Science and Technology, 2 (12) P534-P538, 2013

[8]   I. Radu et al., Novel Low Temperature 3D Wafer Stacking Technology for High Density Device Integration, ESSDERC 2013, Bucharest.

[9]   K. Kuhn et al., Considerations for Ultimate CMOS Scaling, Transactions of Electron Devices, IEEE, 59 – 7, pp. 1813-1828, 2012.

[10] J. H. Yoon, Flash and DRAM Si Scaling Challenges, Emerging non-volatile memory technology, Flash Memory Summit, 2013, Santa Clara.

[11] P. Batude et al., Advances, challenges and opportunities in 3D CMOS sequential integration, Electron Devices Meeting (IEDM), IEEE 5-7 Dec. 2011, Washington, DC.

Through Silicon Via to FinFET noise coupling in 3-D integrated circuits

A. Rouhi Najaf Abadi[1,2], W. Guo[1], X. Sun[1], K. Ben Ali[3], J.P Raskin[3], M. Rack[3], C. Roda Neve[1], M.Choi[4], V. Moroz[4], G. Van der Plas[1], I. De Wolf[1], E. Beyne[1], P. Absil[1]

[1] *imec, 3001 Leuven, Belgium*
[2] *Dept. Materials Engineering, Fac. Engineering, KU Leuven, 3001 Leuven, Belgium*
[3] *ICTEAM, Université Catholique de Louvain, 1348 Louvain-la-Neuve, Belgium*
[4] *SYNOPSYS, Mountain View, CA*

Abstract— **High speed TSV signals can penetrate through the dielectric liner material, transfer in the silicon substrate and degrade the performance of FEOL devices. In this paper we investigate TSV noise coupling to active device including both FinFET and planar transistors. Calibrated TCAD models are used to perform time domain analysis and understand the mechanisms of substrate noise interaction with active device. Parametric simulations are performed in order to understand the tradeoffs among different design parameters. The results demonstrate superior substrate noise immunity of FinFETs over equivalent planar transistors. In addition we show that a scaled TSV diameter, a novel TSV architecture with thick polymer liner, placing the substrate contact closer to active device and a TSV guard ring helps to mitigate the TSV noise. Finally the importance of electromagnetic coupling effects on Keep Out Zone (KOZ) extraction is illustrated.**

Keywords— 3-D integration, Through Silicon Via, noise, FinFET, TCAD

I. INTRODUCTION

Three dimensional (3-D) integration technology is a promising approach to merge Moore's and More than Moore concepts. Through Silicon Vias (TSV) and FinFETs are the key components to enable this combination. FinFET technology offers technology scaling beyond technology nodes achievable with planar devices and enables increasing transistor density to follow Moor's law. However FinFET technology integration into 3D stacked systems may degrade FinFET device performance due to thermo-mechanic and electromagnetic interactions between device and 3D system components e.g. TSVs, microbumps, etc [1-2]. TSVs are considered as the most likely solution for vertical connection between stacked dies. TSV is a copper filled via through the silicon die that is surrounded by liner material. This dielectric liner avoids DC contact to the substrate. However when the TSV carries a high frequency signal, noise may penetrate into the conductive substrate and affect the neighborhood device performance.

Fig. 1 illustrates three main elements of TSV noise coupling to active device. High frequency noise penetrates from TSV to substrate through the metal-oxide-semiconductor capacitor formed by TSV metal, dielectric liner, and the silicon substrate. This high frequency noise then is transferred through the lossy silicon substrate and eventually affects the performance of the

FEOL device (planar or FinFET). While the first two elements are well-studied and various experimental data and simulation studies have been reported on TSV noise mitigation techniques, the effect of substrate noise on FinFET device performance has not been adequately studied yet. Existing literature on TSV noise coupling includes planar FETs [3] and diode structures [4] as victim devices.

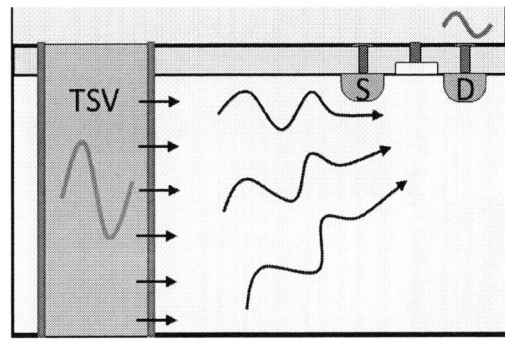

Figure 1. Main elements of TSV noise coupling to active device

The aim of this paper is to investigate the impact of TSV high frequency noise on FinFET and planar devices. Our approach is to use Synopsys Technology Computer Aided Design (TCAD) device simulator to build a calibrated model for TSV noise coupling to active device. This calibrated model enables to understand the mechanism of substrate noise coupling to FinFET drain current and to compare this mechanism to the one for planar devices. The impact of FinFET device channel length and fin width on noise coupling level is studied and explained based on these mechanisms. Furthermore the impact of TSV diameter, liner material and thickness on coupling level is investigated. The effect of substrate contact position and TSV guard ring on TSV noise mitigation is studied. Last but not least, it is shown that the coupling induced variation of transistor current should be considered together with thermo-mechanical stress impact for KOZ extraction.

II. TCAD MODEL CALIBRATION

The TCAD Sentaurus package from Synopsys is used to build a model for TSV-active device noise coupling. The model is built based on RF test structures available on imec FUJI test vehicle to allow further calibration of the model. It consists of a state of the art 5µm/50µm TSV covered by 200nm of oxide liner material and a 32nm nFinFET/ 65nm planar nMOSFET device (Fig. 2) placed at 10µm away from TSV. The specifications of FinFET and planar devices are listed in Table 1. Time domain analysis is performed to apply high frequency sine signals to the TSV and extract the active device saturation current variations. These variations are converted to S-parameter and compared with 2-port (Drain-to-TSV) S-parameter measurements performed on RF test structures to calibrate the models. The result of the model calibration is shown in Fig. 3 which shows a good agreement between simulation and measurement data. The FinFET device shows a better TSV noise immunity compared to the equivalent planar device. TSV noise coupling to FinFET device is 20-30 decibel smaller than the equivalent planar device at 100 MHz and 4-8 decibel smaller at 40 GHz.

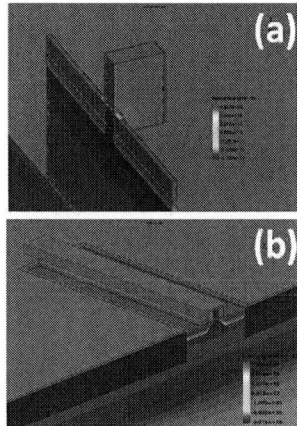

Figure 2. TCAD simulation structure for FinFET (a) and planar FET (b)

Table 1. FinFET and planar transistor specifications based on RF test structures avilable on IMEC FUJI test vehicle

Parameter	FinFET	Planar
Gate length (nm)	40	50
Total width (µm)	6.4	15
Fin width (nm)	20	x
Fin height (nm)	40	x

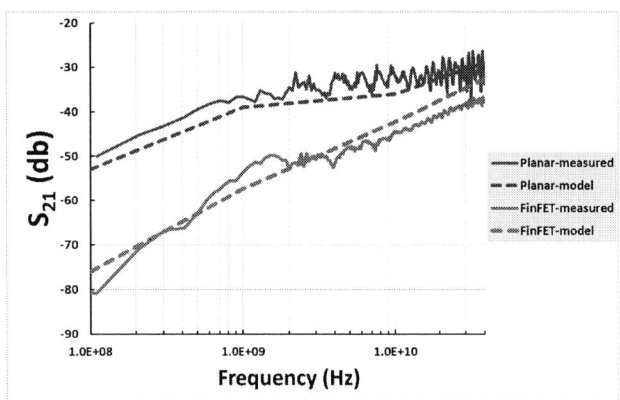

Figure 3. Results of model calibration for planar and FinFET devices mentioned in Table 1 for V_g=1V, V_d=1V, V_{TSV}=0V

III. RESULTS AND DISCUSSION

The calibrated TCAD models are used to investigate the TSV noise coupling mechanism to FinFET and planar devices and to study the impact of TSV, substrate and active device parameters on the noise coupling level.

A. Substrate noise to transistor drain current coupling mechanism

In order to evaluate the mechanism of substrate noise coupling to saturation drain current, the bulk electrostatic potential of the active device is extracted for TSV signal frequencies between 100 MHz and 40 GHz. The amplitude of bulk potential variation is a measure for TSV noise penetration and transfer through the silicon substrate. However we focus on the relative phase change between bulk potential and drain current of the active device. The coupling path between transistor bulk and drain is modelled with a parallel RC circuit (Fig. 4) in which R represents the resistive coupling path through bulk transconductance (gmb) and C represents the capacitive coupling path through drain-bulk junction capacitance (Cdb). The transfer function of such a circuit consists of 3 main parts: A low frequency range dominated by gmb, a high frequency range dominated by Cdb, and a transition frequency range. Fig. 5 (a-d) illustrates the phase change at different frequencies for a planar and a FinFET device in off state. A 90 degree phase shift at all frequencies for both planar and FinFET devices shows that capacitive coupling through the drain-bulk junction capacitance is the coupling mechanism in off-state. When looking at a FinFET device in on-state (Fig. 6-b) at 100 MHz there's no phase shift between bulk potential and drain current, suggesting that there is a resistive coupling path through bulk transconductance. The phase shift increases to 80 degrees when increasing the frequency to 40 GHz (Fig. 6-d) which shows a capacitive coupling path through Cdb at high frequencies. While for planar device coupling through gmb is still the dominant mechanism up to 40 GHz due to a much larger bulk transconductance (Fig. 6-a,c). For the particular FinFET device mentioned in Table 1 the extracted gmb (57.6 µS) and Cdb (1.90 fF) values from the model predict a transition frequency of 5.4 GHz. This is in close agreement with a transition frequency of 4 GHz obtained from the phase shift between bulk potential and drain current curves (Fig. 7).

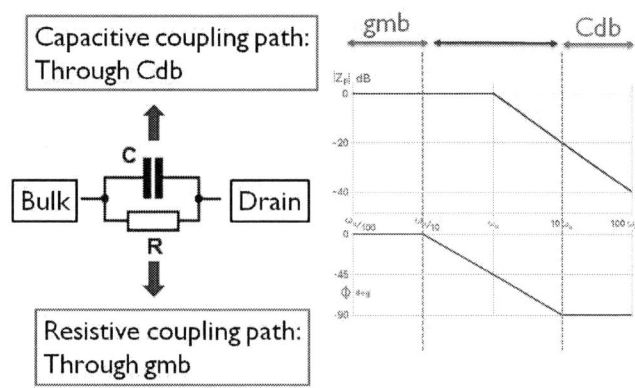

Figure 4. RC coupling path model between bulk potential and drain current.

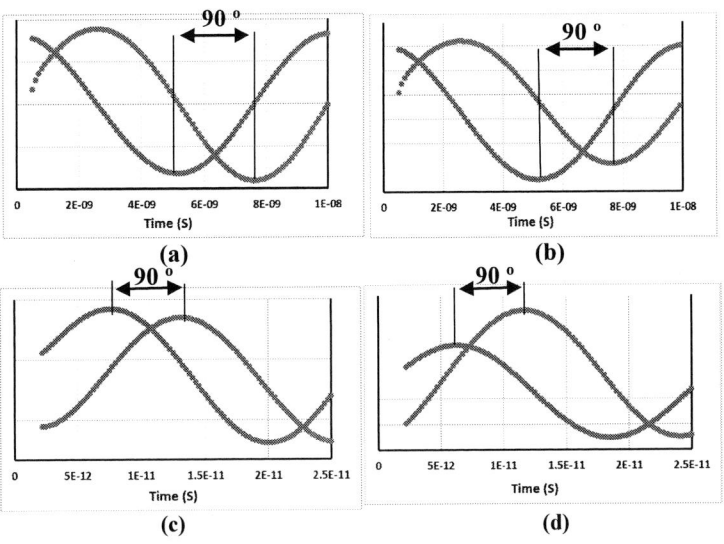

Figure 5. Phase shift between drain off current and bulk electrostatic potential for (a) planar at 100 MHz (b) FinFET at 100 MHz (c) planar at 40 GHz (d) FinFET at 40 GHz

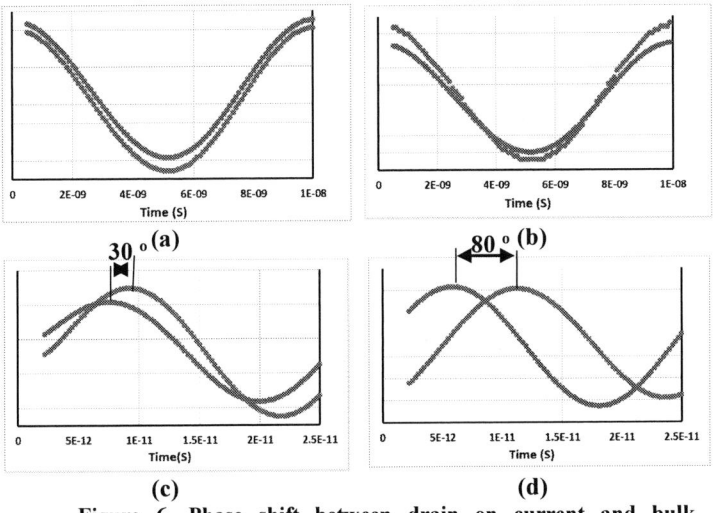

Figure 6. Phase shift between drain on current and bulk electrostatic potential for (a) planar at 100 MHz (b) FinFET at 100 MHz (c) planar at 40 GHz (d) FinFET at 40 GHz

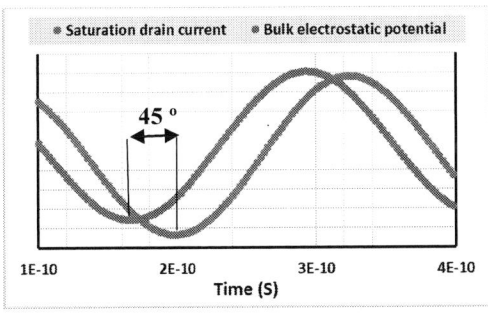

Figure 7. A 45 degree phase shift between drain current and bulk potential of FinFET device at 4 GHz. This is in good agreement with 5.4 GHz value obtained from RC model.

B. *Impact of active device parameters on noise coupling*

Fig. 8 shows the impact of FinFET channel length and fin width on noise coupling when the device is in on-state. A long channel FinFET device (Lg=1μm) exhibits a 15 decibel smaller coupling level at 100 MHz compared to a short channel device (Lg=40nm). This is due to a smaller bulk transconductance of the long channel FinFET which is the dominant coupling path at low frequencies. Table 2 shows the extracted and measured gmb values for short and long channel FinFET which predicts a 13-15.5 dB smaller coupling level for a long channel device. For the case of planar devices there's a constant 7 dB difference for short and long channel devices at all frequency ranges (Fig. 8). This is due to a larger gmb value for the short channel device which is still the dominant coupling mechanism up to 40 GHz. A FinFET device with 40 nm fin width couples 6 dB more than a device with 20nm fin width at 100 MHz and 2.6 dB at 10 GHz. At low frequencies coupling is mainly through gmb and extracted gmb values from the model predict the 6 dB difference (Table 2). At high frequencies Cdb mainly contributes to coupling. Equation 1 is used to calculate the capacitance of the drain-bulk junction. The sidewall capacitance per unit length ($C_{j0,sidewall}$) is negligible since STI technology is implemented. While the area of fin bottom ($AD(S)$) becomes twice by increasing the fin width from 20nm to 40nm, the capacitance per unit area ($C_{j0,bottom}$) decreases by a factor of 1.4 due to the shape of the HDD doping profile. As a result Cdb is increased only by a factor of 1.42 ($\approx$ 3 dB) which matches quite well with the simulation data.

Table 2. Extracted and measured bulk transconductance values for FinFET device.

Gmb (µS)	Lg=40nm finW=20nm	Lg=1µm FinW=20nm	Lg=40nm finW=40nm
Extracted from model (1 fin)	1.2	0.2	2.4
Measured (10 fins)	7.80	1.69	x

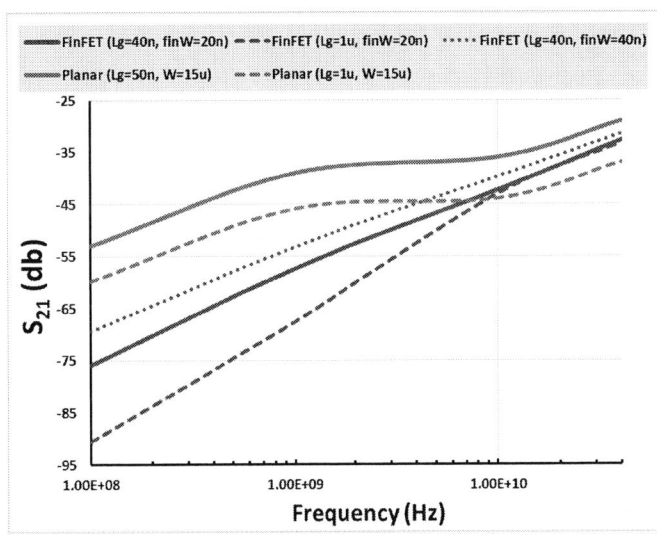

Figure 8. Impact of device channel length and fin width on noise coupling level.

Equation 1

$$Cj = AD(S) \cdot \frac{C_{j0,bottom}}{\left(1 + \frac{V_R}{\emptyset_0}\right)^{m_b}} + PD(S) \cdot \frac{C_{j0,sidewall}}{\left(1 + \frac{V_R}{\emptyset_0}\right)^{m_s}}$$

C. Noise mitigation and impact of TSV architecture on noise coupling

In addition to the impact of substrate noise on active device performance, it is important to investigate the effect of TSV architecture on noise penetration and transfer through the substrate. Fig. 9 illustrates the TSV to planar/FinFET noise coupling level for 3 different TSV architectures: 5µm/50µm and scaled 3µm/50µm via-middle TSVs with 200nm of oxide liner, and via-last TSV architecture with thick (3µm) polymer liner ("donut" TSV). Both scaled and donut TSV architectures reduce the level of coupling mainly at low frequencies- 4 and 20 decibel at 100 MHz respectively- where TSV capacitance is the dominant factor in noise transfer function. The amount of coupling reduction is in close agreement with calculated TSV capacitance ratios considering the TSV as a cylindrical shape capacitor (equations 2-4). Scaling the TSV diameter, thicker liner materials and polymer based liner materials with smaller dielectric constant reduce TSV noise penetration to the substrate. In order to mitigate the TSV noise impact on active device two methods are studied: reducing the distance between substrate contact and active device, and adding a guard ring to the TSV. Fig. 9 (black lines) shows the impact for TSV to FinFET devices noise coupling when these mitigation techniques are implemented. Placing the substrate contact closer to the active device and adding the guard ring makes it easier for the substrate noise to be sunk to the ground. This leads to a 4-6 db noise reduction by placing the substrate contact closer to the active device and 15-17 db noise reduction by adding a guard ring.

Equations 2-4

$$C_{cylinder} = \frac{2 \pi \varepsilon_0 \varepsilon_r h_{TSV}}{\ln\left(\frac{R_{TSV,Total}}{R_{TSV,Copper}}\right)}$$

$$C_{5/50} \approx 1.62 \, C_{3/50} \quad (\sim 4.2 \, db)$$

$$C_{5/50} \approx 15.4 \, C_{Donut} \quad (\sim 23 \, db)$$

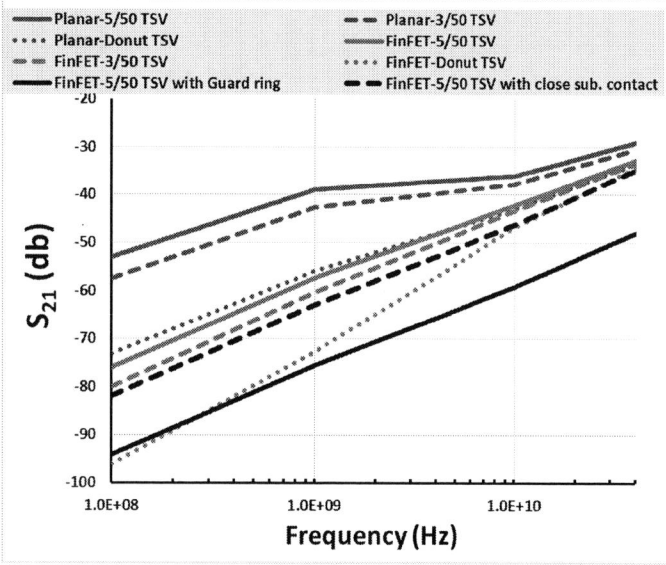

Figure 9. Impact of TSV architecture and noise mitigation techniques on noise coupling.

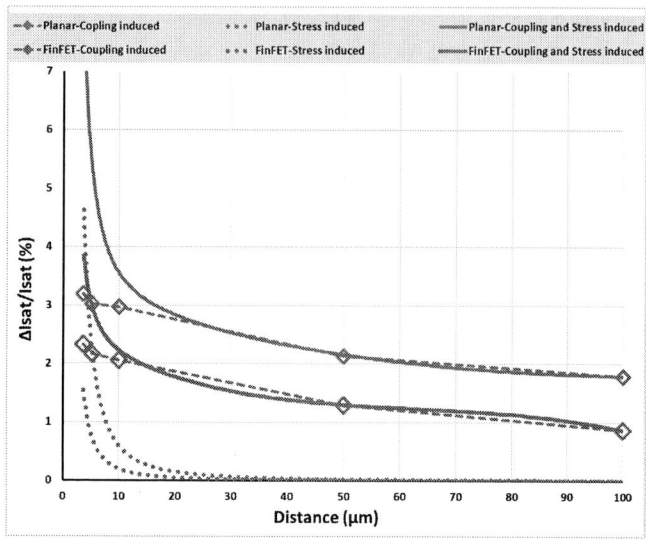

Figure 10. Coupling and stress induced saturation current variation for a 10GHz, 1V TSV signal. No TSV noise mitigation technique is implemented in the model.

D. Coupling induced Keep Out Zone (KOZ)

In order to extract the KOZ for a certain device and TSV technology, coupling induced current variations must be considered together with stress induced current changes. Fig. 10 shows the relative change of the saturation current induced by TSV coupled noise (at 10 GHz, 1 Volt TSV signal) and TSV stress for both FinFET and planar device. When the active device is located at 10µm away from the TSV, the TSV noise changes the saturation current by 3% and 2% for planar and FinFET devices respectively. The coupling induced current change decreases much slower with distance than the stress induced current change. Fig. 10 illustrates that it is important to add up TSV coupling and stress induced current changes especially for larger TSV to device distances. For example for the specific simulated FinFET device located 10µm away from the TSV, the current change induced by TSV stress is only 0.2%. This value increases to 2.2% when the impact from TSV noise is added. Depending on the level of allowed current variation, this may have a big impact on KOZ extraction.

IV. CONCLUSIONS

Calibrated TCAD models are used to investigate TSV noise coupling to FinFET and planar devices. It is shown for FinFETs, that TSV noise couples through gmb at low frequencies and Cdb at high frequencies. While for planar FETs coupling through gmb is still dominant up to 40 GHz due to a much higher gmb. This results in a better noise immunity for FinFET devices. The impact of transistor channel length and fin width on noise coupling is studied and described based on the proposed model for substrate noise to transistor drain coupling. It is shown that scaled and donut TSV architectures, placing the substrate contact closer to active device and adding guard ring around the TSV are efficient methods for TSV noise mitigation.

REFERENCES

[1] W. Guo et al., IEEE-IEDM 2013, pp. 12.8.1-12.8.4. [2] V. Cherman et al., ECTC2014. [3] H. Chaabouni et al., IEEE-IEDM 2010, pp. 35.1.1–35.1.4. [4] M. Brocard et al., Microelectronic Engineering, proceedings of Material for Advanced Metallization.

Simple Wafer Stacking 3D-FPGA Architecture

Motoki Amagasaki, Qian Zhao, Masahiro Iida, Morihiro Kuga, Toshinori Sueyoshi

Graduate School of Science and Technology, Kumamoto University

2-39-1 Kurokami, Chuo-ku, Kumamoto 860-8555, Japan

Email: {amagasaki, iida, kuga, sueyoshi}@cs.kumamoto-u.ac.jp, {cho}@arch.cs.kumamoto-u.ac.jp

Abstract—A three-dimensional (3D) integration based on wafer-to-wafer bonding using through-silicon vias (TSVs) has been developed for the fabrication of new 3D large-scale integrated chips. To balance between cost and performance, and to explore 3D field-programmable gate array (FPGA) with realistic 3D integration processes, we propose spatially distributed and functionally distributed types of 3D FPGA architectures. The goal of this paper is to elucidate the advantages and disadvantages of these two types of 3D FPGAs. According to our evaluation, when only two layers are used, the functionally distributed architecture is more effective. When higher performance is achieved by using more than two layers, the spatially distributed architecture achieves better performance.

I. INTRODUCTION

Conventional 3D FPGAs are classified into spatially distributed types and functionally distributed types on the basis of the distribution of die stacking. Spatially distributed 3D FPGAs are realized by stacking a set of similar silicon dies. Such 3D FPGAs employ a number of through-silicon vias (TSVs) in 3D switch boxes (SBs) to ensure high routability[1][2]. The relation between TSVs and SBs is such that when the channel width (CW) is 50, 100 inter-layer connections will be necessary in each SB. In light of the size of microbumps and TSVs, such architectures will be infeasible to scale down in the near future due to the area overhead. In contrast, functionally distributed types specialize each layer to one function. Existing FPGAs [3] have a structure in which logic circuits and configuration memory bits are placed on different layers. In this type, circuits on each layer are optimized separately. However, the connections between layers are specialized to each design, and so the generalization of connections is difficult. Therefore, although 3D stacking technology is very attractive, effective 3D FPGA architectures with good cost and performance are yet to be introduced.

To balance cost and performance, and to explore 3D FPGA architectures with realistic 3D LSI processes, we present two facile 3D FPGA architectures to distinguish between the features of functionally distributed and spatially distributed approaches. **(1) Functionally distributed approach**: This FPGA consists of two wafers (a logic layer and a routing layer) and mitigates the side effects of vertical wires[4]. Since vertical wires pass through microbumps by using face-down stacking, no TSVs are needed. By dividing routing resources into two layers, a smaller tile can be achieved. Smaller tile sizes result in shorter routing wires and faster signal transport, which improves routing performance. **(2) Spatially distributed approach**: This FPGA is divided into multiple layers that have the same structure, unlike in the functionally distributed type. This architecture can be expanded to more than two layers.

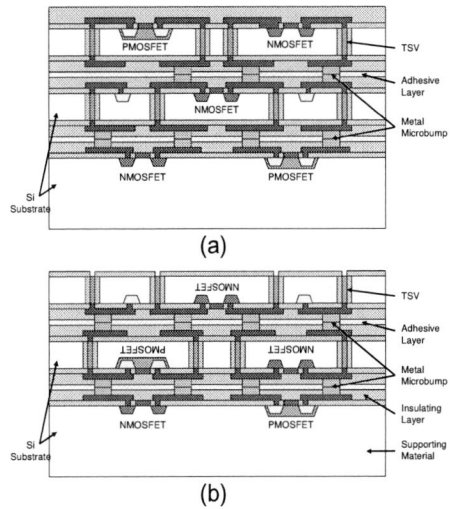

Fig. 1. Cross-sectional structure of 3D LSI architecture: (a) face-up stacking; (b) face-down stacking.

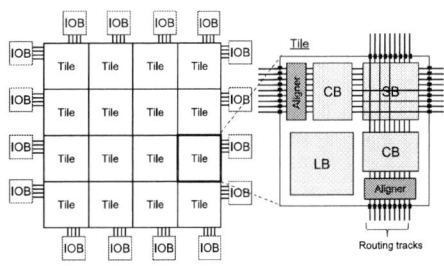

Fig. 2. Completely homogeneous routing architecture.

To decrease total number of vertical connections, vertical wire between layers is limited to outputs of logic cluster.

II. PROPOSED 3D FPGAS

A. 3D Integration and Basic Tile Structure

Figure 1 shows the cross-sectional structure of a 3D LSI circuit fabricated by Koyanagi's 3D integration technology [5]. The thinned upper layers are stacked face-up onto a thick LSI wafer that is face-up (Fig. 1(a)). Figure 1(b) shows the arrangement when the thinned upper layers are stacked face-down onto the thick LSI wafer. If the number of layers is limited to two, no TSVs are needed.

It is important to balance cost and performance when deciding on a 3D FPGA architecture. To this end, we consider the homogeneous (uniform) tile structure[6]. To treat various

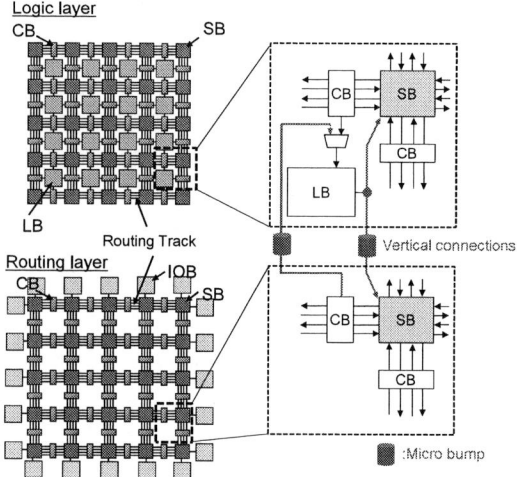

Fig. 3. Functionally distributed 3D architecture.

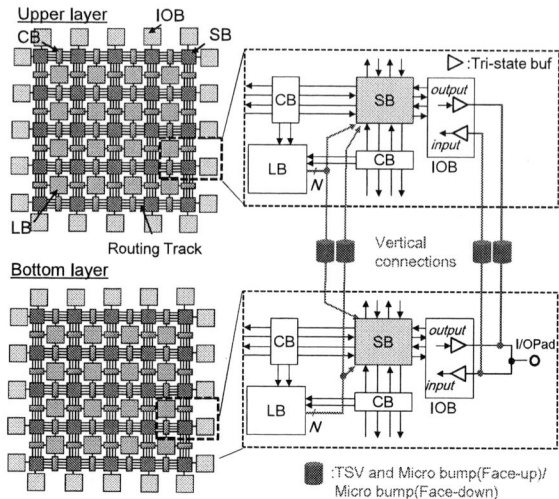

Fig. 4. Spatially distributed 3D architecture.

array sizes in a similar manner, it is important to simplify the tile structure. In this paper, the proposed 3D FPGAs are based on a homogeneous tile structure (Fig. 2).

B. Functionally Distributed 3D FPGA Architecture

Figure 3 shows details of the structures in the functionally distributed 3D FPGA. There are two layers in the architecture examined in the current research: a logic layer and a routing layer. The tiles on the logic layer have an LB and a small part of the routing resources; the tiles on the routing layer have only routing resources. The difference between conventional and proposed 3D SBs is that the 3D connections are made at the LB input and output pins. The number of inter-layer connections within one tile is equal to the total number of LB input and output pins. The number of LB inputs I is determined by the following formula [7].

$$I = \frac{K}{2} \times (N + 1) \quad (1)$$

Here, K is the number of logic cell inputs, and N is the cluster size. The number of vertical wires per one tile are $I + N$. For example, when $K = 6$ and $N = 4$, the numbers of LB inputs and outputs are 15 and 4, respectively. Therefore, the number of vertical connections per tile is 19.

We next discuss a method for determining the minimum width of the routing track channel for the two layers. To find this, we first set the initial CW of the logic layer to 1.5 times the number of LB input pins. Then, the areas used by the small part of connection blocks (CBs) and SB on logic layer can be calculated as the routing area of the logic layer; the tile area of the logic layer is the sum of the logic and small routing resource to connect neighbor LBs. The CW of the routing layer is calculated by allocating the size of the routing area as the size of the logic layer tile area. We next perform routing. If the routing is successful, then the next trail of the logic layer will have its CW set to half the current one. If routing fails, the CW of the next trail of the logic layer will be set to twice the current one. This process is repeated until the minimum CW that can lead to successful routing is found.

By dividing routing resources into two layers, we can achieve a smaller tile. Smaller tiles allow shorter routing wires and thereby enable faster signal transport, which improves the routing performance. The router can choose a network route on the logic layer or the routing layer. Although the number of layers in this type of approach is limited to two, when face-down stacking is used, no TSVs are necessary.

C. Spatially Distributed 3D FPGA Architecture

Figure 4 shows details of the structures in the spatially distributed 3D FPGA. There are two layers in the architecture examined in this example: an upper layer and a lower layer. Both layers have the same structure, which allows using a uniform mask set. The IOBs are connected to the two layers by vertical wires. One difference between this structure and that in our functionally distributed 3D FPGA (Fig. 3) is that vertical wires for LB inputs have been eliminated. The number of inter-layer connections within each tile is equal to twice the number of LB output pins. The total number of vertical wire connections, T_{VC}, is determined by the following formula.

$$T_{VC} = (Arraysize)^2 \times 2N + 4 \cdot Arraysize \times 2C_{IO} \quad (2)$$

In this formula, $Arraysize$ is the side length of the FPGA array, N is the cluster size, and C_{IO} is the IO capacity. Compared with the functionally distributed architecture, which requires $\frac{LUTinputs}{2} \times (N + 1) + N$ vertical connections per tile, this architecture reduces the required number of inter-layer connections. In addition, this architecture can be expanded to more than two layers by stacking dies of the same type. When the number of layers are 3 and 4, the maximum number of TSVs per one tile are $4N$ and $8N$, respectively.

III. DESIGN FLOW AND CAD TOOLS

In this section, we introduce a design flow and CAD tools that can be used for designing spatially distributed architectures. We developed the 3D FPGA design flow by using VTR 7.0 [8], which is the most current version as of this writing. As is done with 2D FPGAs, the circuit (in 'BLIF' format) is first technology-mapped with ABC [9] and then clustered with AAPack, which is included as part of VPR 7.0.

978-1-4799-7670-6/15 $31.00 © 2015 IEEE

Next, the clustered netlist is partitioned by using hMETIS [10], which can efficiently group LBs into a specified number of layers of similar size and with a minimal number of interconnections. These constraints are important for 3D FPGA partitioning. Allocating a similar number of LBs on all layers ensures that the resources on each layer can be fully used. Simultaneously, in order to minimize the overhead from TSVs, partitions with fewer interconnections are preferable. We wrote a script that can generate hypergraphs of LBs from the clustered netlist. The hMETIS tool is used to perform partitioning on the hypergraph generated by the script. Finally, layer allocation information is added back to the clustered netlist as LB attributes.

We created a 3D placement tool that uses VPR 7.0 placer as a base. The tool operates as follows. First, the layer allocation information is read from the clustered and partitioned netlist. Then, a conventional placement by simulated annealing is performed. During the placement process, LB blocks are freely swapped within each layer. The algorithm used is the bounding box method, which focuses on minimizing the bounding-box wire length of the circuit. We plan to implement timing-driven algorithms in future work.

Finally, routing was performed with our novel tool, the EasyRouter [11]. EasyRouter implements a pathfinder routing algorithm similar to that in VPR; however, EasyRouter simplifies the implementation of new FPGA architectures with various routing topologies. In addition, EasyRouter combined with VLSI CAD can provide highly accurate reports on area and critical path delays for FPGA designs that are based on standard cells. Routing is performed in two main steps. First, the router explores the minimum channel-width for each circuit. Next, the CW is fixed at 1.2 times the CW of the circuit with the highest minimum CW (i.e., 1.2-fold the maximum width), and the results are evaluated. This method ensures that all circuits are fairly evaluated on the same device with sufficient resources.

IV. EVALUATION

This section compares the architectures of a functionally distributed 3D FPGA (type 1) with that of the proposed spatially distributed 3D FPGA (type 2), evaluating the area, the critical path delay, and the area delay product. An island-style 2D FPGA(2D_Island) and 2D homogeneous FPGA(2D_Homo) are used as the baseline for evaluations. An analysis of the proposed 3D FPGAs from evaluation results is also given in this section.

A. Evaluation Conditions

All implemented 3D FPGA architectures are homogeneous FPGAs with a lookup-table (LUT) size of 6 and a cluster size of 8. For the routing architecture, we used a Wilton-type SB with $Fs = 3$. The Fc_{in} parameter was set to 0.5, which means that half of the tracks in the routing channel are connected to an LB input through a CB. We set the area of each TSV to $96 \mu m^2$[1] and the delay to 2.2 ps, taking these values from the report in [5]. The type 1 3D FPGA has only two layers, which is denoted by type 1_L2. The type 2 3D FPGAs were

implemented with from 2 to 4 layers, and these are denoted by type 2_L2, type 2_L3 and type 2_L4 for 2, 3, and 4 layers, respectively. The face-up stacked architectures are marked as "(face-up)", and the face-down stacked architectures are noted as "(face-down)".

The 19 largest MCNC benchmarks were used as the evaluation test suite. The device was designed to use 65-nm CMOS technology. The tile was synthesized with a Synopsys Design Compiler F-2011.09-SP2. For the area calculation, tile areas of all architectures were from synthesized results. The total area was calculated by multiplying the number of tiles with the area of one tile. For the delay evaluation, all CBs and SBs are considered to be composed of 2-to-1 multiplexers (MUXes) for both area and delay. The MUXes' physical parameters are incorporated by referenced to the standard cell library.

B. Area and Delay Results

The results of comparing chip areas between type 1 and type 2 3D FPGAs are shown in Fig. 5. We first implemented all benchmarks on each architecture and normalized the area results by the area of the 2D_Island FPGA. All results shown in Fig. 5 are an average of the normalized results for the corresponding architecture. We evaluated face-down and face-up 3D stacking methods for 3D FPGAs. The face-up counts include TSV area in the total area, and the face-down counts do not (more specifically, the area used by TSVs is 0 for face-down stacking).

First, we compare two-layer 3D FPGAs. The face-down stacked type 1_L2 (face-down) and type 2_L2 (face-down) reduce area by about 48.2% and 45.6% from the area of 2D_Island. In contrast, the reduction from the face-up stacked type 2_L2(face-up) is about 43.4%. Next, we compare type 2 3D FPGAs with different numbers of layers. We can see the trend of area reduction from the results of type 2 3D FPGAs having from 2 to 4 layers. Relative to the area of 2D_Island, the type 2_L2 (face-up), type 2_L3 (face-up), and type 2_L4 (face-up) designs reduce area by 43.4%, 56.1%, and 61.7%, respectively. The type 2_L4 (face-up) design offers the best performance of all examined architectures.

To summarize, for a two-layer 3D FPGA, type 1 with face-down stacking has the best performance on area. However, when implementing 3D FPGAs with more than two layers, where face-up stacking is necessary, the type 2 architecture offers a much smaller area.

The results of testing the type 1 and type 2 architectures for critical path delay are shown in Fig. 5, where the delays are normalized by the critical path delay of the 2D_Island architecture. Delay results are not significantly different between (all are within 10% of one another). This is because the MCNC benchmark circuits are not very large, and so the critical path delay is mainly from the LB rather than from routing. However, the results still show some trends. The type 1_L2 (face-down) has 5.3% better delay performance than the 2D_Island FPGA. The improvement is a result of the type 1 3D FPGA having more routing channels on the routing layer. When comparing the type 2_L2 (face-down) and type 2_L2 (face-up) designs, we can see that the TSV overhead in the critical path delay is very small. The TSV delay overhead is caused mainly by

---

[1]In [5], four poly-Si TSVs with a size of $2 \times 12 \ \mu m^2$ are connected in parallel in one vertical interconnection.

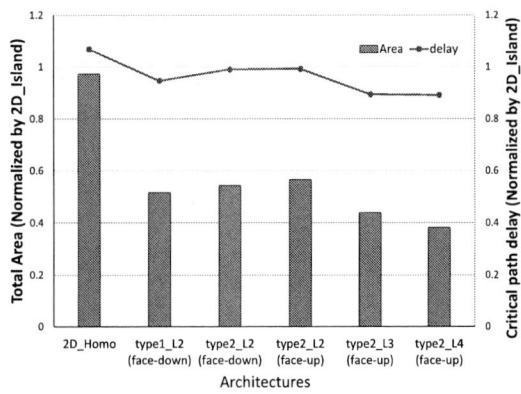

Fig. 5. Area and Delay Results.

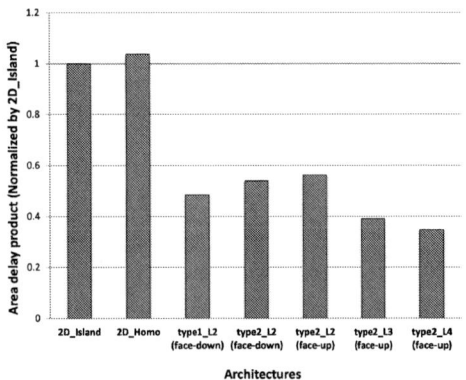

Fig. 6. Area delay product.

increased routing-wire delay due to the larger tile sizes and the delays of the TSVs themselves.

For the type 2 3D FPGA, the critical path delay was mainly affected by two factors. More LBs are allocated closer together as more layers are stacked, which improves the delay. Opposing this, the delay from MUXes of the SBs increase when stacking more layers. In the proposed type 2 3D FPGA, all LB outputs are connected to an SB of tiles at the same coordinate across all layers. When the number of MUX inputs per SB is m, each SB will have $(m - 1)$ 2-to-1 MUXes with logic depth $log_2(m)$. As a result, each type 2 SB MUX has more levels in its logic depth, which lengthens the total routing path. The performance of the type 2_L3 (face-up) architecture is similar to that of the type 2_L4 (face-up) architecture, which is because the increased SB MUX depth offsets the advantages of 3D allocation.

Additionally, partitioning and placement algorithms affect delay performance. In future work, we plan to develop a timing-driven partitioning and placement algorithm for the proposed 3D FPGA.

C. Area Delay Product Results

Finally, the area delay product results are shown in Fig. 6. For the face-down stacking method, the type 1_L2 (face-down) design performs 51.5% better than 2D_Island. For the face-up stacking method, the 4-layer type 2_L4 (face-up) performs 65.2% better than 2D_Island. When limiting the analysis to

two layers, the functionally distributed type of architecture (type 1) is most effective. However, if we prioritize performance and use more than two layers, the spatially distributed architecture (type 2) offers better performance.

V. CONCLUSION

In this paper we proposed a functionally distributed type and a spatially distributed type of 3D FPGA architecture to allow simple die stacking. According to the evaluation results, the functionally distributed type is more effective when limiting designs to two layers, but the spatially distributed architecture with more than two layers is a better choice when performance is prioritized. In this paper, we used relatively large TSVs. However, since more smaller TSVs are been developed in recent researches, functionally distributed type is more effective using these smaller TSVs. In future work, we are planning to stack multiple functionally distributed FPGAs with the face-up method and propose inter-layer high-speed communications that use TSV serial connections [12]. We also intend to improve the CAD toolsets to support algorithms that consider power consumption and timing in order to achieve better power and delay performance.

REFERENCES

[1] M. J. Alexander, J. P. Cohoon, J. L. Colflesh, J. Karro, G. Robins, and C. Science, "Three-Dimensional Field-Programmable Gate Arrays," Proc. of the Eighth Annual IEEE International ASIC Conference and Exhibit, pp.253-256, Sept. 1995.

[2] A. Gayasen, V. Narayanan, M. Kandemir, and A. Rahman, "Designing a 3-D FPGA: Switch Box Architecture and Thermal Issues," IEEE Trans. on VLSI Systems, Vol.16, Issue.7, pp.882-893, July 2008.

[3] T. Naoto, T. Ishida, T. Onoduka, M. Nishigoori, T. Nakayama, Y. Ueno, Y. Ishimoto, A. Suzuki, W. Chung, R. Madurawe, S. Wu, S. Ikeda and H. Oyamatsu, "World's first monolithic 3D-FPGA with TFT SRAM over 90nm 9 layer Cu CMOS," Proc. of VLSIT, pp.219-220, June 2010.

[4] T.Hamada, Q.Zhao, M.Amagasaki, M.Iida, M.Kuga and T.Sueyoshi, 'Three-Dimensional Stacking FPGA Architecture Using Face-to-Face Integration," Proc. of VLSI-SoC, pp.196-201, Oct. 2013.

[5] M. Koyanagi, T. Fukushima, and T. Tanaka, "High-Density Through Silicon Vias for 3-D LSIs," IEEE Journals & Magazines, Vol. 97, Issue 1, pp.49-59, Jan. 2009.

[6] K.Inoue, H.Yosho, M.Amagasaki, M.Iida, and T.Sueyoshi, "An Easily Testable Routing Architecture And Effecient Test Technique," Proc. of FPL, pp.291-294, Aug. 2011.

[7] E. Ahmed and J. Rose, "The Effect of LUT and Cluster Size on Deep-Submicron FPGA Performance and Density," Proc. of FPGAs, pp.3-12, Feb. 2000.

[8] J. Luu, J. Goeders, M. Wainberg, A. Somerville, T. Yu, K. Nasartschuk, M. Nasr, S. Wang, T. Liu, N. Ahmed, K. B. Kent, J. Anderson, J. Rose and V. Betz, "VTR 7.0: Next Generation Architecture and CAD System for FPGAs," ACM TRETS, Vol. 7, No. 2, pp. 6:1 - 6:30, June 2014.

[9] A. Mishchenko et al., "ABC: A System for Sequential Synthesis and Verification," http://www.eecs.berkeley.edu/ alanmi/abc/, 2009.

[10] N. Selvakkumaran and G. Karypis, "Multi-Objective Hypergraph Partitioning Algorithms for Cut and Maximum Subdomain Degree Minimization," IEEE Trans. on CAD of Integrated Circuits and Systems, Vol.25, Issue 3, pp.504-517, Feb. 2006.

[11] Q. Zhao, K. Inoue, M. Amagasaki, M. Iida, and T. Sueyoshi, "FPGA Design Framework Combined with Commercial VLSI CAD," IEICE Trans. on Information and Systems, Vol.E96-D, No.8, pp.1602-1612, Aug. 2013.

[12] T.Kajiwara, Q. Zhao, M. Amagasaki, M. Iida, M. Kuga and T. Sueyoshi, "A Novel Three-Dimensional FPGA Architecture with High-speed Serial Communication Links," Proc. of ICFPT, Dec. 2015.

Design of a Low-power Fixed-point 16-bit Digital Signal Processor Using 65nm SOTB Process

Duc-Hung Le[1*], Nobuyuki Sugii[2], Shiro Kamohara[2], Xuan-Thuan Nguyen[1], Koichiro Ishibashi[1], Cong-Kha Pham[1]

[1] The University of Electro-Communications, Tokyo, Japan

[2] Low-power Electronics Association & Project (LEAP), Japan

Abstract—**In this paper, a design of 16-bit fixed-point digital signal processor (DSP) is proposed. This DSP is based on the Harvard architecture, having two buses for ALU and a pipeline multiply accumulator (MAC). It composes of 16 general purpose 24-bit registers together with 41 four-cycle instruction sets. The DSP has a simple structure which is compact and flexible. The DSP is designed for low-power consumption, and implemented on ASIC using SOTB 65nm process which is a kind of SOI devices. The DSP chip consumes very low-power consumption 282μW at the operation voltage 0.55V and operation frequency 200MHz.**

Keywords— DSP, Fixed-point, Low- power, SOTB.

I. INTRODUCTION

Digital signal processing has become widely available in a large number of applications such as audio and speech signal processing, digital image processing, signal processing for communications, biomedical signal processing, etc. However, standard computers designed for general applications are not optimized for digital signal processing algorithms (i.e. digital filtering, Fourier analysis). Therefore, digital signal processors (DSP), kinds of specialized microprocessors, have been designed to handle specific digital signal processing tasks. Currently, beside the purpose-built hardware such as Application-Specific Integrated Circuits (ASICs), there are additional methods used for implementing digital signal processors such as general-purpose microprocessors, digital signal controllers, Field-Programmable Gate Arrays (FPGAs), and so on.

Research and design on DSP architecture is highly desirable due to its vital role. Ishikawa et al. [1] described a 16-bit fixed-point DSP for telecommunication applications. The DSP was fabricated with 0.5μm CMOS process. Kabuo et al. [2] designed a 16-bit fixed-point DSP with optimized multiply-accumulate (MAC) unit, memories, and instruction sets, which reduced the area and power consumption significantly. The DSP, which was fabricated on 0.5μm double-metal-layer CMOS process, could attain double speed MAC performance. O'Malley et al. [3] proposed a novel and highly versatile Reduced Instruction Set Computer (RISC) based fixed-point DSP which was optimized for digitally controlled switched mode power converters (SMPCs). The integrated circuit was built on a standard 0.35μm digital CMOS process, which occupied less than 1.5mm², and dissipated approximately 5mW at 3.3V. Lee et al. [4] represented an FPGA implementation of 16-bit fixed-

point DSP. The DSP included 211 instructions and 40-bit ALU, 6 level pipelines, 17-bit × 17-bit parallel multiplier for single-cycle MAC operation, 8 addressing modes, 8 auxiliary registers, 2 auxiliary register arithmetic units, 2 of 40-bit accumulators and 2 address generators. The DSP core carried out three test vector sets which were tested on FPGA at the 106 MHz clock rates.

Low-power consumption in DSP is very essential for a variety of applications. There are many approaches for low-power consumption such as sub-threshold region, multi threshold-voltage, multi-V_{DD}, clock gating, silicon-on-insulator (SOI) device, and so on. Among these methods, SOI device is a key approach for low-power consumption designs. Nowadays, design of DSP which is robust, efficient, and low-power is very necessary in high-performance data processing. The proposed 16-bit DSP will be implemented on SOI process for low-power consumption.

II. SOTB TECHNOLOGY

The fully depleted silicon-on-insulator (FD-SOI) MOSFET with an ultrathin buried oxide (BOX) named "Silicon On Thin Box (SOTB)" on 65nm process has been developed [5]-[6]. In this SOTB device, body bias voltage V_{BB} is applied through thin BOX layer. The wide-range back-gate controllability of this structure enables optimization of both performance and power after fabrication. This FD-SOI structure realizes high immunity from short-channel effects and enables a multiple threshold-voltage (V_{TH}) design, because of the high doping concentration in the substrate just beneath the thin BOX. The intrinsic channel suppresses the V_{TH} fluctuation by half of conventional bulk devices with the same V_{TH}. The very low leakage current is achieved on this 65nm SOTB process for low-power consumption.

III. DSP DESIGN

A. Features

The DSP, which is based on RISC, has a simple structure with Harvard architecture. It is compact and flexible in ASIC implementation on 65nm SOTB process for low-power consumption. The DSP is easily integrated with the other signal processing applications on a chip.

B. Architecture

The DSP architecture supports separate instruction and data ports, classifying it as the Harvard architecture to

(*) The author is now affiliated with The University of Science – VNUHCM, Vietnam.

978-1-4799-7670-6/15 $31.00 © 2015 IEEE

accelerate speed of accessing and processing data. The data port connects to both memory and peripheral components, while the instruction port only connects to the memory components. This architecure allows two memory blocks can access data in one clock cycle. The instruction and data memory size are 4K × 16 bits and 64K × 16 bits, respectively. Furthermore, the DSP has 16 24-bit general-purpose registers indexed from r0 to r15, which can be used in many arithmetic logic unit (ALU) operations, except for r0 which is not used as an operand. The registers from r0 to r15 are SRAM with synchronous writing and asynchronous reading. ALU operations take one or two inputs from registers, and store the results back in the registers with up to 24-bit precision. Although the control of subroutines is similar to other processors, a register (generally r0) must be dedicated to protecting the returning address. The same register, subsequently, can be utilized to return to the calling point.

The RISC-based architecture of DSP utilizes 2 24-bit buses, one for operands and one for results, as depicted in Fig. 1. When the DSP is reset, 16-bit program counter (PC) is set to address 0. At the start of each instruction, PC is used to read an instruction from the memory as an address register. The 16-bit instruction is stored in IR (Instruction Register) and the PC is automatically incremented to point to the next instruction. The bank of registers, that maintain the values of r0 to r15, is a memory with synchronous writing and asynchronous reading.

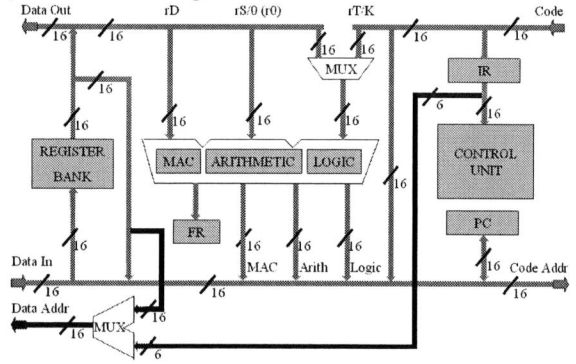

Fig. 1. Block diagram of DSP.

By using two different buses to exchange data to external elements (i.e. one to send data and one to receive them), the design avoids additional tri-state buffers that, indeed, are not necessary in designing chip. Although three or more buses can allow the processor to realize all the instruction operations in only one clock cycle, this DSP only employs two buses to avoid using dual-port memories for the bank of registers. Through these buses, rD (destination registers) and rS|T (source registers) can be observed during the first and last two phases of each instruction cycle, respectively.

The ALU, as depicted in Fig. 2, has three dedicated units and three buffered outputs: the MAC unit for multiplications, with or without accumulation; the arithmetic unit that operates the additions and subtractions, conditionally or not; and the logic unit. After each instruction, the zero (Z), sign (S), and overflow (V) flags are updated. This allows the DSP to carry out the corresponding allocations and conditional jumps. It is worthy to notice that this processor does not

require the carry flag (C) because of the single precision operations. The bus of results is dedicated to collecting the outputs of ALU and other sources (i.e. the external data upon reading from a port, the value of the PC when calling a subroutine, etc.)

Furthermore, the central operation of this DSP is the multiplication, which operates on two 16-bit values and generates a 24-bit result. Besides, all accumulations, additions, and subtractions are designed with a resolution of 24 bits. In other words, the operands and results are presented in <1.15> and <1.23> formats, respectively. In order to build the faster multiplier, a pipeline structure is used by dividing the operation into four stages, multiplying in each step a 16-bit operand by another one of only four bits, emitting a 20-bit intermediate result. This operation can be made by only four adders and an intermediate segmentation register as shown in Fig. 3. This saves execution time. If the operation is implemented in a single clock cycle, 15 adders are employed instead, and the execution time would have been greater due to the segmentation registers.

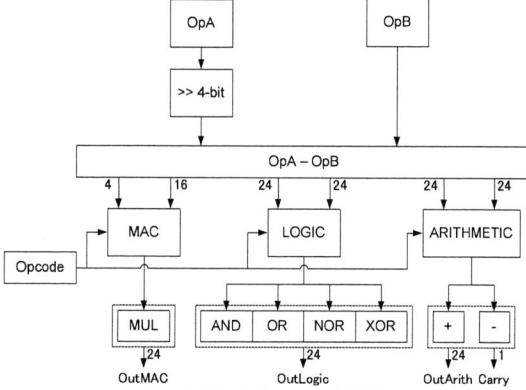

Fig. 2. Block diagram of ALU.

The most beneficial operation of this processor is the fixed-point product, with or without accumulation. There are two kinds of multiplications: those executed in "normalized format" (denoted as rD=rS*rT) and those in "integer format" (denoted as rD=rS×rT). Both use the same physical multiplier and introduce the same one-instruction-cycle latency, but the results are taken in different ways: the first one uses a <1.15> format for both operands (one bit for +/- sign, and fifteen bits for fraction: numbers between +0.999969 and -1.00); the other one, used mainly for shifts, takes the second operand in <8.8> format (8 bits for integer, 8 bits for fraction: numbers between +127.9961 and -128.00).

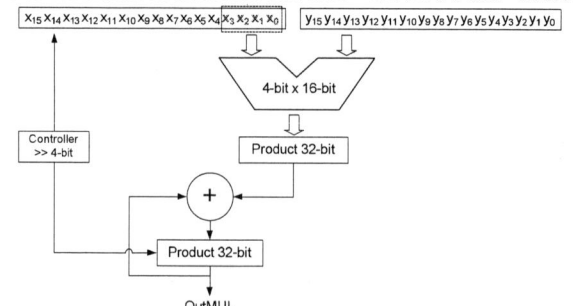

Fig. 3. Block diagram of a multiplier.

978-1-4799-7670-6/15 $31.00 © 2015 IEEE

C. Instruction sets

Most instructions of the DSP use three operands (rD = rS op rT). They can be registers (for example, r12 = r6 + r15), or constants (r12 = r6 + 0,9327). The DSP supports up to 41 instructions by combining 16 different opcodes (operation codes) with 8 flag types, which is divided into three types: R-type, I-type, and J-type.

R-type: The defining characteristic of the R-type instruction word format is that all arguments and results are specified as registers. R-type instructions contain a 4-bit opcode field OP and three 4-bit register fields D, S, and T. Fields S and T specify the source operands, and field D specifies the destination register. R-type instructions include arithmetic and logical operations; its format is illustrated as follows.

15 14 13 12	11 10 9 8	7 6 5 4	3 2 1 0
OP	D	S	T

I-type: The defining characteristic of the I-type instruction word format is that it contains an immediate value embedded within the instruction word. I-type instructions words contain a 4-bit opcode field (OP), a 4-bit register field D/S, and a 16-bit immediate data field IMM16. IMM16 is considered as signed and unsigned comparison operations. I-type instructions also include load and store operations; its format is depicted as follows.

First sub-instruction

15 14 13 12	11 10 9 8	7 6 5 4 3 2 1 0
OP	D/S	N/A

Last sub-instruction

15 14 13 12 11 10 9 8 7 6 5 4 3 2 1 0
IMM16

J-type: J-type instructions contain a 4-bit opcode field OP, 4-bit flag field FLAG, and 8-bit address field IMM8. J-type instructions include absolute jump such as CALL instruction and conditional jumps; its format is shown as follows.

15 14 13 12	11 10 9 8	7 6 5 4 3 2 1 0
OP	FLAG	IMM8

IV. EXPERIMENT RESULTS

The DSP is successfully implemented on 65nm SOTB process. A layout of the DSP chip is shown in Fig. 4. It costs 3,873 cells and the layout area of 161µm×158µm. With the power supply of 0.55V, the power consumption at the operation frequency 200MHz is 282µW. Compared with this same DSP on the 65nm bulk CMOS process, the power consumption on the SOTB process is much lower 6 times. The summary results are depicted in Table I. The leakage current measurement of the DSP is shown in Fig. 5. In this figure, the leakage current values are varied when body bias voltage values V_{BB} are applied to the DSP. At $V_{DD} = 400mV$, the leakage current values $I(V_{DD})$ at body bias voltage $V_{BB} = 1.2V, 0V, -1.2V$ are 0.8mA, 10µA, 0.1µA respectively. The low leakage current values are varied within 2 orders of magnitude in logarithm. Meanwhile, on bulk process, the leakage current is not varied. The comparison results are also described in Table II. The proposed DSP consumes the lowest power consumption among the other designs. The low power is achieved due to employing 65nm SOTB process and taking its advantage of the low leakage current.

DSP function verifications are carried out as follows. A normalized product <1.15>*<1.15> is illustrated in Fig. 6a.

In this experiment, two operands 0x0200 and 0x0400 are multiplied. As a result, the multiplication result is 0x0800; meanwhile, a product of <8.8>×<8.8> is carried out by the two same operands above. The result of this multiplication is 0x0010 (Fig. 6b). Verification of an adder is also implemented. The two operands are 0x0011 and 0x0008 added to each other. The adder result is 0x0019 (Fig. 6c). An example of logic operation is also executed. AND logic is performed by 0x0008 and 0x0007. As a result, AND logic operation is 0x0000 (Fig. 6d).

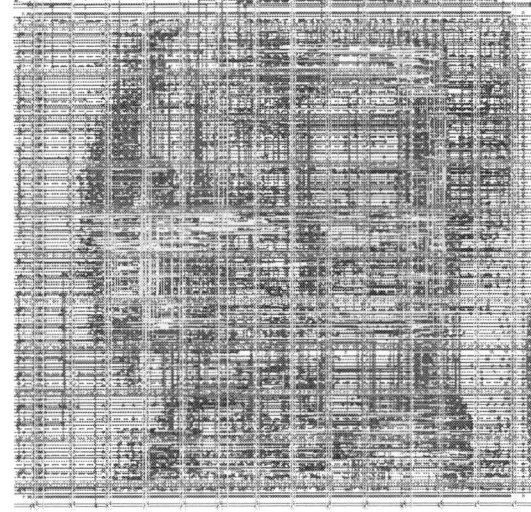

Fig. 4. Layout of the proof-of-concept DSP chip.

TABLE I.   RESULTS OF THE DSP ON CHIP.

Parameters	SOTB (This work)	Bulk CMOS
Process	SOTB 65nm	CMOS 65nm
Voltage	0.55V	1.2V
Core size	161µm × 158µm	161µm × 158µm
Logic cells	3,873 cells	3,024 cells
Max. frequency	200MHz	200MHz
Power consumption	282µW at 200MHz	1.7mW at 200MHz

TABLE II.   COMPARISON OF THE DSPS.

References	Ref. [8]	Ref. [9]	This work
Process	CMOS 28nm	CMOS 0.18µm	SOTB 65nm
Voltage	0.4V	1.8V	0.55V
Logic cells	N/A	2.411	3,873
Power	720µW	1.5mW	282µW
Frequency	33MHz	50MHz	200MHz

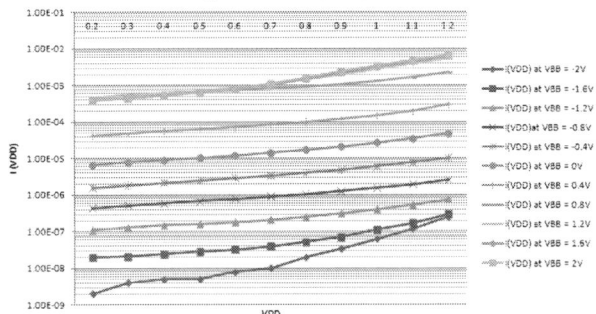

Fig. 5. Leakage current measurement results of DSP on 65nm SOTB process.

V. Conclusion

The proposed 16-bit fixed-point DSP is successfully implemented on 65nm SOTB process. This DSP supports up to 41 instructions due to the combination of opcodes and flags. The central operation of this chip is MAC instruction, which can be completed in four clock cycles. The DSP is designed and implemented on ASIC by using a digital IC design flow. The DSP also consumes very low power by using 65nm SOTB process. The test results show that the chip operates reliably at the voltage 0.55V. At this voltage and the operation frequency 200MHz, the power dissipation is approximately 282µW. The DSP performance is also improved by using optimization techniques to enhance the speed as well as pipelining structures to reduce the power dissipation.

Acknowledgements

This work was performed as "Ultra-Low Voltage Device Project" funded and supported by the Ministry of Economy, Trade and Industry (METI) and the New Energy and Industrial Technology Development Organization (NEDO).

References

[1] T. Ishikawa, H. Suzuki, H. Taki, K. Homma, H. Kabuo, M. Okamoto, K. Uedal, R. Asah, "A 16 bit Low-Power-Consumption Digital Signal Processor for Portable Terminals," Proc. 4th IEEE Int. Conf. Universal Personal Communications, pp. 798-802, Nov. 1995.

[2] H. Kabuo, M. Okamoto, I. Tanaka, H. Yasoshima, S. Marui, M. Yamasaki, T. Sugimura, K. Ueda, T. Ishikawa, H. Suzuki, R. Asahi, "An 80-MOPS-Peak High-Speed and Low-Power Consumption 16-bit Digital Signal Processor," IEEE Journal of Solid-State Circuits, Vol. 31, No. 4, pp. 494 - 530, Apr. 1996.

[3] E. O'Malley, K. Rinne, "A 16-bit fixed-point digital signal processor for digital power converter control," Proc. 20th Annual IEEE Applied Power Electronics Conference and Exposition (APEC), Vol. 1, pp. 50 - 56, Mar. 2005.

[4] D. Lee, C. Ryu, J. Park, K. Kwon, W. Choi, "Design and implementation of 16-bit fixed point digital signal processor," Proc. Int. SoC Design Conf. (ISOCC), pp. II-61 - II.64, Nov. 2008.

[5] H. Makimiya et al., "Design Consideration of 0.4-Operation SOTB MOSFET for Super Low Power Application," IEEE IMFEDK, p.p. 42-43, 2011.

[6] Y. Yamamoto, et al., "Ultralow-Voltage Operation of Silicon-on-Thin-BOX (SOTB) 2Mbit SRAM Down to 0.37V Utilizing Adaptive Back Bias," Symposium on VLSI Technology, p.p.212-213, 2013.

[7] T. Ishigaki, R. Tsuchiya, Y. Morita, H. Yoshimoto, N. Sugii, T. Iwamatsu, H. Oda, Y. Inoue, T. Ohtou, T. Hiramoto, S. Kimura, "Silicon on thin BOX (SOTB) CMOS for ultralow standby power with forward-biasing performance booster," Proc. 38th European Solid-State Device Research Conference (ESSDERC 2008) pp. 198-201, Sep. 2008.

[8] G. Gammie et al., "A 28nm 0.6V low-power DSP for mobile applications," Proc. IEEE International on Solid-State Circuits Conference (ISSCC 2011), pp. 132 - 134, Feb. 2011.

[9] X-T. Nguyen, T-T. Bui, H-T. Huynh, C-K. Pham, D-H. Le, "An ASIC implementation of 16-bit fixed-point Digital Signal Processor", Journal of Science and Technology (Special Issue), Vietnam Academy of Science and Technology, Vol. 51 (4B), pp. 282-289, Oct. 2013.

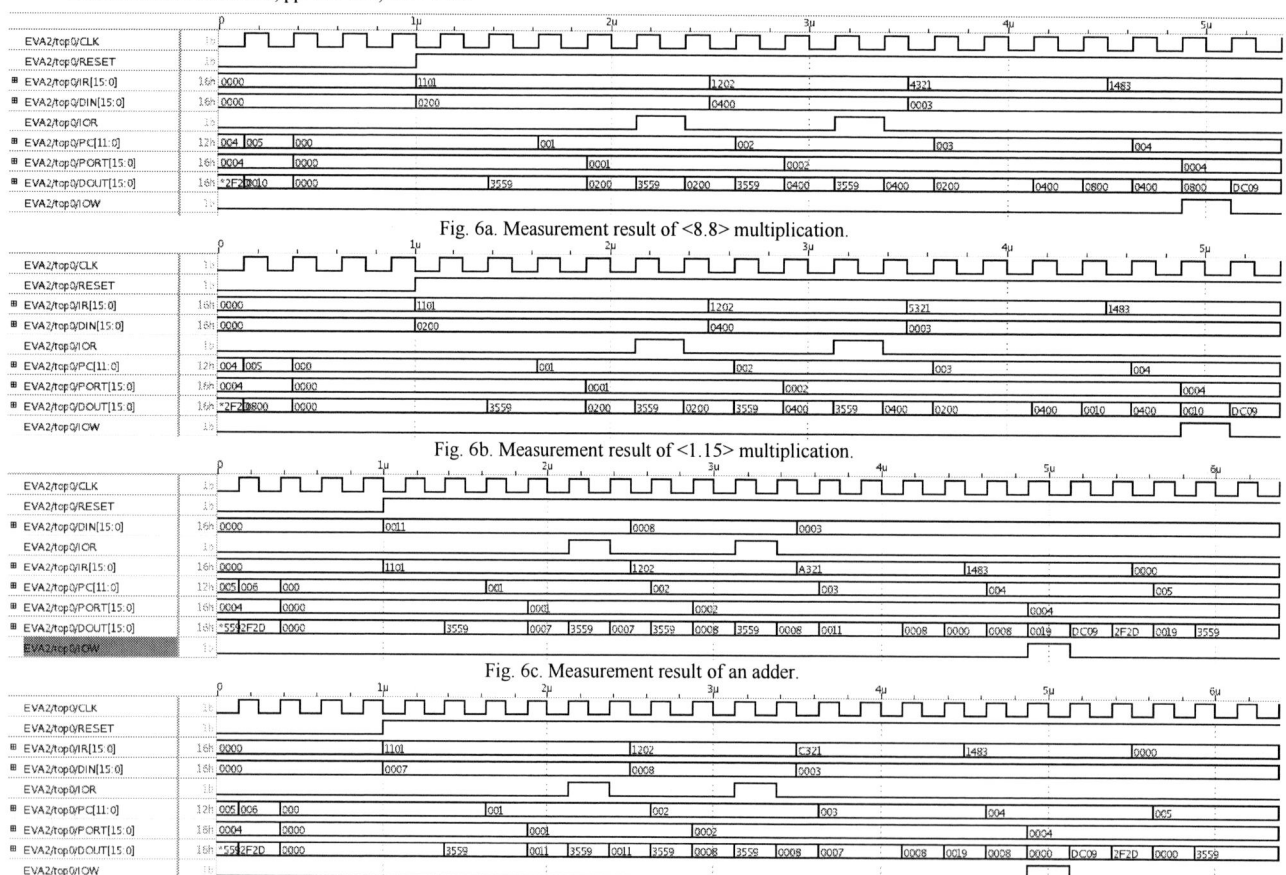

Fig. 6a. Measurement result of <8.8> multiplication.

Fig. 6b. Measurement result of <1.15> multiplication.

Fig. 6c. Measurement result of an adder.

Fig. 6d. Measurement result of AND logic.

Power measurements and cooling of the DOME 28nm 1.8GHz 24-thread ppc64 µServer compute node

Ronald P. Luijten	Matteo Cossale	Rolf Clauberg	Andreas Doering
Cloud & Computing dpt	Cloud & Computing dpt	Cloud & Computing dpt	Cloud & Computing dpt
IBM Research - Zurich	IBM Research - Zurich	IBM Research - Zurich	IBM Research - Zurich
Rüschlikon, Switzerland	Rüschlikon, Switzerland	Rüschlikon, Switzerland	Rüschlikon, Switzerland
lui@zurich.ibm.com			

Abstract— **Within the IBM / ASTRON DOME µServer project, we are finishing two types of memory DIMM-like sized compute nodes. The first is based on the 4-core / 4-thread 2.2 GHz P5040 45nm SoC and the second on a 12-core / 24-thread 1.8 GHz T4240 28nm SoC. Both SoCs employ the 64-bit Power instruction set. We show T4240 Specbench performance results, power consumption, describe the packaging of our first P5040 based 8-way hot-water cooled cluster and privide an outlook.**

Keywords—microserver, SOC, packaging, power-efficiency, DOME, SKA, embedded systems, hot-water cooling

I. INTRODUCTION

In [1] we introduced the ASTRON / IBM µServer[1] project with its key objective to build the world's highest density and most energy-efficient µServer cluster. In summer 2014 we received our first T4240[2] SoC (Server On a Chip) based

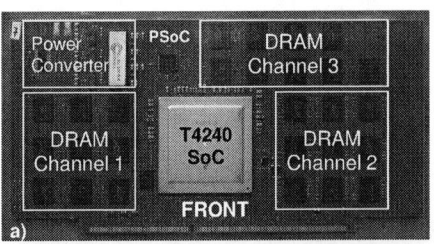

Figure 1: T4240 compute node board a) top view and b) lab setup on single node carrier board

---

[1] This work is supported by the Dutch Government DOME P4 grant
[2] Freescale Semiconductor

DOME prototype board, shown in Fig 1a. This compute node, mounted on our 'mini Base Board' (miniBB) used for standalone, single node aircooled operation is shown in Fig 1b. To conserve cost and time, we manufactured our revision-1 board in a somewhat larger form-factor than the DOME DIMM-like size, and mounted all chips on the front side. We use the same netlist and card-edge connector as for the small form-factor. Figure 2 shows the block diagram of our T4240 node board. The CPU and DRAM frequencies are limited given the first prototype. The miniBB, which requires only a 12V supply, contains the power converters needed for all required voltages (except the 1.0V for T4240), a 1GbE PHY chip (88E1111), an mSATA slot and various connectors. To provide 30A at 1V, the high wattage 1V domain converter is located on the T4240 node board. Our revision-1 T4240 board reliably runs Linux Fedora and various applications. Our demo setup has a management console and root file system mounted on the PC on the bottom left of Fig 2 and the VNC client shown on the PC on the bottom right. As we described in [1], the PSoC[3] collapses 6 key support functions into a single small chip, using USB as the management interface.

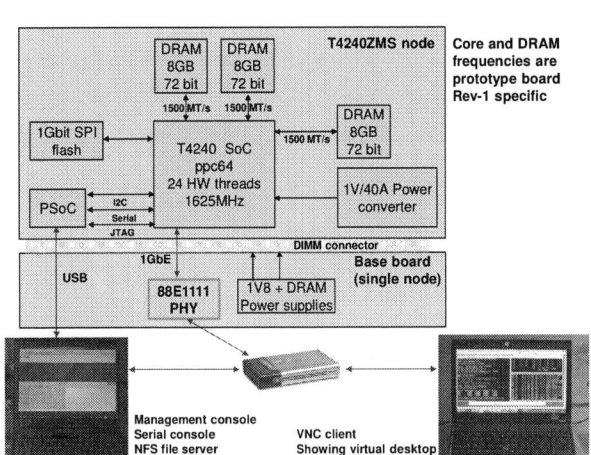

Figure 2: T4240 system under test

---

[3] PsoC® is a trademark of Cypress Semiconductor.

II. PERFORMANCE MEASUREMENTS ON RDB

In order to validate the software stack for the T4240 we have procured a number of T4240RDB (reference design boards) systems from Freescale. The cores run at 1.667GHz. We installed 48Gbyte of DRAM in these systems which run at 1.8667GT/s. Large DIMMs are needed to achieve maximum memory bandwidth using controller- and bank- interleaving. Our RDBs use 2nd revision T4240 samples. We installed Fedora 20, DB2, CPMD and other applications from scratch and have a 4-way cluster, connected with a GbE switch, running IBM DB2-BLU with the workload multi-user driver. We furthermore installed the SPECint® 2006 suite [2]. Figure 3 shows the integer Specbench and Coremark results on our T4240RDB compared to an energy optimized XEON® E3-1230Lv3, earlier published in [3]. We disable turbo mode on the XEON as this CPU will throttle itself down when running at maximum performance for extended duration, and thus we can compare single- and multithread performance at the same core clock rate. As XEONs are optimized for high single thread performance (see Stream results in [4]) we see that single thread performance of the T4240 is around 3 times slower than the XEON. However, in aggregate performance (SPECint rate) the T4240 is 40% faster.

III. POWER CONSUMPTION MEASUREMENTS ON ZMS

From early 2015, we can run the same software stack on our own revision-1 board as on the RDB. Our board has 8GB DRAM in 2 ranks per channel totalling 24GB. Due to a number of limitations, the core clock and DRAM speed is 1625MHz and 1.5GT/s respectively, so we cannot obtain exactly the same test conditions as on our RDB. Furthermore, our revision-1 board only contains 24GB DRAM which implies we cannot run some SPECint rate programs due to memory size requirements. The SoC label is T4240NSE7TTB / UYQCJ1450. The average input current measurements are performed at the 12V inputs of the 1V8 I/O voltage converter, the DRAM 1V35 / 0V67 converter as well as the 1V0 T4240

System under test	Freescale **T4240** 12 cores; 24 thr. 28nm Bulk	Intel Xeon **E3-1230L v3** 4 cores; 8 threads 22nm FinFet
CPU2006 Benchmark Test Environment	System: T4240RDB-PB **1.666 GHz core clock, 1.866 GT/s 6GB DRAM, 3 channels** Fedora 20, Kernel 3.12.19 GCC 4.7.2 gcc options: -O3 -mcpu=powerpc64	System: Supermicro X10SAE **1.8 GHz core clock; Turbo disabled 1.666 GT/s 8 GB DRAM, 2 channels** Fedora 19, Kernel 3.13.9 GCC 4.8.2 gcc options: -O3 -march=native -mtune=native
CINT-base – 1 thread	6.86	20.7
CINT-base – all threads	**109.34 (24 threads)**	**77.6 (8 threads)**
Coremark - all threads	**188K (24 threads)**	**65K (8 threads)**
System Photo		

Figure 3: Performance measurement results

Power measurement on rev 1 board #5, on 7 + 8 april 2015; PSoC firmware 2-mar-15							
current measurements at 12V input of power converters, T4240 temp < 65C							
voltage domain		1V8 I/O		DRAM		1V0 core		total node
current measured @ 12V input								
condition	mA	W	mA	W	A	W	W
PSOC only power	3.4	0.0408	74	0.888	0.0008	0.0096	0.9384
T4240 power on, kept in reset	75	0.9	152	1.824	0.32	3.84	6.564
u-boot prompt (idle)	77.6	0.9312	350	4.2	1.48	17.76	22.8912
Linux prompt, idle system	77.6	0.9312	315	3.78	1.58	18.96	23.6712
BW_MEM 512M, 24 thr	77.3	0.9276	450	5.4	1.65	19.8	26.1276
stream, 24 thread	77.3	0.9276	470	5.64	1.65	19.8	26.3676
BW_MEM 512, 24 thr	77.7	0.9324	320	3.84	2.53	30.36	35.1324
idle at XCFE desktop	77.7	0.9324	320	3.84	1.6	19.2	23.9724
SpecInt PerlBench, 24 thr	77.8	0.9336	400	4.8	2.63	31.56	**37.2936**
SpecInt PerlBench, 12 thr	78	0.936	355	4.26	2.2	26.4	31.596
SpecInt gcc, 12 thr	78	0.936	416	4.992	1.7	20.4	26.328

Notes:

PSOC = compute node controller
'PSOC only power': this means that 1V8,DRAM and 1V domains are switched off
BW_MEM 512M runs out of DRAM memory, stressing the memory system
BW_MEM 512 runs out of cache, stressing the caches and the cores, but not memory
STREAM is run with a 40'000'000 elements and 24 threads, and also stresses the FPUs
SpecInt Perlbench is run at 24 and 12 threads
SpecInt gcc can only be run at 12 threads due to DRAM memory size limitations

Current measurements are +/- 5%, averaged over 15 seconds
These measurements are first and preliminary results on pre-production T4240 silicon

Figure 4: power usage measurement results

core voltage converter. The first two converters are located on the base board or miniBB. Figure 4 shows the preliminary power consumption of our T4240ZMS (Zurich Micro Server) compute node for these voltage domains including the total, for various test conditions. Subtracting the power losses from the external converters, the worst case T4240 node level power consumption (running Specbench) is ~36.3 Watt, which is 70% of our XEON E31230Lv3 test node. Hence our node achieves a factor of 2 more operations per Watt. The 26.4W node power consumption for the Stream benchmark is more representative for an average cloud computing workload. Note that the T4240 I/O power on the 1V8 domain does not change much with workload and is less than 1 Watt.

Figure 5: Hot-water cooled P5040 cluster

IV. COMBINED POWER DELIVERY AND COOLING

Our hot-water cooling solution, earlier demonstrated at scale with the "SuperMUC" (nr.4 / June 2012 TOP500 HPC) system [5], is used to manage the thermal load of a fully populated µServer chassis enabling high packaging density of our compute node boards. An 8-node cooling prototype of our rack-unit based on the P5040[4] SoC is shown in Figure 5.

In the DOME µServer module, the main heat source is the SoC, which is in a BGA package, mounted in the center of the module (Fig 6). In our packaging approach, the lid of the original SoC package is removed and replaced with our custom designed heatspreader. The direct attachment of our heatspreader to the chip reduces the overall thermal resistance, as the BGA-lid thermal interface is removed. A lower thermal resistance allows an increase of the coolant temperature, enabling warm/hot water cooling. When the µServer is in operation, heat generated from the SoC is transferred to the sides of the heatspreader (Fig 7). These sides are thermally connected to an active heat sink; the heat is then transferred to the liquid coolant, which is kept at a constant temperature by the data-center cooling installation.

The electric supply current for a 128-node rack-unit is estimated to be in the order of several hundred Amps at 12V. To ease the design of the PCBs and connectors, the high current delivery is also performed by cooling infrastructure. We provide an electrical connection between the power supply and the compute node, storage, power and switch modules. Taking advantage of the large copper cross section of the heatspreader, the power supply is electrically connected to the heat sink and also electrically connected to the heatspreaders of each module. A heatspreader consists of three copper layers laminated together in an FR4 manufacturing process: one

a) Front

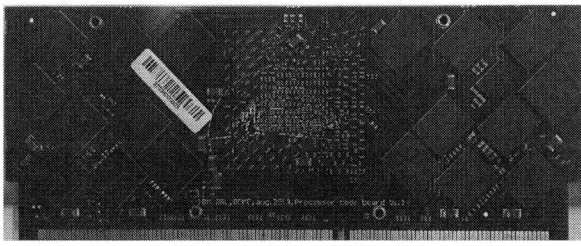

b) Back

Figure 6: P5040 compute node a) Front (without LID) b) Backside

---
[4] Freescale Semiconductor

2mm and two 0.2mm thick layers. They are electrically insulated from each other with 0.1mm dielectric Arlon 92ML type 106, a prepreg with enhanced thermal conductivity. The inner layer acts as +12V supply voltage conductor while the outer layers act as ground conductors. The outer layers also provide a shield against electromagnetic interference carried by the +12V plane. The 2mm thick layer acts as the main thermal path and it is in direct contact with the chip. The heatspreader is mechanically attached to the µServer module with metal rivets, which also carry the supply current. The electrical connection of the heat-spreader to the ground potential is established with the first and last layer. The heat sink has a power rail, which is insulated from the rest of the heat sink and connected to +12V supply. The power rails are electrically connected via spring contacts to the inner layer of the heatspreader, thus establishing the electric contact of the +12V.

V. THERMAL CHARACTERIZATION

We built a thermal test vehicle to characterize the thermal performance. It has the same form-factor as the µServer but instead of the SoC it has a thermal test chip equipped with heaters (resistors) and temperature sensors (diodes). First results show a thermal resistance of 0.85 K/W, which allows a power dissipation (P in Fig 7c) of 36W per module by cooling with 45°C coolant (T_{in}) and ensuring a SoC junction temperature (T_j) below 85°C. Further characterization using the P5040 is in progress and the T4240 revision-2 board will also fit in the same cooling package.

VI. LESSONS LEARNED

Figure 8 summarizes the main bringup steps of our T4240ZMS board. This task was mainly performed by the first author with support from Volterra and Freescale. The u-Boot adaptation was performed by an intern student. We had two boards procured without the T4240 SoC and DRAM chips to validate the power supply. We needed 6 weeks to bringup the

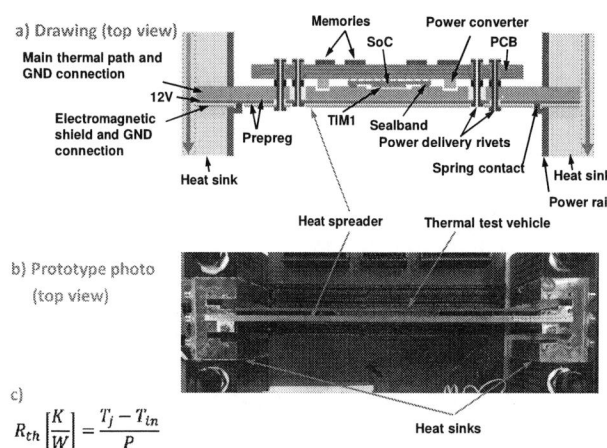

Figure 7: a) Compute node board assembly, b) Thermal test section , c) Thermal resistance formula

Date	step
7 Jul 2014	Revision 1 board received
1 Sep 2014	First power on T4240 module
6 Oct 2014	First assembly program running out of L3 cache
10 Oct 2014	First u-boot prompt (1 DRAM channel working)
6 Nov 2014	Brought up 2nd revision 1 board
11 Nov 2014	Travel to US for demo at OSS14 and SC14
16 Nov 2014	Start linux kernel (missing root FS)
7 Jan 2015	Ethernet working within u-boot
16 Jan 2015	Ethernet working with poky linux with Ram based rootFS
17Jan 2015	Booted Fedora 20 with NFS mounted rootFS

Figure 8: T4240ZMS bringup progress

Volterra based 1.0V supply circuitry. A PSoC programming bug uncovered a bug in Volterra VT1175MA SMBus implementation, causing significant delay. The VT1175MA part has been discontinued in the mean time. The entire month of September was needed to get the T4240 to load the RCW (Reset Configuration Word) and PBL (Pre Boot Loader). In October we got the first program to execute out of L3 cache (mapped as memory), the first DRAM channel configured and u-Boot running. Later that month we discovered that the power supply filter was connected to the wrong PLL for SATA operation, as this PLL did not lock. In January 2015 we procured the revision-1b board. A significant effort was needed to configure the DTB (device tree blob) to boot linux Fedora using a network mounted rootFS (file system). Unfortunately, revision-1b had one more design error for SATA and a revision-1c is currently underway.

In comprehensive internal and external design reviews the teams missed the two SATA mistakes. The complexity of an SoC comprises all chips on a typical server motherboard and all of this needs to be fully understood to get a 100% working board. We underestimated the magnitude of this. We benefited from the choice to make revision-1,1b and 1c in the larger form factor.

We have started the design of the 2U rack unit carrier board. The compute nodes will be networked using 10 GbE. Each T4240 has four 10 GbE ports and we concluded that we cannot route the Ethernet wiring to the 128 nodes boards when operating 10 GbE with XAUI. We concluded we need to run these interface in KR mode (10Gbps per lane), reducing Ethernet wiring fourfold. A regular memory DIMM connector cannot support signalling at 10 Gbps. We therefore decided to change to the 3M SPD08 connector, similar in size as the DIMM connector. SPD08 supports speeds in excesss of 25Gbps. The final T4240 node board in the DOME form factor will use this connector.

Finally, we were too aggressive in reducing component count. The current revision uses a single 125MHz oscilator to source both core and DRAM PLLs. This yields a suboptimal choice of frequencies for the DRAM and we will employ a separate 133MHz oscillator for DRAM on revision-2.

VII. OUTLOOK

Once our rev-1c T4240ZMS has been fully validated, we will manufacture the compute node in the small DOME form-factor. We expect these revision-2 boards around mid 2015. Furthermore, we are currently designing a 64-way cluster in a 2U 19" rack-unit as an intermediate step towards the full 128-way cluster. The bringup of the 64-way unit is planned for year-end 2015. Finally, our plan is to build the next compute node with 64-bit ARMv8 architecture using a Freescale Layerscape™ [5] SoC.

VIII. CONCLUSIONS

The latest result demonstrates that the desired μServer energy-efficiency advantage has been fully achieved. With well chosen system design objectives and a matching SoC, we have shown that we can obtain 40% better aggregate performance as well as 2x more operations per Joule than an energy optimized XEON® E31230Lv3. Thus, the T4240 delivers best of breed performance per Watt and our 2U unit is expected to deliver best performance per Watt per unit of volume. Note that we achieve this with an SoC in 28nm bulk versus a 22nm Finfet CMOS process, and significant further improvements are possible.

ACKNOWLEGDEMENTS

The work described in this paper is the result of a large collaboration across the globe [6]. We thank Freescale Austin, Belgium and Germany and also IBM Zurich, Toronto, Italy and Austin. We thank Astron, Transfer and TPC Electronics in the Netherlands, Volterra in California and Dimema and Tecnomaster in Italy. Special thanks go to Peter van Ackeren for his invaluable help (from PBL and DRAM to Fedora), our intern student Aris Ioannou for his work on u-Boot, and Stephan Paredes and Bruno Michel for the cooling design.

REFERENCES

[1] "The DOME embedded 64 bit microserver demonstrator", R. Luijten and A. Doering, ICICDT 2013, Pavia, Italy, May 2013

[2] "https://www.spec.org/

[3] "Energy-Efficient Microserver Based on a 12-Core 1.8GHz 188K-CoreMark 28nm Bulk CMOS 64b SoC for Big-Data Applications with 159GB/s/L Memory Bandwidth System Density," R. Luijten et al, ISSCC15, San Francisco, CA, Feb 2015

[4] Dual function heat-spreading and performance of the IBM / Astron DOME 64-bit μServer demonstrator", R. Luijten, A. Doering and S. Paredes, ICICDT14, Austin TX, May 2014.

[5] "Waste heat recovery in supercomputers and 3D integrated liquid cooled electronics", Tiwari, et al., Thermal and Thermomechanical Phenomena in Electronic Systems (ITherm), 2012 13th IEEE Intersociety Conference

[6] http://www.zurich.ibm.com/microserver

---

[5] Layerscape™ is a trademark from Freescale Semiconductor

AUTHOR INDEX

Abadi, A. Rouhi Najaf 177
Absil, P. 177
Ackaert, Jan 142
Akbal, M. 114
Alburaikan, Abdullah 43
Ali, K. Ben 177
Ali, S. Z. 31
Amagasaki, Motoki 181
Amouri, Emna 102
André, N. 31
Aoulaiche, M. 13, 74, 86, 146
Aqeeli, Mohammed 43
Athanasiou, S. 166
Baert, Rogier 94
Bardon, M. Garcia 47, 63
Beyne, E. 138, 177
Bhasin, Shivam 102
Boschke, Roman 158, 162
Cabrini, Alessandro 5
Caillat, C. 146
Catthoor, F. 98
Chen, Chien-Ju 35
Chen, Shih-Hung 158, 162
Chen, Yin-Nien 35
Cherman, V. 138
Chiarella, Thomas 51
Cho, M. 74, 86, 146
Choi, M. 177
Chuang, Ching-Te 35
Ciubatoru, Florin 39
Clauberg, Rolf 189
Collaert, N. 63, 106
Colpaert, Tony 142
Cossale, Matteo 189
Cristoloveanu, S. 166
D'Amico, Stefano 126
Danger, Jean-Luc 102
Dao, Thuy 150
Daud, Mohammad 17
De Keersgieter, A. 106
De Meyer, K. 63
De Vos, J. 138
De Wolf, I. 177
Dehan, Morin 51
Doering, Andreas 189
Endo, K. 70
Eneman, G. 13, 63, 106
Eriguchi, Koji 118, 122
Fan, Hua 130
Fazan, P. 13, 74, 86, 146
Feng, Philip X.-L. 25
Fievet, Nathalie 94
Firouzi, Farshed 90
Flandre, D. 31
Francis, L. A. 31

Fuketa, Hiroshi 134
Fukuda, K. 70
Galy, P. 154, 166
Gaudin, Gweltaz 174
Ger, Mu-Ling 150
Gérard, P. 31
Giraud, Bastien 17
Gonzalez, Mario 142
Govoreanu, Bogdan 1
Graba, Tarik 102
Groeseneken, G. 98, 162
Grover, Anuj 17
Guillermet, M. 114
Guo, W. 177
Haond, M. 59
Hellings, Geert 51, 158, 162
Hiraga, Keizo 51
Horiguchi, N. 13, 146
Horiguchi, Naoto 51, 74, 86
Hu, Vita Pi-Ho 35
Hu, Zhirun 43
Huang, Xianjun 43
Huynh-Bao, Trong 9
Iida, Masahiro 181
Ikeda, Taro 118
Ishibashi, Koichiro 185
Ishikawa, Y. 70
Janardan, Dhori Kedar 21
Jang, D. 47, 51, 63
Josse, E. 59
Jurczak, Malgorzata 1
Kaczer, B. 78, 86, 98
Kamohara, Shiro 185
Kasai, Shigeru 118
Kim, Minsoo 51
Kiouseloglou, Athanasios 5
Kobayashi, Kazutoshi 78
Kuga, Morihiro 181
Kumar, Ashish 21
Kumar, Promod 17
Kumar, Vinay 21
Lauwereins, Rudy 39
Le, Duc-Hung 185
Li, G. 31
Li, Qiang 130
Lin, Wei 170
Linten, Dimitri 86, 158, 162
Liu, Airong 130
Liu, Kehong 130
Liu, Y. X. 70
Luijten, Ronald P. 189
Luque, Maria Toledano 74
Lv, Lishan 130
Maggioni, F. 138
Malik, Aditi 142

AUTHOR INDEX

Mallik, A. 63
Masahara, M. 70
Mathieu, Yves..........................102
Matsukawa, T............................70
Matsumoto, Takashi 78
Matsuzawa, Kazuya 82
Mazure, Carlos.........................174
Mazurier, J...............................59
Meneghesso, Gaudenzio 110
Meneghini, Matteo 110
Mercha, A. 9, 47, 51, 55, 63, 94
Mertens, H................................63
Migita, S..................................70
Mitani, Yuichiro........................82
Miyaguchi, Kenichi 51, 55
Mizubayashi, W. 70
Mocuta, A. 13, 51, 55, 67, 106, 146
Moens, Peter.............................110
Mori, T.....................................70
Morita, Y. 70
Moritz, Guillaume......................17
Moroz, V..................................177
Navarro, Gabriele 5
Neve, C. Roda...........................177
Nguyen, Bich-Yen......................174
Nguyen, Xuan-Thuan 185
Noel, Jean-Philippe.....................17
O'Uchi, S..................................70
Oboril, Fabian...........................90
Okada, Yukimasa.......................122
Ono, Kouichi 118, 122
Oprins, H..................................138
Oshima, Azusa..........................78
Ota, H......................................70
Parthasarathy, Chittoor 17
Parvais, Bertrand........................55
Perniola, Luca 5
Pham, Cong-Kha........................185
Poncelet, O................................31
Pourghaderi, M. Ali 67
Qiao, Zhiliang...........................130
Rack, M....................................177
Radu, Ionut..............................174
Radu, Iuliana.............................39
Raghavan, P.............39, 47, 55, 63, 94
Ragnarsson, Lars-åke 55
Raskin, J. P...............................177
Ribes, G...................................114
Ritzenthaler, R. 13, 74, 86, 146
Robert, Frederic.........................94
Roussel, J..................................98
Ryckaert, Julien...........................9
Saha, Kaushik 21
Sakhare, Sushil............................9
Sakurai, Takayasu.......................134

Schoenmaker, Wim 154
Scholz, Mirko 158, 162
Schram, T............13, 74, 86, 146
Schuddinck, P. 47, 63
Sherazi, Yasser 39
Soree, Bart...............................39
Spagnolo, Annachiara 126
Spessot, A. 13, 74, 86, 146
Su, Pin....................................35
Sueyoshi, Toshinori 181
Sugii, Nobuyuki 185
Sun, X....................................177
Tahoori, Mehdi B. 90
Takamiya, Makoto 134
Tanihara, Akira 118
Thean, A.........9, 13, 39, 51, 55, 47, 63, 67, 94, 106, 146
Tomida, Kazuyuki 51
Torelli, Guido 5
Tsai, Tsung-Yi 170
Tsukada, J. 70
Turgis, David 17
Udrea, F...................................31
Vallier, L..................................114
Van Der Plas, G. 138, 177
Vanmeerbeek, Piet 110
Vaysset, Adrien..........................39
Verbruggen, Bob 126
Verheyen, Peter.........................162
Verkest, D. 9, 47, 51, 55, 63, 94
Visweswaran, G. S. 17, 21
Wambacq, Piet.....................9, 55, 126
Wang, Shengcheng 90
Wangy, Chua-Chin......................170
Weber, O...................................59
Weckx, P. 78, 98
Wu, Tse-Ching...........................35
Yakimets, D.........................47, 63
Yakimets, Dmitry 9
Yamamoto, Nobuhiko 118
Yamauchi, H. 70
Yamauchi, Yoshitaka 134
Yanagihara, Yuki 134
Zanoni, Enrico...........................110
Zeng, Y....................................31
Zhang, Leqi 1
Zhao, Qian...............................181
Zografos, Odysseas 39
Zuo, Jiangkai............................150

9781479976706